Variation in Leaf Structure

An Ecophysiological Perspective

W0018646

Variation in Leaf Structure

An Ecophysiological Perspective

Edited by

E. Garnier, J. F. Farrar, H. Poorter and J. E. Dale

Centre d'Écologie Fonctionnelle et Évolutive (CNRS-UPR 9056)
1919 Route de Mende, 34293 Montpellier Cedex 5, France

School of Biological Sciences
University College of North Wales, Bangor
Gwynedd LL57 2UW, UK

Department of Plant Ecology and Evolutionary Biology
Utrecht University
PO Box 800.84, 3508 TB Utrecht
The Netherlands

Institute of Cell and Molecular Biology
Daniel Rutherford Building
The King's Buildings, Edinburgh EH9 3JH, UK

This book is reprinted from a *New Phytologist* Special Issue,
New Phytologist **143**, 1, 1999, and published for
the *New Phytologist* Trust by Cambridge University Press

New
Phytologist

International Journal of Plant Science

CAMBRIDGE
UNIVERSITY PRESS

PUBLISHED BY THE PRESS SYNDICATE OF THE UNIVERSITY OF CAMBRIDGE
The Pitt Building, Trumpington Street, Cambridge, United Kingdom

CAMBRIDGE UNIVERSITY PRESS
The Edinburgh Building, Cambridge CB2 2RU, UK   http://www.cup.cam.ac.uk
40 West 20th Street, New York, NY 10011–4211, USA   http://www.cup.org
10 Stamford Road, Oakleigh, Melbourne, 3166, Australia
Ruiz de Alarcón 13, 28014 Madrid, Spain

© Trustees of *New Phytologist* 1999

First published 1999

Printed in the United Kingdom at the University Press, Cambridge

ISBN 0 521 77678 3

Contents

Preface

This book explores the impacts of variation in leaf structure on processes occurring from the level of the leaf itself to that of ecosystems and biomes. The introductory review by Vince Gutschick assesses the basic consequences of differences in leaf structure, and this is followed by both reviews and research papers that expand on this theme, providing up-to-date accounts of the environmental, genotypic and whole-plant controls over leaf structure, and how that structure influences such diverse processes as gas exchange, litter decomposition and interactions among organisms.

For a long time, ecophysiologists have focused strongly on the photosynthetic properties of the leaf, and how these relate to its structure and chemical composition. For example, a recent survey (Reich *et al.*, 1997) has established that the maximum photosynthetic rate of a wide range of terrestrial plants can be accurately predicted from a combination of leaf structural (specific leaf area (SLA), the ratio of leaf area to leaf mass) and biochemical (total nitrogen concentration) parameters. However, there is increasing awareness that the impact of leaf structure on plant function goes far beyond gas exchange. In particular, SLA has been shown to correlate positively with traits that allow the plant to acquire external resources rapidly, and consequently to grow fast (Poorter & Garnier, 1999). There are also close relationships between the structure of a leaf and its life span: leaves with a high dry matter percentage (and/or low SLA) tend to be long-lasting (Ryser, 1996; Reich *et al.*, 1997), which increases the time that nutrients remain within the plant (Aerts & Chapin, 1999). Leaf structure therefore appears to play a central role in what can be interpreted as a trade-off between a rapid production of biomass and an efficient conservation of nutrients.

Clearly, the structure of leaves has important implications for the performance of plants in specific habitats. For this reason, it has been argued that it could be used to characterize species, define functional groups and predict the response of species to various environmental factors (Westoby, 1998; Weiher *et al.*, 1999). At the ecosystem level, the impacts of differences in leaf structure of component species on primary productivity and cycling of nutrients are currently under intensive investigation (Reich *et al.*, 1997; Cornelissen & Thompson, 1997).

The concluding overview by Malcolm Press draws attention to the difference between function and functional significance, and how this distinction can apply to the many strands of research explored in the rest of the volume. Ranging from basic plant physiology to ecology and remote sensing, the book demonstrates the fundamental implications of leaf structure for plant and ecosystem functioning, and identifies key issues and challenges for research in the next decade.

ACKNOWLEDGEMENTS

The articles in this book are all rigorously peer-reviewed contributions to *New Phytologist*, but have been brought together – and then developed – as a result of the 4th *New Phytologist* Symposium, 'At the Crossroads of Plant Physiology and Ecology: Causes and Consequences of Variation in Leaf Structure', which was organized by Marie-Laure Navas (École Nationale Supérieure Agronomique (ENSA-M), Montpellier, France), Eric Garnier (Centre d'Écologie Fonctionnelle et Évolutive (CEFE)–CNRS, Montpellier) and Catherine Roumet (CEFE–CNRS) (Mousseau, 1999). This meeting was held in October 1998 at ENSA-M, and was a part of the events organized to celebrate the 150th anniversary of the institution. It was made possible through the generosity of the *New Phytologist* Trust, and complementary subsidies from the Languedoc-Roussillon Région and ENSA-M itself. We warmly thank staff from ENSA-M and CEFE for sustained help during the preparation and course of the conference.

Editors
ERIC GARNIER *CEFE–CNRS, Montpellier, France*
JOHN FARRAR *University of Wales – Bangor, UK*
HENDRIK POORTER *University of Utrecht, The Netherlands*
JOHN DALE *Edinburgh, UK*

REFERENCES

Aerts R, Chapin FS III. 1999. The mineral nutrition of wild plants revisited: a re-evaluation of processes and patterns. *Advances in Ecological Research.* (In press.)

Cornelissen JHC, Thompson K. 1997. Functional leaf attributes predict litter decomposition rate in herbaceous plants. *New Phytologist* **135**: 109–114.

Mousseau M. 1999. At the crossroads of plant physiology and ecology. *Trends in Plant Science* **4**: 1.

Poorter H, Garnier E. 1999. Ecological significance of inherent variation in relative growth rate and its components. In: Pugnaire FI, Valladares F, eds. *Handbook of functional plant ecology.* New York, USA: Marcel Dekker. 81–120.

Reich PB, Walters MB, Ellsworth DS. 1997. From tropics to tundra: global convergence in plant functioning. *Proceedings of the National Academy of Sciences, USA.* **94**: 13730–13734.

Ryser P. 1996. The importance of tissue density for growth and life span of leaves and roots: a comparison of five ecologically contrasting grasses. *Functional Ecology* **10**: 717–723.

Weiher E, van der Werf A, Thompson K, Roderick M, Garnier E, Eriksson O. 1999. Challenging Theophrastus: a common core list of plant traits for functional ecology. *Journal of Vegetation Science.* (In press.)

Westoby M. 1998. A leaf-height-seed (LHS) plant ecology strategy scheme. *Plant and Soil* **199**: 213–227.

New Phytol. (1999), **143**, 3–18

Research review
Biotic and abiotic consequences of differences in leaf structure

VINCENT P. GUTSCHICK

Department of Biology, MSC 3AF, New Mexico State University, Las Cruces, NM 88003, USA (tel +1 505 646 5661; fax +1 505 646 5665; e-mail vince@nmsu.edu)

Received 5 October 1998; accepted 9 April 1999

SUMMARY

Both within and between species, leaves of plants display wide ranges in structural features. These features include: gross investments of carbon and nitrogen substrates (e.g. leaf mass per unit area); stomatal density, distribution between adaxial and abaxial surfaces, and aperture; internal and external optical scattering structures; defensive structures, such as trichomes and spines; and defensive compounds, including UV screens, antifeedants, toxins, and silica abrasives. I offer a synthesis of selected publications, including some of my own. A unifying theme is the adaptive value of expressing certain structural features, posed as metabolic costs and benefits, for (1) competitive acquisition and use of abiotic resources (such as water, light and nitrogen) and (2) regulation of biotic interactions, particularly fungal attack and herbivory. Both acclimatory responses in one plant and adaptations over evolutionary time scales are covered where possible. The ubiquity of trade-offs in function is a recurrent theme; this helps to explain diversity in solutions to the same environmental challenges but poses problems for investigators to uncover numerous important trade-offs. I offer some suggestions for research, such as on the need for models that integrate biotic and abiotic effects (these must be highly focused), and some speculations, such as on the intensity of selection pressures for these structures.

Key words: leaf anatomy, leaf morphology, gas exchange, herbivory, leaf nutrients, chemical defence, optimization, trade-off.

INTRODUCTION

Both between and within plants (individuals or species), leaves are diverse in structure. Variations are prominent in, for example, linear dimensions, dissection of the margins, dry mass per unit area, nutrient content, cell size, optical scattering and absorption, stomatal density and apertures, presence or absence of trichomes, and cuticle composition. This diversity is demonstrably under genetic control, including the plastic responses such as leaf size and mass per unit area in response to the light environment. We presume that the differences are largely adaptive, either for acquiring resources (mostly in photosynthesis) or in biotic interactions, such as retarding leaf herbivory. In some cases, the adaptive value of a structural feature can be demonstrated directly, by using some innovative methods. Constraining leaf angle displays can demonstrate the value of leaf angles for photosynthesis (Ehleringer & Hammond, 1987); removing trichomes can demonstrate their effectiveness against herbivory (Kanno, 1996). Many other quantitative benefits (and costs) have been investigated. In this review, I concentrate on the costs, benefits and associated trade-offs for distinct structural features, above the level of genes and biochemistry. Many features, such as trichomes or stomata, show adaptive value for multiple purposes (for an example, see Press, 1999). For example, trichomes can function both in energy balance and in defence, and in defence against both microbes and herbivores. Trade-offs between costs and benefits, particularly as marginal costs and benefits, are expected to be close if function is well optimized by natural selection. Indeed, close balances are often observed. This closeness supports the variability of structures, and it explains partly how very divergent structures can provide adaptation to the same selection pressures in different plant species.

LEAF STRUCTURE AFFECTS RESOURCE ACQUISITION AND USE

The cuticle

The cuticle, present in almost every land plant, is foremost a barrier against water loss as well as against pathogen invasion. The cuticle also offers much protection against a loss of solutes to rain (leaching by throughfall), although some does occur, especially in acid conditions (Pearcy & Baker, 1991). Because leaves flex in the wind and other stresses, the cuticle must be flexible or layered. Abrasion of the cuticle in high wind can increase water loss by transpiration (Grace, 1974; Pitcairn & Grace, 1985). The layer of wax or cutin is commonly rather thin (several μm), except in many xeromorphic plants, where it can reach 60 μm (Ihlenfeldt & Hartmann, 1982). Consequently, the metabolic cost of constructing the cuticle is typically a few percent of total leaf construction cost. (This accounting excludes surface resins, which can make up half the leaf mass in species such as *Larrea tridentata*.) The thickness does vary, and so does the corresponding water permeability, by about one hundredfold (Schreiber & Riederer, 1996). This variation is often in acclimation to water regimes (Turner, 1994).

Stomata

These dynamic pores are present at densities of several hundred per square millimetre. Not all plants have them; indeed, Woodward (1998) poses the question, 'Are stomata necessary?' All major taxa do possess stomata now, having evolved increasing densities of them though the Upper Carboniferous period. At the least, stomata are required for control of the exchange of CO_2 for water vapour, which is also inherently related to transpirational cooling. Woodward (1998) begins with cooling as a need in full sunlight and notes that cooling is effective even for plants of short stature, poorly coupled to the atmosphere (as defined by Jarvis & McNaughton, 1986). This occurs because the stirred part of the troposphere (the convective boundary layer) typically maintains humidities well below saturation, even in the presence of much evapotranspiration. Much more attention has been given to CO_2–water-vapour exchange, as will be discussed shortly.

As a population, stomata can be described by their areal density or by the fraction of epidermal cells that they represent (stomatal index). They are further described by their distribution of apertures. Although the area-averaged effect of stomatal opening is to confer a conductance that is controlled physiologically, not all stomata are open equally. A typical histogram of apertures is unimodal, but transients in light or humidity can induce broader and even multimodal distributions (Buckley *et al.*,

1997). Aperture responds to at least three major physiological state variables (Tardieu, 1994; Tardieu & Simmoneau, 1998): (1) photosynthetic metabolites, so that conductance keeps pace with need for CO_2 substrate (in the leaf interior; Mott, 1988); (2) hormones or regulators, particularly abscisic acid or ABA (as a water-stress signal), primarily from the roots (Blackman & Davies, 1985; implicating cytokinins; Zhang & Davies, 1990; Tardieu *et al.*, 1993, 1996); and (3) hydraulic linkages. Hydraulic linkage is overall to the bulk water status of the plant but local linkages to the epidermal cells are responsible for stomatal responsiveness to humidity (Haefner *et al.*, 1997). Bunce (1997) describes how the three physiological variables can be linked and how the linkages can be deduced experimentally. It is notable that the upper and lower (adaxial and abaxial) leaf surfaces can differ markedly in stomatal density and in physiological responsiveness (Pospísilová & Solárová, 1980).

Jones (1998) distinguishes between three major adaptive functions of stomata: optimizing the trade-off in taking up CO_2 while losing water; controlling the risk of dehydration, particularly poising the leaf water potential above the point of catastrophic xylem cavitation (Tyree & Sperry, 1988); and regulating of temperature by transpirational cooling. These functions will be discussed individually here. A uniform framework to explain all these functions simultaneously is not yet available, either mechanistically or evolutionarily (that is, demonstrating the adaptiveness, or cruder optimality, of observed behaviour).

Optimization of assimilation rate/transpiration rate (A/E) by stomata. Stomata cost almost nothing to develop. Similarly, they cost little metabolic energy to operate (Assmann & Zeiger, 1987). Consequently, the costs and benefits in their operation are almost wholly those of resource use (water, CO_2, nutrients) that they modulate. The most important trade-off is that of photosynthetic CO_2 gain against water loss, or A against E. Cowan & Farquhar (1977) proposed an optimization principle, that stomata should maintain a constant ratio of the marginal increase in CO_2 gain to marginal increase in transpiration:

$$\partial A/\partial E = \text{constant}. \qquad \text{Eqn 1}$$

The basis of this principle is that there is a metabolic cost (proportional to E) of maintaining the magnitude of E required for a given A: the cost of constructing and maintaining roots and other tissue. In an exceptional combination of theory and experiment, Givnish (1986) showed how the constant could be evaluated for particular plants and growth conditions. He included the effects of water stress, not just water use, and extended the theory to allocation of root and shoot.

How do stomata (or conductance per leaf area,

g_s) regulate both A and E? We must specify the micrometeorological environment: flux densities of photosynthetically active radiation PAR, near-infrared (NIR) and thermal infrared (TIR); air temperature, humidity, CO_2 concentration, and windspeed; and the resistance of any canopy boundary layer between the leaf and our point of measuring the micrometeorological variables. Three major processes then must be modelled: leaf assimilation, leaf energy balance, and stomatal response to the leaf environment. The equations can be formulated to a very good level of accuracy, although the simultaneous solution is mathematically challenging (Collatz *et al.*, 1991; V. P. Gutschick, unpublished). A simpler, partly qualitative viewpoint can be taken, to show that increasing g_s increases A but decreases water-use efficiency (WUE): WUE $= A/E$. Consider first the transport of water vapour, through the conductance of the stomata (g_s) and the conductance of the boundary layers (leaf and canopy taken together, with total conductance g_b). The total conductance is $g_{bs} = 1/(1/g_s + 1/g_b)$, using the addition of series resistances (1/conductance). Clearly, stomata control only part of this conductance, a point to which we must return. The transpiration rate is simply this total conductance multiplied by the difference in water vapour 'concentration' from leaf interior to stirred air outside the boundary layers: $E = g_{bs}(e_{leaf} - e_{air})/P$. This formula incorporates the total air pressure, P, so that E is formulated in terms of the water-vapour mole fraction. Conductance is used in the familiar molar units (mol m^{-2} s^{-1}), which are less dependent on temperature and pressure than the old velocity units (m s^{-1}; see Jones (1992), pp. 54–55). Next, consider the transport of CO_2, through similar physical paths, followed by its biochemical reaction in mesophyll cells. The physical-path conductance of CO_2 from its partial pressure in free air (C_a) to that in the substomatal cavity (C_s) is very similar to that for water vapour: $g'_{bs} = 1/(1.6/g_s + 1.37/g_b)$; the factors 1.6 and 1.37 account for the lower diffusibility of CO_2 than of water vapour. The reaction rate, which is simply A, can be approximated (from full enzyme kinetics, as in Farquhar *et al.* (1980)) as being proportional to the CO_2 mole fraction in the substomatal cavity, C_s/P, yielding $A = g_m C_s/P$ in units of mol m^{-2} s^{-1} (commonly quoted in micromoles, not moles). Empirically and enzyme-kinetically, g_m is simply the slope of A against C_s/P, which is fairly constant over modest ranges of C_s.

With these definitions, we obtain

$$A = g_m C_s = g'_{bs}(C_a - C_s) \qquad \text{Eqn 2a}$$

Solving for C_s and expressing A in terms of free-air CO_2 level and the conductances, we have:

$$C_s = C_a g'_{bs}/(g'_{bs} + g_m) \qquad \text{Eqn 2b}$$
$$A = C_a g_m g'_{bs}/(g'_{bs} + g_m) = C_a g'_{bs}/(1 + g'_{bs}/g_m) \qquad \text{Eqn 2c}$$

We thus obtain an expression for instantaneous water-use efficiency of the leaf:

$$
\begin{aligned}
\text{WUE} &= \frac{A}{E} \\[6pt]
&= \frac{C_a g'_{bs}}{1 + g'_{bs}/g_m} \frac{e_{leaf} - e_{air}}{g_{bs}} \\[6pt]
&= \frac{C_a}{e_{leaf} - e_{air}} \frac{g'_{bs}}{g_{bs}} \frac{1}{1 + g'_{bs}/g_m} \qquad \text{Eqn 2d}
\end{aligned}
$$

In the final formula, the first factor is dependent mostly on the external environment: $C_a/(e_{leaf} - e_{air})$. (Of course, leaf-interior vapour pressure is a function of leaf temperature, which is affected by the water-vapour conductance, a complication that we dismiss for now and that has an equally complex resolution in theory and experiment.) The second factor is almost a constant, having a narrow range from 0.62 to 0.72. The final factor expresses the physiological control of WUE by the factor g'_{bs}/g_m. Physical conductance and biochemical capacity act in completely complementary fashions. In any environment with boundary layers fixed by leaf dimensions and windspeed, a higher stomatal conductance increases g'_{bs} and confers both high A (Eqn 2c) and low WUE (Eqn 2d). The trade-off is so sharp that most plants control g_s so as to keep C_s/C_a in a very narrow range, *c.* 0.7 for plants with the C_3 pathway (Bell, 1982; Wong *et al.*, 1985).

Stomata have somewhat different leverage over A and WUE than in this simple model, because there are biophysical feedbacks: (1) an increase in g_s leads to decreased leaf temperature, which flattens the rate of decrease of WUE; (2) within the leaf and canopy boundary resistances, increased g_s humidifies the air, similarly flattening the drop of WUE (an equivalent statement is that stomatal control is diluted by the boundary-layer resistance, as is apparent in the formula for g_{bs} already described); (3) losses of water to soil evaporation or to competitors decrease as g_s increases. This last feedback arises from two effects; (a) a more humid canopy decreases the gradient in water-vapour pressure from soil to canopy air, and (b) in the long term, high g_s confers higher A and faster growth. The canopy closes earlier and suppresses soil evaporation (analogous results were reported as a function of crop planting density by Richards (1991)). In certain sets of conditions, WUE can even rise with modest increases in g_s; a case in a field experiment is reported by Meinzer *et al.* (1997). In general, these feedbacks dilute the control of both A and E or WUE by stomata, for the canopy as a whole. For individual leaves, and especially for individual plants competing with the group, many feedbacks such as changes in canopy humidification do not apply, and stomatal control retains much value.

Discussions of the value of stomatal control in balancing the instantaneous rates of carbon (C) gain and water loss are incomplete. Much more general models can be constructed. If we consider only water: (1) water costs can vary with soil water status (more root mass is needed as this declines); (2) water availability rates can be constrained by the decrease in water potential that threatens to cause xylem cavitation; (3) the rate of water use, and thus WUE, can be less relevant than constraints on total water availability (water volume in the potential rooting volume). Consider the change from conditions of unlimited water bearing a cost of acquisition (root function) to a limitation in volume. The value or weighting of WUE then increases relative to A. Models with varied degrees of inclusiveness and complexity exist, predicting varied optimal programmes for stomatal control (Berninger *et al.*, 1996; Haxeltine & Prentice, 1996; Santrucek & Sage, 1996). Many predicted (and observed) changes in stomatal conductance, from both aperture control and stomatal density development, apply over the span of leaf development time, not just as a response to the immediate environment. In a viewpoint covering the longest time spans, we must consider the evolution of developmental controls over stomatal density and physiological controls over aperture. Robinson (1994) argues that plant families or taxa that evolved earlier were constrained, and therefore did not develop as 'efficient' stomatal control. A precise definition of efficiency must be developed to clarify Robinson's point fully. One must also ask in what trade-offs these older taxa excel, so that they are not extinguished by modern taxa.

We should also consider other resources, such as N, as changing the cost–benefit structure for stomatal control. The simple arguments already discussed assumed that mesophyll conductance, or some other measure of photosynthetic capacity or investment, was given *a priori*. The relative values and availabilities of water and N must actually be balanced, so that g_s and g_m (or leaf N content) are optimized together. This will be discussed further in a later section.

For stomata to (nearly) optimize A/E, they must respond appropriately to environmental signals. To respond to A, they must respond to a photosynthetic metabolite. This metabolite must be near or in the guard cells (Jarvis & Davies, 1998). No metabolite has yet been identified, although it has been demonstrated that stomata respond to internal CO_2 partial pressure in the leaf (Mott, 1988). Stomata must also respond to E or to the atmospheric humidity (absolute or relative) that helps to determine E. Empirically, the response of g_s to the full set of environmental conditions, including humidity, is often well approximated by the Ball–Berry model (Ball *et al.*, 1987), but not always (Jarvis & Davies, 1998; Dewar, 1995):

$$g_s = mAh_s/C_s + b \qquad \text{Eqn 3}$$

(h_s and C_s are, respectively, the relative humidity and the CO_2 concentration at the leaf surface, beneath any leaf boundary layer; the slope m and the intercept b are both measures of commitments to use water and to favour assimilation over water-use efficiency). Stomata that respond according to this form do decrease g_s as evaporative demand (closely proportional to $1 - h_s$) rises. The apparent response to humidity actually derives from a direct, mechanistic response of g_s to transpiration rate (Mott & Parkhurst, 1991) and, more specifically, to epidermal transpiration rate (Saliendra *et al.*, 1995; Haefner *et al.*, 1997). However, as leaf temperature rises, the resupply of water is also activated, such that the net response of g_s is close to a response in $(e_{leaf} - e_{air})/(a$ temperature function closely paralleling $e_{leaf})$; thus, a response to e_{air}/e_{leaf}, or relative humidity. Haefner *et al.* (1997) demonstrated the realism of the full hydraulic model, not only for bulk leaf conductance but also for its patchy behaviour and for its transient behaviour, opposite in direction to the final response.

Regulation of leaf water potential by stomata. Stomatal conductance and transpiration increase together, and the water potential decrease from soil to leaf, $\psi_s - \psi_L$, increases with E. Models of varying complexity and inclusiveness show how ψ_L responds to g_s (Jones, 1992, especially p. 158 *et sqq.*). Many of these models are used to argue that g_s can be set to maintain ψ_L above the point of catastrophic xylem cavitation (Tyree & Sperry, 1988), which is very expensive in lost function in leaf and stem. (Partial cavitation, as is often observed in field conditions (Meinzer *et al.*, 1997) might nevertheless be within the optimal behaviour.) However, stomata do not respond directly to bulk ψ_L but to particular combinations of epidermal and guard-cell water potentials (Haefner *et al.*, 1997). A specific structural (hydraulic) linkage enforces this form of response. Leaves also respond to ABA (Tardieu & Simmoneau, 1998) as a signal of root or soil water status; perhaps the ultimate response is more directly to soil mechanical strength (Tardieu, 1994; Masle, 1998) than to water potential alone, with the former as a better indicator of future prospects of water extraction. The resultant combination of response to hydraulic signals and to ABA can result in ψ_L that is stable, or at least kept above a 'floor' value, in many plants called isohydric (Saliendra *et al.*, 1995; Tardieu & Simmoneau, 1998).

Stomatal regulation in response to both A/E and water status can be joined via mechanistic responses of g_s to photosynthetic metabolite(s), transpiration and ABA, as already noted.

Regulation of leaf temperature by stomata. Leaf temperature affects all manner of resource use. Increasing temperature (T), up to an optimum, is desirable for activating CO_2 assimilation and increasing the photosynthetic N-use efficiency. It can also aid photosynthate transport and organ development. Extremes are to be avoided. At high T, thermal damage can occur; stomata do open at high T, perhaps adaptively to limit T. Low T can lead to chilling and freezing injuries; in some plants, chilling alone is not damaging, but it is when combined with high light levels (Ball *et al.*, 1991, 1997).

Both stomatal conductance and leaf geometry affect leaf T. At steady state, leaf T adjusts to balance the energy fluxes per unit leaf area:

$$0 = Q_{SW}^+ + Q_{TIR}^+ - Q_{TIR}^- - Q_E - Q_{cc} \qquad \text{Eqn 4}$$

(Q_{SW}^+ is the net rate of absorption of shortwave radiant energy, which depends on spectral absorptivity and leaf display angle). The influx of thermal infrared radiant energy (Q_{TIR}^+) is almost independent of leaf structure or of display angle, depending only on surrounding temperatures and the nearly invariant leaf thermal absorptivity. The next three terms for energy losses all depend on leaf T, increasing (mostly nonlinearly) with T. Thermal infrared losses (Q_{TIR}^-) increase as the fourth power of T but are essentially independent of leaf structure, display angle or physiology. Evaporative cooling (Q_E) is simply the molar transpiration rate times the heat of vaporization of water. This cooling rate responds to stomatal conductance (a component of total conductance g_{bs}'; see the discussion around Eqns 2a–2d) and to T (because internal vapour pressure in the leaf rises exponentially with T). Convective and conductive cooling to air (Q_{cc}) increases linearly with T. As stomatal conductance increases, Q_E increases. This affords stomata a modest control of T, over a range in the order of 10°C, depending on, for example, windspeed. I use the term 'modest' because transpiration changes less than proportionally with g_s, because stomatal resistance is diluted in boundary-layer resistance and there is a negative feedback (increasing g_s also decreases the vapour-pressure deficit, $e_{leaf} - e_{air}$ in Eqn 2a). Empirically, there is little evidence of a direct response to T under normal environmental conditions, as expressed in the general success of the Ball–Berry and related models; in these models, the only effects of T are on A. The adaptive reasons are not clear.

Acclimation of stomatal density and control programme to long-term environmental conditions. Stomatal density decreases as atmospheric CO_2 concentration rises; the stomatal index decreases even more regularly (Morison, 1998). This decrease is arguably very adaptive (Kurschner *et al.*, 1998), given a link to stomatal conductance (note that the conductance contribution of individual stomata can also change, e.g., if the size or maximal aperture changes). Simply, assimilation is enhanced without cost, or the

E to support a given A decreases. This developmental response is seen over evolutionary time as well as in the lifetime of single plants (Wagner *et al.*, 1996). Of course, it is variable with leaf position in the canopy; Poole *et al.* (1996) caution that this must be accounted for in interpreting data, especially palaeontological. The response of stomatal density seems to be independent of life form (e.g., herb or tree) but dependent on exposure and on initial stomatal density (Beerling & Kelly, 1997).

Changes in other environmental variables such as humidity or PAR flux density over the duration of a leaf's growth also affect the stomatal density as well as the stomatal control programme (the short-term responsiveness of g_s to environmental variables such as radiative fluxes and humidity, as is expressed in Eqn 3) (Bunce, 1998). The control programme (expressed, e.g., as magnitudes of slope m and of intercept b in Eqn 3) acclimates at elevated [CO_2]. A significant part of the acclimation might be to altered water status (Morison, 1998). Typically, the acclimation preserves the ratio of internal to ambient CO_2 partial pressures, C_s/C_a (Morison, 1998). We can rewrite Eqn 2d, for water-use efficiency, using A as in the far right-hand side of Eqn 2a, to obtain

$$\text{WUE} = \frac{g_{bs}' C_a (1 - C_s/C_a)}{g_{bs}(e_{leaf} - e_{air})} \qquad \text{Eqn 5}$$

Given that g_{bs}'/g_{bs} varies little (see previously), we see that as ambient CO_2 pressure C_a rises at constant C_s/C_a, there is an increase in WUE (and in water status).

Trichomes

These leaf hairs function in defence (see later), but also affect gas exchange and temperature. In many plants, trichomes decrease the absorption of shortwave radiation by leaves and keep them cooler (Ehleringer, 1981; Baldocchi *et al.*, 1983). The silversword plant on Mt Haleakala, Hawaii, is an exception: it uses partly focused light reflected from trichomes to keep its apical meristem very much warmer than ambient air, to aid its development (Melcher *et al.*, 1994). The principal cost of pubescence to alter leaf T is decreased light interception. This might be a negligible cost in high-light (light-saturated) environments, or even a benefit, from the avoidance of photoinhibition (see Press (1999) for a further discussion of leaf pubescence).

Trichomes also keep water droplets off the leaf surface and the stomata (Brewer & Smith, 1997), which helps to maintain leaf gas exchange (Smith & McClean, 1989; Brewer & Smith, 1995). In calcicole species, trichomes act as sinks for excess calcium that would otherwise cause stress to the plants and result in stomatal closure (De Silva *et al.*, 1996).

Overall leaf size, shape and display

Leaves range widely in linear dimensions, from millimetres to nearly 1 m; they also vary in shape, from nearly circular with entire margins to deeply lobed or serrated margins. A highly dissected leaf margin decreases the effective size of the leaf. Small size or dissection thus increases the boundary-layer conductance, which is proportional to $\sqrt{[\text{windspeed}/(\text{linear dimension})]}$. The fractional control of A and E by stomata is kept higher than in large or entire leaves. Also, heat transport is facilitated, so that leaf temperatures are held close to air temperature. This can bear a cost, in that it decreases leaf cooling and WUE gains at high transpiration rates. As a benefit, leaves suffer less extremes of temperatures, neither high T in high sun and low transpiration, nor low T at night under radiative cooling (radiation frosts are a hazard; Leuning, 1988). All these effects are modulated by leaf position within the canopy, of course. Deeper in the canopy, wind penetration is decreased, as is radiative input both in shortwave and thermal radiation. As a result, the trade-offs vary with position and so can the leaf shape.

Size as linear dimension and thickness affects the efficiency of resource use. Larger, thicker petioles are demanded for broader and thicker leaves. 'Sun' leaves are thicker than shade leaves, for example. Costs of petioles actually make sun plants less effective than shade plants of the same size, for intercepting PAR (Sims & Pearcy, 1994). Why, then, does sun architecture occur? Vallardes & Pearcy (1998) propose that shade plants would be more damaged by photoinhibition, from high PAR interception on leaves of limited electron-transport capacity. Niklas (1992) argues that in one species of plant, petiole investment is excessive for light interception, summed over the day. I suggest that the architecture might be closer to optimal if one were to account for sunlight interception being more valuable early and late in the day, when vapour-pressure differences are smaller and WUE is larger.

Compound leaves require a higher investment in support (rachis plus petioles) than do simple leaves of the same area. Givnish (1978) argues that compound leaves are nevertheless a cheap disposable structure in seasonally dry tropics. They decrease water loss from (absent) branches in the dry season. Niinemets (1998a) also notes that petioles in compound leaves are low in N content and are therefore cheaper to construct in terms of the most limiting resource.

Leaf display angle presents a richness of trade-offs in leaf function. Angles that favour high light interception (normal to the sun, and perhaps actively tracking the sun) favour high efficiencies in using nutrients and N, but low efficiencies in using light (much light is intercepted at irradiances far exceeding the light-saturation point of the leaf) and in

using water. An extensive discussion, pointing to the abundant detail in the literature, has been given elsewhere (Gutschick, 1997). Several interesting phenomena can be summarized here. One is that the optimal leaf angle varies with depth in the canopy (Loomis & Williams, 1969; Duncan, 1971; Niinemets, 1998b) and with the relative importance of WUE over assimilation and growth rate (plants can change solar tracking modes with changes in water status (Forseth & Ehleringer, 1983; Reed & Travis, 1987)). Superior light-use efficiency is achieved with leaves that are more erect. Crops have been bred for this trait (Trenbath & Angus, 1975), which apparently gives up the stronger shading of competitors afforded by more planophile leaves. Canopies as a whole show different (higher) light-use efficiencies than individual leaves (Monteith, 1994). Finally, it is inadequate to characterize a leaf's light interception and light-use (or N-use) efficiency on the basis of the total of direct and diffuse interception. The distribution between direct and diffuse light is important, both for single leaves (fine leaves that give diffuse solar shadows or penumbras are beneficial; Gutschick, 1991) and for whole canopies (Leuning *et al.*, 1998).

Mesophyll structure, particularly total mass and nitrogen investments

The development of the palisade layer(s) of cells, as number and length, is most responsive to light levels. Sun leaves are markedly thicker and can have additional palisade layers compared with shade leaves (Nobel & Hartsock, 1981; Thompson *et al.*, 1988). In terms of resource use, development of the mesophyll (palisade plus spongy mesophyll) represents foremost the investment in N and in total construction costs. Therefore, the remaining discussion will focus on N and on total dry mass per leaf area.

Nitrogen is allocated to leaves increasingly as the opportunity for photosynthesis increases, especially with increasing PAR flux density. The declining investment in old leaves can be reversed (Johnston *et al.*, 1969). This is clearly adaptive, at least for plants in which fast growth is valuable, and the patterns along the gradient of microenvironments in a canopy have been so analysed (Hirose & Werger, 1987); for a first-principles derivation of the optimum investment in overall mass, see Gutschick & Wiegel (1988). Accompanying the increasing mass of N per leaf area in high sunlight is an increasing partitioning to carboxylation enzymes at the expense of light-capturing chlorophyll complexes (Cowan, 1986; Evans, 1989), another clearly adaptive pattern. In general, the patterns tend to maximize canopy photosynthetic rate per total mass or per total mass of N. These rates per mass are much more directly related to relative growth rate than are rates per leaf area (Gutschick, 1987).

To maintain an optimal distribution of N between leaves, N in old leaves that have been overtopped must be moved to new leaves. Such resorption and remobilization of N is widely observed, whereas the fraction reabsorbed is moderate, as it is for most nutrients (one-half or less, on a mass basis; Vitousek, 1982), notably less than the optimality models predict. A fundamental limitation on the remobilization of all nutrients is that enough of the biochemical and phloem-transport systems must be maintained to catabolize cell contents and to export nutrients in low-molecular-mass compounds.

The distribution of N investment, or total mass per leaf area, between different leaves on a plant is only part of the pattern. What sets the absolute magnitude of N per leaf area, N_a, or the related fractional content of N, f_N? There are substantial differences in N_a or f_N between plants in similar microenvironments (say, trees compared with grasses in the same geographic location). One determinant is certainly functional balance between root and shoot. Strong photosynthetic function of N in the shoot, as in high light, dilutes N, whereas strong root function in acquiring N increases f_N. A 'passive' balance sets f_N in this accounting (Gutschick, 1993; Gutschick & Kay, 1995). Rate limitations on the uptake of N (or P) lead to low f_N. High $[CO_2]$ increases shoot function and leads similarly to low f_N, as is widely observed, both in current experiments and in comparisons of leaves from earlier centuries with present-day leaves (Peñuelas & Matamala, 1990). Low-N soils often lead to thick or sclerophyllous leaves, which not only have low f_N but great thickness, or dry mass per unit area, and long leaf lifetimes. Givnish (1979) explained the combined pattern. He demonstrated, for example, that greater rates of increase of lifetime A occur at greater leaf thickness; only with great thickness does the marginal benefit equal the marginal cost of constructing the leaf and supporting its function. This pattern should nevertheless be re-examined in the light of recent findings showing that construction costs do not vary systematically with leaf structure (Poorter & Villar, 1997).

The optimum for average mass (or N) per leaf area in the whole canopy is rather broad and flat. Consequently, even substantial deviations from the optimum bear little cost in photosynthetic performance. Gutschick & Wiegel (1988) proposed that canopies develop with a mass per unit area that is well below the optimum. The extra leaf area developable per unit mass can decrease light availability to competing plants. The hypothesis has not been tested yet. The optimum mass per leaf unit area increases with leaf area index; indeed, canopies do follow this trend. A broadly related hypothesis is that leaves absorb excess light (by having higher PAR absorptivity than is optimal) to deprive competing plants of light. Conversely, plants with decreased chlorophyll levels were proposed as being superior in photosynthesis (Gutschick, 1984); field tests bore this out in soybean (Pettigrew *et al.*, 1989).

A second determinant is niche differentiation between life forms, such as trees versus grasses. Trees commonly have low A per leaf area (Wullschleger, 1993) (the exceptions cited by Nelson (1984) remain exceptions). Their strategy of development and of N use in particular differs from that of grasses. In regions with seasonal leafing-out, established trees use N that was reabsorbed to the trunk (Ryan & Bormann, 1982) for an early leaf flush that establishes a superior claim to light interception. However, young trees are often outcompeted by grasses. It remains puzzling to me why juvenile tree foliage is not programmed developmentally to attain high N_a and A. Certainly, such plasticity is possible, at least in shape, and might extend to N_a. Developmental plasticity can have little or no cost (J. Schmitt, pers. comm.). One possibility is that grasses are always superior in A per unit mass, given their very low investment in supporting structures. (A similar argument can be made for vines.) Thus, the competition for A per unit mass is lost, and trees, committed to a woody structural base, however modest at first, must express superiority in other resource use or later in time.

A third determinant is the trade-off between the instantaneous photosynthetic N-use efficiency (PNUE) and WUE. High N content, absolute or as a mass fraction, can confer high WUE. It increases the mesophyll conductance and decreases g_s/g_m, which increases WUE (see Eqn 2d). However, assimilation per mass of leaf is a modestly declining function of mass per unit area (Gutschick & Wiegel, 1988), at all magnitudes of mass per unit area. Thus, assimilation per unit mass of N declines similarly. Furthermore, if high g_m decreases C_s, the carboxylation rate per mass of Rubisco enzyme declines, also decreasing A per mass of N in the whole leaf. The magnitude of N_a that is optimal depends on the relative costs and benefits of water and N. High N availability favours high WUE and low PNUE. Such trade-offs are seen in different shrub species (Field *et al.*, 1983), but their generality is questionable (Meinzer *et al.*, 1992, Poorter & Farquhar, 1994).

Kranz anatomy

The specialization of leaf cells into mesophyll and bundle sheath is of most marked value in allowing the C_4 path of photosynthesis. Numerous articles have reviewed the advantages of C_4 photosynthesis in the efficiencies of using water, N, light, and even C substrates themselves (see, e.g., Ehleringer & Monson (1993), who also present the arguments that the pathway evolved in response to low $[CO_2]$ in the

Miocene epoch). The assimilation rates of C_4 plants are far less sensitive to ambient $[CO_2]$ than those of C_3 plants, because C_4 plants have a CO_2-concentrating biochemical pump. The relative performances of the two pathways in the currently increasing CO_2 levels are of intense interest. I still regard it as difficult to explain why C_4 plants have not replaced C_3 plants even more extensively. A common argument is that C_4 plants are more down-regulated by low temperatures. However, some C_4 plants function well at low T; if this can evolve in a few families, why has it not done so in all the families in which the C_4 path evolved independently? (This topic is also discussed by Press, 1999.)

Joint 'optimization' of nitrogen content and leaf lifetime

A variety of arguments have been developed to relate these traits. One early argument is that leaves with high N content per unit mass have high costs of construction and should require longer lifetimes to pay these costs back. A more careful analysis of payback rates, and field measurement, shows the reverse: high-N leaves are short-lived (Williams *et al.*, 1989; Reich *et al.*, 1997). At the other extreme, evergreens have long lifetimes that are commonly correlated with low N_a. However, this trait need not simply optimize PNUE evaluated over the whole life cycle. Jonasson (1989) found that, in five shrub species over widely different locations, leaf lifetimes were not markedly long, nor was the efficiency of reabsorbing N from senescing leaves high. He proposed that evergreenness was an adaptation to low rates of N supply from soil.

Extensive reviews have been presented by Garnier & Aronson (1998) and by Grime *et al.* (1997); and for a discussion of nutrient utility as affected by both leaf lifetime and nutrient resorption see Eckstein *et al.* (1999).

Leaf optical structure, relevant to PAR absorption

Many structures within the leaf, particularly the cell walls, scatter light (Fukshansky, 1991). The combined scattering and absorption of light lead to a steep gradient of total flux density with depth in the leaf. Thus, the photosynthetic rate also varies steeply with depth. However, novel leaf-sectioning experiments reveal that photosynthesis does not fall off with the same profile as light (Nishio *et al.*, 1993). This is true in particular at high light flux densities. At low light, photosynthesis does follow the light absorption profile, so that no light is wasted in excess absorption and the quantum yield reaches its limiting value near 0.05 mol CO_2 mol^{-1} photons. The adaptive value of the manner in which light absorption (pigment concentration) and enzyme capacity are distributed across the leaf thickness has

been investigated several times (Gutschick, 1984; Parkhurst, 1994). Modelling demonstrates that the exact shape of these profiles is not critical, but having an appreciable gradient allows a high actual quantum yield in high light. Interestingly, the whole-leaf rate of photosynthesis is not strongly altered by the distribution of stomatal conductance between top and bottom surfaces, even though CO_2 is constrained to enter the leaf at the surface with the lowest light level (Gutschick, 1984).

More details of leaf optical structure and the effects on leaf performance are given by Evans (1999) and Han *et al.* (1999).

Leaf optical structure for UV protection

Ultraviolet B and C flux is damaging to DNA in all cells, including plant cells. Leaves deploy UV-absorbing compounds (especially flavonoids) in the epidermis and throughout the mesophyll. As one might expect, the degree of protection is greatest in sun leaves most exposed to UV, and in longest-lived leaves having the greatest time-integrated dose (Day *et al.*, 1993). The cost of UV screens is currently being assessed. The demand for screening varies with depth in the leaf and also laterally, because plant cells both scatter and focus radiation. Consequently, screening is being investigated as a fully three-dimensional phenomenon (Alenius *et al.*, 1995).

LEAF STRUCTURE AFFECTS INTERACTIONS WITH HERBIVORES AND PATHOGENS

Mechanical defences

Waxy cuticle. This is the primary defence against microbial and viral invasion, as well as a barrier to water and solute loss (Hadley, 1980). Thickness is quite variable, as already noted; chemical composition and fine mechanical structure are perhaps more important in functions such as providing a barrier against water loss (Kerstiens, 1996). Fungi had a key role in the evolution of cuticle properties (Taylor & Osborn, 1996). Penetrating the cuticle remains a key step in the fungal invasion of leaves (Mendgen *et al.*, 1996). Fungi can degrade the cuticle (Commenil *et al.*, 1998; Sugui *et al.*, 1998), but leaves can detect the products (Schweizer *et al.*, 1996) and initiate other defences such as the hypersensitive reaction.

Trichomes. These leaf 'hairs' may be extensions of single epidermal cells. They may also themselves be multicellular (Esau, 1965). Trichomes occur in almost every plant family (Johnson, 1975) and commonly have a defensive value (Levin, 1973) because they impede herbivores mechanically (including making attachment difficult) or irritate them.

Trichomes can also secrete defensive compounds, as discussed later. Other functions of trichomes are known: (1) acquiring resources (leaf trichomes hold water and absorb both water and nutrients in some bromeliads (Raven *et al.*, 1992), while root trichomes are commonly known as root hairs, performing the same function); (2) limiting the interception of UV radiation or of total radiation and hence limiting leaf temperatures (Ehleringer, 1981; Baldocchi *et al.*, 1983); and (3) excreting excess salt, as in mangroves and many other species in diverse biomes.

Trichomes can deter feeding by small invertebrates (Letourneau, 1997). Kanno (1996) established this for soybeans attacked by false melon beetles. In addition to correlating the extent of herbivory with hairiness, he manipulated hairiness by shaving leaves. Finally, he showed that trichomes were responsible, rather than solvent-extractable chemical defences, by applying the latter in reciprocal treatments. In other plant–herbivore systems, the protection afforded by trichomes is less marked. Gannon & Bach (1996) found that the development of bean beetle larvae on soybeans was variously retarded or accelerated by trichome density, according to the specific larval stage. Nevertheless, the hairiest leaves induced a much higher mortality in larvae. Interestingly, stinging trichomes on two herbaceous plant species (*Urtica dioica* and *Laportea canadensis*) did not deter four species of invertebrate herbivores (Tuberville *et al.*, 1996). Such stinging trichomes are known to be more effective against large (mammalian) herbivores.

Trichomes bear costs. Substrates are required for constructing them, perhaps several percent of leaf construction costs. Another cost is a 'lost-opportunity' cost, the decrease in light interception. Typically, the effect on photosynthesis of directly sunlit leaves is very small. These leaves are light-saturated, and the principal effect might be a lowering of leaf temperature. A 1°C decrease might be caused by a 10% decrease in PAR absorptivity (depending on boundary-layer and stomatal conductances). This T decrease can decrease light-limited photosynthesis by approx. 6%. However, a water-limited sun plant also improves its water-use efficiency, to a degree dependent on air humidity and stomatal responses, for example. Returning to lost photosynthesis as a cost, this will be most significant for shade-lit leaves. A 10% decrease in PAR absorptivity for these leaves translates to a 10% decrease in photosynthesis. However, shade-lit leaves do not contribute heavily to photosynthesis in a whole plant canopy (as in crop monocultures). Even in a dense canopy, only approx. 20% of photosynthesis is by shade-lit leaves, as indicated in many model studies.

Because trichomes bear costs, their density seems to be regulated (inducible by threats) in some species

(Baur *et al.*, 1991). Trichomes can also alter the susceptibility of plants to fungal infection. For example, Wilson & Hanna (1998) found a positive correlation of pubescence with severity of smut (*Moesziomyces penicillariae*) in pearl millet. Enhanced trapping of fungal or oomycotal spores might be the cause.

Estimating the total cost of trichome presence is therefore somewhat involved. The combined cost of trichomes and related chemical defences can be moderately significant and therefore subject to measurable selection pressure. Mauricio & Rausher (1997) compared stands of *Arabidopsis thaliana* with and without natural insect enemies. The presence of insects selected for greater production of trichomes and defensive chemicals, the glucosinolates. Both the benefits of resistance and the costs in growth were resolvable. On the scale of a single plant generation (that is, showing plastic response, not natural selection), Wilkens *et al.* (1996) related trichome density of the tomato to resource availability and to caterpillar foraging on leaves. They tested a model of trade-offs of defence costs against benefits. This model focuses on net growth potential, and it predicts the highest levels of defence at an intermediate availability of resources. The model held for variations in availability of light, but less closely for water. Overall, trichomes seem to have modest defensive value (quantitative, not absolute) and modest cost. They are clearly not required, given that many plants have few or no leaf trichomes, yet they survive. Plants have other methods of defence or avoidance or leaf herbivory to use in various combinations, as we shall see later. The defences are effective, in that leaf herbivory is estimated at only 5–20% in some representative ecosystems (Golley, 1977; Schowalter *et al.*, 1981). The costs overall can be substantial, particularly for perennials. Tropical forests seem to suffer higher herbivory rates than do temperate forests, especially in young leaves and understory leaves, and more in seasonally dry forests than in wet forests (Coley & Barone, 1996).

Spines. Plants have a variety of sharp structures for defence: spines, which are modified leaves or stipules; thorns, which are modified branches; and prickles, such as sand burrs, which are epidermal growths. Here discussion will be limited to spines, but the cost–benefit analyses apply quite directly to thorns and prickles. Spines deter large herbivores, primarily mammals. Mammals have two basic methods of eating leaves: wholesale, with branches (pruning), and selectively, for leaves alone (picking). The formidable spines of *Acacia tortilis* in East Africa seem to protect neighbouring true leaves from pruning by goats, but not from picking (Gowda, 1996). They preserve not the leaves themselves but the potential to regenerate leaves from meristems. Also preserved are carbohydrate and nutrient reserves in the branch tissues. These are smaller but critical benefits. The cost of the spines was more than repaid by the shoot biomass saved. An attractive prospect is constructing and testing a model for the whole life cycle, to assess deferred benefits and costs such as these.

Tough tissues. Within leaves, the tissues with the highest mechanical resistance or toughness include the large veins. Toughness is conferred by thick cell walls and lignification in xylem vessels and in non-conducting fibre cells. Choong (1996) found quantitative relations of cell-wall volumetric content and fibre content to toughness in *Castanopsis fissa*. Toughness protected older leaves, whereas younger leaves had more protection by phenolic compounds. (The protective value of lower N content in old leaves was not evaluated.)

Abrasives. The most widespread abrasive compound is silica, effectively restricted to the grasses and some minor taxa such as *Equisetum*. The silica in grasses is hydrated, that is, opal (Baker, 1960). Its abrasiveness (Esau, 1965) limits herbivory on grasses, which otherwise are highly attractive for their absence of woody tissue and of toxins. One of the most remarkable hypotheses, being examined currently (Kaiser, 1998) is that the widespread replacement of other vegetation by grasses in the Miocene era drove the extinction of North American horses, which had lower rates of tooth growth than horses on other continents.

Stomatal design. Does the distribution of stomata, or their individual structure, help to decrease the risk of fungal entry into leaves? I once speculated that the distribution of stomata favouring abaxial over adaxial surfaces might be so protective (Gutschick, 1984). However, the macroevolution of plants seems not to favour a particular pattern of adaxial:abaxial ratio in stomatal density (Beerling & Kelly, 1996). It seems that the structure of individual stomata is more important in deterring fungal invasion. The topography of stomata can induce fungi to form (or not to form) appressoria, the hyphae that are specialized to penetrate other cells (Read *et al.*, 1997). Protection can be offered by a thick wax layer (Rubiales & Niks, 1996) or by waxy plugs that bear a cost of decreased photosynthesis (Brodribb & Hill, 1997).

Other structures, and overall architecture. A number of plant species have small structures (domatida) that harbour mites, which in turn can clean away fungal spores (O'Dowd & Willson, 1989). More generally, grasses protect their meristems from most grazing animals by their position at or below soil level. That is, aboveground, grasses are almost all leaf. In addition, the bunchgrass growth habit allows the rapid recovery of tissue growth after grazing; carbohydrates are mobilized rapidly from roots (Richards & Caldwell, 1985). Clonal growth is another way of protecting a sufficient number of meristems.

Chemical defences

Classes of compound. Chemical classifications of defensive compounds are readily found in both the introductory literature (Salisbury & Ross, 1992) and the specialist literature (Rosenthal & Berenbaum, 1991). In addition to chemical structure, an important distinction (Feeny, 1975) is between the *quantitative* defences present in large quantities and the high-potency *qualitative* defences. In the former group, tannins and resins have high metabolic costs, both for the producing leaf and for the herbivore that detoxifies the compounds. The latter group includes cardiac glycosides (Boppre, 1978) and insect hormones or analogues (Slama, 1980, 1987; Bowers, 1991). These toxins are present at levels $<1\%$ in dry mass but are highly effective at deterring herbivory (digitoxin) or at killing herbivores (insect juvenile hormones such as juvabione, which prevent normal metamorphosis). The potent chemicals trade off their high effectiveness against (1) limited range (e.g., cardiac glycosides, unlike tannins, do not affect invertebrates) and (2) a negligible cost to the herbivore once a biochemical adaptation evolves to avert the defence. It is well known that some insects can even turn the glycosides into protectants for themselves against their own predators (Boppre, 1978; Holzinger & Wink, 1996). Tannins act broadly against both invertebrates and vertebrates but require large investments of C.

More traditional categorizations by mode of action resolve the antifeedants, the toxins, the anti-digestants and the phytohormones. One antifeedant shared by several plant species is 2,4-dihydroxy-7-methoxy-1,4-benzoaxazin-3,1 (DIMBOA) (Barry *et al.*, 1994); resistance to corn borers, for example, is quantitatively related to DIMBOA concentration. Potent toxins that are effective against almost all herbivores (Rosenthal & Janzen, 1979) include alkaloids, glucosinolates and cardiac glycosides. These first two classes of chemical protectants are most common in crop plants and their relatives (Letourneau, 1997). As with all defences, they can be overcome by some herbivore species or races. Herbivores must expend both energy and nutrients in detoxifying plant protectants (Foley, 1992), which gives the chemicals a remanent value as defences even against capable herbivores.

The energetic and N costs of N-based defences are significant. However, as N availability to plants rises, these costs are diluted in a larger flux of metabolic energy. One might hypothesize that defences are most supportable at high N, given this dilution of costs and the greater benefit of defence (plants are more at risk at high N content because they are nutritionally more attractive to herbivores). The null hypothesis, by contrast, is that high-N plants are more damaged than low-N plants. Letourneau (1997) has reanalysed 135 published studies that purported to demonstrate that high N is correlated with high herbivory. She found very mixed correlation in true field conditions (not in pots or glasshouses). One alternative to high N attracting herbivores is that some insects are 'powerful' rather than discriminating feeders, and can consume more of low-N leaves to extract sufficient protein (Slansky & Feeny, 1977; Moran & Hamilton, 1980; Feeny, 1991; Woodward *et al.*, 1991;). One can also compare different genotypes under the same conditions to estimate whether the cost of defence is significant. Hoffland *et al.* (1996) found that radish cultivars that were more protected against a fungus had to bear significant costs of more cell wall in leaves (but less wall and more protein in roots). Darrow & Bowers (1997) found no apparent cost (as decreased growth rate) for more potent defences, namely the iridioid glycosides, in *Plantago* species that were better protected against lepidopteran insects. Costs of defence can be decreased by the defences' being inducible, as in systemic acquired resistance to fungi, with chitinase to break down fungal cell walls (Enkerli *et al.*, 1993). Infrequent expression of defences also decreases the selection pressure on insects to overcome defences, which could be more important than saving metabolic costs.

Attractants for enemies of plant pests. A classic association is that of ants with acacia trees, both in the neotropics and in East Africa. The classic analysis of Janzen (1966) has ramified greatly A recent analysis by Young *et al.* (1997) shows that five ant species colonize East African acacias in succession. Some trees provide nectar and young proteinaceous leaves for the ants to eat. The ants, in turn, protect the tree leaves from both mammalian and insect herbivores by aggressive attacks. The trade-offs vary with ant species. Some inhibit flowering, while making other leaves healthier. The association is obligatory; the acacias do not survive long without ants. Ant–plant associations also extend well beyond the acacias (see, e.g., Koptur, 1992; Davidson & McKey, 1993; Fonseca, 1994). The patterns of defences according to leaf age and the trade-offs of defence with leaf longevity do not fit a single, simple model (Fonseca, 1994).

A variety of plants exude chemical attractants for insects that prey upon othe insect species attacking the plants (Turlings *et al.*, 1995; Margolies *et al.*, 1997). Conversely, some plants exude repellents for insect herbivores when the leaves are attacked (Bernasconi *et al.*, 1998). The cost of these attractants or repellents relative to their benefits have yet to be assessed, although final yields with and without predators can be assessed.

Nutritional unattractiveness from low nutrient content. This topic has been discussed, in part, already. In general, low N content in leaves makes them unattractive to herbivores (Fox & Macauley, 1977; McNeil & Southwood, 1978). At very low N contents, in sclerophyllous leaves, herbivory is limited to those animals that have evolved special digestive strategies (i.e. leaf fermentation in the gut; Cork, 1996). These strategies allow the herbivore to extract sufficient N from quantities of leaf matter that would otherwise exceed intake capacity. Fermentation also decreases the passive loss of N in faeces.

Balancing defence costs and benefits. Theories of optimal defence have been predicated on several grounds, such as the balance of photosynthetic C supplies relative to N supplies (Bryant *et al.*, 1983). Assuming that secondary metabolism is plastic and scales up or down adaptively, high N availability favours the use of C supplies for growth relative to the production of C-based defences (i.e. tannins). As Herms & Mattson (1992) review in detail, the premises of plasticity and adaptiveness merit testing, as do the predictions of shifts in defence with nutrient status. The authors also present a wide-ranging discussion of complementary theories of defence, such as environmental constraint models (Bryant *et al.*, 1985), which postulate that other stresses such as drought might militate against growth, making the diversion of C to defence free from additional costs. The broadest framework is the model in which growth is traded off against differentiation (structural and metabolic) (Loomis, 1932, 1953). It should be noted that evidence for all the theories remains mixed at present.

A general premise is that the more valuable leaves should be better protected (e.g. by spines and chemicals). The comparative adjective 'more' requires elaboration. The youngest leaves on any one plant commonly have the highest photosynthetic rate (per mass as well as per area). Harper (1989) argues cogently that the earliest productivity of leaves (as exported photosynthate) is the most valuable and the most important to defend. The argument applies primarily in exponential growth, in which reinvestment of photosynthate in the ability to make more photosynthate is a critical determinant of final reproductive biomass and thus of fitness. Indeed, higher investments in defensive alkaloids can be found in younger leaves (Iwasa *et al.*, 1996). However, older leaves can contain more tannins, which cannot be retranslocated to younger leaves (Oleksyn *et al.*, 1997). Another sense of 'more valuable' arises in comparing plants of different nutritional status. Leaves of low-N plants in tropical forests have more tannins than leaves of trees in adjacent open areas (disturbed sites of high N) (Coley *et al.*, 1985). Are the low-N leaves more valuable to the low-N tree than high-N leaves are to the high-N tree? Or are high-N leaves better protected with N-based alkaloids that are cheaper for them to make? It is critical to make the proper comparison. A second consideration here is that some defences compromise other defences. Letourneau (1997, Table 1) gives some intriguing examples, such as trichomes hindering predators of mites as well as herbivorous mites themselves. Another example is an alkaloid, tomatine, decreasing the effectiveness of parasitoids of a leaf herbivore. These second-order interactions make it more challenging to assess costs and benefits. Reviews by Pasteels *et al.* (1983) and Gauld & Gaston (1994) are informative.

CONCLUSIONS

Many aspects of leaf structure have been studied for their adaptive value in permitting efficient resource use and in defence against herbivory. Within limited spheres (limited ranges of species to compare, or a limited range of environments), the studies are very enlightening about how leaf structure is moulded in many details (i.e. thickness, nutrient content and hairiness). Some larger challenges remain. First, some analyses must be joined. There are traits such as stomatal conductance that affect the use of light and water, and also N. Similarly, trichomes affect both defence and the use of resources such as light. Costs and benefits must be assessed for all these resources, in a single metabolic 'currency' or, better, a currency of ultimate Darwinian fitness. Economic analogies for multiple resource inputs do exist that are promising (Bloom *et al.*, 1985), but implementation is very difficult. I have attempted to use such a framework for evaluating the optimum in g_s and root:shoot ratio jointly, given specified availabilities of water and N. Problems arise from the beginning: for example, root investment is a resource use, but g_s is a modulator of resource-use efficiency, not a resource. The effect of g_s on leaf temperature, which also modulates performance, is further removed. More general frameworks, such as those in Gutschick (1987), only propose the existence of objective functions (goals) and of numerical searches for an optimum. Numerical values, however, are less informative and extensible than are true formalisms, such as equations for marginal gains.

Second, we must explain why different solutions exist for the same challenges: for example, different combinations of leaf N, stomatal conductance and leaf lifetime occur in coexisting plants. One general approach to explaining this fact is that multiple solutions are very nearly equally close to optimal. A second approach is that the different combinations represent evolutionarily stable strategies (Maynard Smith & Price, 1973) in a game-theory analysis of competition. Tilman (1982) elaborated some specific

behaviour in resource use alone, called resource partitioning, but greater generality is demanded. A third approach is to consider that selection pressures to evolve various traits and combinations of traits are much decreased by ubiquitous trade-offs, and very many in number, such that they all dilute each other. Regarding trade-offs, consider dry mass per leaf area. Gains in light-use efficiency are compromised by, or traded off against, losses in N-use efficiency. There are also trade-offs in time; WUE is more valuable in one growth stage, PNUE in another. This is the analogue of alternating selection in genetics. Trade-offs are so common and so closely matched that the theoretical optimum for this trait, and many others, is very broad. The concept of dilution arises from population genetics theory. To eliminate variant phenotypes, variant genotypes must be eliminated. Strong selection pressure can do so, over a certain number of generations, or weak selection over many generations. However, there cannot be a large number of strong selection pressures. Each strong selection requires a large number of 'genetic deaths' (a large proportion of propagules dying); many strong selection pressures eliminate the entire population or species. One can readily think of hundreds or perhaps thousands of selection pressures on a comparable number of traits. One concludes that most of the net selection pressures, after trade-offs are accounted for, are very weak or are multimodal. The challenge of explaining leaf structure then is largely inverted, to demonstrate that trade-offs are closely balanced and net selection pressure is weak, for all traits showing variability.

ACKNOWLEDGEMENTS

My wife, Dr Lou Ellen Kay, provided many helpful discussions, particularly on biotic interactions. Laura Huenneke of our faculty also provided useful comments. Eric Garnier and two anonymous reviewers made many important suggestions. The remaining deficiencies in the presentation on biotic interactions are my own. In apology, I offer that this topic lies outside my research experience. Nevertheless, the topic has been rewarding to learn, and for that I thank Eric Garnier, the organizer of the Fourth *New Phytologist* Symposium. My own research, on which part of this presentation was based, was supported by grant TUL-022-94/95 from the US Department of Energy, National Institutes for Global Environmental Change; grant DEB-94111971 from the National Science Foundation, Long-Term Ecological Research Program; and grant DE-FG02-97ER62332 from the US Department of Energy, Terrestrial Ecology Program.

REFERENCES

Alenius CM, Vogelmann TC, Bornmann JF. 1995. A three-dimensional representation of the relationship between penetration of UV-B radiation and UV-screening pigments of *Brassica napus*. *New Phytologist* **131**: 297–302.

Assmann S, Zeiger E. 1987. Guard cell bioenergetics. In: Zeiger E, Farquhar GD, Cowan IR, eds. *Stomatal function*. Stanford, USA: Stanford University Press, 163–193.

Baker G. 1960. Hook-shaped opal phytoliths in the epidermal cells of oats. *Australian Journal of Botany* **8**: 69–74.

Baldocchi D, Verma SB, Rosenberg NJ, Blad BL, Garay A, Specht JE. 1983. Leaf pubescence effects on the mass and energy exchange between soybean canopies and the atmosphere. *Agronomy Journal* **75**: 537–542.

Ball JT, Woodrow IE, Berry JA. 1987. A model predicting stomatal conductance and its contribution to the control of photosynthesis under different environmental conditions. In: Biggins J, ed. *Progress in photosynthesis research.* Dordrecht, The Netherlands, M. Nijhoff, vol. 4, 221–224.

Ball MC, Egerton JJG, Leuning R, Cunningham RB, Dunne P. 1997. Microclimate above grass adversely affects spring growth of seedling snow gum (*Eucalyptus pauciflora*). *Plant, Cell and Environment* **20**: 155–166.

Ball MC, Hodges V, Laughlin CP. 1991. Cold-induced photoinhibition limits regeneration of snow gum at treeline. *Functional Ecology* **5**: 663–668.

Barry D, Alfara D, Darrah LL. 1994. Relation of the European corn borer (Lepidoptera: *Pyralidae*) leaf-feeding resistance and DIMBOA content in maize. *Environmental Entomology* **23**: 177–182.

Baur R, Binder S, Benz G. 1991. Nonglandular leaf trichomes as short-term inducible defense of the grey alder, *Alnus incana* (L.), against the chrysomelid beetle, *Agelastica alni* L. *Oecologia* **87**: 219–226.

Beerling DJ, Kelly CK. 1996. Evolutionary comparative analyses of the relationship between leaf structure and function. *New Phytologist* **134**: 35–51.

Beerling DJ, Kelly CK. 1997. Stomatal density responses of temperate woodland plants over the past 7 decades of CO2 increase: a comparison of Salisbury (1927) with contemporary data. *American Journal of Botany* **84**: 1572–1583.

Bell CJ. 1982. A model of stomatal control. *Photosynthetica* **16**: 486–495.

Bernasconi ML, Turlings TCJ, Ambrosetti L, Bassetti P, Dorn S. 1998. Herbivore-induced emissions of maize volatiles repel the corn leaf aphid, *Rhopalosiphum maidis*. *Entomologia Experimentalis et Applicata* **87**: 133–142.

Berninger F, Makela A, Hari P. 1996. Optimal control of gas exchange during drought: empirical evidence. *Annals of Botany* **77**: 469–476.

Blackman PG, Davies WJ. 1985. Root to shoot communication in maize plants of the effects of soil drying. *Journal of Experimental Botany* **36**: 39–48.

Bloom AJ, Chapin FS III, Mooney HA. 1985. Resource limitation in plants – an economic analogy. *Annual Review of Ecology and Systematics* **16**: 363–392.

Boppre M. 1978. Chemical communication, plant relationships, and mimicry in the evolution of danaid butterflies cardiac glycosides, pyrrolizidine alkaloids. *Entomologia Experimentalis et Applicata* **24**: 64–77.

Bowers WS. 1991. Insect hormones and antihormones in plants. In: Rosenthal GA, Berenbaum MR, eds. *Herbivores: their interactions with secondary plant metabolites, 2nd edn.* San Diego, CA, USA: Academic Press, vol. 1, 431–456.

Brewer CA, Smith WK. 1995. Leaf surface wetness and gas exchange in the pond lily *Nuphar polysepalum* (Nymphaeaceae). *American Journal of Botany* **82**: 1271–1277.

Brewer CA, Smith WK. 1997. Patterns of leaf surface wetness for montane and subalpine plants. *Plant, Cell and Environment* **20**: 1–11.

Brodribb T, Hill RS. 1997. Imbricacy and stomatal wax plugs reduce maximum leaf conductance in southern hemisphere conifers. *Australian Journal of Botany* **45**: 657–668.

Bryant JP, Chapin FS III, Klein DR. 1983. Carbon/nutrient balance of boreal plants in relation to herbivory. *Oikos* **40**: 357–368.

Bryant JP, Chapin FS III, Reichardt P, Clausen T. 1985. Adaptation to resource availability as a determinant of chemical defense strategies in woody plants. *Recent Advances in Phytochemistry* **19**: 219–237.

Buckley TN, Farquhar GD, Mott KA. 1997. Qualitative effects of patchy stomatal conductance distribution features on gas-exchange calculations. *Plant, Cell and Environment* **20**: 867–880.

Bunce JA. 1997. Does transpiration control stomatal responses to water vapour pressure deficit? *Plant, Cell and Environment* **19**: 131–135.

Bunce JA. 1998. Effects of environment during growth on the sensitivity of leaf conductance to changes in humidity. *Global Change Biology* **4**: 269–274.

Choong MF. 1996. What makes a leaf tough and how this affects the pattern of *Castanopsis fissa* leaf consumption by caterpillars. *Functional Ecology* **10**: 668–674.

Coley PD, Barone JA. 1996. Herbivory and plant defenses in tropical forests. *Annual Review of Ecology and Systematics* **27**: 305–335.

Coley PD, Bryant JP, Chapin FS III. 1985. Resource availability and plant anti-herbivore defense. *Science* **230**: 895–899.

Collatz GJ, Ball JT, Grivet C, Berry JA. 1991. Physiological and environmental regulation of stomatal conductance, photosynthesis and transpiration: a model that includes a laminar boundary layer. *Agricultural and Forest Meteorology* **54**: 107–136.

Commenil P, Belingerhi L, DeHorter B. 1998. Antilipase antibodies prevent infection of tomato leaves by *Botrytis cinerea*. *Physiological and Molecular Plant Pathology* **52**: 1–14.

Cork SJ. 1996. Optimal digestive strategies for arboreal herbivorous mammals in contrasting forest types: why koalas and colobines are different. *Australian Journal of Ecology* **21**: 10–20.

Cowan IR. 1986. Economics of carbon fixation in higher plants. In: Givnish TJ, ed. *On the economy of plant form and function.* Cambridge, UK: Cambridge University Press, 133–170.

Cowan IR, Farquhar GD. 1977. Stomatal diffusion in relation to leaf metabolism and environment. *Symposia of the Society for Experimental Biology* **31**: 471–505.

Darrow K, Bowers, MD. 1997. Phenological and population variation in iridoid glycosides of *Plantago lanceolata* (Plantaginaceae). *Biochemical and Systematic Ecology* **25**: 1–11.

Davidson DW, McKey D. 1993. Ant–plant symbioses: stalking the Chyachaqui. *Trends in Ecology and Evolution* **8**: 326–332.

Day TA, Martin G, Coleman TC. 1993. Penetration of UV-B radiation in foliage: evidence that the epidermis behaves as a non-uniform filter. *Plant, Cell and Environment* **16**: 735–741.

DeSilva DLR, Hetherington AM, Mansfield TA. 1996. Where does all the calcium go? Evidence of an important regulatory role for trichomes in two calcicoles. *Plant, Cell and Environment* **19**: 880–886.

Dewar RC. 1995. Interpretation of an empirical model for stomatal conductance in terms of guard cell function: theoretical paper. *Plant, Cell and Environment* **18**: 365–372.

Duncan WG. 1971. Leaf angles, leaf area, and canopy photosynthesis. *Crop Science* **11**: 482–485.

Eckstein RL, Karlsson PS, Weih M. 1999. Life span and nutrient resorption as determinants of plant nutrient conservation in temperate-arctic regions. *New Phytologist* **143**: 177–189.

Ehleringer J. 1981. Leaf absorptances of Mojave and Sonoran Desert plants. *Oecologia* **49**: 366–370.

Ehleringer JR, Hammond SD. 1987. Solar tracking and photosynthesis in cotton leaves. *Agricultural and Forest Meteorology* **39**: 25–35.

Ehleringer JR, Monson RK. 1993. Evolutionary and ecological aspects of photosynthetic pathway variation. *Annual Review of Ecology and Systematics* **24**: 411–439.

Enkerli J, Gisu U, Mosinger E. 1993. Systemic adquired resistance to *Phytophthora infestans* in tomato and the role of pathogenesis-related proteins. *Physiological and Molecular Plant Pathology* **43**: 161–171.

Esau K. 1965. *Plant anatomy, 2nd edn.* New York, USA: Wiley.

Evans JR. 1989. Photosynthesis and nitrogen relationships in leaves of C_3 plants. *Oecologia* **78**: 9–19.

Evans JR. 1999. Leaf anatomy enables more equal access to light and CO_2 between chloroplants. *New Phytologist* **143**: 93–104.

Farquhar GD, von Caemmerer S, Berry JA. 1980. A biochemical model of photosynthetic CO_2 assimilation in leaves of C_3 species. *Planta* **149**: 78–90.

Feeny P. 1975. Biochemical coevolution between plants and their insect herbivores. In: Gilbert LE, Raven PH, eds. *Coevolution of animals and plants.* Austin, TX, USA: University of Texas Press, 3–19.

Feeny P. 1991. The evolution of chemical ecology: contributions from the study of herbivorous insects. In: Rosenthal GA, Berenbaum MR, eds. *Herbivores: their interactions with secondary plant metabolites, 2nd edn.* San Diego, CA, USA: Academic Press, vol. 2, 1–44.

Field C, Merino J, Mooney HA. 1983. Compromises between water-use efficiency and nitrogen-use efficiency in five species of California evergreens. *Oecologia* **60**: 384–389.

Foley WJ. 1992. Nitrogen and energy retention and acid–base status in the common ringtail possum (*Pseudocheirus peregrinus*) – evidence of the effects of absorbed allelochemicals. *Physiological Zoology* **65**: 403–421.

Fonseca CR. 1994. Herbivory and the long-lived leaves of an Amazonian ant tree. *Journal of Ecology* **82**: 833–842.

Forseth I, Ehleringer JR. 1983. Ecophysiology of two solar-tracking desert winter annuals. III. Gas-exchange responses to light, CO_2, and VPD in relation to long-term drought. *Oecologia* **57**: 344–351.

Fox LR, Macauley BJ. 1977. Insect (*Paropsis atomaria*) grazing on eucalyptus in response to variation in leaf tannins and nitrogen. *Oecologia* **29**: 145–162.

Fukshansky L. 1991. Photon transport in leaf tissue: applications in plant physiology. In: Myneni RB, Ross J, eds. *Photon–vegetation interactions: applications in optical remote sensing and plant ecology.* Berlin, Germany: Springer Verlag, 253–302.

Gannon AJ, Bach CE. 1996. Effects of soybean trichome density on Mexican bean beetle (Coleoptera, Coccinellidae) development and feeding preference. *Environmental Entomology* **25**: 1077–1082.

Garnier E, Aronson J. 1998. Nitrogen-use efficiency from leaf to stand level: clarifying the concept. In: Lambers H, Poorter H, Van Vuuren MMI, eds. *Inherent variation in plant growth. physiological mechanisms and ecological consequences.* Leiden, The Netherlands: Backhuys Publishers, 515–538.

Gauld ID, Gaston KJ. 1994. The taste of enemy-free space. In: Hawkins BA, Sheehan W, eds. *Parasitoid community ecology.* Oxford, UK: Oxford University Press, 281–299.

Givnish TJ. 1978. On the adaptive significance of compound leaves, with particular reference to tropical trees. In: Tomlinson PB, Zimmermann MH, eds. *Tropical trees as living systems.* Cambridge, UK: Cambridge University Press, 351–380.

Givnish T. 1979. On the adaptive significance of leaf form. In: Solbrig OT, Jain S, Johnson GB, Raven PH, eds. *Topics in plant population biology.* New York, USA: Columbia University Press, 375–410.

Givnish TJ. 1986. Optimal stomatal conductance, allocation of energy between leaves and roots, and the marginal cost of transpiration. In: Givnish TJ, ed. *On the economy of plant form and function.* Cambridge, UK: Cambridge University Press, 171–213.

Golley FB. 1977. Insects as regulators of forest nutrient cycling. *Tropical Ecology* **18**: 116–123.

Gowda JH. 1996. Spines of *Acacia tortilis*: what do they defend and how? *Oikos* **77**: 279–284.

Grace J. 1974. The effect of wind on grasses. I. Cuticular and stomatal transpiration. *Journal of Experimental Botany* **25**: 542–551.

Grime JP and 34 others. 1997. Integrated screening validates primary axes of specialization in plants. *Oikos* **79**: 259–281.

Gutschick VP. 1984. Photosynthesis model for C_3 leaves incorporating CO_2 transport, radiation propagation, and biochemistry. 2. Ecological and agricultural utility. *Photosynthetica* **18**: 569–595

Gutschick VP. 1987. *Functional biology of crop plants.* London, UK: Croom Helm/Portland, OR, USA: Timber Press.

Gutschick VP. 1991. Joining leaf photosynthesis models and canopy photon-transport models. In: Myneni RB, Ross J, eds. *Photon–vegetation interactions: applications in optical remote sensing and plant ecology.* Berlin, Germany: Springer Verlag, 501–535.

Gutschick VP. 1993. Nutrient-limited growth rates: roles of nutrient-use efficiency and of adaptations to increase uptake rate. *Journal of Experimental Botany* **44**: 41–51.

Gutschick VP. 1997. Photosynthesis, growth rate, and biomass allocation. In: Jackson LE, ed. *Ecology in agriculture.* San Diego, CA, USA: Academic Press, 39–78.

Gutschick VP, Kay LE. 1995. Nutrient-limited growth rates: quantitative benefits of stress responses and some aspects of regulation. *Journal of Experimental Botany* **46**: 995–1009.

Gutschick VP, Wiegel, FW. 1988. Optimizing the canopy photosynthetic rate by patterns of investment in specific leaf mass. *American Naturalist* **132**: 67–86.

Hadley NF. 1980. Surface waxes and integumentary permeability. *American Scientist* **68**: 546–553.

Haefner JW, Buckley TN, Mott KA. 1997. A spatially explicit model of patchy stomatal responses to humidity. *Plant, Cell and Environment* **20**: 1087–1098.

Han, T, Vogelmann T, Nishio J. 1999. Profiles of photosynthetic oxygen-evolution within leaves of *Spinacia oleracea. New Phytologist* **143**: 83–92.

Harper JL. 1989. The value of a leaf. *Oecologia* **80**: 53–58.

Haxeltine A, Prentice IC. 1996. A general model for light-use efficiency of primary production. *Functional Ecology* **10**: 551–561.

Herms DA, Mattson WJ. 1992. The dilemma of plants: to grow or defend. *Quarterly Review of Biology* **67**: 283–335.

Hirose T, Werger MJA. 1987. Maximizing daily canopy photosynthesis with respect to the leaf nitrogen allocation pattern in the canopy. *Oecologia* **72**: 520–526.

Hoffland E, Niemann GJ, Vanpelt, JA, Pureveen JBM, Eijkel GB, Boon JJ, Lambers H. **1996.** Relative growth rate correlates negatively with pathogen resistance in radish: the role of plant chemistry. *Plant, Cell and Environment* **19**: 1281–1290.

Holzinger F, Wink M. 1996. Mediation of cardiac glycoside insensitivity in the monarch butterfly (*Danaus plexippus*): role of an amino acid substitution in the ouabain binding site of Na^+,K^+-ATPase. *Journal of Chemical Ecology* **22**: 1921–1937.

Ihlenfeldt H-D, Hartmann HEK. 1982. Leaf surfaaces in Mesembryanthemaceae. In: Cutler DF, Alvin KL, Price CE, eds. *The plant cuticle.* San Diego, CA, USA: Academic Press, 397–423.

Iwasa Y, Kubo T, VanDam N, deJong TJ. 1996. Optimal level of chemical defense decreasing with leaf age. *Theoretical Population Biology* **50**: 124–148.

Janzen DH. 1966. Coevolution of mutualism between ants and acacias in Central America. *Evolution* **20**: 249–275.

Jarvis AJ, Davies WJ. 1998. The coupled response of stomatal conductance to photosynthesis and transpiration. *Journal of Experimental Botany* **49**: 399–406.

Jarvis PG, McNaughton KG. 1986. Stomatal control of transpiration: scaling up from leaf to region. *Advances in Ecological Research* **15**: 1–49.

Johnson HB. 1975. Plant pubescence: an ecological perspective. *Botanical Review* **41**: 233–258.

Johnston TJ, Pendleton JW, Peters DB, Hicks DR. 1969. Influence of supplemental light on apparent photosynthesis, yield, and yield components of soybeans (*Glycine max* L.). *Crop Science* **9**: 577–581.

Jonasson S. 1989. Implications of leaf longevity, leaf nutrient reabsorption and translocation for the resource economy of five evergreen plant species. *Oikos* **56**: 121–131.

Jones HG. 1992. *Plants and microclimate: a quantitative approach to environmental plant physiology,* 2nd edn. Cambridge, UK: Cambridge University Press.

Jones HG. 1998. Stomatal control of photosynthesis and transpiration. *Journal of Experimental Botany* **49**: 387–398.

Kaiser J. 1998. Tracking vanishing mammals and elusive nitrogen. *Science* **281**: 1274–1275.

Kanno H. 1996. Role of leaf pubescence in soybean resistance to the false melon beetle, *Atrychya menetriesi* Faldermann (Coleoptera, Chrysomelidae). *Applied Entomology and Zoology* **31**: 597–603.

Kerstiens G. 1996. Signaling across the divide: a wider perspective of cuticular structure structure–function relationships. *Trends in Plant Science* **1**: 125–129.

Koptur S. 1992. Plants with extrafloral nectaries and ants in Everglades habitats. *Florida Entomologist* **75**: 38–50.

Kurschner WM, Stulen I, Wagner F, Kuiper PJC. 1998. Comparison of paleobotanical observations with experimental data on the leaf anatomy of Durmast oak [*Quercus petraea* (Fagaceae)] in response to environmental change. *Annals of Botany* **81**: 657–664.

Letourneau DK. 1997. Plant-arthropod interactions in agroecosystems. In: Jackson LE, ed. *Ecology in agriculture.* San Diego, CA, USA: Academic Press, 239–290.

Leuning, R. 1988. Leaf temperatures during radiation frost. Part II. A steady state theory. *Agricultural and Forest Meteorology* **42**: 135–155.

Leuning R, Kelliher FM, De Pury DGG, Schulze E-D. 1998.

Leaf nitrogen, photosynthesis, conductance and transpiration: scaling from leaves to canopies. *Plant, Cell and Environment* **18**: 1183–1200.

Levin DA. 1973. The role of trichomes in plant defense. *Quarterly Review of Biology* **48**: 3–15.

Loomis RS, Williams WA. 1969. Productivity and the morphology of crop stands: patterns with leaves. In: Eastin JD, Haskins FA, Sullivan CY, van Bavel CHM, eds. *Physiological aspects of crop yield.* Madison, WI, USA: American Society of Agronomy, 27–51.

Loomis WE. 1932. Growth–differentiation balance vs. carbohydrate–nitrogen ratio. *Proceedings of the American Society of Horticultural Science* **29**: 240–245.

Loomis WE. 1953. Growth and differentiation – an introduction and summary. In: Loomis WE, ed. *Growth and differentiation in plants.* Ames, IA, USA: Iowa State University Press, 1–17.

Margolies DC, Sabelis MW, Boyer JE Jr. 1997. Response of a phytoseiid predator to herbivore-induced plant volatiles: selection on attraction and effect on prey exploitation. *Journal of Insect Behavior* **10**: 695–709.

Masle J. 1998. Growth and stomatal responses of wheat seedlings to spatial and temporal variation in soil strength of bi-layered soils. *Journal of Experimental Botany* **49**: 1245–1257.

Mauricio R, Rausher MD. 1997. Experimental manipulation of putative selective agents provides evidence for the role of natural enemies in the evolution of plant defense. *Evolution* **51**: 1435–1444.

Maynard Smith J, Price GR. 1973. The logic of animal conflict. *Nature* **246**: 15–18.

McNeil S, Southwood TRE. 1978. The role of nitrogen in the development of insect/plant relationships. In: Harborne JB, ed. *Biochemical aspects of plant and animal coevolution.* London, UK: Academic Press, 77–98.

Meinzer FC, Andrade JL, Goldstein G, Holbrook NM, Cavelier J, Jackson P. 1997. Control of transpiration from the upper canopy of a tropical forest: the role of stomatal, boundary-layer and hydraulic architecture components. *Plant, Cell and Environment* **20**: 1242–1252.

Meinzer FC, Rundel PW, Goldstein G, Sharifi MR. 1992. Carbon isotope composition in relation to leaf gas exchange and environmental conditions in Hawaiian *Metrisideros polymorpha* populations. *Oecologia* **91**: 305–311.

Melcher PJ, Goldstein G, Meinzer FC, Minyard B, Giambelluca TW, Loope LL. 1994. Determinants of thermal balance in the Hawaiian giant rosette plant *Argyroxiphium sandwicense. Oecologia* **98**: 412–418.

Mendgen K, Hahn M, Deising H. 1996. Morphogenesis and mechanisms of penetration by plant-pathogenic fungi. *Annual Review of Phytopathology* **34**: 367–386.

Monteith JL. 1994. Validity of the correlation between intercepted radiation and biomass. *Agricultural and Forest Meteorology* **68**: 213–220.

Moran N, Hamilton WD. 1980. Low nutritive quality as defense against herbivores. *Journal of Theoretical Biology* **86**: 247–254.

Morison JIL. 1998. Stomatal response to increased CO_2 concentration. *Journal of Experimental Botany* **49**: 443–452.

Mott KA. 1988. Do stomata respond to CO_2 concentrations other than intercellular? *Plant Physiology* **86**: 200–203.

Mott KA, Parkhurst DF. 1991. Stomatal responses to humidity in air and helox. *Plant, Cell and Environment* **14**: 509–515.

Nelson ND. 1984. Woody plants are not inherently low in photosynthetic capacity. *Photosynthetica* **18**: 600–605.

Niinemets U. 1998a. Are compound-leaved woody species inherently shade-intolerant: an analysis of species ecological requirements and foliar support costs. *Plant Ecology* **134**: 1–11.

Niinemets U. 1998b. Adjustment of foliage structure and function to a canopy light gradient in two coexisting deciduous trees: variability in leaf inclination angles in relation to petiole morphology. *Tree Structure and Function* **12**: 446–451.

Niklas KJ. 1992. Petiole mechanics, light interception by lamina, and 'economy in design'. *Oecologia* **90**: 518–526.

Nishio JN, Sun J, Vogelmann TC. 1993. Carbon fixation gradients across spinach leaves do not follow internal light gradients. *Plant Cell* **5**: 953–961.

Nobel PS, Hartsock TL. 1981. Development of leaf thickness for *Plectanthrus parviflorus* – influence of photosynthetically active radiation. *Physiologia Plantarum* **51**: 163–166.

O'Dowd DJ, Willson MF. 1989. Leaf domatia and mites on

Australasian plants: ecological and evolutionary implications. *Biological Journal of the Linnean Society* **37**: 191–236.

Oleksyn J, Tjoelker MG, Lorenplucinska G, Konwinska A, Zytkowiak R, Karolewski P, Reich PB. 1997. Needle CO_2 exchange, structure and defense traits in relation to needle age in *Pinus heldreichii* Christ: a relic a Tertiary flora. *Tree Structure and Function* **12**: 82–89.

Parkhurst DF. 1994. Diffusion of CO_2 and other gases inside leaves. *New Phytologist* **126**: 449–479.

Pasteels JM, Gregoire J-C, Rowell-Rahier M. 1983. The chemical ecology of defense in arthropods. *Annual Review of Entomology* **28**: 263–289.

Pearcy KE, Baker EA. 1991. Effects of simulated acid rain on needle wettability and rain retention by two Sitka spruce (*Picea sitchensis*) clones. *Canadian Journal of Forest Research* **21**: 694–697.

Peñuelas J, Matamala R. 1990. Changes in N and S leaf content, stomatal density and specific leaf area of 14 plant species during the last three centuries of CO_2 increase. *Journal of Experimental Botany* **41**: 1119–1124.

Pettigrew WT, Hesketh JD, Peters DB, Woolley JT. 1989. Characterization of canopy photosynthesis of chlorophyll-deficient soybean isolines. *Crop Science* **29**: 1025–1029.

Pitcairn CER, Grace J. 1985. Wind and surface damage. *Progress in Biometeorology* **2**: 115–126.

Poole I, Weyers JDB, Lawson T, Raven JA. 1996. Variations in stomatal density and index: implications for paleoclimatic reconstructions. *Plant, Cell and Environment* **19**: 705–712.

Poorter H, Farquhar GD. 1994. Transpiration, intercellular carbon dioxide concentration and carbon-isotope discrimination in 24 wild species differing in relative growth rate. *Australian Journal of Plant Physiology* **21**: 507–516.

Poorter H, Villar R. 1997. The fate of acquired carbon in plants: chemical composition and construction costs. In: Bazzaz FA, Grace J, eds. *Plant resource allocation.* San Diego, CA, USA: Academic Press, 39–72.

Pospísilová J, Solárová J. 1980. Environmental and biological control of diffusive conductances of adaxial and abaxial leaf epidermes. *Photosynthetica* **14**: 90–127.

Press MC. 1999. The functional significance of leaf structure: a search for generalizations. *New Phytologist* **143**: 213–219.

Raven PH, Evert RF, Eichhorn SE. 1992. *Biology of plants,* 5th edn. New York, USA: Worth.

Read ND, Kellock LJ, Collins TJ, Gundlach AM. 1997. Tole of topography sensing for infection-structure differentiation in cereal rust fungi. *Planta* **202**: 163–170.

Reed R, Travis RL. 1987. Paraheliotropic leaf movements in mature alfalfa canopies. *Crop Science* **27**: 301–304.

Reich PB, Walters MB, Ellsworth DS. 1997. From tropics to tundra: global convergence in plant functioning. *Proceedings of the National Academy of Sciences, USA* **94**: 13730–13734.

Richards JH, Caldwell MM. 1985. Soluble carbohydrates, concurrent photosynthesis and efficiency in regrowth following defoliation: a field study with *Agropyron* species. *Journal of Applied Ecology* **22**: 907–920.

Richards R. 1991. Crop improvement for temperate Australia: future opportunities. *Field Crops Research* **26**: 141–169.

Robinson JM. 1994. Speculations on carbon dioxide starvation, Late Tertiary evolution of stomatal regulation and floristic modernization. *Plant, Cell and Environment* **17**: 345–354.

Rosenthal GA, Berenbaum MR, eds. 1991. *Herbivores; their interactions with secondary plant metabolites,* 2nd edn, vol. 1 (*The Chemical Participants*). San Diego, CA, USA: Academic Press.

Rosenthal GA, Janzen DH. 1979. *Herbivores: their interaction with secondary plant metabolites.* New York, USA: Acadmic Press.

Rubiales D, Niks RE. 1996. Avoidance of rust infection by some genotypes of *Hordeum chilense* due to their relative inability to induce the formation of appressoria. *Physiological and Molecular Plant Pathology* **49**: 89–101.

Ryan DF, Bormann FH. 1982. Nutrient resorption in northern hardwood forests. *BioScience* **32**: 29–32.

Saliendra NZ, Sperry JS, Comstock JP. 1995. Influence of leaf water status on stomatal response to humidity, hydraulic conductance, and soil drought in *Betula occidentalis. Planta* **196**: 357–366.

Salisbury FB, Ross, CW. 1992. *Plant physiology,* 4th edn. Belmont, CA, USA: Wadsworth.

Santrucek J, Sage RF. 1996. Acclimation of stomatal conductance to a CO_2-enriched atmosphere and elevated temperature in *Chenopodium album*. *Australian Journal of Plant Physiology* **23**: 467–478.

Schowalter TD, Warren Webb J, Crossley DA Jr. 1981. Community structure and nutrient content of canopy arthropods in clearcut and uncut forest ecosystems. *Ecology* **62**: 1010–1019.

Schreiber L, Riederer M. 1996. Ecophysiology of cuticular transpiration: comparative investigation of cuticular water permeability of plant species from different habitats. *Oecologia* **107**: 426–432.

Schweizer P, Felix G, Buchala A, Muller C, Metraux JP. 1996. Perception of free cutin monomers by plant cells. *Plant Journal* **10**: 331–341.

Sims DA, Pearcy RW. 1994. Scaling sun and shade photosynthetic acclimation of *Alocasia macrorrhiza* to whole-plant performance. I. Carbon balance and allocation at different daily photon flux densities. *Plant, Cell and Environment* **17**: 881–887.

Slama K. 1980. Animal hormones and antihormones in plants. *Biochemie und Physiologie der Pflanzen* **175**: 177–193.

Slama K. 1987. Insect hormones and bioanalogues in plants. *Series Entomologica* **41**: 9–16.

Slansky F, Jr., Feeny P. 1977. Stabilization of the rate of nitrogen accumulation by larvae of the cabbage butterfly (*Pieris rapae*) on wild and cultivated food plants. *Ecological Monographs* **47**: 209–228.

Smith WK, McClean TM. 1989. Adaptive relationship between leaf water repellency, stomatal distribution, and gas exchange. *American Journal of Botany* **76**: 465–469.

Sugui JA, Pascholati SF, Kunoh H, Howard RJ, Nicholson RL. **1998.** Association of *Pestalotia malicola* with the plant cuticle: visualization of the pathogen and detection of cutinase and nospecific esterase. *Physiological and Molecular Plant Pathology* **52**: 213–221.

Tardieu F. 1994. Growth and functioning of roots and of root systems subjected to soil compaction: towards a system with multiple signaling. *Soil Tillage Research* **30**: 217–243.

Tardieu F, Lafarge T, Simmoneau Th. 1996. Stomatal control by fed or endogenous xylem ABA in sunflower: interpretation of correlations between leaf water potential and stomatal control in anisohydric species. *Plant, Cell and Environment* **19**: 75–84.

Tardieu F, Simmoneau T. 1998. Variability among species of stomatal control under fluctuating soil water status and evaporative demand: modelling isohydric and anisohydric behaviours. *Journal of Experimental Botany* **49**: 419–432.

Tardieu F, Zhang J, Gowing DJG. 1993. Stomatal control by both [ABA] in the xylem sap and leaf water status: a test of a model for droughted or ABA-fed field-grown maize. *Plant, Cell and Environment* **16**: 413–420.

Taylor TN, Osborn JM. 1996. The importance of fungi in shaping the paleoecosystem. *Reviews of Palaeobotany and Palynology* **90**: 249–262.

Thompson WA, Stocker GC, Kriedemann PE. 1988. Growth and photosynthetic response to light and nutrients of *Flindersia brayleyana* F. Muell., a rainforest tree with broad tolerance to sun and shade. *Australian Journal of Plant Physiology* **15**: 299–315.

Tilman D. 1982. *Resource competition and community structure.* Princeton, NJ, USA: Princeton University Press.

Trenbath BR, Angus JF. 1975. Leaf inclination and crop production. *Field Crops Abstracts* **28**: 231–244.

Tuberville TD, Dudley PG, Pollard AJ. 1996. Responses of invertebrate herbivores to stinging trichomes of *Urtica dioica* and *Laportea canadensis*. *Oikos* **75**: 83–88.

Turlings TCJ, Loughrin JH, McCall PJ, Rose USR, Lewis WJ, Tumlinson JH. 1995. How caterpillar-damaged plants protect themselves by attracting parasitic wasps. *Proceedings of the National Academy of Sciences, USA* **92**: 4169–4174.

Turner IM. 1994. A quantitative analysis of leaf form in woody plants from the world's major broadleaved forest types. *Journal of Biogeography* **21**: 413–419.

Tyree MT, Sperry JS. 1988. Do woody plants operate near the point of catastrophic xylem dysfunction caused by dynamic water stress? Answers from a model. *Plant Physiology* **88**: 574–580.

Vallardes F, Pearcy RW. 1998. The functional ecology of shoot architecture in sun and shade plants of *Heteromeles arbutifolia* M. Roem., a California chaparral shrub. *Oecologia* **114**: 1–10.

Vitousek P. 1982. Nutrient cycling and nutrient use efficiency. *American Naturalist* **119**: 553–572.

Wagner F, Below R, Deklerk R, Dilcher DL, Joosten H, Kurschner WM, Visscher H. 1996. A natural experiment on plant acclimation: lifetime stomatal frequency response of an individual tree to annual atmospheric CO_2 increase. *Proceedings of the National Academy of Sciences, USA* **93**: 11705–11708.

Wilkens RT, Shea GO, Halbreich S, Stamp NE. 1996. Resource availability and the trichome defenses of tomato plants. *Oecologia* **106**: 181–191.

Williams K, Field CB, Mooney HA. 1989. Relationships among leaf construction cost, leaf longevity, and light environment in rain-forest plants of the genus *Piper*. *American Naturalist* **133**: 198–211.

Wilson JP, Hanna WW. 1998. Smut resistance and grain yield of pearl millet hybrids near-isogenic at the *tr* locus. *Crop Science* **38**: 649–651.

Wong S-C, Cowan IR, Farquhar GD. 1985. Leaf conductance in relation to rate of CO_2 assimilation. 1. Influence of nitrogen nutrition, phosphorus nutrition, photon flux density, and ambient partial pressure of CO_2 during ontogeny. *Plant Physiology* **78**: 821–825.

Woodward FI. 1998. Do plants really need stomata? *Journal of Experimental Botany* **49**: 471–480.

Woodward FI, Thompson GB, McKee IF. 1991. The effects of elevated concentrations of carbon dioxide on individual plants, populations, communities and ecosystems. *Annals of Botany* **67** (Suppl. 1): 23–38.

Wullschleger SD. 1993. Biochemical limits to carbon assimilation in C_3 plants: a retrospective analysis of the A/C_i curves from 109 species. *Journal of Experimental Botany* **44**: 907–920.

Young TP, Stubblefield CH, Isbell LA. 1997. Ants of swollen-thorn acacias: species coexistence in a simple system. *Oecologia* **109**: 98–107.

Zhang J, Davies WJ. 1990. Does ABA in the xylem control the rate of leaf growth in soil-dried maize and sunflower plants? *Journal of Experimental Botany* **41**: 1125–1132.

New Phytol. (1999), **143**, 19–31

Research review
A mechanical perspective on foliage leaf form and function

KARL J. NIKLAS

Section of Plant Biology, Cornell University, Ithaca, New York 14853, USA (tel +1 607 255 8727; fax +1 607 255 5407; e-mail kjn2@cornell.edu)

Received 11 November 1998; accepted 24 March 1999

SUMMARY

The mechanical behaviour of large foliage leaves in response to static and dynamic mechanical forces is reviewed in the context of a few basic engineering principles and illustrated in terms of species drawn from a variety of vascular plant lineages. When loaded under their own weight or subjected to externally applied forces, petioles simultaneously bend and twist, and thus mechanically operate as cantilevered beams. The stresses that develop in petioles reach their maximum intensities either at their surface or very near their centroid axes, where they are accommodated either by living and hydrostatic tissues (parenchyma and collenchyma) or dead and stiff tissues (sclerenchyma and vascular fibres) depending on the size of the leaf and the species from which it is drawn. Allometric analyses of diverse species indicate size-dependent variations in petiole length, transverse shape, geometry and stiffness that accord well with those required to maintain a uniform tip-deflection for leaves with laminae differing in mass. When dynamically loaded, the laminae of many broad-leaved species fold and curl into streamlined objects, thereby reducing the drag forces that they experience and transmit to their subtending petioles and stems. From a mechanical perspective, the laminae of these species operate as stress-skin panels that distribute point loads more or less equally over their entire surface. Although comparatively little is known about the mechanical structure and behaviour of foliage leaves, new advances in engineering theory and computer analyses reveal these organs to be far more complex than previously thought. For example, finite-element analyses of the base of palm leaves reveal that stresses are decreased when these structures are composed of anisotropic as opposed to isotropic materials (tissues).

Key words: biomechanics, finite element analyses, leaves, plants.

INTRODUCTION

'Apart from the mere mechanical support of its own weight in still air, the leaf will have to resist undamaged the pressure of winds, and maintain its tissues functionally unimpaired by them.'

F. O. Bower (1925)

Botanists have long recognized that, despite their diversity in size, shape and internal construction, the petioles and laminae of foliage leaves operate as cantilevered beams and stress-skin panels respectively (Bower, 1925; Gibson *et al.*, 1988; Vogel, 1989; Niklas, 1992a) that must be sufficiently stiff to support their own weight mechanically against the pull of gravity yet flexible enough to bend, twist or fold without typically breaking when subjected to large externally applied dynamic forces (Raupach & Thom, 1981; Cannell & Coutts, 1988; Vogel, 1989). When viewed in this highly abstract way, leaves manifest a basic mechanical structure and behaviour that can be quantitatively understood in terms of a

comparatively few, albeit basic, engineering principles. The objective of this paper is to review these principles and to illustrate them with examples of leaves drawn from a variety of vascular plant species.

However, no attempt is made to be synoptic. Phyletic differences in leaf morphology and anatomy cannot be discounted, however much leaves might share the same basic mechanical structure. Leaves have evolved many times in different clades (Stewart & Rothwell, 1993; Taylor & Taylor, 1993), and, although they invariably develop exogenously as appendicular structures produced by shoot apical meristems (Esau, 1967; Mauseth, 1988), it is reasonable to infer that leaves of tracheophytes are not invariably homologous structures. If so, then the general similarity in the mechanical structure of leaves, which seems to hold for most vascular plant species, is a result of convergent evolution. This implies that only the most general model for leaf biomechanics can be adduced.

Because all form–function relationships are size-dependent, an allometric as well as an engineering approach is also taken, one that emphasizes and evaluates the extent to which the physical or mechanical properties of leaves scale with respect to size. Many of the engineering principles governing the behaviour of leaf biomechanics require size-dependent variations, so the hypothesis that leaves comply with a general biomechanical model can be tested in part on the basis of whether they manifest size-dependent variations predicted on the basis of this model.

For convenience, the features of the leaf mechanical model are presented in four separate sections: the mechanics of cantilevered structures, the internal stresses and strains that develop in them, the mechanical structure of the leaf lamina, and, finally, the responses of entire leaves to wind-induced drag. Emphasis is placed on the mechanics of the petiole for two reasons. First, the lamina of fern, cycad and most dicot leaves is attached to the stem by means of this structure, the leaves of many monocot species have prominent and well developed petiole-like structures (e.g. *Dieffenbachia*, *Philodendron* and palm species), referred to as pseudopetioles (Mauseth, 1988), and the strap-shaped leaves of grasses, the sessile leaves of many dicot species, and the needle-like leaves of many conifer species are also cantilevered mechanical 'devices'. Thus, most leaves possess a petiole or petiole-like structure. Second, much of what is currently known about leaf biomechanics is based on the behaviour of petioles because these structures are easily modelled from an engineering perspective. In contrast, the behaviour of thin laminae is largely anecdotally known because these structures are very difficult to model analytically.

THE CANTILEVERED LEAF

The foliage leaves of most species fall into one of three categories of mechanical structure: (a) an untapered cantilevered beam (the petiole) anchored at one end (by a phyllopodium) and supporting a mass (the lamina) at the other (e.g. simple and palmately compound leaves), (b) a tapered cantilevered beam (rachis) anchored at its base (by a petiole) supporting a series of masses (leaflets) along its length (e.g. pinnately compound leaves), and (c) a cantilevered beam bearing a more or less uniform load distribution along its length (e.g. grass and pine leaves). Despite their differences, all three categories manifest the same behaviour: they bend under their own weight and they bend or twist in response to the movement of the fluid in which they are submerged (i.e. static and dynamic bending, respectively) (Fig. 1). Because leaf orientation to direct sunlight is governed in part by the flexure resulting from self-loading and because the response of leaves to high wind speeds or falling debris is dictated by their

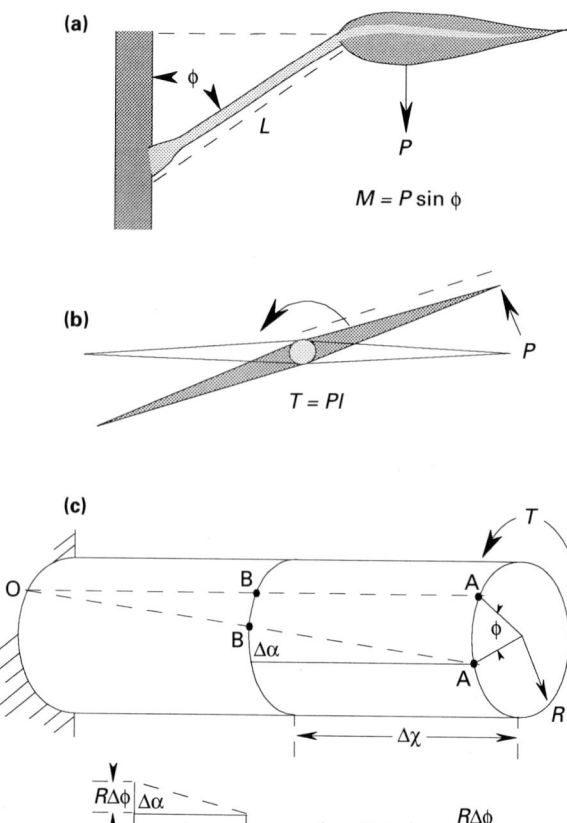

Fig. 1. Bending and torsional moments experienced by leaves. (a) A petiole of length L experiences a bending moment M equal to the combined load (weight) P resulting from its mass and that of the lamina to which it is attached × the sine of the inclination angle ϕ from the vertical. (b) When dynamically loaded by wind pressure, resolved here as a point load P acting the edge of a lamina, a leaf experiences a torque T equal to the load P × the lever arm l through which the load P acts. (c) Longitudinal variation in the torsional shear strain γ as a function of the change in distance Δx from the tip of a petiole subjected to a torque T and with an angle of twist ϕ. The point A will rotate to point A′ through the angle ϕ, whereas the point B will rotate to point B′ through the angle $\phi - \Delta\phi$. The small element ABB′A′, will therefore experience a shear strain γ equal to $\Delta\alpha \approx \tan\Delta\alpha = R\Delta\phi/\Delta x$, where R is the distance from the centre of the petiole. Accordingly, the torsional shear strain increases from the centre to the surface of the beam-like petiole and increases from the base to the free end of the petiole.

ability to twist as well as to bend, the mechanical principles governing the physical behaviour of leaves in response to forces are physiologically and ecologically important.

Although bending and twisting forces are mechanically different phenomena (Wainwright *et al.*, 1976), the formulae describing the response of any cantilevered device to these forces are mathematically simple and similar (Bisshop & Drucker, 1945; Timoshenko & Gere, 1961). Specifically, the resistance offered to a bending force (P) is inversely proportional to the cube of the cantilevered length (L) and directly proportional to its flexural rigidity (EI), which is a measure of the resistance of a

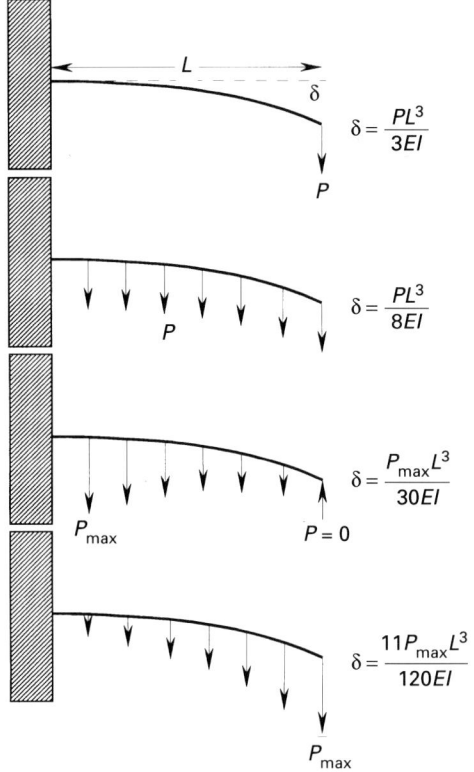

$$\delta = \frac{PL^3}{3EI}$$

$$\delta = \frac{PL^3}{8EI}$$

$$\delta = \frac{P_{max}L^3}{30EI}$$

$$\delta = \frac{11P_{max}L^3}{120EI}$$

Fig. 2. Tip-deflection formulas for cantilevered beams experiencing small elastic deformations resulting from different distributions of mechanical force P. In each case, the vertical deflection, δ, measured at the free end of the cantilever is proportional to the force P and the cube of beam length L and is inversely proportional to the flexural rigidity of the beam EI. Taken from Niklas (1992a, p. 171).

structure to bending (Fig. 2). Similarly, the ability of a cantilever to resist a twisting force is inversely proportional to L and directly proportional to its torsional rigidity (GJ), which is a measure of the resistance of a structure to twisting (Wainwright *et al.*, 1976; Niklas, 1992a). In each case, the resistance to a mechanical force is inversely proportional to some function of L and directly proportional to the product of a measure of the stiffness of a material (i.e. in bending, stiffness is numerically expressed by the Young modulus E; in twisting, stiffness is quantified by the torsional shear modulus G) and a parameter called the second moment of area, which describes how the geometry and the absolute size of a material contribute to the ability to cope with bending or twisting (i.e. in bending, the moment is called the axial second moment of area I; in torsion, the moment is called the polar second moment of area J).

Traditional engineering literature assumes that solids are linearly elastic materials, that is, that the strains produced in a solid that result from any externally applied force are always proportional to the magnitude of the force acting though a material normalized with respect to the cross-sectional area through which the force acts. Because the magni-

tudes of the elastic moduli E and G are the slopes of the respective stress–strain curves for a solid material and because stress is calculated such that it is independent of the size of the material being tested (stress = force divided by the area through which it acts), the assumption of linear elasticity means that E and G are assumed to be size-independent and constant properties for each material. Although this assumption greatly simplifies engineering theory, it does not hold for most plant materials. The available data indicate that the magnitudes of E and G increase to a limit, as the cross-sectional area of tissue samples increases. A partial explanation for this phenomenon is that the stiffness of most tissue types depends on the number of cells in a tissue sample, because adjoining cell walls assist one another in resisting deformation (Niklas, 1992a).

Numerous studies also show that the elastic moduli of even an anatomically well defined tissue type vary across species as a consequence of inter-specific differences in cell-wall thickness, shape or chemistry (Wainwright *et al.*, 1976; Vincent, 1990; Niklas, 1992a). Other studies indicate that the elastic moduli of a tissue type can vary within a single plant owing to physiological changes. This is especially true for thin-walled, living tissues (e.g. parenchyma and collenchyma) whose stiffness can change in response to cell turgor (Falk *et al.*, 1958; Greenberg *et al.*, 1989; Niklas, 1989; Niklas & Paolillo, 1997). Because the stiffness of organic structures, such as leaves, also depends on which tissues (and how much of each) are present, ontogenetic and phylogenetic variations in leaf stiffness are expected regardless of whether fundamental anatomical differences exist across clades or whether transient physiological changes occur.

Similarly, the numerical values of the second moments of area I and J depend on cross-sectional shape and absolute size (Timoshenko & Gere, 1961; Wainwright *et al.*, 1976; Niklas, 1992a), and can therefore vary markedly within and across species as a consequence of ontogenetic or phyletic variation. Indeed, the magnitudes of I and J can change along the length of a single leaf as a result of variations in the cross-sectional shape or size of a petiole, rachis or leaf lamina. The second moments of area are also physiologically dependent parameters because cross-sectional size and shape can be altered as a consequence of water stress and tissue dehydration. Because bending or twisting rigidity is the product of bulk tissue stiffness and the axial or polar second moment of area, and because each of these parameters can vary between and within species, petioles are subtle mechanical devices and an individual plant can produce a large number of leaves whose mechanical behaviours vary widely depending on the ambient environmental conditions attending the growth and development of individual leaves (Tomlinson, 1962; Vogel, 1992; Niklas, 1996).

Nevertheless, if transient physiological changes are neglected, the mechanical behaviour of leaves tends to scale with respect to size in a fairly predictable and uniform way across and within species. More importantly, the scaling of leaf biomechanics complies well with the basic engineering principles governing cantilevered beams.

This is illustrated by first considering a petiole bending under the action of a single force (P) acting on its free end (the weight of the lamina of a simple or palmately compound leaf) and whose deflection (δ) at any distance (x) from the free end ($x = 0$) is given by the formula:

$$\delta_x = \left(\frac{-PL^3}{3EI}\right)\left(1 - \frac{3x}{2L} + \frac{x^3}{2L^3}\right) \qquad \text{Eqn 1}$$

(L is the length of the cantilever (Bisshop & Drucker, 1945; Timoshenko & Gere, 1961) (Fig. 2)). For very small elastic deflections (i.e. $\delta = 0.10L$), the tip deflection of the petiole is approximated by the formula:

$$\delta_{x=0} = PL^3/3EI \qquad \text{Eqn 2}$$

From these equations it is obvious that, regardless of the materials used in their construction, short petioles have smaller tip-deflections than longer petioles. It is also apparent that for equivalent lengths, petioles composed of stiff materials have smaller tip-deflections than those composed of less stiff materials. Parenchyma is the least stiff, wood the stiffest known plant tissue;. Between these two tissues, in ascending order of stiffness, are collenchyma, primary xylem and sclerenchyma (Wainwright *et al.*, 1976; Niklas, 1992a, 1993). Because wood is generally unavailable for the fabrication of leaves, any or all of the remaining 'building materials' serve as the principal stiffening agent in a petiole.

The influence of the axial second moment of area on tip-deflection is also obvious because an increase in I can have the same effect as an increase in E regardless of the tissues used to construct a leaf. A large increase in I actually requires no addition of mass (and thus no increase in self-loading) because I is shape- as well as size-dependent. For example, the second moment of area for a petiole with an elliptical cross section depends on the orientation of the major axis of the elliptical section with respect to the vertical. Assuming that the major axis of the ellipse is twice the length of the minor axis, I increases fourfold when the major axis is vertical as opposed to horizontal to the Earth's surface. Naturally, very long petioles with elliptical cross sections are not rigid structures because they bend easily when loaded dynamically (Grace, 1978).

As noted, the biomechanical properties of leaves seem to scale in accordance with Eqn 2. Allometric analyses of 193 leaves differing in size drawn from 19 dicot and monocot species reveal that petiole flexural rigidity scales with respect to petiole length such that

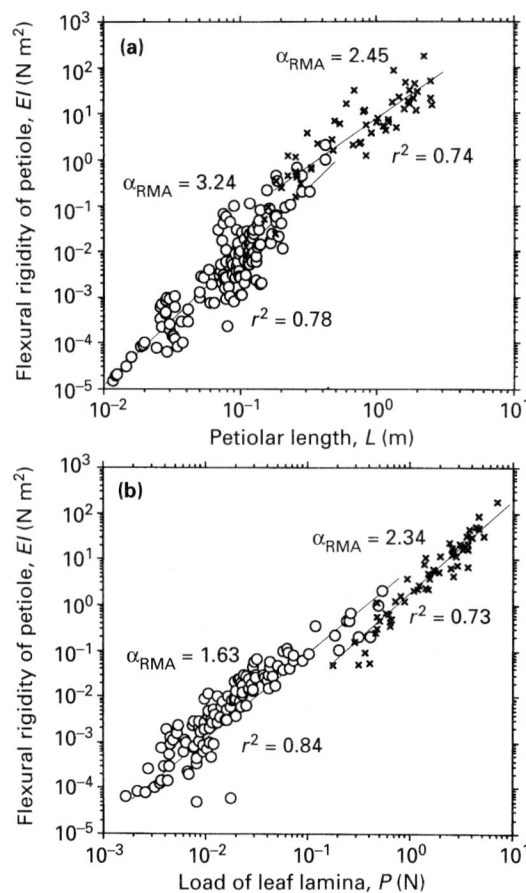

Fig. 3. Allometric relationships between petiole flexural rigidity EI and length L (a), and between petiole flexural rigidity EI and lamina load P (b) for simple and palmately (open circles) and pinnately (crosses) compound leaves ($n = 119$) drawn from different species. Solid lines denote reduced major-axis regression curves with slopes α_{RMA}. Data taken from Niklas (1994).

$EI \propto L^{3.24}$ ($r^2 = 0.78$) for simple and palmately compound leaves, and $EI \propto L^{2.45}$ ($r^2 = 0.74$) for pinnately compound leaves (Fig. 3a), whereas Eqn 2 predicts $EI \propto L^3$. Similarly, petiole rigidity scales with respect to the load P generated by laminae such that $EI \propto P^{1.63}$ ($r^2 = 0.84$) for simple and palmately compound leaves, and $EI \propto P^{2.34}$ ($r^2 = 0.73$) for pinnately compound leaves (Fig. 3b), whereas Eqn 2 predicts that the minimum scaling factor should be $EI \propto P$. Importantly, these allometric relationships are also obtained when the mean values of EI, L and P for each of the 19 species are used in regression analyses (Niklas, 1994). Consequently, when attempts are made to correct for the 'phyletic effect' of over-representing one or a few species, scaling relationships remain intact, suggesting that tip-deflections of leaves differing markedly in size can be held more or less constant regardless of difference in size by virtue of proportionate increases in petiole length, bulk tissue stiffness, transverse geometry or shape with respect to lamina weight.

In terms of torsion, most structures twist when they bend, and cantilevered leaves are no exception.

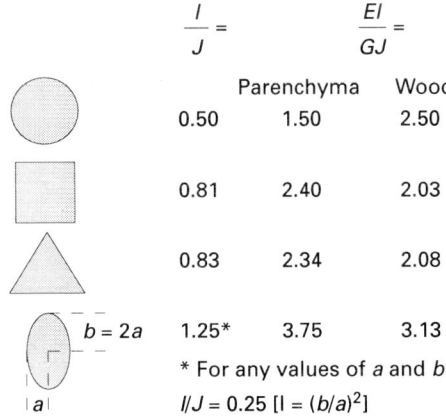

$$\frac{I}{J} = \qquad \frac{EI}{GJ} =$$

	Parenchyma	Wood
0.50	1.50	2.50
0.81	2.40	2.03
0.83	2.34	2.08
1.25*	3.75	3.13

$b = 2a$

* For any values of a and b

$I/J = 0.25$ [$I = (b/a)^2$]

Fig. 4. Quotients of axial and polar moments of area I/J and of flexural and torsional rigidity EI/GJ for beam-like petioles with different cross-sectional geometries composed of either pure parenchyma or wood. An elliptical cross section whose major axis is twice its minor axis is the only transverse geometry for which $I > J$; in each case, $EI > GJ$ because $E \gg J$ for most plant tissues.

The biomechanics and allometry of twisting are as stimple as those of bending. When subjected to a torque at its free end, the angle of petiole twist is given by the formula:

$$\phi = TL/GJ \qquad \text{Eqn 3}$$

(T is the torque, L is petiole length and GJ is the petiole's torsional rigidity (Timoshenko & Gere, 1961; Wainwright *et al.*, 1976; Vincent, 1990; Niklas, 1992a.). Eqn 3 is analogous to the tip-deflection relationship for a cantilevered beam (see Eqn 2) because it shows that the deformation ϕ resulting from a twisting force T is proportional to length L and inversely proportional to torsional rigidity GJ. Once again, we see that resistance to twisting can be increased by decreasing the length of the cantilever or increasing either its torsional stiffness G or its polar moment of area J, which is shape- and size-dependent.

From a purely geometrical perspective, most beams resist torsion better than bending because the quotient of I and J for the most commonly encountered cross sections tends to be significantly less than one (Niklas, 1992a; Vogel, 1992). There are, naturally, exceptions to this generalization. An elliptical beam resists bending better than twisting when its major axis is aligned in the vertical direction (e.g. $I/J = 1.25$ when the major axis is twice the minor axis) (Fig. 4). However, the same geometry and orientation permit large lateral deflections when mechanical forces, such as wind pressure, act from side to side. The petiole and rachis of *Rhus typina* leaves have elliptical cross sections whose major axes are aligned in the vertical direction, and thus the leaves of this species resist bending when loaded from top to bottom but deflect easily when exposed to horizontally moving wind.

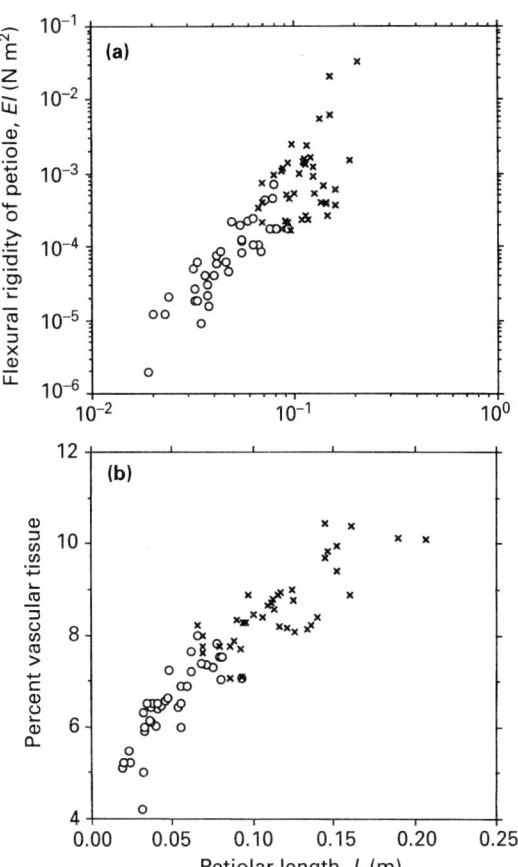

Fig. 5. Flexural rigidity EI (a) and percentage vascular tissues (b) of petioles of *Acer saccharum* plotted against petiole length L for leaves drawn from trees grown in an exposed (windy) (open circles) site and a sheltered (calm) (crosses) site. Data taken from Niklas (1996).

Nevertheless, the effect of the material properties of tissues typically overwhelms the effect of geometry: all plant tissues are far less stiff in torsion than in bending (Niklas, 1992a; Vogel, 1992, 1995). Indeed, on the basis of the stiffness of parenchyma and wood measured in twisting and bending, it is easy to show that beams differing in their cross-sectional shape resist bending much more than twisting (Fig. 4). This is likely to be highly adaptive for large foliage leaves whose laminae would otherwise experience large and potentially damaging drag forces if their petioles were unusually stiff and unyielding (Ennos, 1993, 1994; Vogel, 1996).

There is ample evidence that leaf development adaptively responds to ambient wind conditions (Venning, 1949; Jacobs, 1973; Niklas, 1992b). For example, the petioles of *Acer saccharum* leaves removed from trees growing in windy sites are shorter but significantly less rigid than those from trees growing in sites sheltered from the wind (Fig. 5). Leaves with shorter petioles are expected to resist bending more than leaves with longer petioles, but less rigid petioles are expected to bend and twist more easily in the wind. The data from *A. saccharum* illustrate the phenomenon known as thigmomorphogenesis, a developmental response to chronic mechanical perturbation (Knight, 1811; Jacobs, 1973; Telewski & Jaffe, 1986; Telewski, 1995). The typical response of plants to persistent mechanical disturbance is to produce shorter and broader stems and petioles. If this 'geometrical response' were not compensated for by means of a decrease in the stiffness of tissues, thigmomorphogenesis would result in mechanically unyielding organs that would experience extremely high drag forces and be liable to breaking. In leaves of *A. saccharum*, the petioles produced in windy sites are shorter but much less stiff than those from trees growing in sheltered sites. Thus, the petioles produced in windy sites are highly flexible and therefore bend and twist easily. Analyses of the volume fractions of petiole tissues from *A. saccharum* show that the decrease in the stiffness of petioles growing in windy sites is a result of a decrease in the percentage of vascular tissue (Fig. 5) and a corresponding increase in the amount of collenchyma, a viscoelastic material capable of considerable extension before it breaks. It is noteworthy that allometry of *A. saccharum* leaves does not differ between windy and protected sites; the differences seen in the physical and material properties of leaves are merely the extremes of a continuous scaling relationship that reflects a conservative yet 'plastic' developmental repertoire.

INTERNAL STRESSES AND STRAINS

Tissues of any organ deform as the organ bends and twists because mechanical forces acting through any structure generate stresses and corresponding strains. The cells within a leaf must therefore either resist deformation by stiffness or must deform without breaking by elasticity or viscoelastic extensibility. Indeed, all evidence indicates that tissues are deployed spatially in the petiole and lamina as effectively to cope with mechanical forces as they are to deal with the physiological imperatives of respiration and photosynthesis.

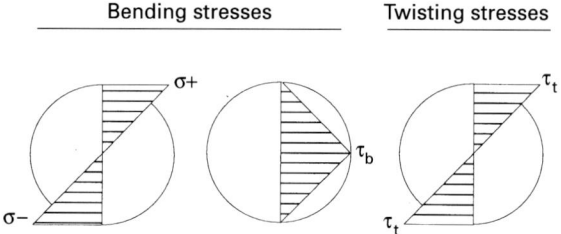

Fig. 6. Distribution patterns of bending and twisting stresses in a representative cross section through a terete petiole. Tensile and compressive bending stresses ($\sigma+$ and $\sigma-$, respectively) and bending shear stresses τ_b reach their maximum intensities at the surface and at the centroid axis, whereas torsional shear stresses τ_t reach their maximum intensities at the surface of the petiole. Taken from Niklas (1992a, p. 159).

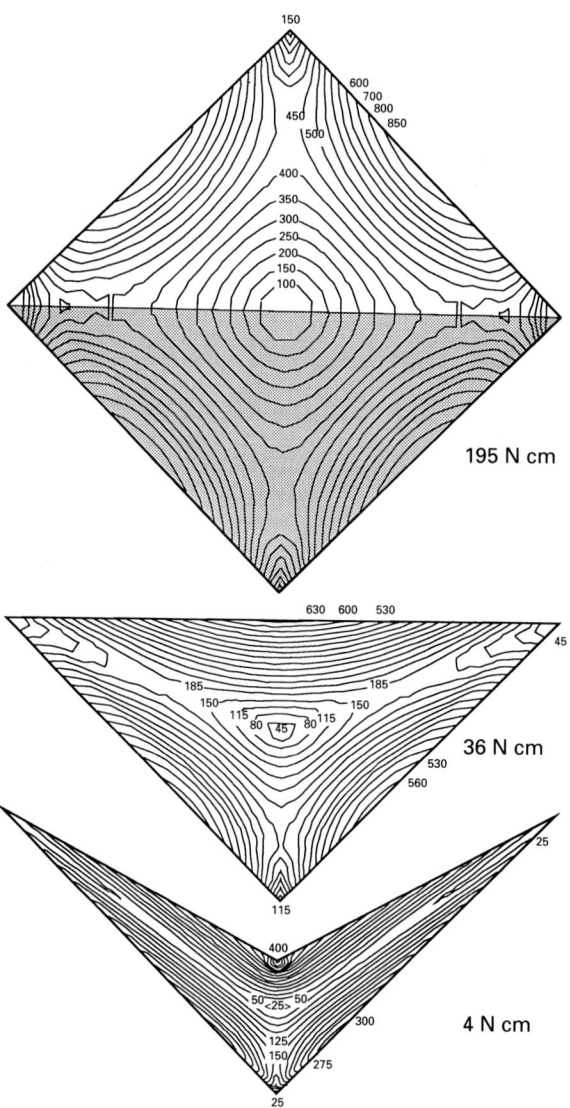

Fig. 7. Finite-element analyses of tensile and compressive bending stresses developing within beams with different transverse geometries subjected to different bending moments (stress magnitudes plotted as contour lines; the magnitudes of bending moments are shown at the right for each geometry). Each simulation assumes that beam materials are isotropic and linearly elastic.

The stress-distribution patterns in beams of circular cross section are easily generalized (Timoshenko & Gere, 1961; Wainwright *et al.*, 1976). For small flexures, the tensile and compressive stresses resulting from a bending force ($\sigma+$ and $\sigma-$, respectively) always reach their highest intensities at the surface of each cross section, as do the torsional shear stresses τ_t, whereas the shear stresses resulting from bending τ_b reach their maximum intensities at the centre of a circular cross section (Timoshenko & Gere, 1961; Wainwright *et al.*, 1976; Niklas, 1992a) (Fig. 6). The transverse stress distribution patterns for other geometries are far more complex and sometimes counter-intuitive. For example, tensile and bending stresses reach their maximum intensities at the centre of the sides (rather than at the corners) of a square or triangular cross section (Fig. 7). The

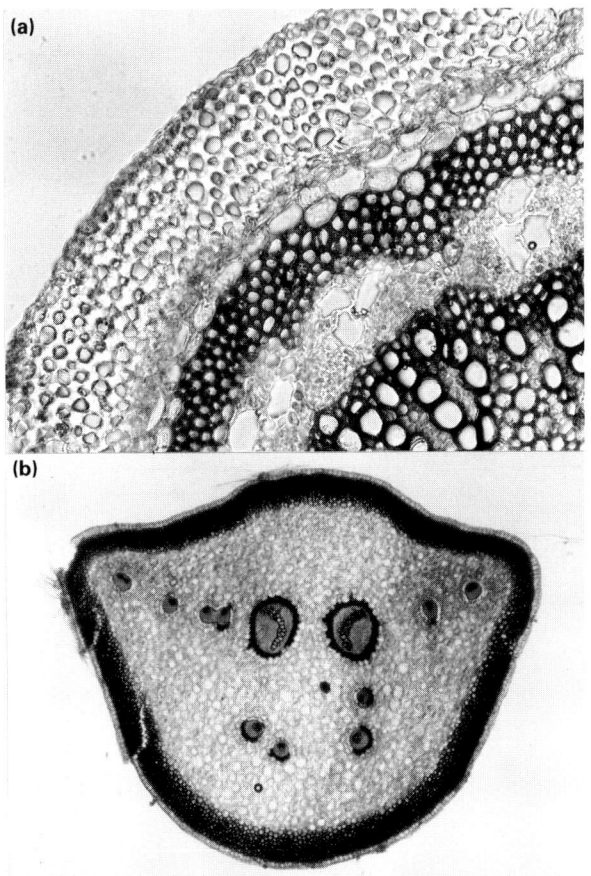

Fig. 8. Transverse petiole anatomy of representative leaves from a dicot, *Acer saccharum* (a) with a collenchymatous hypodermis and a fern, *Cyrtomium falcatum* (b) with a sclerenchymatous hypodermis. Tissues treated with HCl–phloroglucinol to stain for lignified cell walls.

same stresses are maximized in the dorsal notch of a chevron-shaped cross section, which incidentally coincides with the typical location of phloem and bundle-sheath fibre strands in cross sections of leaf laminae. Nevertheless, in each of these cases, the minimum shearing stresses occur very near or at the centre of the petiole cross section. The issue is therefore how best to deploy different tissue types to accommodate these stress patterns and thus avoid tissue breakage.

As previously mentioned, most plant tissues are far weaker in torsion than in either compression or tension, and, with the exception of wood, most are stronger in tension than in compression (Niklas, 1992a). Thus, the shear stresses generated by bending and twisting are likely to cause most damage to a leaf, by contrast with the tensile and compressive stresses resulting from bending. One biomechanical solution that copes with high shear stresses is reliance on a material that behaves very much like a fluid, that is, one that can change shape easily but that can conserve its volume. Parenchyma and collenchyma have these properties by virtue of their high Poisson ratios. Collenchyma has the added advantage of being very extensible and yet capable of elastically

restoring itself after external mechanical forces have been removed (Esau, 1936; Jarvis *et al.*, 1984). Parenchyma and collenchyma are also both highly viscoelastic tissues; that is, their deformations are time-dependent such that they act like a rigid solid when rapidly stressed and behave more like a fluid when slowly stressed (Jaccard & Pilet, 1975, 1977). These properties collectively make parenchyma and collenchyma suitable building materials for the central and peripheral regions in the cross sections through petioles that must normally bend and twist (Fig. 8a). Parenchyma located at the centre of the cross section of the petiole can deal with the shear stresses resulting from bending forces by virtue of its high Poisson ratios (i.e. $v = 5$), whereas collenchyma located just beneath the epidermis can deal with the high torsional shear stresses by means of the elastic extension of its cell walls. Both tissues confer remarkable stiffness when petioles are mechanically loaded rapidly, but they deform and recover without breaking when loaded and unloaded slowly. An additional advantage of these tissues is that both operate as hydrostatic materials; that is, their stiffness in bending or twisting increases (to an upper limit) as a function of their turgor pressure. When supplied with water, these tissues are rigid, but when deprived of water, both 'deflate' and become more flexible. Thus, slightly wilted petioles can bend and twist and thus decrease drag forces more easily than those which are fully turgid, and wilted petioles can reorient laminae such that light interception (and thus further water loss) is decreased.

Thick-walled nonhydrostatic tissue with low Poisson ratios, for example sclerenchyma, are an alternative to the use of hydrostatic, viscoelastic tissues with large Poisson ratios. Sclerenchyma is almost as stiff as wood in bending or twisting. Thus, location of this tissue just beneath the epidermis can rigidify and stiffen petioles, especially because it maximizes the contribution of a stiff material to the axial and polar second moments of area of a structure as a whole. By contrast with the petioles of many arborescent dicot species, which are exposed to potentially high and damaging winds (Raupach & Thom, 1981) and typically develop collenchyma in their hypodermis (Esau, 1967), the petioles of many herbaceous fern species, especially those living near the ground, where wind speeds tend to be low, develop sclerenchyma just beneath their epidermis (Fig. 8b). Although this suggestion is highly speculative, it is possible that the difference in the hypodermal petiole anatomies of arborescent dicot and herbaceous fern species is an adaptation to the differences in the ambient wind speeds that attend the two different growth habits of these species. This speculation is reinforced somewhat by the observation that the petioles of many tree fern species develop more collenchyma and less sclerenchyma just beneath their epidermis.

Even though the stress distribution pattern in each cross section through a petiole is the same (Fig. 6), the magnitudes of the highest bending stresses increase towards the base of the petiole, whereas those of torsional stresses decrease basipetally from the tip of the petiole. Thus the maximum tensile stresses resulting from bending occur along the upper surface at the base of the petiole, whereas the maximum torsional shear stresses occur along the surface of the petiole just beneath its point of attachment to the lamina. Morphometric analyses of the volume fractions of different tissue types along the lengths of many dicot and monocot petioles indicate that the proportion of parenchyma and collenchyma increases basipetally towards the base, whereas the proportion of vascular tissues increases acropetally towards the free end of the petiole. In fact, the absolute amount of vascular tissues in each cross section varies little along the petiole length such that the changes in the proportion of the ground tissues are a consequence of an increase in the absolute size of these tissues. For dicot species, the basipetal increase in the absolute amount of ground tissue is manifested by the presence of the phyllopodium, an inflated region at the base of the petiole and typically composed of hydrostatic and viscoelastic tissues. Its shape and location are well suited to the high bending stresses that occur at the base of any cantilevered petiole because its expanded cross section decreases the magnitudes of bending and twisting stress.

THE LAMINA

The lamina also experiences bending and twisting forces, and, when dorsiventral in morphology and thin in cross section, it can be conveniently modelled as a stress-skin panel or a polylaminated sandwich board (Allen, 1969; Gibson *et al.*, 1988) whose facings (lined above and below with epidermis) are rigidly bonded to interconnecting stringers (the vascular bundles running throughout the lamina) (Fig. 9). This structural configuration spreads any force acting normal to its surface over a large surface area and produces small lateral deflections that are easily accommodated by the stringers holding the webbing together (Niklas, 1992a). The advantage of this behaviour is that lateral loadings are resisted primarily by in-plane membrane forces that develop within the webbing or 'skin' rather than by the usual bending forces seen typically in structures made up of columns or beams. Curved folded plates have the additional mechanical advantage of supporting lateral loads by arch compression (a curvilinear deformation spanning the width and breadth of the whole structure in which all components share a common load by distributing strains as equally as possible). In this way, no single portion of the construction bears a load per surface area that is

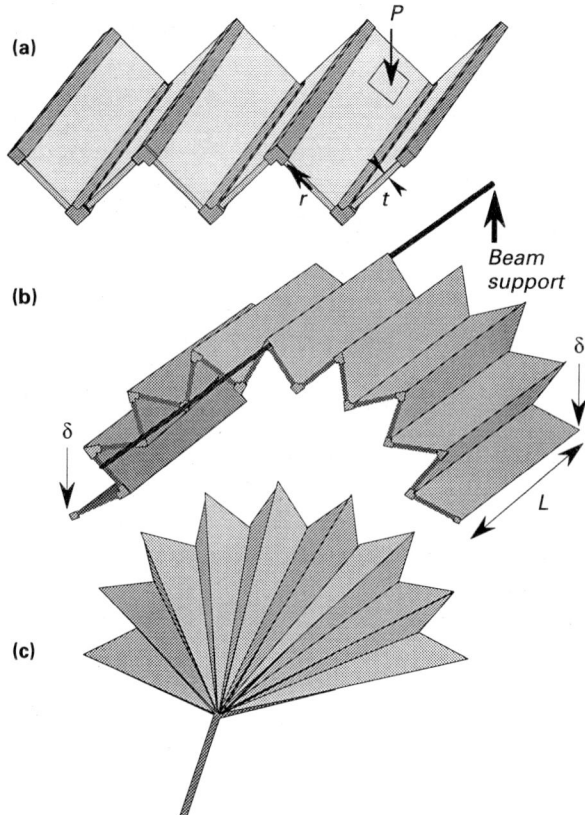

Fig. 9. Mechanical structure of plicated stress-skin panels composed of webbed runners with thickness t and length L. (a) A load P acting normal to the panel web generates tangentially acting stresses r in runners whose magnitudes are dispersed throughout the structure. (b) A beam-supported plicated stress-skin panel experiences broad intra-marginal arching with a maximum vertical deflection, δ. This configuration minimizes tensile stresses experienced throughout the panel. (c) A plicated stress-skin panel with convergent runners attached at one point to a cantilevered beam. This structure is mechanically stable under its own weight and can fold upon itself to become a streamlined shape when subjected to airflow passing along the petiole and the length of its runners.

greater than average. Stress-skin panels and thin polylaminated sandwich boards are remarkably flexible when bent or twisted and can rapidly deform elastically without damage, especially if the stringers radiate outwards from (and converge on) a single point. The thin laminae of many leaves easily fold in rapidly moving wind, thereby decreasing their drag and the probability of mechanical damage (Vogel, 1989).

The disadvantage of very thin stress-skin panels is that they are prone to ripping at their edges when folded excessively, like fabrics that are stretched or pulled. For woven structures, the solution is a selvedge or woven border, hem or gusset. The hem is used to strengthen a continuous margin by doubling it up; the gusset is inserted as a triangular patch at the base of some indentation to decrease the risk of tearing inwards from a weakened point. These familiar mechanical solutions coincide with the marginal strengthening of the laminae of many

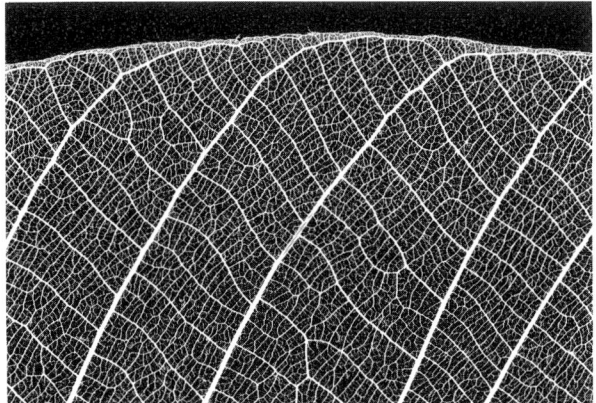

Fig. 10. A network of thicker and thinner vascular stands describing broad intra-marginal archings along the margin of a leaf of *Ficus*.

leaves (Fig. 10). Arched selvedges in leaves are common among dicot species. A network of vascular stands consisting of stronger and weaker stands describe broad intra-marginal archings succeeded by others of smaller curve nearer the margin. The curve of each arch runs for some distance nearly parallel to the leaf margin. The parallel venation of monocot leaves acts in much the same way because the most marginal strand and the sclerified leaf margin act as a single long strengthening arch. Another effective design is to place a strong band of stiff tissue along the entire margin of the lamina, as seen in the leaves of *Ulmus* and *Platanus*. Finally, the mechanical equivalent of gussets is seen in the form of hardened patches of tissue located at the base of deep indentations at the margins of leaves, as in those of many ferns (e.g. *Asplenium horridum*).

The presence of a polylaminated construction with very stiff and less stiff layers also decreases the risk of a crack or tear passing through the whole structure because the compliant (less stiff) layer can deform extensively and thereby decrease the stress concentration at the tip of the advancing crack (Vincent, 1990). The sharp end of a crack acts as a force concentrator and can be stopped by running into either a void or a material that deforms rapidly and easily. The greater the difference between the material properties of the adjoining layers of material, the less likely it is that a crack will propagate through the two layers. A decrease in the size of the components in the stiffer of the two layers is also helpful because every crack requires a minimum size before it will propagate. Increasing the surface contact made between the stiff and less stiff layers also increases the probability that a crack will run into the less stiff layer and thus run its course. The parallel venation of grass leaves is an example. If a grass leaf is slit at its margin, it tears more easily but only in direct proportion to its reduced cross section; it tears jaggedly as the crack passes up and down between the parallel veins and intervening softer tissues of the leaf (Vincent, 1982, 1983).

The leaves of grasses and other monocots provide a special case because their sheaths clasp stem internodes and thus contribute mechanically to the support of the shoot as a whole (Tomlinson, 1962; Niklas, 1990, 1992a, 1998). Developmental studies show that clasping leaf sheaths are stiffer and more rigid than the internodes that they surround during the early development of the stem, when internodes are not fully extended, and that internode tissues stiffen gradually over the course of development and become as stiff as the sheaths once they extend beyond their enclosing leaves. However, the sheaths continue to support fully extended stem internodes mechanically because they externally impart rigidity to internodes against local (Brazier) buckling when shoots bend. The leaf sheath also helps internodes to resist torsional deformations.

DRAG

The lamina imposes a bending moment or a torque on its petiole proportional to its surface area, but this proportionality seems to differ between simple and compound leaves: larger simple leaves seem to benefit slightly by the surface area that they gain with respect to the loads that their laminae produce (Fig. 11). Although far too few species have been examined, allometric analyses of three species with simple leaves indicate that the lamina surface area scales as the 1.12 power of the weight of the lamina, or $S \propto P^{1.12}$ ($r^2 = 0.985$), whereas the relationship for three species with palmately or pinnately compound leaves is $S \propto P^{0.90}$ ($r^2 = 0.983$). These scaling exponents differ at the 1% level, thus suggesting a statistically significant difference between the two

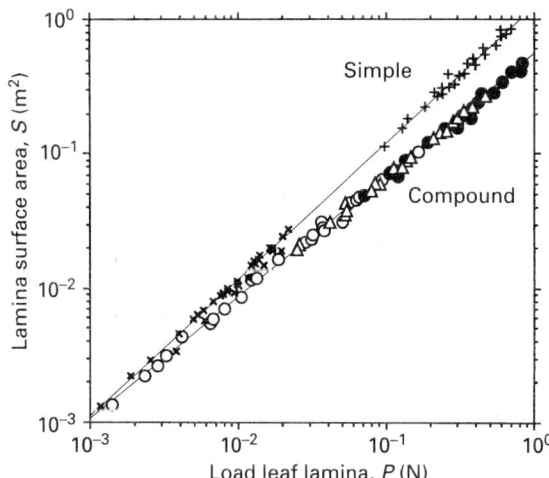

Fig. 11. Lamina surface area S plotted against lamina mass-force (load P) for two species with simple leaves and three species with pinnately compound leaves. Statistical comparisons between the slopes of the reduced major-axis regression curves (solid lines) indicate that the slopes differ at the 1% level such that the allometry of S with respect to P differs significantly for the two leaf morphologies.

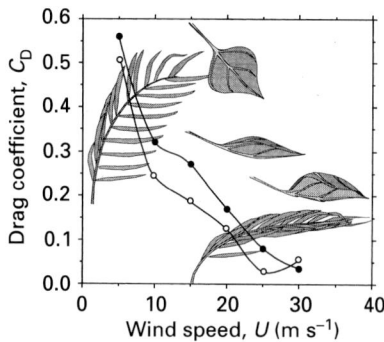

Fig. 12. Drag coefficient C_D plotted against wind speed U for a *Catalpa* leaf (closed circles) and a *Chamaedorea* leaf (open circles) with equivalent surface areas ($S = 227$ cm²; see Eqn 5) placed in a wind tunnel. The illustrated reconfigurations of leaves for progressively higher wind speeds are drawn from photographs taken orthogonal to the direction of airflow.

general leaf morphologies that is the result of differences in either density of lamina tissue or in lamina thickness.

Whether this difference is functionally meaningful remains speculative, but it is clear that the drag force F_D experienced by any object obstructing the flow of a fluid (water or air) is equal to the product of the density ρ of the fluid, the square of the fluid velocity U, the area S_P that the object projects in the direction of flow, and the drag coefficient C_D (Vogel, 1994):

$$F_D = 0.5\rho U^2 S_P C_D \qquad \text{Eqn 4}$$

In instances in which the fluid flows over the length of a streamlined object, the drag force is due almost entirely to skin friction and it is more accurate to substitute the wetted surface area of the object S_W for the projected surface area S_P, such that:

$$F_D = \rho U^2 S C_D \qquad \text{Eqn 5}$$

(For leaves, S is either the upper or lower lamina surface area such that the total area of the lamina equals $2S$.) Regardless of which surface area is deemed the more appropriate, both these formulae show that the drag force experienced by a leaf is proportional to the lamina surface area.

Importantly, regardless of published statements, the drag coefficient is not constant (Mayhead, 1973), especially for biological objects that can twist, bend and fold onto themselves and so reconfigure their shape in response to a moving fluid (Vogel, 1989, 1994). For example, wind-tunnel experiments with portions of intact shoots show that the laminae of leaves with very different morphologies curl and fold when subjected to high winds and thereby decrease their projected surface areas and their drag coefficients (Vogel, 1989). Because the drag force is proportional to the product of S_P and C_D (Eqns 4, 5) and because the stems of plants experience the cumulative drag force exerted by the wind on all their leaves, elastic changes in lamina shape confer obvious benefits (Fig. 12). The laminae of the simple

leaves of many species, such as *Catalpa*, roll into streamlined tubes whose overlapping margins decrease the skin drag of the leaf as a whole. When combined with the effects of the bending of stems, this configuration greatly decreases the drag forces acting on entire shoots. Wind reconfiguration of simple leaves is not observed for all species (the leaves of *Quercus* spp. do not appreciably reshape themselves even in very high winds (Vogel, 1989)), but it is almost always a consequence of the flexure of the abaxial leaf surface such that the adaxial surface of the leaf folds upon itself and becomes concave. This might be a result of the high extensibility of the spongy mesophyll that can stretch and deform and thus permit the upper surface of the leaf to roll up. The reconfiguration of pinnate leaf morphologies, such as those of many palm species, involves the elastic flexure of the rachis and the pivoting of the individual leaflets at their attachment points to the rachis (Fig. 12). Not all pinnate leaves behave in this way; the leaves of cycads, such as *Dioon* and *Zamia*, do not reconfigure to any appreciable extent. However, for many species, the entire pinnate leaf operates mechanically as a flexible branch bearing numerous simple leaves with elastic hinges at their bases.

The base of the palm pinnate leaf, which clasps its subtending stem, is also mechanically complex. Tests of surgically isolated tissue samples show that the leaf base is extremely anisotropic, such that its response to forces depends on the direction in which these forces act (Spatz *et al.* (1995) discuss stem tissue anisotropy). Finite-element analyses of the morphology of the palm leaf base show that the mechanical anisotropy of its tissues decreases the magnitudes of the bending and torsional shear stresses that develop in the structure as a whole in comparison with the magnitudes of the stress that would develop in the same structure if it were composed of isotropic materials (Fig. 13). Thus, significantly higher bending or torsional moments are required to break the leaf base. Finite-element analyses also show that the regions within the leaf base that experience the highest intensities of stresses are structurally reinforced by concentrations of thick-walled tissue and cell types that are typically aligned parallel to the direction of the forces generating these stresses. These tissues and cell types are concentrated along the rim and the abaxial surface of the chevron-shaped petiole where it clasps the shoot. Computer simulations show further that the geometric and material properties of the palm leaf base confer remarkable rigidity in bending yet significant flexibility in twisting. Finally, computer simulations show that the acropetal changes in the transverse shape and size of the rachis of the pinnately compound palm leaf keep the magnitudes of bending and torsional shear stresses fairly uniform along the length of the leaf. These features col-

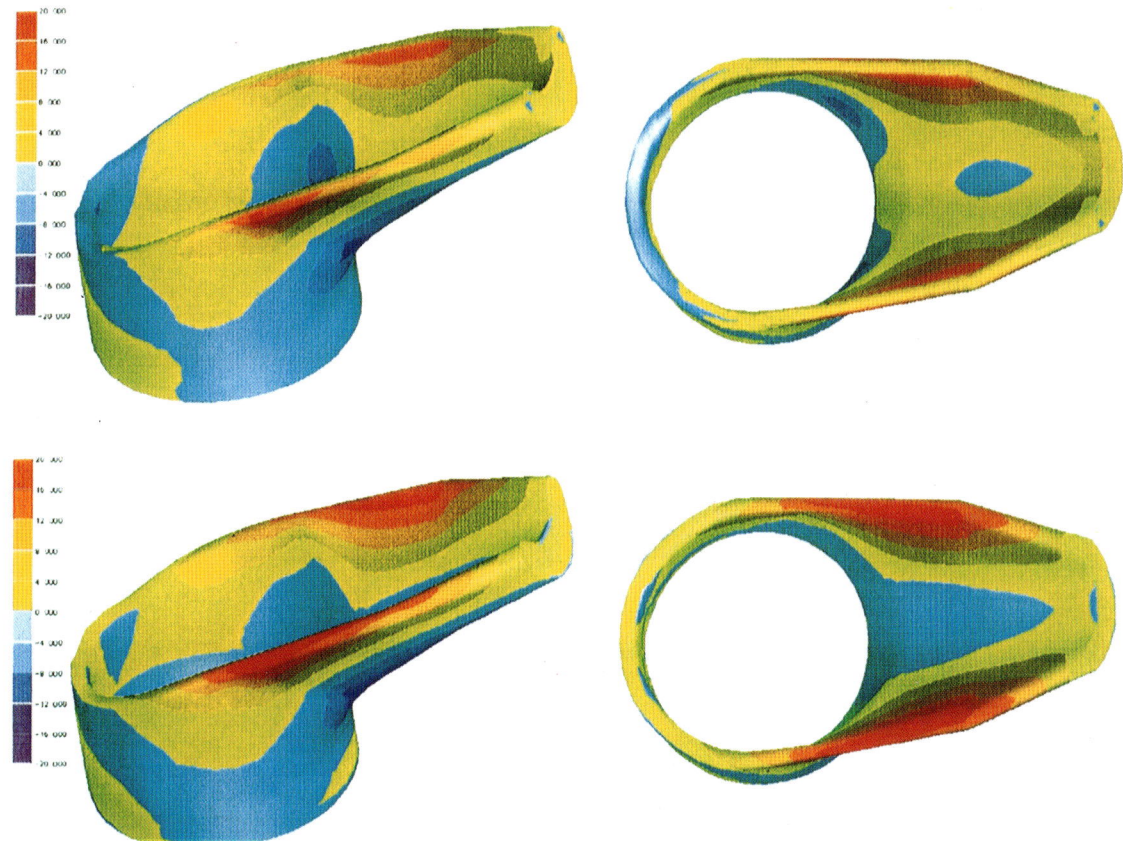

Fig. 13. Finite-element analyses of the longitudinal principal bending stresses generated within the closed base of a palm leaf subjected to a bending moment equivalent to that generated by the leaf mass-force acting along its length (distal rachis and leaflets not shown). The space created by the tubular portion of these simulations is occupied by the palm stem in real plants. Tensile and compressive stress intensities are equivalent (see the scales at the left of each pair of views). Upper (lateral and abaxial) views of leaf base are based on the assumption that the leaf materials (tissues) are anisotropic in their mechanical behaviour; bottom lateral and abaxial views of the leaf base assume that the leaf tissues are isotropic in their behaviour. Bulk leaf tissue stiffness and Poisson ratios were empirically determined on the basis of bending tests of five *Chamaedorea* leaves. Finite-element analyses indicate that the anisotropic behaviour of tissues results in lower global stress intensities than in simulations with isotropic materials.

lectively permit the palm leaf to reconfigure its shape and to bend and twist without breaking, even in gale-force winds.

CONCLUSION

We still know surprisingly little about the mechanical design of leaves, especially how they respond to large externally applied dynamic forces that can change direction and duration of application in a short time. However, the general picture that is emerging is that the morphology and anatomy of leaves reflect a compromise between the obvious requirement to sustain the static loads that leaf tissues impose on themselves and the requirement to twist and bend easily when subjected to large and potentially devastating mechanical forces, such as wind-induced pressure. Nevertheless, this biomechanical perspective still requires integration with what is known about leaf physiology and allometry. For example,

we know that the dimensional limit on the diameter of an internode in primary growth is dictated by the simultaneous functional obligations for the mechanical support and the supply of water by vascular tissues to the leaves supported by the stem. Thus, the dimensional limit on primary stem diameter must relate to SLA, which is correlated with many ecologically and physiologically important features (e.g. relative growth and gas exchange rates), yet varies enormously within and across species as a function of irradiance levels, wind conditions and precipitation (Salisbury & Ross, 1992). Unfortunately, our ability to identify and quantify the intrinsic allometric, biomechanical and ecophysiological constraints on leaf size and number is limited largely by a dearth of integrative studies.

Similarly, much of the engineering theory and practice applied to understanding leaf and stem biomechanics is devoted to the design of rigid and strong structures that deform little when loaded. As emphasized throughout this paper, this literature is not irrelevant to plant biomechanics because it provides some general design principles. Nevertheless, the traditional engineering literature provides an incomplete perspective in terms of organic structures that typically violate the basic assumptions that lie at the heart of traditional solid mechanics. Unlike their fabricated counterparts, biological structures change size, shape and internal organization as they develop and mature, and the materials out of which they are constructed rarely if ever behave as linearly elastic isotropic materials. In addition, organic structures rarely experience simple loading regimes; nor do they function in only one or a very few capacities.

Recently, however, a discernible shift in the emphasis of the engineering literature has occurred, one that focuses attention on the mechanics of thin flexible shells, beams and shafts that achieve large elastic displacements by small strains (Axelrad, 1984). This shift has been made possible in part by developments in finite-element analyses and the availability of high-speed computers that permit rigorous quantitative treatments of complex structures composed of nonlinear elastic and anisotropic materials. These new tools, combined with recent advances in engineering theory (Gibson & Ashby, 1988) and increased attention to biomechanical constraints on ecophysiological phenomena, hold out a promise that we might be soon able to dissect organic form–function relationships quantitatively and understand more fully why leaves are shaped and structured as they are.

REFERENCES

Allen HG. 1969. *Analysis and design of structural sandwich panels.* Oxford, UK: Pergamon Press.

Axelrad EL. 1984. Flexible shells. In: Axelrad EL, Emmerling, FA, eds. *Flexible shells.* Berlin, Germany: Springer Verlag, 44–63.

Bisshop KE, Drucker DC. 1945. Large deflection of cantilever beams. *Quarterly Applied Mathematics* 3: 272–275.

Bower FO. 1925. *Plants and man.* London, UK: Macmillan.

Cannell M, Coutts MP. 1988. Growing in the wind. *New Scientist* 21: 42–46.

Ennos AR. 1993. The scaling of root anchorage. *Journal of Theoretical Biology* 161: 61–75.

Ennos AR. 1994. The biomechanics of root anchorage. *Biomimetics* 2: 129–137.

Esau K. 1936. Ontogeny and structure of collenchyma and of vascular tissues in celery petioles. *Hilgardia* 10: 431–476.

Esau K. 1967. *Plant anatomy.* New York, USA: John Wiley.

Falk S, Hertz H, Virgin H. 1958. On the relation between turgor pressure and tissue rigidity. I. Experiments on resonance frequency and tissue rigidity. *Physiologia Plantarum* 11: 802–817.

Gibson LJ, Ashby MF. 1988. *Cellular solids, structure and properties.* Oxford, UK: Pergamon Press.

Gibson LJ, Ashby MF, Easterling KE. 1988. Structure and mechanics of the iris leaf. *Journal of Materials Science* 23: 3041–3048.

Grace J. 1978. The turbulent boundary layer over a flapping *Populus* leaf. *Plant, Cell, and Environment* 1: 35–38.

Greenberg AR., Mehling A, Lee M, Bock JH. 1989. Tensile behavior of grass. *Journal of Materials Science* 24: 2549–2554.

Jaccard M, Pilet PE. 1975. Extensibility and rheology of collenchyma cells. I. Creep relaxation and viscoelasticity of

young and senescent cells. *Plant and Cell Physiology* 16: 113–120.

Jaccard M, Pilet PE. 1977. Extensibility and rheology of collenchyam cells. II. Low-pH effect on the extension of collocytes isolated from high- and low-growing material. *Plant and Cell Physiology* 18: 883–891.

Jacobs MR. 1973. Thigmomorphogenesis: the response of plant growth and development to mechanical stress. *Planta* 114: 143–157.

Jarvis MC, Logan AS, Duncan HJ. 1984. Tensile characteristics of collenchyma cell walls at different calcium contents. *Physiologia Plantarum* 61: 81–86.

Knight TA. 1811. On the causes which influence the direction of the growth of roots. *Philosophical Transactions of the Royal Society of London* 1811: 209–219.

Mauseth JD. 1988. *Plant anatomy.* Menlo Park, CA, USA: Benjamin Cummings.

Mayhead GJ. 1973. Some drag coefficients for British forest trees derived from wind tunnel studies. *Agricultural Meteorology* 12: 123–130.

Niklas KJ. 1989. Mechanical behavior of plant tissues as inferred from the theory of pressurized cellular solids. *American Journal of Botany* 76: 929–937.

Niklas KJ. 1990. The mechanical significance of clasping leaf sheaths in grasses: evidence from two cultivars of *Avena sativa. Annals of Botany* 65: 505–512.

Niklas KJ. 1992a. *Plant biomechanics: an engineering approach to plant form and function.* Chicago, USA: University of Chicago Press.

Niklas KJ. 1992b. Gravity-induced effects on material properties and size of leaves on horizontal shoots of *Acer saccharum* (Aceraceae). *American Journal of Botany* 79: 820–827.

Niklas KJ. 1993. The scaling of plant height: a comparison among major plant clades and anatomical grades. *Annals of Botany* 72: 165–172.

Niklas KJ. 1994. *Plant allometry: the scaling of form and process.* Chicago, USA: University of Chicago Press.

Niklas KJ. 1996. Differences between *Acer saccharum* leaves from open and wind-protected sites. *Annals of Botany* 78: 61–66.

Niklas KJ. 1998. The mechanical roles of clasping leaf sheaths: evidence from *Arundinaria téctata* (Poaceae) shoots subjected to bending and twisting forces. *Annals of Botany* 81: 23–34.

Niklas KJ, Paolillo DJ, Jr 1997. The role of the epidermis as a stiffening agent in *Tulipa* (Liliaceae) stems. *American Journal of Botany* 84: 735–744.

Raupach MR, Thom AS. 1981. Turbulence in and above plant canopies. *Annual Review of Fluid Mechanics* 13: 97–129.

Salisbury FB, Ross CW. 1992. *Plant Physiology.* Belmont, CA, USA: Wadsworth.

Spatz H-C, Beismann H, Emanns A, Speck T. 1995. Mechanical anisotropy and inhomogeneity in the tissues comprising the hollow stem of the giant reed *Arundo donax. Biomimetics* 3: 141–155.

Stewart WN, Rothwell GW. 1993. *Paleobotany and the evolution of plants.* Cambridge, UK: Cambridge University Press.

Taylor TN, Taylor EL. 1993. *The biology and evolution of fossil plants.* Englewood Cliffs, NJ, USA: Prentice-Hall.

Telewski FW. 1995. Wind induced physiological and developmental responses in trees. In: Coutts MP, Grace J, eds. *Wind and trees.* Cambridge, UK: Cambridge University Press, 237–263.

Telewski FW, Jaffe MJ. 1986. Thigmomorphogenesis field and laboratory studies of *Abies frazei* in response to wind or mechanical perturbation. *Physiologia Plantarum* 66: 211–218.

Timoshenko SP, Gere JM. 1961. *Theory of elastic stability.* New York, USA: McGraw-Hill.

Tomlinson PB. 1962. The leaf base of palms. Its morphology and mechanical biology. *Journal of the Arnold Arboretum* 43: 23–50.

Venning FD. 1949. Stimulation by wind motion of collenchyma formation in celery petioles. *International Journal of Plant Sciences* 110: 511–514.

Vincent JFV. 1982. The mechanical design of grass. *Journal of Materials Science* 17: 856–860.

Vincent JFV. 1983. The influence of water content on the stiffness and fracture properties of grass leaves. *Grass and Forage Science* 38: 107–114.

Vincent JFV. 1990. *Structural biomaterials.* Princeton, NJ, USA: Princeton University Press.

Vogel S. 1989. Drag and reconfiguration of broad leaves in high winds. *Journal of Experimental Botany* **40**: 941–948.

Vogel S. 1992. Twist-to-bend ratios and cross-sectional shapes of petioles and stems. *Journal of Experimental Botany* **43**: 1527–1532.

Vogel S. 1994. *Life in moving fluids.* Princeton, NJ, USA: Princeton University Press.

Vogel S. 1995. Twist-to-bend ratios of woody structures. *Journal of Experimental Botany* **46**: 981–985.

Vogel S. 1996. Blowing in the wind: storm-resisting features of the design of trees. *Journal of Arboriculture* **22**: 92–98.

Wainwright SA, Biggs WD, Currey JD, Gosline JM. 1976. *Mechanical design in organisms.* Princeton, NJ, USA: Princeton University Press.

New Phytol. (1999), **143**, 33–43

Research review
Modelling leaf expansion in a fluctuating environment: are changes in specific leaf area a consequence of changes in expansion rate?

F. TARDIEU*, C. GRANIER AND B. MULLER

Institut National de la Recherche Agronomique, Laboratoire d'Ecophysiologie des Plantes sous Stress Environnementaux, 2 Place Viala, 34060 Montpellier, France

Received 5 October 1998; accepted 27 March 1999

SUMMARY

Leaf expansion rate varies with leaf temperature, photon flux density (PPFD), evaporative demand and soil water status. In most simulation models, it is calculated every day by multiplying the amount of carbohydrate available to leaves by specific leaf area (SLA). However, leaf expansion rate is considerably reduced by mild water deficits which do not affect photosynthesis, and is not affected by a reduction in the PPFD intercepted during rapid leaf expansion. Specific leaf area undergoes a several-fold variability depending on PPFD, soil water status and time of day. It is increased when environmental conditions have a greater depressive effect on expansion rate than on photosynthesis, and is decreased in the opposite case. It is therefore appropriate to model leaf expansion independently of the plant carbon budget. Consistent characteristics can be deduced from a series of experiments, allowing a model of leaf expansion to be proposed. (i) Time courses of relative leaf expansion rate and of epidermal cell division rate are well conserved within a plant and across a large range of environmental conditions, provided that durations and rates are expressed in thermal time. Maximum relative rates are common to all zones of a leaf and to all leaves of a plant, in maize and sunflower. (ii) A water deficit, or a reduction in intercepted PPFD, imposed in the first half of the period of leaf development affects the relative expansion rate in the deficit only, but permanently affects the absolute expansion rate. In contrast, a reduction in PPFD causes no effect on leaf expansion if imposed in the rapid expansion period when the leaf is autotrophic. (iii) Expansion rate is related to evaporative demand and to the concentration of ABA in the xylem sap with relationships that apply under both field and laboratory conditions. (iv) Tissue expansion and epidermal cell division behave as independent processes which determine epidermal cell area at each time.

Key words: leaf expansion, cell division, specific leaf area, sunflower, maize.

INTRODUCTION

Leaf area largely determines light interception and transpiration of plants (Monteith, 1977). Its increase with time depends on environmental conditions, such as water deficit (Randall & Sinclair, 1988; Lecoeur *et al.*, 1995), PPFD (Milthorpe & Newton, 1963; Dengler, 1980), temperature (Ben Haj Salah & Tardieu, 1995; Granier & Tardieu, 1998b) and evaporative demand (Ben Haj Salah & Tardieu, 1996). In the literature, three approaches have been reported to the interpretation of temporal changes in leaf area and to accounting for environmental effects.

Leaf expansion can be considered to depend on carbon (C) availability. The resulting equation relates the increase in plant leaf area on a given day ($\Delta A/\Delta t$, mm² d⁻¹) to C gain on that day (difference between photosynthesis, $\mathcal{J}$, and respiration, R, g d⁻¹), to the proportion of C allocated to leaves on the same day (p_l, g g⁻¹), and to specific leaf area (SLA, leaf area per unit d.wt, mm² g⁻¹):

$$\Delta A/\Delta t = (\mathcal{J} - R) \times p_l \times SLA. \qquad \text{Eqn 1}$$

This approach is used in most current simulation models (e.g. CERES, EPIC, GOSSYM). It implies either that p_l and SLA are constant for a genotype, or that their changes with plant age and with environmental conditions can be predicted in a simple way.

*Author for correspondence (tel +33 4 99 61 26 32; fax +33 4 67 52 21 16; e-mail tardieu@ensam.inra.fr).

Leaf expansion can also be considered as linked to the deformation of epidermal cell walls, which are the less plastic tissues of leaves during a deformation (Kutschera, 1992). In the equation of Lockhart (1965), the relative expansion rate (RER) is linked to epidermal cell turgor (P, Pa) and to the rheological properties of epidermal cells, namely wall extensibility (ϕ, mm^2 mm^{-2} d^{-1} Pa^{-1}) and minimum turgor allowing epidermal cell expansion (Y, Pa):

$$RER = \Delta A / A \, \Delta t = \phi(P - Y). \qquad \text{Eqn 2}$$

Turgor is well maintained in plants subjected to water deficit (Rhizopoulou & Davies, 1991), to changes in temperature (Pritchard *et al.*, 1990) or to nitrogen deprivation (Palmer *et al.*, 1996). Changes in expansion rate are therefore essentially linked to cell-wall rheological properties, via the activities of enzymes such as xyloglucan endotransglycosylase (XET) (Pritchard *et al.*, 1993), expansins (Mac-Queen-Mason & Cosgrove, 1995; Reinhardt *et al.*, 1998) or peroxidases (Thompson *et al.*, 1998), under the control of chemical signals such as ABA (Ben Haj Salah & Tardieu, 1997) and/or of hydraulic signals (Chazen & Neuman, 1994; Ben Haj Salah & Tardieu, 1997).

Leaf expansion can finally be considered as the result of both epidermal cell division and epidermal cell expansion. Leaf area is the product of epidermal cell number (N) and mean area of epidermal cells (A_c, mm^2) so, at each time interval (dt):

$$RER = dA / A \, dt = dN / N \, dt + dA_c / A_c \, dt. \qquad \text{Eqn 3}$$

The two terms in the right part of Eqn 3 are frequently considered as independent, so any increase in epidermal cell division rate should result in an increase in leaf expansion rate (LER). Consistently, manipulation of enzymes of the cell cycle affect leaf area (Hemerly *et al.*, 1995) and root length (Doerner *et al.*, 1996). If this view is correct, epidermal cell division should be the most important process because epidermal cell number of a leaf located at a given position of the stem is usually much more variable than individual cell size in a range of environmental conditions (Dale, 1992). Conversely, this view has been questioned by Green (1976) because an increase in epidermal cell division rate does not cause increase in leaf area unless accompanied by increase in epidermal cell-wall expansion rate.

Eqns 1–3 are mathematically correct and appear to be based on physiologically sound processes. However, they are mutually incompatible, since each is based on a unique process. For instance, the use of Eqn 1 implies that changes in SLA are considered a key mechanism in the control of leaf area. In contrast, the use of either Eqn 2 or Eqn 3 implies that C gain is not directly a mechanism of leaf expansion, so SLA should be considered as a variable (output of a model) rather than as a parameter (intrinsic plant property provided to the model). Similarly, cell division is considered a key mechanism if Eqn 3 is used, but not in Eqn 2 if Green's view that cell division *per se* does not provoke tissue expansion is accepted. A mathematical combination of Eqns 1–3 could allow all processes to be taken into account by expressing a parameter of any equation as a function of the variables of the other two. This noncontroversial data-processing increases the number of parameters to be estimated, thereby decreasing the predictive value of a model. Furthermore, the resulting parameters could lose any relevance if processes are not independent.

We suggest that an appropriate way of integrating the processes involved by Eqns 1–3 is to determine which equations are sufficiently robust to hold true in a range of experimental conditions without necessitating the reestimation of parameters, and which equations result in an unacceptable variability of parameters. We present here a synthesis of results which analyse leaf expansion and cell division in a wide range of environmental conditions and determine which processes can be considered well conserved.

CAN LEAF EXPANSION BE MODELLED USING A CARBON BALANCE EQUATION?

Equations are used frequently in crop modelling for prediction, despite not directly reflecting a physiological mechanism (Passioura, 1996). The crucial question is therefore whether Eqn 1 allows prediction of leaf expansion when plants are subjected to changing temperature, PPFD or water availability.

Is leaf expansion rate related to leaf carbon budget?

A rapid analysis could lead to the conclusion that Eqn 1 is acceptable because a change in the C balance of the plant due to a decrease in PPFD causes a reduction in leaf area (Wilson, 1966; Dengler, 1980). We have shown recently that sunflower leaves located in positions 15–25 on the stem are up to 10 times larger in plants subjected to a mean PPFD of 55 mol m^{-2} d^{-1} during leaf expansion, than in those subjected to 11 mol m^{-2} d^{-1} (O. Turc *et al.*, unpublished data). However, an analysis of the time course of expansion does not support the hypothesis that on any one day the expansion rate of a leaf is related to C balance on that day, calculated either at plant or a leaf level.

A short period of mild water deficit, which does not affect photosynthesis per unit leaf area, causes considerable reduction in absolute expansion rate for the whole development of sunflower leaf 8 (Fig. 1a,b). This reduction in expansion rate continues to

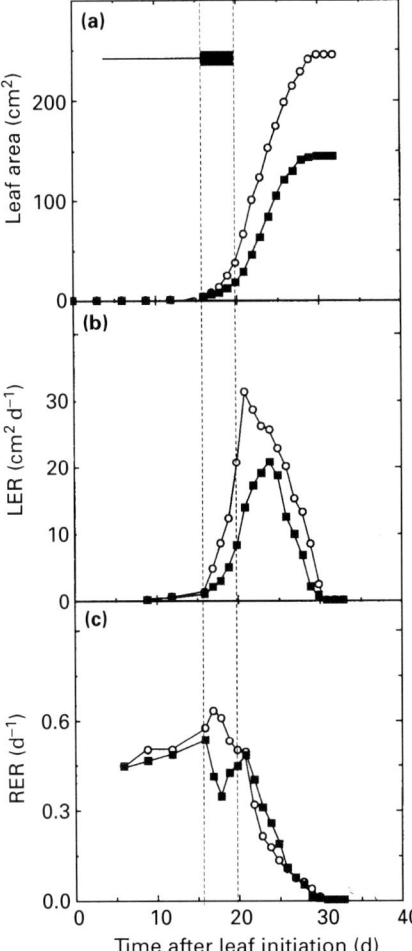

Fig. 1. Effects of water deficit on leaf expansion. The graphs show changes with time in area (a), in absolute expansion rate (LER, b) and in relative expansion rate (RER) (c) of leaf 8 of sunflower plants grown in well watered conditions (open circles) or with early mild water deficit (closed squares). In a glasshouse experiment, plants were subjected to a 5-d water deficit in which available soil water content was maintained constant at 23% of retention capacity. Predawn leaf water potential was −0.5 MPa and concentration of ABA in the xylem sap was 50 μmol m⁻³ in the deficit period. Plants were otherwise well watered (predawn leaf water potential: −0.3 MPa and near-zero xylem (ABA)). Mean daily incident photosyntheitc photon flux density (PPFD) was 28 mol m⁻² d⁻¹ and leaf temperature was 25°C during leaf development. The horizontal thick bar indicates the period of constant water deficit (redrawn from Granier & Tardieu, 1999).

be observed several weeks after the plants have been rewatered and have recovered maximum photosynthetic rate (Lecoeur *et al.*, 1995, Granier & Tardieu, 1999). A treatment which reduces photosynthesis neither during water deficit nor after rewatering can, therefore, reduce final leaf area by 45%. Leaves subjected to this treatment have higher concentrations of soluble sugars and starch (Granier, 1998) and lower SLA, because the same amount of C gain occurs in a reduced leaf area.

Covering 40% of plant leaf area with aluminium foil does not affect the expansion rate of sunflower

leaf 8, if the reduction in photosynthetic leaf area is imposed during the period of maximum absolute expansion rate of leaf 8, i.e. while carbon demand of the leaf is at maximum (Fig. 2a, intercepted PPFD reduced by 40% days 20–27). Similarly, leaf elongation rate of maize is unaffected by changes in PPFD during the period when elongation rate is at a maximum, in the range of PPFD 8–50 mol m⁻² d⁻¹ (Ben Haj Salah & Tardieu, 1996). By contrast, reducing intercepted PPFD by 40% in the early stages of the development of sunflower leaf 8 severely reduces expansion rate (Fig. 2b), although the level of C demand necessary for expansion of this young leaf is still very low.

Leaf expansion rate observed on one day is, therefore, neither linked to plant carbon balance nor to leaf carbon balance on the same day. This does not imply, of course, that leaves do not need carbon to grow. During the period when absolute expansion rate is maximum, leaves are autotrophic so growing tissues are close to a carbon source. This results in a high sink priority (Minchin *et al.*, 1993) so leaf expansion can be sustained even with low PPFD. In contrast, root elongation rate is closely dependent on intercepted PPFD since roots have a lower sink priority because they are further from a carbon source (Aguirrezabal *et al.*, 1994; Muller *et al.*, 1998). The same reasoning probably applies to young heterotrophic leaves whose expansion is reduced by a decrease in intercepted PPFD (Fig. 2).

Are changes in SLA in growing leaves a consequence of changes in leaf expansion rate and in photosynthesis rate?

If leaf expansion rate on a given day is independent of the carbon budget of the leaf, SLA is not an independent variable when leaf carbon balance or leaf expansion rate are changing with time. Eqn 1 can be rewritten as:

$$SLA = [dA/dt]/[(J-R)\,p].\qquad\text{Eqn 4}$$

This implies that SLA decreases when expansion is more affected than leaf carbon budget, and increases in the opposite case. We do not discuss here the role of carbon export by the leaf (see Gunn *et al.*, 1999), but concentrate on the respective roles of net photosynthesis (J − R) and expansion (dA/dt).

The water deficit presented in Fig. 1 had no effect on sunflower photosynthesis, but reduced LER (Fig. 3a,b). Consistent with Eqn 4, SLA was decreased by 35% on average in all leaf zones after 3 d of deficit (Fig. 3c). Conversely, a water deficit which had similar effects on photosynthesis and on expansion resulted in an unchanged SLA in maize (Fig. 4; B. Muller & M. Stosser, unpublished data).

The early reduction in intercepted PPFD presented in Fig. 2b caused a larger reduction in net photosynthesis (40%) than in relative expansion rate

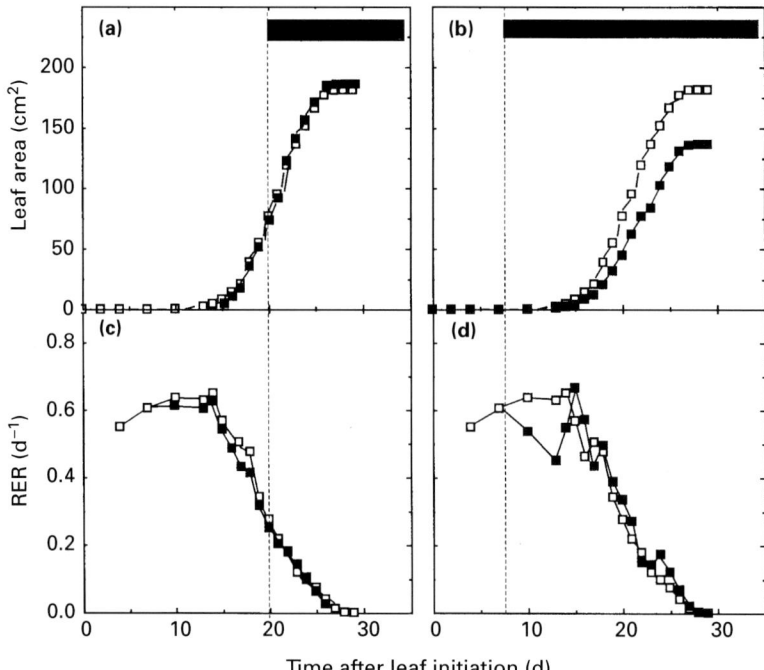

Fig. 2. Effects of PPFD on leaf expansion. The graphs show changes with time in leaf area (a,b) and in leaf relative expansion rate (RER) (c,d) of leaf 8 of sunflower in control plants (open squares) and plants with late (a,c) or early (b,d) reduction in intercepted PPFD (closed squares). Plants were grown in the field with a mean incident photosynthetic photon flux density (PPFD) of 45 mol m^{-2} d^{-1} and mean leaf temperature of 21°C in leaf development. Leaves 1, 3 and 5 were covered with aluminium foil to reduce photosynthetic leaf area by 40%. Horizontal thick bars represent the periods of reduction in intercepted PPFD. Vertical bars indicate confidence intervals at $P = 0.05$ (C. Granier & F. Tardieu, unpublished data).

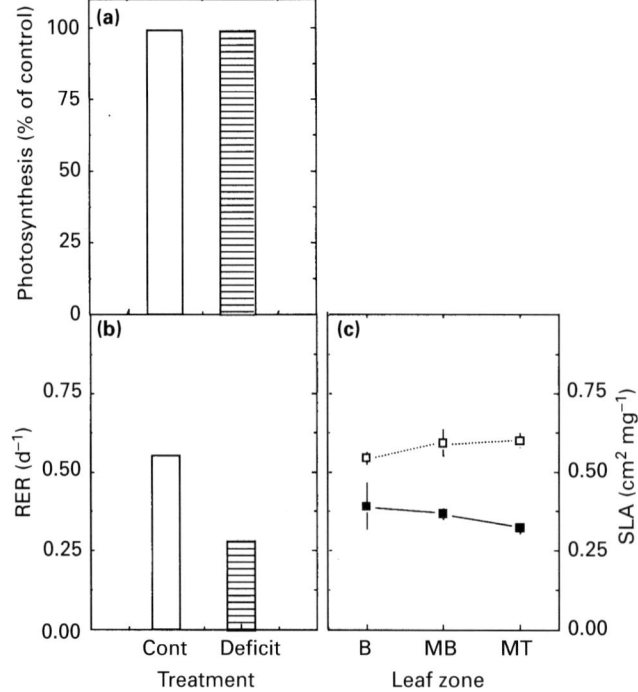

Fig. 3. Effects of mild water deficit on photosynthesis (a), relative expansion rate (RER) of leaf 8 (b) and specific leaf area (SLA) in three zones of leaf 8 (c) in sunflower. Leaf base, B; middle-base, MB; and middle-tip, MT. The experiment is that described in the legend to Fig. 1. Control plants, Cont; plants subjected to water deficit, Def. Vertical bars indicate confidence intervals at $P = 0.05$ (C. Granier & F. Tardieu, unpublished data).

(21%), resulting in a large increase in SLA (Fig. 5c). The same effect was observed in maize when incident PPFD was reduced by 75%. This reduced photo-synthesis but not leaf expansion, thereby increasing SLA (Fig. 4). Finally, changes in SLA between morning and afternoon also followed the tendency

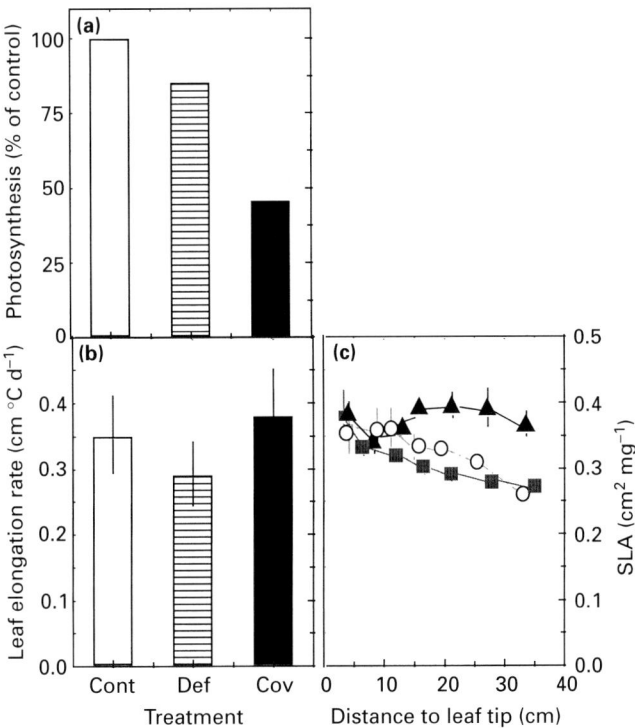

Fig. 4. Effects of a reduction in incident photosynthetic photon flux density (PPFD) and of mild water deficit on photosynthesis (a), absolute leaf elongation rate (b) and specific leaf area (SLA) along leaf 4 (c) in maize. Plants were grown in the field with a mean PPFD of 45 mol m^{-2} d^{-1} and a mean meristem temperature of 24°C during leaf development. The treatments were imposed for 100°d (7 d) when leaf 4 was *c.* 10 cm long. Control plants, Cont; plants subjected to water deficit, Def; plants grown with an incident PPFD reduced by 75%, Cov. Predawn leaf water potential was −0.3 MPa in water-stressed plants and −0.1 MPa in well watered plants. SLA is presented as a function of the distance of the considered leaf zone from the leaf tip. Vertical bars indicate confidence intervals at $P = 0.05$ (B. Muller & M. Stosser, unpublished data).

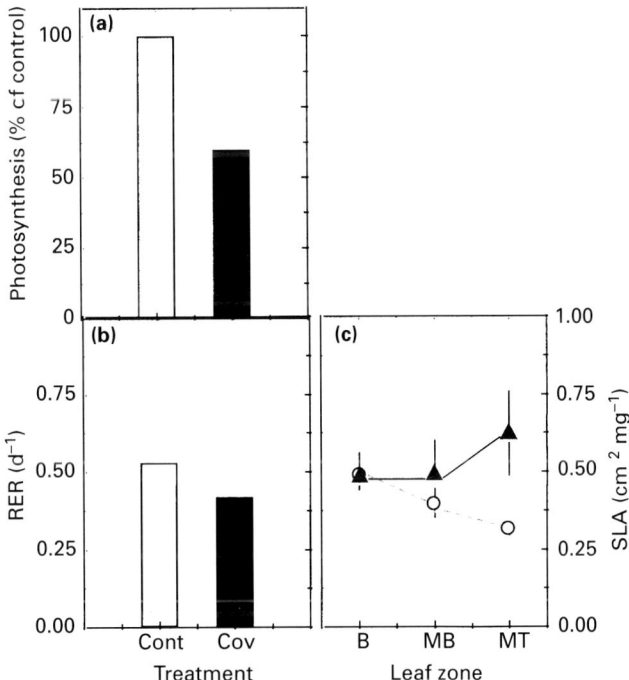

Fig. 5. Effect of a reduction in intercepted PPFD on photosynthesis (a), relative expansion rate (RER) of leaf 8 (b) and specific leaf area (SLA) (c) in three zones of sunflower leaf 8 base (B), middle-base (MB) and middle-tip (MT). The experiment is that described in the legend to Fig. 2. Control plants, Cont; plants grown with a reduced intercepted PPFD, imposed by covering 40% leaf area with aluminium foil (C. Granier & F. Tardieu, unpublished data), Cov.

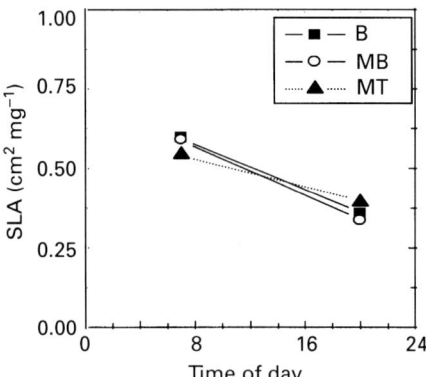

Fig. 6. Temporal changes in specific leaf area (SLA) in three zones of sunflower leaf 8. Base (B), middle-base (MB) and middle-tip (MT) leaves were harvested at the beginning and end of the photoperiod in the field experiment described in the legend to Fig. 2 (C. Granier & F. Tardieu, unpublished data).

suggested by Eqn 4. SLA decreased by one-third in all leaf zones (Fig. 6) because leaf expansion rate occurs at similar rates during day and night if expressed in thermal time (see below), but photosynthesis only occurs during the day. In all experiments, changes in SLA were partly accounted for by changes in concentrations in soluble carbohydrates and in starch which followed the same trends (Granier, 1998). However, a part of the variability of SLA still remains unexplained.

A MODEL OF LEAF EXPANSION BASED ON THERMAL TIME THAT HOLDS FOR BOTH MONOCOTYLEDONS AND DICOTYLEDONS

Results presented above suggest that tissue expansion and epidermal cell division can be analysed and modelled independently of C balance, changes in which would be expected to result in changes in SLA instead of changes in LER.

Time course of expansion and epidermal cell division in sunflower leaves

A common time course was observed in a series of 13 experiments carried out in the field, in the glasshouse and in the growth chamber at contrasting temperatures and PPFDs (Granier & Tardieu, 1998b). A sunflower leaf expands for several weeks after its initiation on the apex, in two phases which each last approx. half the period of leaf development.

During the first phase (days 1 to 15 in Fig. 7), absolute expansion rate is slow and relative expansion rate (RER) is high. Expansion is exponential (linear relationship between time and the logarithm of leaf area) and RER is constant and maximum if leaf temperature is constant (Fig. 7c; the transient decrease in RER on day 13 is due to a lower temperature on this day; see also control treatments

in Figs 1, 2). This constant RER is uniform over the whole leaf in this phase (Fig. 7d, day 14) and similar in leaves at different positions on the stem (Granier & Tardieu, 1998a for leaves 8 and 16; O. Turc *et al.*, unpublished data for leaves 6 to 32).

The second phase is characterized by a fast absolute expansion rate and a decreasing RER (Fig. 7b,c). Although the increase in leaf area appears to be broadly linear during this phase (Fig. 7a), the absolute expansion rate begins to decrease 1–2 d after reaching its maximum so no 'linear phase' can be defined. The decrease in RER occurs first at the leaf tip and progresses towards the leaf base (Fig. 7d, day 18). The greater final area of the base is, therefore, linked to a longer exponential period and not to a difference in RER. It is noteworthy that this exponential period, which determines leaf shape and final leaf area, occurs while the leaf is not visible and has a low absolute expansion rate.

This temporal pattern is essentially conserved for epidermal cell division in a leaf. Increase in epidermal cell number per leaf is initially slow, while the relative cell division rate (RDR, number of new epidermal cells per cell and per unit time) is at a maximum, and constant with time in the first period if leaf temperature is constant (Fig. 8b,d). During the phase with exponential increase in epidermal cell number, RDR and mean cell-cycle duration are common to all leaf zones and to several leaves of a plant (Granier & Tardieu, 1998a). At the end of this period, the absolute rate of increase in epidermal cell number reaches a maximum, while the relative cell-division rate decreases (i.e. cell-cycle duration increases). The reduction in relative cell-division rate occurs first at the leaf tip, progressing towards the leaf base.

Time course of leaf expansion and epidermal cell division in maize

The increase in maize leaf length follows a pattern similar to that found for leaf area in sunflower, but with a period with linear elongation. During a first period, lasting approx. half the duration of leaf development (days 1 to 12 in Fig. 5), the relative elongation rate is maximum and constant (Fig. 9b and c) while the absolute elongation rate is slow. It is common to all leaves in a plant, except for the first 5 leaves preformed in the seed (Muller, Reymond & Tardieu unpublished data). This phase is followed by a period with linear expansion (Fig. 9b) that starts shortly before the leaf emerges from the whorl. The absolute elongation rate is then constant with time and is common for all leaves (except for the first five leaves). Similar results are conserved in sorghum (Lafarge *et al.*, 1998). The duration of the period with linear elongation is short by comparison with that of leaf development (1/3 of total development duration in the experiment presented in Fig. 9).

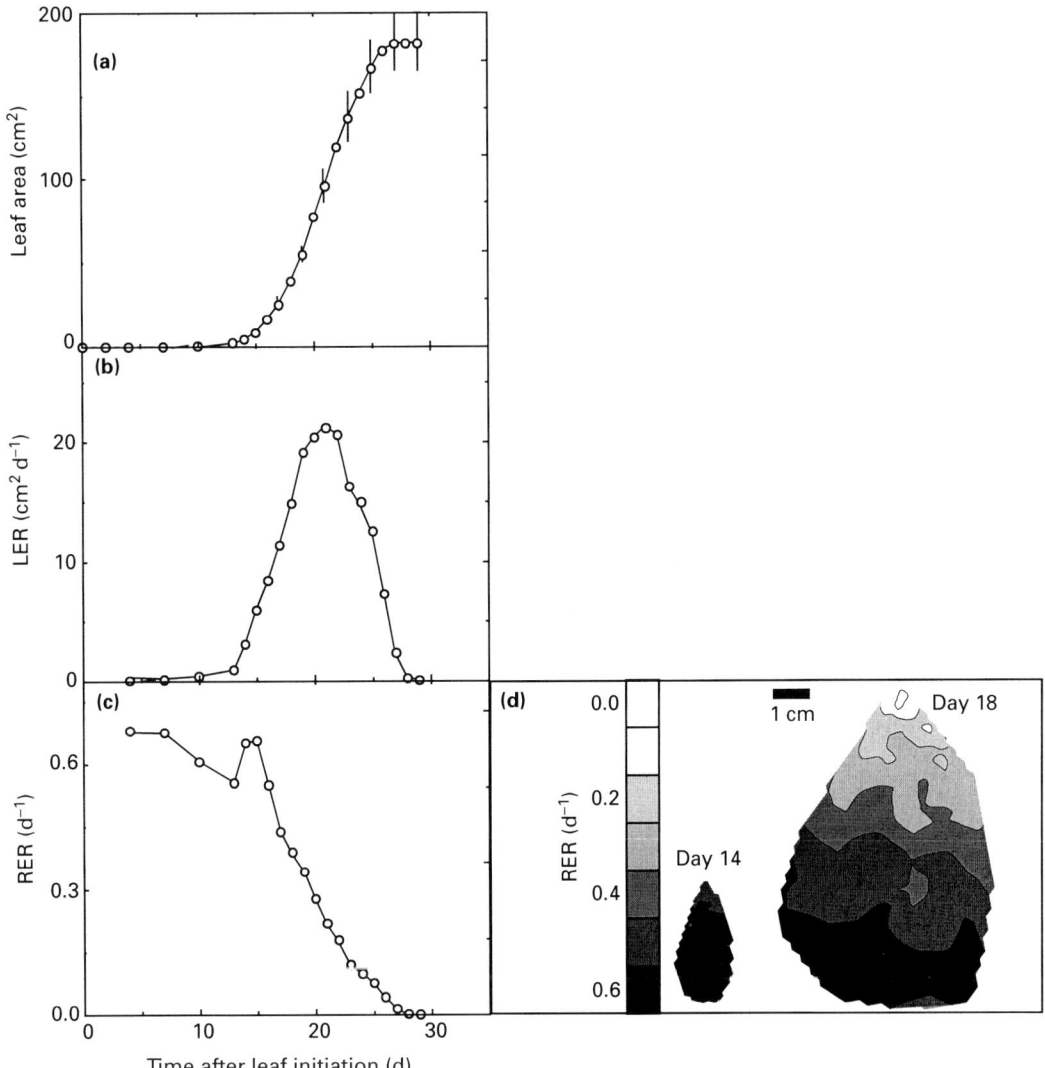

Fig. 7. Change with time in leaf area (a), absolute leaf expansion rate (LER) (b) and relative leaf expansion rate (RER) (c) of leaf 8 of sunflower in a field experiment similar to that described in the legend to Fig. 2. Spatial distributions of RER on days 14 (left) and 18 (right) are shown in panel (d). They were calculated in 120 triangles delimited by ink dots drawn on the leaf surface on day 12 (redrawn from Granier & Tardieu, 1998a).

During this period, both epidermal cell division and tissue elongation are restricted to the base of the leaf. Tissue expansion occurs in the first 8 cm and epidermal cell division in the first 2 cm behind the insertion point (Fig. 10a). The periods during which a group of epidermal cells divides and expands are short compared to those in sunflower (Ben Haj Salah & Tardieu, 1995). A new cell, pushed forward by younger cells, quits the meristematic region after 30–60 h and leaves the expansion zone in 40–90 h (vs 20 and 30 days respectively in sunflower).

How are these patterns altered by temperature? The use of thermal time

Rates of leaf expansion and of epidermal cell division are linearly related to leaf temperature in a wide range of temperatures (typically 10–30°C) in sunflower (Granier & Tardieu, 1998b), maize (Ben Haj Salah & Tardieu, 1995) and sorghum (Lafarge *et al.*, 1998). Common relationships hold for both constant and fluctuating temperatures in the field and the growth chamber. The reciprocals of the durations of expansion and of cell division are also linearly related to leaf temperature with stable relationships. All these relationships have a common x-intercept for a given species (T_0: 5°C for sunflower, 10°C for maize and 11°C for sorghum), and apply to temperatures ranging from this x-intercept to *c.* 30°C. In this range:

$$\mathrm{d}P/\mathrm{d}t = \mathrm{a}(T - T_0) \qquad \text{Eqn 5}$$

and

$$1/\mathrm{d} = \mathrm{b}(T - T_0) \qquad \text{Eqn 6}$$

($\mathrm{d}P/\mathrm{d}t$ is the rate of the studied process (expansion, epidermal cell division or leaf initiation), T is current temperature, a and T_0 are the slope and the x-

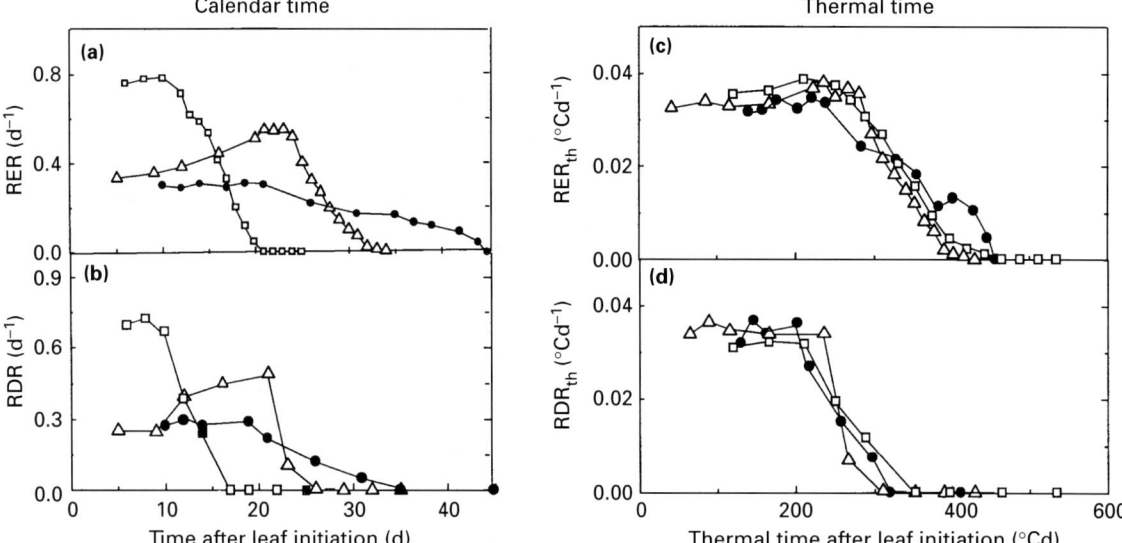

Fig. 8. Change with time in relative expansion rate (RER) (a) and in relative division rate (RDR) (b) in leaf 8 of sunflower grown in three experiments at contrasting temperatures: in the glasshouse with fluctuating temperature, mean temperature 26°C (open squares), in the field with fluctuating temperatures, mean temperature 18.5°C (open triangles) and in the growth chamber at a constant temperature of 14°C (closed circles). Each point represents the mean of five leaves. The same data are presented in panels (c) and (d) as a function of thermal time (redrawn from Granier & Tardieu, 1998b).

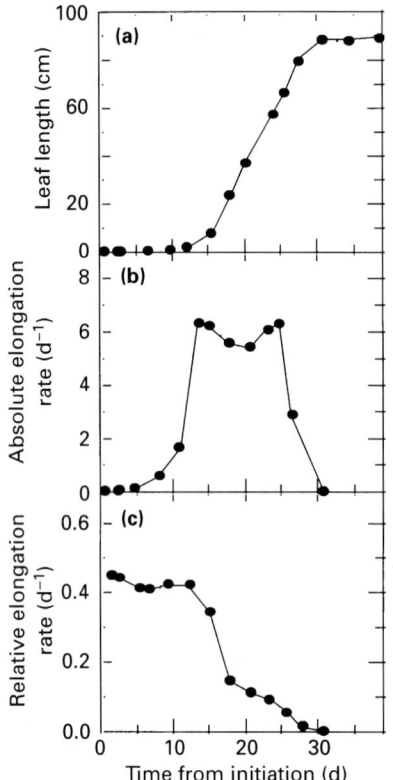

Fig. 9. Changes with time in leaf length (a), absolute leaf elongation rate (b) and in relative leaf elongation rate (c) of leaf 14 in maize. Plants were grown in a field experiment with a mean incident PPFD of 46 mol m^{-2} d^{-1} and a mean apex temperature of 23°C (B. Muller *et al.*, unpublished data).

intercept of the relationship between dP/dt and T, d is the time in which expansion (respectively epidermal cell division) occurs in a leaf, or the time in which leaf initiation occurs on the apex.) It follows that, at time d:

$$P = a \int_0^d (T - T_0) \, . \, dt \qquad \text{Eqn 7}$$

$\int_0^d (T - T_0) \, . \, dt$ is thermal time (unit °Cd when calculated with a daily timestep), which has already been used for a long time in the calculation of plant cycle duration (Reaumur 1735, Arnold 1959). Its use intuitively implies that time, as sensed by plants, elapses more rapidly at high than at low temperatures. We show here that a common calculation of thermal time can apply to epidermal cell division, to tissue expansion and to leaf initiation because x-intercepts (T_0) are common to these three processes.

The relevance of expressing durations and rates in thermal time can be seen in Fig. 8. Time courses of relative cell division rate and of relative expansion rate of sunflower leaves are presented for three experiments carried out at contrasting leaf temperatures, either fluctuating (field or greenhouse) or constant (growth chamber). The durations of periods with exponential cell division and tissue expansion, and total durations of expansion and division differed if expressed in clock time. However, they were unified if expressed in thermal time, regardless of the temperature imposed to leaves. In the same way, the rates of epidermal cell division and of tissue expansion increased with temperature if expressed per unit clock time but were common to all situations when expressed per unit thermal time. It follows that the duration of epidermal cell cycle, of 67, 31 and 23 h at 14, 22 and 26°C, respectively, was 22°Cd, regardless of leaf temperature when expressed in thermal time, in the field as well as in the laboratory (Granier & Tardieu, 1998b).

Similar results were observed in maize and sorghum. Durations and rates expressed in thermal time were common to several experiments at different temperatures in different experimental conditions (Ben Haj Salah & Tardieu, 1995; Lafarge *et al.*, 1998). For instance, the time for a cell to cross the elongation zone of maize was 30°Cd regardless of temperature if expressed in thermal time, while it ranged from 34 to 195 h if expressed in clock time (Ben Haj Salah & Tardieu, 1995).

A MODEL OF LEAF EXPANSION, EPIDERMAL CELL DIVISION AND INDIVIDUAL CELL EXPANSION IN THE ABSENCE OF WATER OR NUTRIENT DEFICITS

The model for sunflower (and presumably any dicot) leaves is summarized in Fig. 11. Three phases can be identified, whose durations are constant among a large range of experiments if expressed in thermal time (Fig. 11c). During the first phase, both relative expansion rate and relative cell division rate are maximum, constant with time and common to all leaves which are in this phase. The duration of this phase is constant in a given zone of a leaf, but differs among zones and among leaves of a plant. A second phase follows, during which relative expansion rate remains constant but relative cell division rate decreases. Relative expansion decreases during the third phase while cell division is almost stopped.

Individual cell area remains small (c. 80 to 120 μm²) and has a constant relative expansion rate while both tissue relative expansion rate and cell division rate are constant and maximum (Fig. 11b,d). It suddenly increases, with a maximum relative expansion rate, during phase 2. Relative cell expansion rate decreases afterwards, together with tissue relative expansion rate. This behaviour can be predicted from Eqn 3 if relative cell division rate and relative expansion rate are considered as independent. If so, relative individual cell expansion rate can be calculated as the difference between relative expansion rate and relative cell division rate. It is therefore constant during phase 1 and equals the difference between RER and RDR. It increases during phase 2 because RDR decreases without change in RER. It decreases in phase 3 as RER decreases.

Maize and sorghum (and presumably any monocot) leaf development follow a model of development similar to that in Fig. 11, but with a linear phase. It applies to African fields as well as to European fields and to growth chamber experiments (Lafarge *et al.*, 1998). Both the relative elongation rate during the exponential phase and the absolute elongation rate during the linear phase are common to all leaves of a plant, except for the first five leaves preformed in the seed, and the last three leaves which usually remain small. Apparent independency of cell division and of tissue expansion are also apparent in maize. Cells remain short close to the insertion point because RDR and RER are similar (Fig. 10b). They begin to increase in length in the zone where RDR decreases and RER increases, and reach their maximum length at the end of the elongation zone.

Effects of water availability and of intercepted PPFD

The durations of epidermal cell division and tissue expansion of sunflower leaves are not affected by mild water deficits or by reductions in intercepted PPFD, which largely reduce final leaf area (Figs 1 and 2). This is also the case for pea under water deficit (Turc & Lecoeur, 1997) and for maize or sorghum exposed to a large range of PPFD (Lafarge *et al.*, 1998; Muller *et al.*, unpublished). Only drastic reductions in incident PPFD (lower than 2 mol m⁻² d⁻¹) or in water availability (leaf predawn water potential lower than −1 MPa) increased these durations (C. Granier & F. Tardieu, unpublished).

Time courses of relative cell division rate and relative leaf expansion rate have common baselines in plants subjected to short water deficits and in watered plants. Relative rates are reduced during the deficit period only (Fig. 1c) but absolute rates are permanently reduced, even after rewatering. This is due to the fact that absolute expansion rate on a day is the product of relative expansion rate and of leaf area on the same day. Leaf area is smaller in droughted plants at the end of the water deficit period, so absolute expansion rate remains smaller. The same reasoning applies to the increase in cell

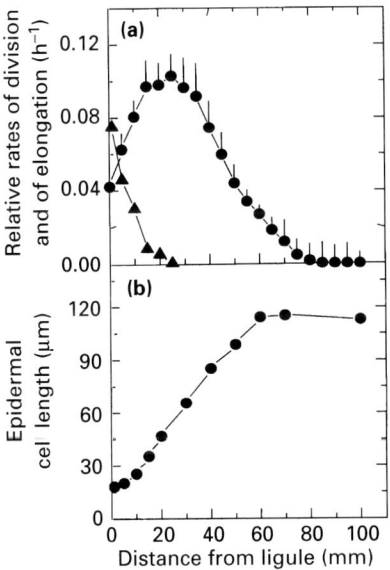

Fig. 10. Spatial distributions of relative elongation rate (closed circles), relative cell division rate (closed triangles) (a) and epidermal cell length (closed circles) (b) in leaf 6 of maize in a field experiment (redrawn from Ben Haj Salah & Tardieu, 1995).

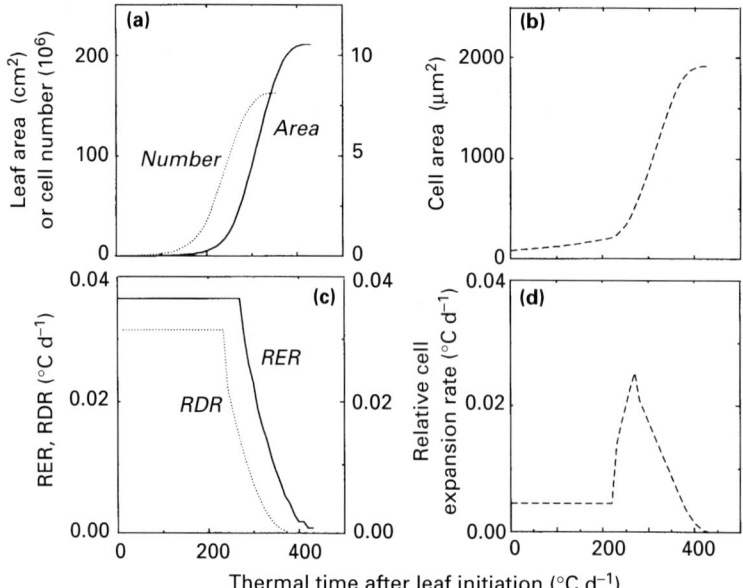

Fig. 11. Modelled time courses of leaf area and epidermal cell number per leaf (a) and epidermal cell area (b) in sunflower leaf 8, averaged over 16 experiments in the field, growth chamber and glasshouse. Corresponding relative leaf expansion rate (RER) and relative cell division rate (RDR) are shown in (c) and relative individual cell expansion rate in (d) (redrawn from Granier & Tardieu, 1998b, 1999).

number, which is permanently affected in droughted plants although relative cell division rate recovers shortly after rewatering. In the same way, a reduction in incident PPFD during the early phase of leaf development causes a decrease in relative expansion rate and division rate (Fig. 2). RER and RDR recover but absolute rates remain lower in shaded plants.

Expansion rate is directly affected by water deficit, via a change in cell wall rheological properties involving cell wall enzymes and the expression of specific mRNAs (Saab & Sharp, 1995). This is accompanied by changes in the activity of the enzyme cdc2 kinase which plays a crucial role in the control of cell division (Schuppler *et al.*, 1998). We have shown that maize leaf elongation rate is reduced by water deficit via two effects, one involving a hydraulic signal and the other linked to ABA signalling (Ben Haj Salah & Tardieu, 1997). The first effect is observed during the day, even in well watered plants, when leaf-to-air vapour pressure difference (VPD) increases, with a linear relationship between elongation rate and VPD. The second is observed during the night in droughted plants. A linear relationship is observed between leaf elongation rate and the concentration of ABA in the xylem sap, with a common relationship for endogenous ABA and for exogenous ABA artificially fed to plants. The effect of water deficit during the day is accounted for by the sum of these two effects. The concentrations of soluble carbohydrates in the leaf are increased in all these cases, so reductions in elongation rate cannot be attributed to a change in carbon availability.

CONCLUSION

Leaf expansion rate observed on a given day is not controlled by the leaf carbon budget on the same day. Specific leaf area, which is considered as a key controlling parameter in many models, behaves as a consequence of simultaneous changes in expansion rate and photosynthesis rate. Environmental conditions which have a greater depressive effect on expansion rate than on photosynthesis cause a decrease in SLA. Those which have a greater depressive effect on photosynthesis cause an increase in SLA. These effects result in large variations in SLA, depending on available PPFD, plant water status and time of day. Models which derive leaf expansion from carbon budget and which use SLA as an input parameter are, therefore, prone to large errors. These errors are frequently hidden by the use of extra equations (with extra parameters estimated *ad hoc*) which increase or decrease SLA depending on environmental conditions. In the same way, SLA should probably not be considered as a direct adaptive process in ecological modelling. It may still be an interesting characteristic for comparing species placed in competition for light, and may be correlated to other plant characteristics of interest in ecological modelling. We suggest that the time at which leaves are collected (50% reduction in SLA between morning and afternoon), and environmental conditions (effects of PPFD and water deficit) must be taken into account prior to data interpretation.

We believe that leaf expansion and cell division can, and should, be modelled as primary processes.

However, neither the Lockhart equation (Eqn 2) nor the equation relating leaf expansion to increases in cell number and in cell area (Eqn 3) can be used in a straightforward way, as they overlook the program of leaf development. These changes can be synthesised as follows:

Time courses of relative leaf expansion rate and of cell division rate are well conserved within a plant and among a large range of environmental conditions, provided that durations and rates are expressed in thermal time. Maximum relative rates, observed during exponential phases, are common to all zones of a leaf and to all leaves of a plant, in maize and sunflower. They are also common to a range of experiments performed at contrasting temperatures, in the absence of water, nutrient or light deficits. They can therefore be considered as stable parameters which characterize a genotype.

A water deficit affects these time courses during deficit only, without appreciable 'after effect', if expansion and division are analysed via relative rates. It is possible to express the effects of evaporative demand and of the concentration of ABA in the xylem with stable relationships which apply to field conditions as well as laboratory conditions.

A reduction in PPFD causes no effect on leaf expansion during the rapid growth period, while the leaf is autotrophic, but considerably reduces the relative expansion rate during early leaf development. As in the case of water deficit, a transient reduction in relative rates of expansion and division results in a permanent reduction in absolute rates.

The most straightforward way of modelling leaf growth is to consider that tissue expansion and cell division are independent processes, and that they determine cell area at each time.

REFERENCES

Arnold CM. 1959. The determination and significance of the base temperature in a linear heat unit system. *American Society of Horticultural Science Proceedings* **74**: 430–445.

Aguirrezabal LAN, Deléens E, Tardieu F. 1994. Root elongation rate is accounted for by intercepted PPFD and source–sink relations in field and laboratory grown sunflower. *Plant, Cell and Environment* **17**: 443–450.

Ben Haj Salah H, Tardieu F. 1995. Temperature affects expansion rate of maize leaves without change in spatial distribution of cell length. Analysis of the coordination between cell division and cell expansion. *Plant Physiology* **109**: 861–870.

Ben Haj Salah H, Tardieu F. 1996. Quantitative analysis of the combined effects of temperature, evaporative demand and light on leaf elongation rate in well watered field and laboratory-grown maize plants. *Journal of Experimental Botany* **47**: 1689–1698.

Ben Haj Salah H, Tardieu F. 1997. Control of leaf expansion rate of droughted maize plants under fluctuating evaporative demand. A superposition of hydraulic and chemical messages? *Plant Physiology* **114**: 893–900.

Chazen O, Neuman PM. 1994. Hydraulic signals from the roots and rapid cell-wall hardening in growing maize (*Zea Mays* L.) leaves are primary responses to polyethylene glycol induced water deficits. *Plant Physiology* **104**: 1385–1392.

Dale JE. 1992. How do leaves grow? *Bioscience* **42**: 423–432.

Dengler NG. 1980. Comparative histological basis of sun and shade leaf dimorphism in *Helianthus annuus*. *Canadian Journal of Botany* **58**: 717–730.

Doerner P, Jorgensen JE, You R, Steppuhn J, Lamb C. 1996. Control of root growth and development by cyclin expression. *Nature* **380**: 520–523.

Durand R. 1969. Signification et portée des sommes de température. *Bulletin technique d'information* **238**: 185–190.

Granier C. 1998. Analyse de l'expansion de feuilles de tournesol en conditions fluctuantes. Effets de la température, du déficit hydrique et du rayonnement. Montpellier, France: Ecole Nationale Supérieure Agronomique de Montpellier.

Granier C, Tardieu F. 1998a. Spatial and temporal analyses of expansion and cell cycle in sunflower leaves. A common pattern of development for all zones of a leaf and different leaves of a plant. *Plant Physiology* **116**: 991–1001.

Granier C, Tardieu F. 1998b. Is thermal time adequate for expressing the effects of temperature on sunflower leaf development? *Plant Cell and Environment*, **21**: 695–703.

Granier C, Tardieu F. 1999. Reductions in area, cell number and cell area in sunflower leaves subjected to short water deficits with different timings. Variability in responses can be simulated using a simple model of leaf development. *Plant Physiology* **119**: 609–619.

Green PB. 1976. Growth and cell pattern formation on an axis: critique of concepts, terminology, and mode of study. *Botanical Gazette* **137**: 187–202.

Gunn S, Farrar JF, Collis BE, Nason M. 1999. Specific leaf area in barley: individual leaves versus whole plants. *New Phytologist* **143**: 45–51.

Hemerly A, Almeida Engler J, Bergounioux C, Van Montagu M, Engler G, Inzé D, Ferreira P. 1995. Dominant negative mutants of the Cdc2 kinase uncouple cell division from iterative plant development. *EMBO (European Molecular Biology Organization) Journal* **14**: 3925–3936.

Kutschera U. 1992. The role of the epidermis in the control of elongation growth in stem and coleoptiles. *Botanica Acta* **105**: 246–252.

Lafarge T, De Raissac M, Tardieu F. 1998. Elongation rate of sorghum leaves has a common response to meristem temperature in diverse African and European environmental conditions. *Field Crops Research* **58**: 69–79.

Lecoeur J, Wery J, Turc O, Tardieu F. 1995. Expansion of pea leaves subjected to short water deficit: cell number and cell size are sensitive to stress at different periods of leaf development. *Journal of Experimental Botany* **46**: 1093–1101.

Lockhart JA. 1965. An analysis of irreversible plant cell elongation. *Journal of Theoretical Biology* **8**: 453–470.

MacQueen-Mason SJ, Cosgrove DJ. 1995. Expansin mode of action on cell walls. Analysis of wall hydrolysis, stress relaxation, and binding. *Plant Physiology* **107**: 87–100.

Milthorpe FL, Newton P. 1963. Studies on the expansion of the leaf surface. III. The influence of radiation on cell division and leaf expansion. *Journal of Experimental Botany* **14**: 483–495.

Minchin PEH, Thorpe MR, Farrar JF. 1993. A simple mechanistic model of phloem transport which explains sink priority. *Journal of Experimental Botany* **44**: 497–955.

Monteith JL. 1977. Climate and the efficiency of crop production in Britain. *Philosophical Transactions of the Royal Society*. London, B, **281**: 277–294.

Muller B, Stosser M, Tardieu F. 1998. Spatial distributions of tissue expansion and cell division rates are related to irradiance and to sugar content in the growing zone of maize roots. *Plant, Cell and Environment* **21**: 149–158.

Palmer SJ, Berridge DM, Mac Donald AJS, Davies WJ. 1996. Control of leaf expansion in sunflower (*Helianthus annuus* L.) by nitrogen nutrition. *Journal of Experimental Botany* **47**: 359–368.

Passioura JB. 1996. Simulation models: science, snake oil, education or engineering? *Agronomy Journal* **88**: 690–694.

Pritchard J, Hetherington PR, Fry SC, Tomos AD. 1993. Xyloglucan endotransglycosylase activity, microfibril orientation and the profiles of cell wall properties along growing regions of maize roots. *Journal of Experimental Botany* **44**: 1281–1289.

Pritchard J, Wyn Jones RG, Tomos AD. 1990. Measurement of yield threshold and cell wall extensibility of intact wheat roots under coherent ionic, osmotic and temperature treatments. *Journal of Experimental Botany* **41**: 669–675.

Randall HC, Sinclair TR. 1988. Sensitivity of soybean leaf development to water deficits. *Plant, Cell and Environment* **11**: 835–839.

Reinhardt D, Wittwer F, Mandel T, Kuhlemeier C. 1998. Localized upregulation of a new expansin gene predicts the site of leaf formation in the tomato meristem. *Plant Cell* **10**: 1427–1437.

Rhizopoulou S, Davies WJ. 1991. Influence of soil drying on root development, water relations and leaf growth of *Ceratonia siliqua* L. *Oecologia* **88**: 41–47.

Saab IN, Ho THD, Sharp RE. 1995. Translatable RNA populations associated with maintenance of primary root elongation and inhibition of mesocotyl elongation by abscisic acid in maize seedlings at low water potentials. *Plant Physiology* **109**: 593–601.

Schuppler U, He PH, John PCL, Munns R. 1998. Effect of water stress on cell division and cell division cycle 2-like cell cycle kinase activity in wheat leaves. *Plant Physiology* **117**: 667–678.

Thompson DS, Davies WJ, HO LC. 1998. Regulation of tomato fruit growth by epidermal cell wall enzymes. *Plant, Cell and Environment* **21**: 589–599.

Turc O, Lecoeur J. 1997. Leaf primordium initiation and expanded leaf production are co-ordinated through similar response to air temperature in pea (*Pisum sativum* L.) *Annals of Botany* **80**: 265–273.

Wilson GL. 1966. Studies on the expansion of the leaf surface. V-Cell division and expansion in a developing leaf as influenced by light and upper leaves. *Journal of Experimental Botany* **7**: 440–451.

New Phytol. (1999), **143**, 45–51

Specific leaf area in barley: individual leaves versus whole plants

S. GUNN*, J. F. FARRAR, B. E. COLLIS AND M. NASON

School of Biological Sciences, University of Wales, Bangor, Gwynedd LL57 2UW, UK

Received 5 October 1998; accepted 8 April 1999

SUMMARY

We have explored the relationships between specific leaf area calculated for a whole plant and its individual leaves. Barley was grown in hydroponics in controlled environment cabinets. Plants were harvested on the basis of physiological age (defined as the number of days after full expansion of leaves on the main stem) and the area and weight of whole, fully expanded, leaves measured and specific leaf area (SLA) of individual leaves or whole plants calculated. Specific leaf area calculated for individual leaves (SLA_L) varied with leaf position and with leaf age after full expansion whereas SLA calculated for whole plants (SLA_P) varied with plant age. The same conclusions were reached whether the results were based on total dry weight or dry weight minus soluble carbohydrates ('structural weight'). Transferring plants to shade on the day of full expansion of the third leaf on the main stem increased the SLA_P, and also SLA_L of leaves 3 and 4 on the main stem (leaf 4 being the younger leaf of the two), because of a decrease in the 'structural weight' of these leaves. However SLA_L of leaf 2 (which was older than leaf 3) was not affected by shading; the effect was confined to leaves developing in the new conditions.

Key words: specific leaf area, barley, dry weight, structural dry weight, shade.

INTRODUCTION

Specific leaf area (SLA) was introduced as a concept in the analysis of whole plant growth and was defined as the total leaf area divided by the total leaf weight (Evans, 1972). Defined in this way, SLA has been used to draw conclusions about the relative thickness of leaves which in turn has led to questions about how leaf structure is affected by environmental conditions or experimental treatments. Therefore there is a conceptual leap from a property of the whole plant (traditional specific leaf area) to specific changes in the structure and chemical composition of individual leaves.

SLA has been correlated with variables as diverse as net photosynthesis (McClendon, 1962), relative growth rate (Atkins & Lambers, 1998; Poorter & Van der Werf, 1998), yield (Singh *et al.*, 1985) and leaf structure (Cambridge & Lambers, 1998; Pyankov *et al.*, 1998). However there rarely seems to be a clear distinction made between whole plant SLA (i.e. a mean of all leaves, SLA_P) and SLA of individual leaves (SLA_L) and how these relate to

actual changes in leaf structure and chemical composition. This problem was noted by Garnier & Freijsen (1994) who alluded to the paucity of data on the relationship between SLA_L and SLA_P.

Furthermore there is the question of what controls SLA_L, and whether there are any mechanistic explanations for effects of experimental treatment. Correlations of SLA_P with different variables between species are descriptive and do not lead to mechanistic explanations (Shipley, 1995). A better method is to manipulate SLA_L within a single species by altering the environment. For this approach to work, it is necessary to define exactly when and where changes in SLA_L occur so that changes can be related to other parameters of plant growth. The partitioning of carbon (C) between different leaves and within leaves between transport carbohydrate (soluble sugars in the cytosol and phloem vessels), storage carbohydrate (fructans in the vacuole and starch in the chloroplast) and structural C (cell wall, lipid and protein) gives a basis for understanding changes in leaf weight and hence SLA_L. These pools of C vary over different timescales (from hours to days or weeks) and all may contribute to changes in SLA_L. Variation in inorganic compounds may also cause changes in SLA_L (Heilmeier & Monson, 1994; Van Arendonk & Poorter, 1994), but the abundance of C make its partitioning and

*Author for correspondence (fax +44 1248 370731; e-mail s.gunn@bangor.ac.uk).

Abbreviations: SLA, specific leaf area; SLA_L, specific leaf area of individual leaf blades; SLA_P, SLA of the whole plant (mean); dae, day after full expansion.

metabolism a strong candidate for controlling changes in SLA_L.

We wanted to determine whether SLA_L and SLA_P can be used interchangeably in drawing conclusions about individual leaves by posing the following questions. (1) Does SLA_L vary with leaf position when leaves are harvested at a given developmental stage? (2) Does the SLA_L of a leaf change with age after it has fully expanded? (3) Can SLA_P be predicted from SLA_L? (4) When plants are transferred from high to low light, what aspect of SLA changes? We also investigated the role of C in the variation of SLA_L with two further questions: (5). If SLA_L changes is this because of changes in leaf area or leaf weight? (6) Do the answers to the previous questions change if the results are expressed on the basis of total weight or structural weight?

MATERIALS AND METHODS

Plant growth

Seeds of *Hordeum vulgare* L. (barley) cv. Klaxon were germinated and grown in controlled environment chambers (Sanyo Gallenkamp PG660, Leicester, UK), at a mean CO_2 concentration of 350 μmol CO_2 mol^{-1}, at 20°C with a 16 h photoperiod, a photon flux density of 400 μmol m^{-2} s^{-1} at plant height, supplied by halogen metal halide bulbs supplemented with tungsten filament bulbs and a vapour pressure deficit of 0.7 kPa. Air was drawn into the cabinets through a modified inlet port from a fan (Type-3MS11, Air Control Installations, Chard, UK) at 60 dm^3 min^{-1} which produced 5.5 changes of air h^{-1}.

Sixty plants were grown in two troughs each containing 7 dm^3 of solution aerated at 1.2 dm^3 min^{-1}. Plants were spaced to minimize shading between plants and there was little shading within each plant. The temperature of the solution was not controlled but was ±1°C of the air temperature. Solutions were changed every 3 or 4 d. The plants were grown in full strength Long Ashton solution (mol m^{-3}, full strength); KNO$_3$ (4), Ca(NO$_3$)$_2$.4H$_2$O (4), NaH$_2$PO$_4$.2H$_2$O (1.33), MgSO$_4$.7H$_2$O (1.5), FeEDTA Na (0.1), MnSO$_4$.4H$_2$O (0.01), CuSO$_4$. 5H$_2$O (0.001), ZnSO$_4$.7H$_2$O (0.001), H$_3$BO$_3$ (0.05), Na$_2$MoO$_4$.2H$_2$O (0.004), NaCl (0.1), Na$_2$SiO$_3$. 5H$_2$O (0.05).

Transfer to low light

On the day of full expansion of leaf three on the main stem, 10 plants were transferred to low light (90 μmol m^{-2} s^{-1} at plant height). Neutral density filters (layers of muslin) were supported on a wire frame such that air could circulate freely over the plants: shaded and unshaded plants were grown in the same trough.

Dry weight and leaf area

Plants were harvested (a) 2 dae of successive leaves on the main stem, (b) 2 dae of leaf 4 (low light) or (c) every day after full expansion of the third leaf on the main stem from the day of full expansion (0 dae) to 5 dae. At each harvest four replicate plants were divided into; main stem leaf blades which were fully expanded, and tiller leaf blades which were fully expanded. Leaf area was measured on a flatbed scanner with computer software (Delta T Devices, Cambridge, UK) before leaves were dried in an oven at 70°C for 48 h.

'Structural weight'

'Structural weight' was calculated as total dry weight minus soluble carbohydrates. Soluble carbohydrates were extracted from dried fully expanded leaf blades in 5 ml 80% ethanol at 60°C overnight, followed by 5 ml 40% ethanol at 60°C for 2 h and then 5 ml distilled water at 60°C for 2 h and the three extracts combined and made up to 20 ml. A further extraction in 5 ml distilled water at 60°C for 2 h did not contain significant amounts of carbohydrate. Soluble carbohydrates were determined by the phenol sulphuric method (Dubois *et al.*, 1956), with sucrose as the standard. Starch was not determined as it is a very small proportion (2.5±0.5%) of the carbohydrate of barley leaves grown under these conditions (B. Collis, unpublished).

Specific leaf area

Specific leaf area of individual leaf blades (leaf SLA, SLA_L) was calculated as:

$$SLA_L = \frac{\text{individual leaf area}}{\text{individual leaf weight}}$$

whereas whole plant SLA (SLA_P) was calculated as:

$$SLA_P = \frac{\Sigma \ (\text{area of all fully expanded leaves})}{\Sigma \ (\text{weight of all fully expanded leaves})}$$

but excluding senescent leaves.

Means were compared by ANOVA using the computer package SPSS (version 7, SPSS, Chicago, US). Levine's test was used to test for the equality of variances and significant interactions were compared using Tukey's honestly significant test. Data are shown as means of four replicates ±1 SE.

Allometric coefficients

Allometric coefficients were calculated for the relationships between the natural logarithms of leaf area and leaf dry weight by geometric mean regression (Gunn *et al.*, 1999). Two relationships were determined for; all fully expanded leaves individually, and total of all fully expanded leaves on a plant. Goodness of fit of the points to a straight line was assessed using the coefficient of determination (r^2) (Zar, 1996). A comparison of the two correlation coefficients was carried out after a Fisher's z transformation of r (Zar, 1996). There was no significant difference between the two correlation coefficients (r for all fully expanded leaves individually was 0.978 and for the total of all fully expanded leaves was 0.989). Comparisons of v were carried out using a modified t-test and results are shown with standard errors (Ricker, 1984). A comparison of the elevations (as opposed to the intercepts) of the regressions (i.e. a comparison of the vertical positions of the lines on the graphs) was carried out using a t-test (Zar 1996).

RESULTS

Specific leaf area of individual leaf blades (SLA_L)

The SLA_L of leaf 4 on the main stem was smaller than for leaves 1, 2 or 3, whether expressed on total ($P < 0.001$) or structural weight basis ($P < 0.001$, Fig. 1) whereas SLA_L of leaf 3 was lower than leaf 2 when expressed on a structural basis. The SLA_L of leaf 3 was lower between 2–5 than 0–1 d after full expansion when expressed on a total ($P < 0.01$) or structural weight basis ($P < 0.01$, Fig. 2a) because of an increase in the structural dry weight of the leaf (Fig. 2b). Leaf area was unaffected by shading (Fig. 2b).

Whole plant (SLA_P)

Plants were harvested on the basis of development (2 dae of successive leaves on the main stem). The

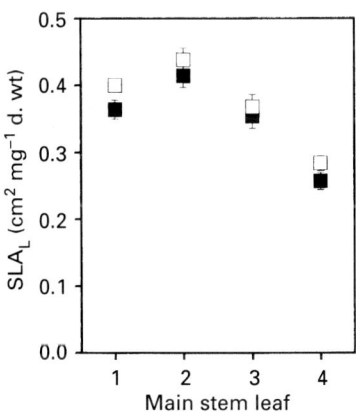

Fig. 1. The effect of leaf position on the specific leaf area of individual leaf blades (SLA_L, cm^2 mg^{-1}) of barley grown in hydroponics. Plants were harvested 2 d after expansion of successive main stem leaves. Values are the mean ± SE of 4 replicates. SLA_L calculated using total dry weight (filled squares); SLA_L calculated using extracted dry weight (open squares).

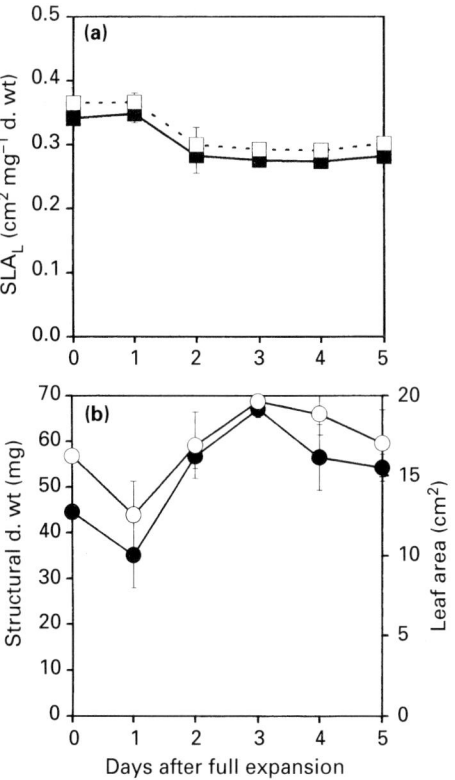

Fig. 2. The effect of leaf age on (a) the specific leaf area (SLA_L, cm^2 mg^{-1}) of leaf 3 on the main stem and (b) the structural dry weight (mg) and leaf area (cm^2) of barley (*Hordeum vulgare*) grown in hydroponics. Plants were harvested every day from full expansion (0 dae) to 5 dae. Values are the mean of 4 replicates ± SE. SLA_L calculated using total dry weight (filled squares); SLA_L calculated using extracted dry weight (open squares); dry weight (filled circles); leaf area (open circles).

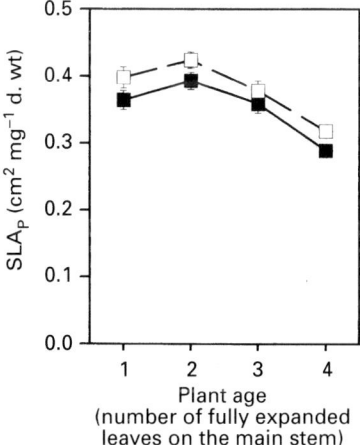

Fig. 3. The effect of plant age (as determined by number of main stem leaves) on the specific leaf area of whole plants (SLA_P, cm^2 mg^{-1}) of barley (*Hordeum vulgare*) grown in hydroponics. Plants were harvested 2 d after expansion of successive main stem leaves. Values are the mean of 4 replicates ± SE. SLA_P calculated using total dry weight (filled squares); SLA_P calculated using extracted dry weight (open squares).

corresponding chronological ages were, 2 d after expansion of leaf 1, 11 d; leaf 2, 15 d; leaf 3, 19 d and leaf 4, 23 d after sowing.

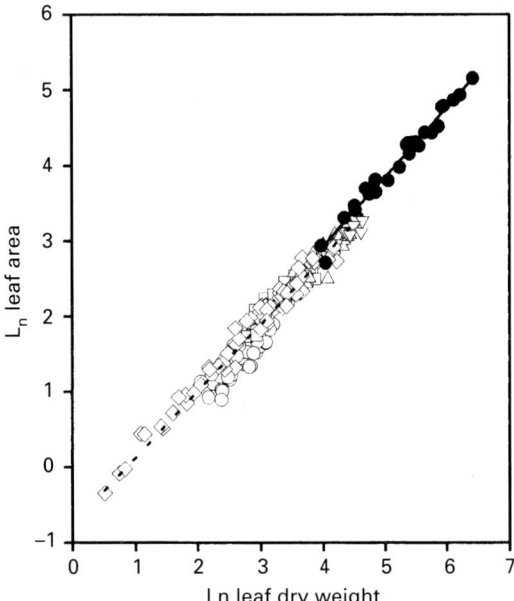

Fig. 4. The allometric relationship between leaf area and leaf weight for barley (*Hordeum vulgare*) grown in hydroponics. Individual leaves are shown by open symbols and the dotted line (leaf 1 on the main stem (circle); leaf 2 on the main stem (square); leaf 3 on the main stem (triangle, apex uppermost); leaf 4 on the main stem (triangle, apex down); tiller leaves (diamond)), and the sum of the individual leaves are indicated by solid circles and the solid line.

When plants were harvested 2 dae of successive leaves SLA_P fell with plant age when expressed on a total ($P < 0.001$) or structural weight basis ($P < 0.001$, Fig. 3).

Can SLA_P be predicted from the SLA_L?

There was no significant difference between the allometric coefficient, calculated by geometric mean regression, for all fully expanded leaves individually ($v = 0.90 \pm 0.01$) and that calculated for the total of all fully expanded leaves ($v = 0.93 \pm 0.03$) whereas both were significantly lower than 1 ($P < 0.001$; Fig. 4). There was no significant difference between the elevations.

Table 2. *Results from a two way ANOVA of the effect of leaf position and shading on specific leaf area of individual leaves (SLA_L), structural dry weight and leaf area*

Source of variation	SLA_L	Structural d. wt	Leaf area
Shade	***	**	ns
Leaf	**	***	***
Shade × leaf	**	ns	ns

***, $P \leqslant 0.001$; **, $P \leqslant 0.01$; ns, no significant difference.

Transfer to low light

On the day of full expansion of leaf 3 on the main stem, plants were transferred from high to low light (400 to 90 μmol m⁻² s⁻¹ at plant height). SLA_P is increased by shading when this is expressed on a total ($P < 0.001$) or structural weight basis ($P < 0.001$, Table 1) over the next 6 d.

An ANOVA (Table 2) of the results from shaded and unshaded plants harvested 2 d after full expansion of the fourth leaf on the main stem showed that there was a significant interaction between leaf and shading such that SLA_L of leaf blades 3 and 4 on the main stem, but not leaf blade 2, was increased by shading when this is expressed on a total (results not shown) or structural weight basis (Fig. 5a). Shading decreased dry weight (both total and structural) but not leaf area (Table 2, Fig. 5b,c).

DISCUSSION

We have investigated the variation in SLA_L within a single plant to test if SLA_P could be related to SLA_L and vice versa. We set out to answer a number of questions.

Does SLA_L vary with leaf position when leaves are harvested at a given developmental stage?

When leaves of barley are harvested at the same physiological age the SLA_L varies with leaf position by as much as 38%. This is in contrast to wheat

Table 1. *The effect of shade on whole plant SLA (SLA_P, cm² mg⁻¹) of barley (Hordeum vulgare) plants grown in hydroponics and harvested 2 d after full expansion of leaf blade 4 on the main stem*

	SLA_P (Total d. wt basis)		SLA_P (Structural d. wt basis)	
	Control	Shade	Control	Shade
2 dae leaf 4	0.288 ± 0.005	0.414 ± 0.012	0.318 ± 0.006	0.434 ± 0.019

Plants were transferred from 400 to 90 μmol m⁻² s⁻¹ at plant height on the day of full expansion of the third leaf on the main stem. SLA_p was calculated on the basis of total dry weight or structural dry weight. Values are the means ± SE of 4 replicates.

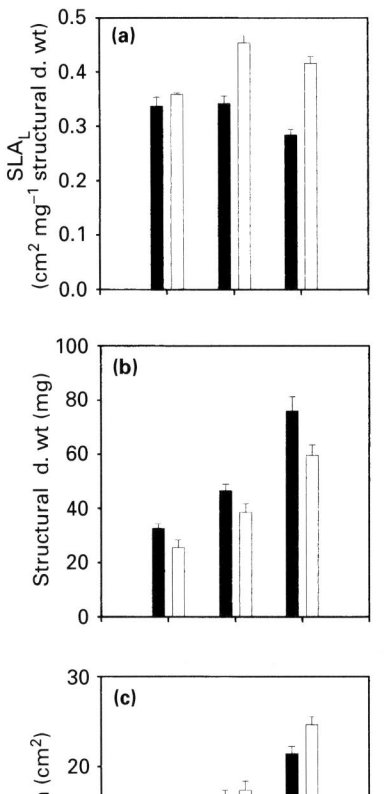

Fig. 5. The effect of shade on (a) specific leaf area of individual leaves on the main stem (SLA_L, cm^2 mg^{-1}), (b) structural dry weight (mg) and (c) leaf area (cm^2) of barley (*Hordeum vulgare*) plants grown in hydroponics and harvested 2 d after full expansion of leaf blade 4 on the main stem. Plants were transferred from 400 to 90 µmol m^{-2} s^{-1} at plant height on the day of full expansion of the third leaf on the main stem. Values are the means of 4 replicates ± SE. Control plants (filled bars); shaded plants (open bars).

Table 3. *A comparison of (A) SLA_P of plants harvested 2 dae leaf 4 on the main stem ($SLA_P = 0.318$, Fig. 3) with SLA_L of leaves 1, 2, 3, 4 on the main stem (see text for assumptions about SLA_L)*

	Leaf 1	Leaf 2	Leaf 3	Leaf 4
(A)	+ 26	+ 38	+ 16	− 11
(B)		+ 6	+ 28	+ 44

($[[SLA_P - SLA_L]/SLA_L] \times 100$) and (B) SLA_L of unshaded plants with shaded plants ($[[SLA_L \text{unshaded} - SLA_L \text{shaded}] \; SLA_L \text{ unshaded}]/ \times 100$) (results from Fig. 5).

Can SLA_P be predicted from SLA_L?

In order to calculate how well SLA_P would be predicted by SLA_L of any individual leaf we shall assume that SLA_L of leaves 1, 2 and 3 on the main stem of plants harvested 2 dae of leaf 4 on the main stem are the same as for leaf 1 harvested 2 dae, leaf 2 harvested 2 dae and leaf 3 harvested 2 dae (Fig. 1) and compare these values for SLA_P of plants harvested 2 dae of leaf 4 on the main stem (Fig. 3). The results show that SLA_P underpredicts SLA_L by as much as 11% and overpredicts by as much as 38% (Table 3). These results are comparable with changes in SLA_L due to shading: in barley SLA_L increased by a maximum of 44% (Table 3) whereas in soybean SLA_L of the first fully expanded trifoliate leaf increased by *c.* 50% after 12 d of shading (Pons & Pearcy, 1994). Therefore we cannot draw any conclusions about individual leaf structure from SLA_P. However when SLA is plotted allometrically (Fig. 4) as the natural logarithum of leaf weight versus leaf area the points of individual leaves and of whole plants lie on the same line. Since the slope of the line < 1, larger leaves have smaller SLA_L than smaller leaves. Hence only if data are expressed allometrically can predictions about SLA_L be made from SLA_P and vice versa.

When plants are transferred from high to low light, what aspect of SLA changes?

Growth at low light or switching plants from high to low light generally causes an increase in SLA_P and SLA_L because of an increase in leaf area (Blackman, 1956; Evans, 1972; Rice & Bazzaz, 1989). However there does not appear to have been any work which determines if all leaves on a plant are affected in the same way by shading of the whole plant. In barley switched to low light on the day of full expansion of the third leaf on the main stem, SLA_P was increased compared with controls. However whole plant shading affected individual SLA_L differentially. Leaves which developed in low light (in this case leaf 4 on the main stem) did indeed have a higher SLA_L than those which developed in high light when both were harvested at the same physiological age.

where, after full expansion, SLA_L for different leaves varied by only *c.* 5% (Rawson *et al.*, 1987). It also contrasts with results of Poorter & de Jong (1999) who studied SLA_L for 70 species in the field. They found on average a difference of only 4% between the youngest fully expanded leaf and the oldest still viable leaf.

Does the SLA_L of a leaf change with age after it has fully expanded?

SLA_L of leaf 3 of barley decreases as the leaf ages, after full expansion, by as much as 20%. Similar results have been found for primary leaves of barley (Sicher *et al.*, 1984). In wheat, however, the SLA of some leaves increased 16–20 d after tip emergence (Rawson *et al.*, 1987), beyond the timescale of the experiment reported here.

Reducing the light available to the whole plant also increased the SLA of leaves which were fully expanded at the time the treatment was imposed (leaf 3). However the SLA_L of the oldest leaf (leaf 2), which was 6 dae when the treatment was imposed, was unaffected by shading.

If SLA_L changes is this because of changes in leaf area or leaf weight?

The variations in SLA_L due to leaf ageing and shading were analysed by parallel investigations of leaf area and leaf weight. The decrease in SLA_L with leaf age was due to an increase in structural dry weight; older leaves were heavier. The increase in SLA_L in both newly expanded and expanding leaves because of shading was due to the structural weight of the leaves being lower. This was similar to results from soybean (Pons & Pearcy, 1994).

Do the answers to the previous questions change if the results are expressed on the basis of total weight or structural weight?

The influence of non-structural carbohydrates on leaf weight was analysed, to exclude the possibility that accumulation or mobilization of storage carbohydrate during leaf ageing, or differential accumulation by leaves at the same developmental stage could influence SLA_L. The decrease in SLA_L with increasing leaf position was not due to differences in accumulation of soluble carbohydrates; SLA_L calculated on total weight or structural weight differed by only 5–10%. Likewise, the decrease in SLA_L because of leaf ageing was not due to an accumulation of soluble carbohydrates with time; structural dry weight increased.

General conclusions

These simple experiments serve to emphasize that there are several routes which can lead to a single value of SLA_L; area may be increased due to a treatment imposed before leaf emergence and expansion, soluble carbohydrate content may increase or decrease, structural weight may change after full leaf emergence. Since SLA_L is so plastic it may be optimistic to try and find meaningful correlations between SLA_P and growth rate or to try and relate changes in SLA_P to changes in the structure of individual leaves. Care must also be taken in assessing the effects of treatment on SLA and leaf structure since each leaf on a single plant may react to the treatment in a different way depending on its position on the plant, age, stage of development and maturity when the treatment was imposed. When making comparisons of SLA_L and leaf structure, intrablade variability must also be taken into ac-

count. Rawson *et al.* (1997) found that SLA of wheat leaves varied from the tip to the base of a leaf and that the mid-vein had more effect on the SLA of the base than the tip. SLA_P as a component of growth analysis remains valid in determining whether any changes in leaf area ratio are due to a greater investment in leaf weight or leaf area for a given plant as it grows. But the assumption that it carries useful information about individual leaves is not supported. SLA_P is determined by the sum of the processes which determine the surface expansion and net weight gain of individual leaves. Surface expansion is considered by Tardieu (1999) and net weight gain wil be considered here.

Net weight gain by a leaf is a complex of many processes. To progress, first consider only C, which constitutes between 38–48% of plant dry weight (Poorter *et al.*, 1997), and concentrate on the system about which we know most: the control of export of C from mature leaves. Export in the phloem is by Munch pressure flow, and is driven by gradients of turgor pressure within the phloem between source and sink (Farrar, 1992). Precisely, export of C is a function of the loading of sucrose and of other turgor-generating solutes in the source leaf and of their removal in sinks, as well as of apoplastic solutes. Export is therefore not a function of the source leaf alone. Rather it is a whole plant property, a conclusion happily converged on by theory (Minchin *et al.*, 1993) and experiment (Moorby & Jarman, 1976; Minchin *et al.*, 1994). It is clear therefore that the rate of export of C in the phloem from mature source leaves is a whole-plant property, dependent on the nature of the transport system and on events in other sources and in sinks. It is certainly not determined wholly by events within the leaf itself. The principles which underlie this conclusion will also underlie the control of import to a young, developing leaf, and indeed any exchange of any substance between a leaf and the remainder of the plant. It follows that the net fluxes of C and other components which together constitute leaf weight are whole-plant properties. SLA_L, and thus SLA_P, is a whole plant property in the sense that its particular value is determined mechanistically by a set of processes that involve parts of the plant remote from the leaf or leaves being considered.

REFERENCES

Atkin OK, Lambers H. 1998. Slow-growing alpine and fast growing lowland species: a case study of factors associated with variation in growth rate among herbaceous higher plants under natural and controlled conditions. In: Lambers H, Poorter H, Van Vuuren MMI, eds. *Inherent variation in plant growth. Physiological mechanisms and ecological consequences.* Leiden, The Netherlands: Backhuys Publishers, 259–288.

Blackman GE. 1956. Influence of light and temperature on leaf growth. In: Milthorpe FL, ed. *The growth of leaves.* London, UK: Butterworths Scientific Publications, 151–169.

Cambridge ML, Lambers H. 1998. Specific leaf area and functional leaf anatomy in Western Australian seagrassses. In:

Lambers H, Poorter H, Van Vuuren MMI, eds. *Inherent variation in plant growth. Physiological mechanisms and ecological consequences.* Leiden, The Netherlands: Backhuys Publishers, 89–99.

Dubois M, Gilles KA, Hamilton JK, Rebus P, Smith F. 1956. Colorimetric method for the determination of sugars and related substances. *Analytical Chemistry* **28**: 350–356.

Evans GC. 1972. *The quantitative analysis of plant growth.* Oxford, UK: Blackwell Scientific Publications.

Farrar JF. 1992. The whole plant: carbon partitioning during development. In: Pollock CJ, Farrar JF, Gordon AJ, eds. *Carbon partitioning within and between organisms.* Oxford, UK: Bios, 163–180.

Garnier E, Freijsen AHJ. 1994. On ecological inference from laboratory experiments conducted under optimum conditions. In: Roy J, Garnier E, eds. *Whole plant perspective on carbon–nitrogen interactions.* The Hague, The Netherlands: SPB Academic Publishing, 267–292.

Gunn S, Bailey SJ, Farrar JF. 1999. Partitioning of dry weight and leaf area within plants of three species grown at elevated CO_2. *Functional Ecology.* (In press.)

Heilmeier H, Monson RK. 1994. Carbon and nitrogen storage in herbaceous plants. In: Roy J, Garnier E, eds. *Whole plant perspective on carbon–nitrogen interactions.* The Hague, The Netherlands: SPB Academic Publishing, 149–171.

McClendon JH. 1962. The relationship between the thickness of deciduous leaves and their maximum photosynthesis rate. *American Journal of Botany* **49**: 320–322.

Minchin PEH, Farrar JF, Thorpe MR. 1994. Partitioning of carbon in split root systems of barley: effect of temperature of the root. *Journal of Experimental Botany* **45**: 1103–1109.

Minchin PEH, Thorpe MR, Farrar JF. 1993. A mechanistic model of phloem transport which explains sink priority. *Journal of Experimental Botany* **44**: 947–955.

Moorby J, Jarman PD. 1976. The use of compartmental analysis in the study of the movement of carbon through leaves. *Planta* **122**: 155–168.

Pons TL, Pearcy RW. 1994. Nitrogen reallocation and photosynthetic acclimation in response to partial shading in soybean plants. *Physiologia Plantarum* **92**: 636–644.

Poorter H, de Jong R. 1999. A comparison of specific leaf area, chemical composition and leaf construction costs of field plants from 15 habitats differing in productivity. *New Phytologist* **143**: 163–176

Poorter H, VanBerkel Y, Baxter R, Den Hertog J, Dijkstra P, Gifford RM, Griffin KL, Roumet C, Roy J, Wong SC. 1997. The effect of elevated CO_2 on the chemical composition and construction costs of leaves of 27 C_3 species. *Plant Cell and Environment* **20**: 472–482.

Poorter H, Van der Werf A. 1998. Is inherent variation in RGR determined by LAR at low irradiance and by NAR at high irradiance? A review of herbaceous species. In: Lambers H, Poorter H, Van Vuuren MMI, eds. *Inherent variation in plant growth. Physiological mechanisms and ecological consequences.* Leiden, The Netherlands: Backhuys Publishers, 309–336.

Pyankov VI, Ivanova LA, Lambers H. 1998. Quantitative anatomy of photosynthetic tissues of plant species of different functional types in a boreal vegetation. In: Lambers H, Poorter H, Van Vuuren MMI, eds. *Inherent variation in plant growth. Physiological mechanisms and ecological consequences.* Leiden, The Netherlands: Backhuys Publishers, 71–87.

Rawson HM, Gardner PA, Long MJ. 1987. Sources of variation in specific leaf area in wheat grown at high temperature. *Australian Journal of Plant Physiology* **14**: 287–298.

Rice SA, Bazzaz FA. 1989. Quantification of plasticity of plant traits in response to light intensity: comparing phenotypes at a common weight. *Oecologia* **78**: 502–507.

Ricker WE. 1984. Computation and uses of central trend lines. *Canadian Journal Zoology* **62**: 1897–1905.

Shipley B. 1995. Structured interspecific determinants of specific leaf area in 34 species of herbaceous angiosperms. *Functional Ecology* **9**: 312–319.

Sicher RC, Kramer DF, Harris WG. 1984. Diurnal carbohydrate metabolism of barley primary leaves. *Plant Physiology* **76**: 165–169.

Singh BB, Shrivastava MK, Lalchand. 1985. Relationships among leaf chlorophyll, bean yield and other characters in field grown soybean cultivars. *Photosynthetica* **19**: 240–243.

Tardieu F, Granier C, Muller B. 1999. Modellling leaf expansion in a fluctuating environment: are changes in specific leaf area a consequence of changes in expansion rate? *New Phytologist* **143**: 33–44

Van Arendonk JJCM, Poorter H. 1994. The chemical composition and anatomical structure of leaves of grass species differing in relative growth rate. *Plant, Cell and Environment* **17**: 963–970.

Zar JH. 1996. *Biostastical analysis, 3rd edn.* New Jersey, USA: Prentice Hall.

New Phytol (1999) 143 53-61

Contribution of carbohydrate pools to the variations in leaf mass per area within a tomato plant

N. BERTIN[1]*, M. TCHAMITCHIAN[1], P. BALDET[2], C. DEVAUX[2], B. BRUNEL[1] AND C. GARY[1]

[1] *INRA, Unité de Bioclimatologie, Domaine St Paul, Site Agroparc, F-84914 Avignon Cedex 9, France*
[2] *INRA, Unité de Physiologie Végétale, Domaine de la Grande Ferrade, F-33883 Villenave d'Ornon Cedex, France*

Received 11 November 1998; accepted 24 March 1999

SUMMARY

The contribution of the starch and soluble carbohydrate pools to the diurnal variations of leaf mass per unit area (LMA) has been investigated in tomato leaves. A glasshouse experiment was carried out with plants pruned to two or five fruits per truss. Leaflets were sampled at sunrise, noon and sunset at different positions within the leaf (basal or terminal), and on different sympods along the stem. Carbohydrate contents and LMA were significantly higher in the terminal than in the basal leaflets, except at sunrise. During the day, differences in starch accumulation between terminal and basal leaflets increased with leaf height on the plant. Among sympods, the soluble carbohydrate content of the terminal leaflets did not vary significantly, whereas at 13.00 h the LMA was minimum in the middle of the plant and maximum at the top, and the leaf starch content significantly increased half-way up the plant. The plant fruit load had only small and non-significant effects on the LMA and carbohydrate contents. The response of LMA and carbohydrate contents to changing source activity was observed under controlled climatic conditions. The starch pool of fully expanded leaves was rapidly filled and emptied under increasing and decreasing source activity. In young expanding leaves, this pool was hardly filled during daylight. On average the soluble carbohydrates did not contribute significantly to the diurnal variations in LMA, whereas fluctuations in starch explained c. 70% and 44% of these variations in the upper and lower leaves, respectively. The results are discussed with respect to the modelling of LMA at the level of individual tomato leaves or sympods.

Key words: tomato, *Lycopersicon esculentum*, SLA, starch, carbohydrate, source:sink ratio.

INTRODUCTION

The ratio between leaf mass and leaf area (LMA) is of great interest for crop growth models since it is often used to relate leaf area expansion to increase of leaf dry weight. No current crop growth model addresses the simulation of LMA in a mechanistic way. Indeed LMA has been considered either as a constant (Van Keulen *et al.*, 1982), as a function of plant developmental stage (Wilkerson *et al.*, 1983) or as depending on climate fluctuations or season (Jones *et al.*, 1991; Heuvelink, 1996). In many of these models, leaf area expansion is calculated from the newly synthesized assimilates partitioned to leaves

multiplied by the average LMA of the plant. Therefore any factor that affects the LMA also directly influences the rate of leaf area expansion. This disagrees with recent observations on glasshouse tomato plants of different ages which suggest that leaf area is hardly affected by the balance between the supply and demand of carbon assimilates (Bertin & Gary, 1998). To account for this result, a mechanistic concept of LMA has been formalized in the tomato growth model TOMGRO (Gary *et al.*, 1996), where both structural leaf mass and mobile carbon (C) assimilates vary in response to the plant source–sink balance, whereas leaf expansion is a function of leaf age and temperature only. This model has been globally validated at the plant level (Bertin & Gary, 1998), but very few data are available to validate it at leaf or sympod levels on plants with

* Author for correspondence (tel +33 4 90 31 64 62; fax +33 4 90 89 98 10; e-mail bertin@avignon.inra.fr).

many growing sinks. The aim of this work was to contribute to the understanding of the variations in LMA within a plant as a function of leaf age, leaf position, plant source–sink balance and environment.

LMA is responsive to many external factors such as CO_2 concentration, light, temperature, and to the resulting internal balance between source and sink activities (Dijkstra, 1989). During tomato plant development, leaves of the successive sympods grow under different source–sink balances because of climatic variations and changes in the number of sinks. At the leaf level, variations in LMA are assumed to depend mainly on the balance between photoassimilate production and export by the leaf, which is partly determined by the sink demand. Variations that could not be attributed to carbohydrates were assumed to represent the variations of the so-called structural LMA, which designates in fact a pool of substances other than carbohydrates. In tomato leaves, the pools of hexose and starch fluctuate in response to the source–sink balance, whereas the pool of sucrose is small and changes little (Ho *et al.*, 1983). For plants reduced to one leaf and one truss, *c.* 60% of the C fixed by photosynthesis remains in the leaf after the light period, mainly as starch, hexose, malic acid and proteins. The first three compounds are further mobilized during the night, but decreased activity of sinks reduces the nocturnal C export by $> 30\%$, and induces an accumulation of starch and hexoses in the leaf (Ho *et al.*, 1983; Hammond *et al.*, 1984). The rate of C export from mature pepper leaves increases with increasing sink demand, and the leaf starch and hexose contents decrease rapidly, whereas photosynthesis is not inhibited in response to a lower demand (Shaw *et al.*, 1986). Exploratory work on the diurnal variations of carbohydrate contents in tomato leaves indicated that they would contribute only partly to the variations in LMA over 24 h, depending on plant fruit load and leaf position within the canopy (Bertin & Gary, 1998).

We investigated the daily variations of LMA, starch and soluble carbohydrate contents in tomato leaves of different ages and positions within the canopy. Measurements were made on plants with different leaf:fruit ratios to modify the size of the different pools. In parallel a 9-d experiment under controlled conditions showed the dynamics of the LMA and carbohydrate pools after a period of high source activity followed by prolonged exposure to darkness.

MATERIALS AND METHODS

Plant material and experimental conditions

Two experiments were carried out at the INRA, Avignon, France, the first in a multispan plastic glasshouse, the second in a climatic chamber, with the same round indeterminate tomato cultivar (Raïssa). Flowers were open pollinated by bees and all side shoots were removed as they appeared. Plant nutrition and chemical pest and disease control followed commercial practices. Plant sympods consisted of one truss, two leaves below and one leaf above this truss; sympods will be referred to as S_i, i indicating the truss number from the base to the top of the plant.

For the glasshouse experiment, seeds were sown in sand and planted on rockwool slabs in the glasshouse after 10 wk at a density of 1 plant m^{-2}. A secondary stem developed below the first truss of all plants giving a density of 2 stems m^{-2}. Half of the plants were pruned to two (T2), and the other half to five (T5), flowers per truss . Soil temperature was 18°C (night) and 20°C (day). Air temperature ranged between 17 and 20°C (night) and between 22 and 27°C (day). Average global radiation increased from *c.* 5 MJ m^{-2} d^{-1} (January) to *c.* 20 MJ m^{-2} d^{-1} (April). Sampling took place at anthesis of the ninth truss, on 15 April, with a global radiation of 24 MJ m^{-2} d^{-1}. Plants used for destructive measurements were located on the same double row and surrounded by border plants.

For the growth chamber experiment, seeds were sown in sand and afterwards pricked out in the compartment of a multispan Venlo-type glasshouse. Tomato seedlings were grown in 10 dm^3-pots filled with a balanced oxygenated nutrient solution at a density of 2 plants m^{-2} (Brunel & Sarrouy, 1993). All inflorescences were pruned to six flowers. Day–night air and solution temperatures were around 23–19°C. At anthesis of the fifth truss, 12 plants were transferred to a 8.75 m^2 growth chamber. The day–night air temperature was 18–15°C and air was enriched with pure CO_2 to *c.* 950 ml CO_2 l^{-1} during the photoperiod from 05.00 to 21.00 hours, with a light intensity of 1000 µmol m^{-2} s^{-1} PAR.

Variation of leaf mass per unit area and total non-structural carbohydrate contents within the leaf (glasshouse)

Tomato leaves are compound leaves comprising basically seven leaflets: one terminal, one top-pair, one middle-pair and one basal-pair. To assess the range of variation within a leaf, LMA and total non-structural carbohydrate (TNC) contents were measured on the terminal leaflet and on one of the basal-pair leaflets. Leaflets were randomly sampled from the three leaves of each of the second, fourth and sixth sympods of five plants from the T5 treatment, at sunrise (05.00 hours GMT), noon (13.00 hours GMT) and sunset (18.00 hours GMT). Influence of the leaflet position on different sympods and at different daytimes was analysed by ANOVA ($n = 84$) (Splus software; Becker *et al.*, 1988). For a given sympod, light intensity was about three–five times lower inside the canopy (basal-pair leaflets) than between rows (terminal leaflets) depending on leaf height.

Daily variation of leaf mass per unit area and total non-structural carbohydrate content within the plant (glasshouse)

The vertical gradient of LMA and TNC within the plant was measured on the terminal leaflets from the T5 plants over 24 h. One terminal leaflet was randomly sampled from the three leaves of each sympod (sympods S_2–S_8) at sunrise and at noon, on five different plants. Vertical variations were analysed in interaction with daytime by ANOVA ($n = 64$) (Splus software; Becker *et al.*, 1988). At the first and last sampling times (05.00 and 18.00 hours), no vertical gradient of PAR was measured since light was mostly diffuse. At 13.00 hours, the vertical gradient at the level of the terminal leaflets indicated a light attenuation of *c*. 50% from the top to the base of the plants.

Leaf mass per unit area and total non-structural carbohydrate content as affected by sink (glasshouse) and source (growth chamber) activity

The influence of the sink number on LMA and TNC content was assessed by comparison of the T2 and T5 treatments. At the sampling date, the difference in plant fruit load between T2 and T5 was about 14 fruits per plant, on both the main and secondary stems. Terminal leaflets were sampled on one of the three leaves from sympods S_2, S_4, S_6 at 13.00 and 18.00 hours on five plants from each treatment (T2 and T5). Influence of the sink number on LMA and TNC content was analysed in interaction with daytime and leaf position by ANOVA ($n = 56$) (Splus software; Becker *et al.*, 1988).

The dynamics of LMA and TNC content in response to the source activity were examined in the growth chamber. The climatic conditions already described (long photoperiod, high light intensity and CO_2 enrichment) were maintained for 4 d in order to favour photosynthesis and replenish the C storage pools. These 4 d were followed by a period of low source activity: 16 h at a light intensity of 500 µmol m^{-2} s^{-1} PAR, followed by 4 d of darkness. During the dark period, air temperature was constant at 19°C, and CO_2 enrichment ceased. Leaflets were randomly sampled at 05.00 and 21.00 hours among the terminal and top-pair leaflets on one of the three leaves of S_4 and S_5, on five different plants.

Measurement of leaf mass per unit area and analysis of carbohydrate

After sampling, whole leaflets were immediately weighed, the area was measured with a leaf area meter, and the sample was quickly frozen in liquid nitrogen and kept at −80°C until analysis. The leaflet dry weight was determined after freeze-drying and the LMA (g m^{-2}) was calculated as the ratio between dry weight and projected area of the leaf laminae.

In the glasshouse experiment, soluble carbohydrates were extracted by an improved method (L. Gomez, unpublished) based on the work of Deelens & Garnier-Dardart (1977) and Dickson (1979). Ground samples were mixed in a water-methanol-chloroform mixture. After centrifugation, two-thirds of the methanol-water fraction was vacuum-dried and water added. After elimination of the phenols by adding polyvinyl pyrrolidone, the aqueous extract was filtered for HPLC analysis of the soluble sugars (column Sugar-pak I, Waters, MA, USA). The remaining third of the methanol-water fraction was washed again with methanol and centrifuged; the mixture was evaporated, the residue was suspended in water, and starch was broken up by autoclaving. After enzymatic hydrolysis of starch by amyloglucosidase, the glucose was measured enzymatically (Boehringer–Mannheim, 1979).

In the second experiment, soluble carbohydrates were extracted according to Brouquisse *et al.*, (1991) and the enzymatic analysis followed the method of Kunst *et al.* (1984).

RESULTS

Leaf mass per unit area, carbohydrate contents and position of leaflet within the leaf

The effect of leaflet position within the leaf was tested on three sympods (S_2, S_4, S_6) from the T5 plants. The ANOVA showed that the LMA and the starch and soluble carbohydrate contents were significantly different ($P < 0.01$) between terminal and basal leaflets, except at sunrise when the carbohydrate pools were minimum (Fig. 1). The differences in LMA and soluble carbohydrates were independent of the sympod on which leaflets were sampled. On average in S_2, S_4 and S_6, the increases of LMA and soluble carbohydrates from sunrise to sunset were, respectively, three and four times higher in the terminal than in the basal leaflets (Fig 1a,c). Starch content was significantly affected by the leaflet position within the leaf, by the sympod level and by the time of sampling. All these factors had significant interactions due to the 20-fold increase of starch in the terminal leaflet of S_6 during the day, much larger than in the other sympods, whatever the leaflet position in the leaf (Fig. 1b).

Leaf area per unit mass and carbohydrate contents as affected by leaf position within the plant

Variation in LMA and carbohydrate content among sympods was analysed on the terminal leaflets from

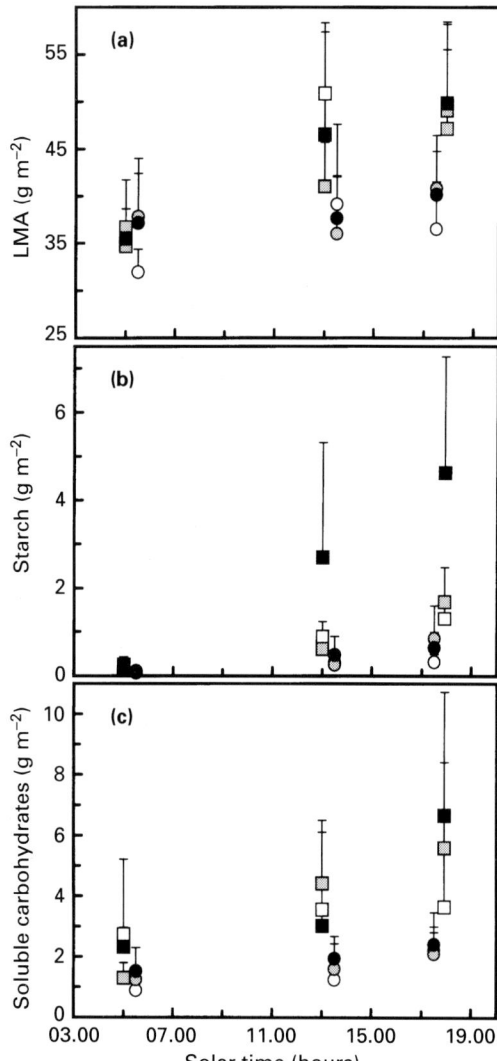

Fig. 1. Changes in (a) leaf mass per unit area (LMA), (b) starch and (c) soluble carbohydrates in terminal (squares) and basal (circles) leaflets of the same leaves during the day. Leaves belonged to S_2 (open symbols), S_4 (grey symbols) or S_6 (closed symbols) sympods. Values are means of five replicates $+1$ SE (vertical bars).

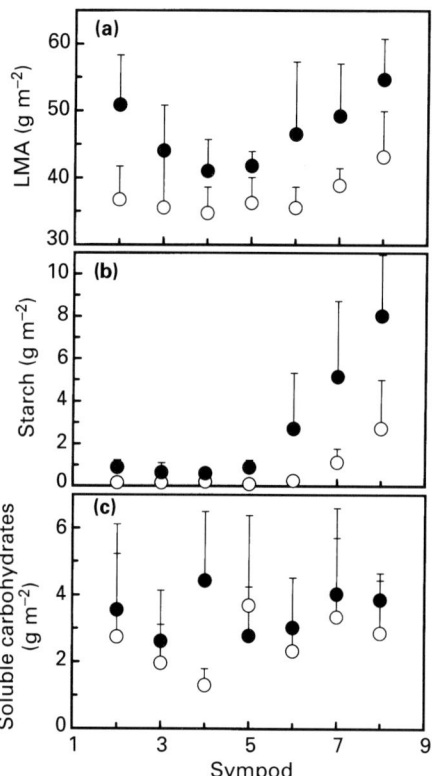

Fig. 2. Changes in (a) leaf mass per unit area (LMA), (b) starch and (c) soluble carbohydrates among seven sympods of the T5 plants. Leaflets were sampled in the terminal position, at 05.00 (open symbols) and 13.00 hours (closed symbols). Values are means of five replicates $+1$ SE (vertical bars).

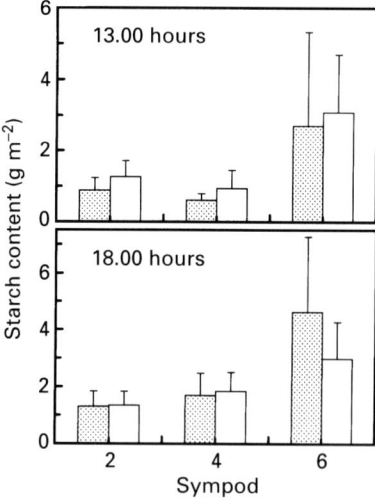

Fig. 3. Effect of plant fruit load on the starch content of the terminal leaflets sampled on S_2, S_4, S_6 at 13.00 and 18.00 hours. Plants were pruned to two (open bars) or five (shaded bars) flowers per truss. Values are means of five replicates $+1$ SE (vertical bars).

the T5 plants at sunrise and noon (Fig. 2). The ANOVA indicated that the soluble carbohydrates did not vary significantly, whereas LMA and starch were significantly ($P < 0.01$) affected by leaf position and by the time of sampling. At sunrise, the LMA was stable *c.* 35 g m^{-2} from S_2 to S_6, and it significantly increased from S_6 to the top at *c.* 40 g m^{-2} (Fig. 2a). At 13.00 hours, variations of LMA within the plant were more pronounced, with a decrease from 50 g m^{-2} at the bottom of the plant to 40 g m^{-2} on the fourth sympod, and a further increase up to the top of the plant to around 55 g m^{-2}. The starch content of the leaf was close to zero until S_5 and increased in the upper half of the plant to *c.* 2.5 g m^{-2} at sunrise and to *c.* 8.0 g m^{-2} at 13.00 hours (Fig. 2b). Because of the large increase of starch between sunrise and noon in the terminal leaflet of the upper sympods, the interaction between sam-

pling time and leaf position had a significant effect on the leaf starch content only. The soluble carbohydrate content was *c.* 3.0 g m^{-2} and, unlike starch, did not vary significantly along the stem or during the day (Fig. 2c).

Leaf mass per unit area and carbohydrate contents as affected by sink and source activities

The effects of plant fruit load on the LMA and leaf carbohydrate content were studied on the terminal leaflets from three sympods (S_2, S_4 and S_6), sampled at 13.00 and 18.00 hours on plants thinned to two or five flowers per truss. Neither the LMA nor the soluble carbohydrates were affected by plant fruit load. Similarly the leaf starch content was not significantly affected by fruit load, except that it interacted with the sampling time ($P < 0.05$). At 13.00 hours, slightly more starch accumulated in leaves of the T2 compared with the T5 plants (Fig. 3a). At 18.00 hours, there was a clear difference at S_6 only, but with a higher starch accumulation in leaves of the T5 plants (Fig. 3b). Because of variability we

did not detect any significant differences on individual sympods between T2 and T5 plants, at either 13.00 or 18.00 hours. The significant interaction between plant fruit load and sampling time may have reflected the increase of starch between 13.00 and 18.00 hours being slightly larger for the T5 than for the T2 plants.

Leaf carbohydrate contents and LMA responded to source activity at S_4 and S_5 in the growth chamber experiment. S_4 was at a developmental stage corresponding to that of the upper sympods (S_8) in the glasshouse experiment, whereas S_5 was younger and probably still expanding (*c.* 30–35 d old). From day 22 to day 26, climatic conditions were expected to promote photosynthesis and fill up the leaf carbohydrate pool. During this period, the LMA of the terminal leaflets of both sympods varied to some extent (Fig 4a). Over the day–night cycle (days 25–26), the LMA of S_4 increased from 31 to 35 g m^{-2} during daytime and decreased to 28 g m^{-2} during the night. On the contrary the LMA of S_5 decreased during the day and increased at night. This course was parallel to that of the soluble carbohydrates (Fig 4c), whereas starch, as expected, increased during daytime and decreased during the night similarly for both sympods. On day 26, the PAR was reduced from 500 to 210 µmol m^{-2} s^{-1} and no starch accumulated during this period. The LMA of S_4 was stable and that of S_5 decreased in parallel to the soluble carbohydrate content. Consequently at the beginning of the dark period, the starch and soluble carbohydrate contents were c. 1–2 g m^{-2}, and the LMA was low, c. 26.5 g m^{-2} for both sympods (Fig. 4). During darkness, the LMA variations were small and not significant. The starch and soluble carbohydrate contents fell rapidly within 10 h in S_4, and more slowly in S_5. After 3 d of darkness, the leaf carbohydrate content was close to zero and the LMA was still around 25 g m^{-2}.

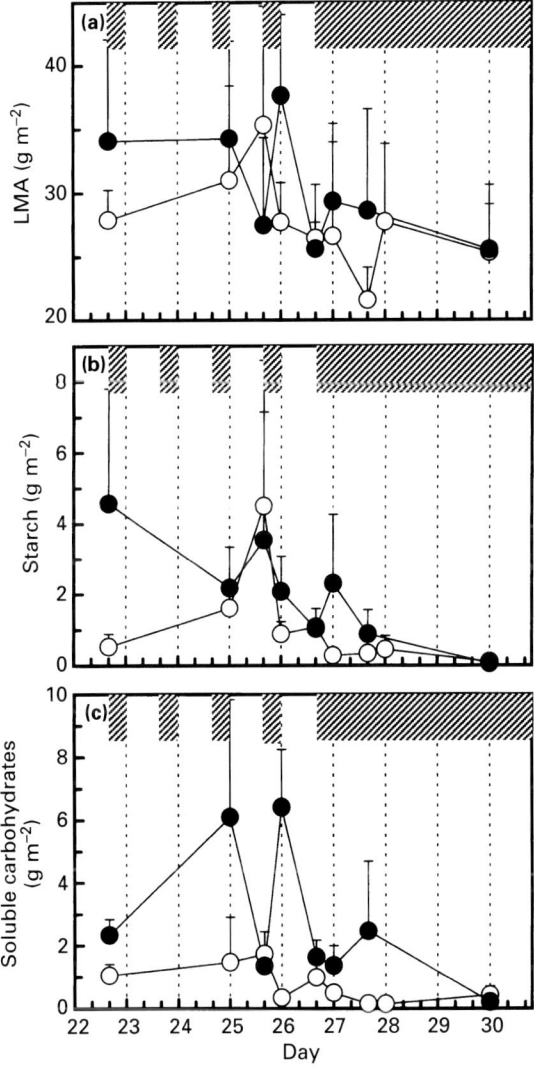

Fig. 4. Changes in (a) leaf mass per unit area (LMA), (b) starch and (c) soluble carbohydrate contents of terminal leaflets of S_4 (open symbols) and S_5 (closed symbols) measured under controlled climatic conditions. Inflorescences were pruned to five flowers. Hatched areas indicate the periods of darkness. Values are obtained at 05.00 and 21.00 hours each day and are the means of five replicates +1 SE (vertical bars).

Contribution of leaf carbohydrates to the daily variations of leaf mass per unit area

The contribution of starch and soluble carbohydrates to the daily variations of LMA was assessed on the terminal leaflets of the different sympods from the T5 plants, since this was the most complete data set. For all sympods, the LMA typically increased from the morning to midday and then levelled off, whereas the starch and soluble carbohydrate contents increased from sunrise to sunset with an amplitude depending on leaf position (Table 1). The soluble carbohydrate content was higher than the starch content in all sympods except in the youngest (S_7 and S_8). The ratio of soluble carbohydrate:starch was high in the morning and decreased during the day by a factor of *c.* 9 in the lower half of the plant, and a factor of only 2 in S_8. This suggested that the contribution of the leaf carbohydrate pools to the

Table 1. *Diurnal courses of leaf mass per unit area (LMA), starch and soluble carbohydrate contents of the terminal leaflets sampled at 05.00, 13.00 and 18.00 hours (solar time) on the fourth, sixth and eighth sympods of tomato plants thinned to five flowers per truss*

	500	1300	1800
LMA	34.7 ± 3.94	41.0 ± 4.68	49.1 ± 6.45
S_4 Starch (g m^{-2})	0.20 ± 0.14	0.61 ± 0.19	1.69 ± 0.78
Soluble carbohydrates (g m^{-2})	1.30 ± 0.50	4.42 ± 2.07	5.60 ± 2.81
LMA	35.5 ± 3.14	46.5 ± 10.9	49.8 ± 8.65
S_6 Starch (g m^{-2})	0.25 ± 0.18	2.70 ± 2.62	4.62 ± 2.64
Soluble carbohdrates (g m^{-2})	2.32 ± 0.65	3.02 ± 1.50	6.66 ± 4.06
LMA	43.1 ± 6.82	54.7 ± 6.15	52.1 ± 8.49
S_8 Starch (g m^{-2})	2.72 ± 2.28	8.02 ± 2.88	9.32 ± 2.76
Soluble carbohydrates (g m^{-2})	2.86 ± 1.58	3.86 ± 0.79	4.98 ± 1.49

Means of five replicates ± SE.

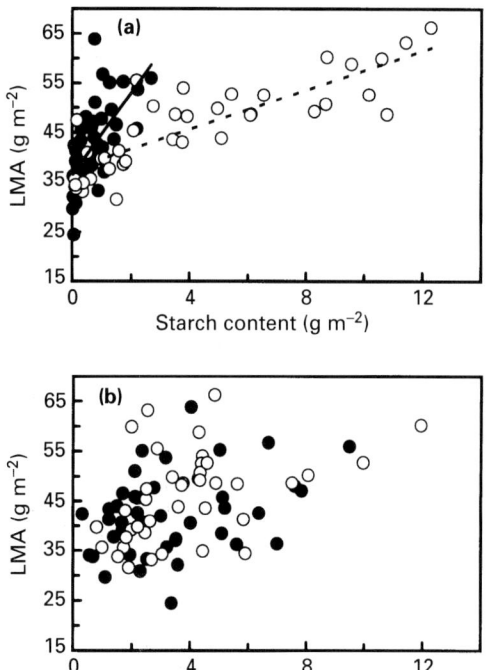

Fig. 5. Relationships between LMA and (a) starch or (b) soluble carbohydrate contents measured from 05.00 to 18.00 hours on the terminal leaflets of the T5 plants. Leaflets were sampled on the lower (S_2, S_3, S_4, S_5, closed symbols) and on the upper (S_6, S_7, S_8, open symbols) half of the plant. In (a), the lines represent the linear regressions fitted to individual measurements for both groups of sympods.

short-term variation of LMA may depend on leaf position within the canopy. Indeed the relation between LMA and starch content was different between the lower (S_2, S_3 S_4, S_5) and the upper (S_6, S_7 and S_8) half of the plant (Fig. 5a). The same range of variation in LMA was observed for all leaves with similar minima and maxima, whereas variations in starch content were much larger in the upper leaves. The range of variation of the soluble carbohydrate content was similar for the two groups of leaves (Fig. 5b). The contribution of the carbohydrates to the daily variations of LMA was calculated by multiple

linear regression. The soluble carbohydrates did not significantly contribute to variation in LMA, whereas starch was highly significant ($P < 0.001$) for both groups of sympods. The determination coefficients (r^2) between LMA and starch content are good estimations of the LMA variability that can be attributed to starch fluctuations, and were 70% and 44% in the upper and lower sympods, respectively.

DISCUSSION

Short term changes in the leaf mass per unit area and carbohydrate pools of leaves at different positions in the canopy

In tomato leaves, starch is accumulated in the light when the photosynthesis rate exceeds translocation rate (Ho *et al.*, 1983, 1989). In the morning, the newly fixed C is preferentially allocated to soluble carbohydrates (sucrose and hexose) in order to fill an export pool, before starch reserves increase during the day. Sucrose content is fairly stable and its transport is maintained at night by the breakdown of starch (Ho *et al.*, 1983, 1989; Shaw *et al.*, 1986). Therefore the decrease in leaf starch of 8 g m^{-2} that we measured at noon from the top to the bottom of the canopy was probably due to the natural vertical profile of light within the canopy. Indeed, at the level of the terminal leaflets, we measured a PAR attenuation of *c.* 50% from the top to the base of the plant at 13.00 hours. Heterogeneity of light may explain why starch did not accumulate in either the terminal leaflets from S_2 to S_5, or in the basal-pair leaflets regardless of their position on the stem, since they received only a low light intensity. Similar observations were reported by Ammerlaan *et al.* (1986) and attributed to the fact that the upper fourth of the canopy accounts for 60% of the whole CO_2 assimilation (Acock *et al.*, 1978).

The general profile of LMA is similar to that reported in other experiments (Bertin & Gary, 1998). It may reflect the C balance at the level of individual sympodial units. Indeed, in tomato plants each inflorescence is mainly supplied by the six nearby leaves; roots are supplied by the lowest leaves and the apex is supplied by the youngest expanded leaves (Bonnemain, 1968). Thus, in our experiment, the lowest LMA in the middle part of the plant would correspond to the leaves that supplied the most consuming sinks, that is the truss bearing the fastest growing fruits. Nevertheless, the high LMA of the lower leaves associated with low starch pools remains unexplained. Ammerlaan *et al.* (1986) measured maximum LMA on the middle third of the plant, and minimum LMA on the upper third of the plant, for plants at similar stage and growing period. They did not mention the time at which they sampled, but they applied high CO_2 enrichment which may have modified the ratio between source and sink activities. Sampling time must be standardized since we measured a variation in LMA of only 5 g m^{-2} among sympods at sunrise and sunset, but a variation of *c.* 15 g m^{-2} at 13.00 hours.

Dawn values of LMA were close to the structural LMA (LMA − TNC) since no or very low starch pools were measured in leaves in the morning. The structural LMA assessed on the successive sympods was rather stable (34±1.7 g m^{-2} for T5 plants) indicating that leaf structure was not much affected by the conditions under which leaves developed. It may be that on long production cycles, the source–sink balance remains fairly stable after ripening of the first fruits, and that variations in the structural LMA actually occur only over short periods, an idea supported by long term measurements of LMA on plants grown with different source:sink ratios (Bertin, 1995; Bertin & Gary, 1998). From the second to the sixth sympods, no starch was found at sunrise, suggesting that in these leaves, photosynthesis was in balance with C export over 24 h. On the contrary in S_7 and S_8 (upper, younger sympods), low amounts of starch were recorded at sunrise indicating a slight positive balance between C fixation and export which may result from a high photosynthetic rate and a low sink demand from the young inflorescences (compared with that of older growing fruits). This excess of C was not translocated to the lower sympods at night. This would support the hypothesis that the source–sink balance for C assimilates usually operates preferentially at sympod level rather than at plant level.

Responses of leaf mass per unit area and carbohydrate pools when sink and source activity changes

The LMA and carbohydrate contents were not significantly affected by fruit load. The leaf starch content was similar on the T2 and T5 plants, so in the long term, the balance between photosynthate production and use has not been significantly affected by sink removal. A small inhibition of photosynthesis by sink removal (Ho *et al.*, 1983) or starch accumulation (Czarnowski & Starzecki, 1989) has been observed in tomato. On plants with one truss and one leaf, fruit removal induces a rapid increase of leaf starch and hexose contents, partly due to the lower export rate during the night (Ho *et al.*, 1983; Hammond *et al.*, 1984). On plants with many sources and sinks, these results were not seen clearly. Perhaps both photosynthesis and sink demand were reduced, resulting in the same leaf C balance; or photosynthesis was unaffected, but in this case other sinks compensated for the reduction of fruit number. For example in the growth chamber experiment, the young expanding leaves acted as sink organs, even after half-expansion. Moreover, the stem may act as a substitute sink when fruits are removed, and non-structural carbohydrates are stored in the stem mostly as starch and sucrose (Hammond *et al.*, 1984). Unfortunately we did not measure the stem carbohydrate contents. Ammerlaan *et al.* (1986) reported a slight, unexpected, decrease of starch content from the bottom to the top of plants with a low fruit load, but this was not confirmed by our measurements.

Carbohydrate pools and LMA respond to changing source activity under controlled climatic conditions. In full expanded leaves (S_4), a pool of starch was rapidly filled during and after transfer to conditions favouring high assimilate supply. This pool was rapidly and totally mobilized in the dark. Starck & Witek-Czuprynska (1991) observed only a partial degradation and export of starch in leaf blades of indeterminate fruiting tomato plants exposed to 36 h darkness, and they measured a decrease of the total LMA much greater than that of the carbohydrates which remained in the blades in high amounts. In determinate tomato plants, LMA and starch content may even increase during 36 h darkness, indicating that assimilates are supplied by other leaf parts, such as midrib or petiole (Starck & Witek-Czuprynska, 1991). Similarly in leaves of Virginia creeper, where starch largely fluctuates in response to storage and mobilization, in prolonged darkness starch content remains high in some favoured leaf parts either from a re-synthesis, or from a lower hydrolysis or an internal translocation of soluble carbohydrates (Vignes, 1972). By contrast, in fully expanded leaves we found no carbohydrates after 24 h darkness. Nevertheless low amounts of starch had been accumulated during the pre-dark period and were nearly depleted before darkness began. This may be due to the developmental stage of plants with five fruiting trusses, which require a lot of assimilates and thus reduce accumulation of starch in the leaves. In the youngest leaves (S_5), during the day–night cycle before prolonged darkness we were surprised to find LMA and soluble carbohydrates decreasing during the day and increasing during the night, with low variations in starch. These leaves seemed to function as sink organs supplied by soluble carbohydrates during the day, but according to Ho & Shaw (1977) the import of C to 25-d-old leaves is very low, and almost nil after 30 d (the age of the S_5 leaves here). Thus our results may be explained by the balance between production and translocation of photosynthates allowing a slight storage of starch. Moreover starch can be rapidly degraded when sink demand exceeds the current C fixation (Ho *et al.*, 1989). The soluble compounds accumulated during the night may come from the small starch pool and from other leaf parts as hypothesized by Starck & Witek-Czuprynska (1991). During the first 24 h darkness, carbohydrate pools emptied with a slight delay, suggesting that indeed other sources may supply leaf blades.

Contribution of the carbohydrate pools to the variations of leaf mass per unit area

The daily variations in LMA are only partly explained by variations in starch and soluble carbohydrate contents of leaves (Bertin & Gary, 1998). The soluble carbohydrates did not contribute significantly to the diurnal variations in LMA, whereas starch fluctuations explained only 70% and 44% of the variation in LMA in the upper and lower sympods, respectively. The contribution of non-structural carbohydrates to the diurnal increase of LMA largely varies among species: 70% in alfalfa and 30% in corn (Chatterton *et al.*, 1972), 31% in basil and 10% in spinach grown under CO_2 enrichment (Holbrook *et al.*, 1993), and *c.* 60% in potato (Cao & Tibbitts, 1997). Chatterton *et al.* (1972) suggested that diurnal fluctuations of the residual (LMA–non-structural carbohydrates) may indicate the variation of components such as free amino acids. Our measurements showed that these residual components must represent a large proportion of the daily increase of dry matter, especially in old leaves. They may be organic acids, amino acids, proteins, minerals or secondary metabolites, but little information is available. In expanded leaves, amino acids, malic acid and hexose phosphates represent *c.* 25%, 6% and 6%, respectively, of the newly fixed C (Ho & Shaw, 1977). These authors demonstrated that syntheses of most proteins and structural polysaccharides occur during the early phase of leaf expansion and would be supplied mainly by imported assimilates, whereas the synthesis of other carbohydrates occurs later from self-fixed assimilates. All the leaves sampled in our experiment were more than half expanded, and thus we can assume that they did not import C and that the percentage of C allocated to structural material was <10% (Ho & Shaw, 1977).

Carbohydrate accumulation in leaves, and the relationship between starch and LMA, appeared to vary according to conditions, developmental stage and fruit load, leaf age and position along the stem that define their contribution to supply the most active sinks. On fruit-producing plants, variations in the LMA and carbohydrate content were attenuated compared with simpler source–sink systems reported in the literature. In this experiment, we did not observe large pools of leaf starch, even under controlled conditions, probably because the assimilate supply met, but did not exceed, the demand from the numerous growing sinks. Although the amount of starch stored in leaves is insignificant compared with tomato fruit yield, a small contribution may be of major importance over short critical periods, such as flower formation or fruit set (Ammerlaan *et al.*, 1986). This justifies the need to improve the modelling of LMA and carbohydrate variation during plant development. This work demonstrated that a simple model of C balance at the leaf level does not alone explain the variations in LMA. In particular, the diurnal variation of non-carbohydrate compounds needs to be better understood.

ACKNOWLEDGEMENTS

We acknowledge the helpful technical assistance of C. Orlando and C. Polizzano. We are grateful to Dr P. Raymond and L. Gomez for carrying out the carbohydrate analysis in their laboratories in Bordeaux and Avignon (INRA). Thanks to Drs H. Gautier and M. Staudt for their valuable comments on the manuscript. The growth chamber experiment was funded by the PDZR programme of the EU.

REFERENCES

Acock B, Charles-Edwards DA, Fitter DJ, Hand DW, Ludwig LJ, Warren Wilson J, Withers AC. 1978. The contribution of leaves from different levels within a tomato crop to canopy net photosynthesis: an experimental examination of two canopy models. *Journal of Experimental Botany* **29**: 815–827.

Ammerlaan AWS, Joosten MHAJ, Grange RI. 1986. The starch content of tomato leaves grown under glass. *Scientia Horticulturae* **28**: 1–9.

Becker RA, Chambers JM, Wilks AR. 1988. *The S language. A programming environment for data analysis and graphics.* CA, USA: Wadsworth & Brooks/Cole Advanced Books & Software.

Bertin N. 1995. Competition for assimilates and fruit position affect fruit set in indeterminate greenhouse tomato. *Annals of Botany* **75**: 55–65.

Bertin N, Gary C. 1998. Short and long term fluctuations of the leaf mass per area of tomato plants - implications for growth models. *Annals of Botany* **82**: 71–81.

Boehringer-Mannheim. 1979. *Méthode enzymatique pour l'analyse Agro-Alimentaire.* Meylan, France: Boehringer Mannheim.

Bonnemain JL. 1968. Transport du ^{14}C assimilé chez les Solanacées. *Revue Générale de Botanie* **75**: 579–610.

Brouquisse R, James F, Raymond P, Pradet A. 1991. Study of glucose starvation in excised maize root tips. *Plant Physiology* **96**: 619–626.

Brunel B, Sarrouy C. 1993. Système automatisé de culture hydroponique en pots dans une serre. *Cahier des techniques de l'INRA* **32**: 19–26.

Cao W, Tibbitts TW. 1997. Starch concentration and impact on specific leaf weight and element concentrations in potato leaves under varied carbon dioxide and temperature. *Journal of Plant Nutrition* **20**: 871–881.

Chatterton NJ, Lee DR, Hungerford WE. 1972. Diurnal change in specific leaf weight of *Medicago sativa* L. and *Zea mays* L. *Crop Science* **12**: 576–578.

Czarnowski M, Starzecki W. 1989. The relationship between photosynthesis inhibition and starch accumulation in tomato leaves. *Bulletin of the Polish Academy of Sciences Biological Sciences* **37**: 7–9.

Deelens E, Garnier-Dardart J. 1977. Carbon isotope composition of biochemical fractions isolated from leaves of *Bryophyllum daigremontianum* Berger, a plant with crassulacean acid metabolism: some physiological aspects related CO_2 dark fixation. *Planta* **135**: 241–248.

Dickson RE. 1979. Analytical procedures for the sequential extraction of ^{14}C-labeled constituents from leaves, bark and wood of cottonwood plants. *Physiologia Plantarum* **45**: 480–488.

Dijkstra P. 1989. Cause and effect of differences in specific leaf area. In: Lambers H, Cambridge ML, Konings H, Pons TL, eds. *Causes and consequences of variation in growth rate and productivity of higher plants.* The Hague, The Netherlands: SPB Academic Publishing, 125–140.

Gary C, Bertin N, Tchamitchian M. 1996. 'TOMGRO' modèle explicatif de fonctionnement des cultures sous serre: un outils pour la recherche et pour l'aide à la décision. In: Baille A, ed. *Actes du séminaire de l'AIP intersectorielle Serres.* Avignon, France: INRA, 90–99.

Hammond JBW, Burton K, Shaw AF, Ho LC. 1984. Source–sink relationships and carbon metabolism in tomato leaves. 2. Carbohydrate pools and catabolic enzymes. *Annals of Botany* **53**: 307–314.

Heuvelink E. 1996. *Tomato growth and yield: quantitative analysis and synthesis.* PhD thesis, Wageningen Agricultural University, The Netherlands.

Ho LC, Grange RI, Shaw AF. 1989. Source/sink regulation. In: Baker DA, Milburn J, eds. *Transport of photoassimilate.* Harlow, UK: Longman, 306–343.

Ho LC, Shaw AF. 1977. Carbon economy and translocation of ^{14}C in leaflets of the seventh leaf of tomato during leaf expansion. *Annals of Botany* **41**: 833–848.

Ho LC, Shaw AF, Hammond JBW, Burton K. 1983. Source–sink relationships and carbon metabolism in tomato leaves. 1. 14C assimilate compartmentation. *Annals of Botany* **52**: 365–372.

Holbrook GP, Hansen J, Wallick K, Zinnen T. 1993. Starch accumulation during hydroponic growth of spinach and basil plants under carbon dioxide enrichment. *Environmental and Experimental Botany* **33**: 313–321.

Jones JW, Dayan E, Allen LH, Van Keulen H, Challa H. 1991. A dynamic tomato growth and yield model (TOMGRO). *Transactions of the American Society of Agricultural Engineers* **34**: 663–672.

Kunst A, Draeger B, Ziegenhorn J. 1984. Method in enzymatic analysis. *Metabolites* **1**: 163–172.

Shaw AF, Grange RI, Ho LC. 1986. The regulation of source leaf assimilate compartmentation. In: Cronshaw J, Lucas WJ, Giaquinta RT, eds. *Plant biology Vol. I, Phloem transport.* New York, USA: Alan R. Liss, 391–398.

Starck Z, Witek-Czuprynska B. 1991. Modifications of source–sink relationships by prolonged darkness in fruiting tomato plants. *Acta Physiologiae Plantarum* **13**: 271–281.

Van Keulen H, Penning de Vries FWT, Drees EM. 1982. A summary model for crop growth. In: Penning de Vries FWT, van Laar HH, eds. *Simulation of plant growth and crop production.* Wageningen, The Netherlands: PUDOC, 87–97.

Vignes D. 1972. Variations et déplacement de l'amidon dans la feuille de vigne vierge. *Comptes rendus de l'Académie des Sciences* **D 275**: 1375–1378.

Wilkerson GG, Jones JW, Boote KJ, Mishoe JW. 1983. Modelling soybean growth for crop management. *Transactions of the American Society of Agricultural Engineers* **26**: 63–73.

New Phytol. (1999), **143**, 63–72

The relationship between leaf composition and morphology at elevated CO$_2$ concentrations

MICHAEL L. RODERICK*, SANDRA L. BERRY AND IAN R. NOBLE

Ecosystem Dynamics Group, Research School of Biological Sciences, Institute of Advanced Studies, The Australian National University, Canberra ACT 0200, Australia

Received 11 November 1998 ; accepted 26 February 1999

SUMMARY

The composition and morphology of leaves exposed to elevated [CO$_2$] usually change so that the leaf nitrogen (N) per unit dry mass decreases and the leaf dry mass per unit area increases. However, at ambient [CO$_2$], leaves with a high leaf dry mass per unit area usually have low leaf N per unit dry mass. Whether the changes in leaf properties induced by elevated [CO$_2$] follow the same overall pattern as that at ambient [CO$_2$] has not previously been addressed. Here we address this issue by using leaf measurements made at ambient [CO$_2$] to develop an empirical model of the composition and morphology of leaves. Predictions from that model are then compared with a global database of leaf measurements made at ambient [CO$_2$]. Those predictions are also compared with measurements showing the impact of elevated [CO$_2$]. In the empirical model both the leaf dry mass and liquid mass per unit area are positively correlated with leaf thickness, whereas the mass of C per unit dry mass and the mass of N per unit liquid mass are constant. Consequently, both the N:C ratio and the surface area:volume ratio of leaves are positively correlated with the liquid content. Predictions from that model were consistent with measurements of leaf properties made at ambient [CO$_2$] from around the world. The changes induced by elevated [CO$_2$] follow the same overall trajectory. It is concluded that elevated [CO$_2$] enhances the rate at which dry matter is accumulated but the overall trajectory of leaf development is conserved.

Key words: elevated [CO$_2$], leaf composition, leaf morphology, global change.

INTRODUCTION

It is now well established that atmospheric [CO$_2$] has increased from pre-industrial levels; current predictions indicate that this increase will continue into the foreseeable future (IPCC, 1996). This increase might have an indirect effect on plant growth via changes in climate. However, there is also a direct effect due to the sensitivity of photosynthesis to [CO$_2$] that is usually called CO$_2$ fertilization.

At the scale of the leaf, a change in [CO$_2$] has an almost instantaneous effect on photosynthesis and existing models can be used to describe this short-term dependence (Farquhar *et al.*, 1980). Elevated [CO$_2$] is also known to change the composition and morphology of leaves. In general, leaves grown at elevated [CO$_2$] show a decrease in the mass of N per unit dry mass (Körner & Miglietta, 1994; McGuire *et al.*, 1995; Drake *et al.*, 1997) and an increase in the

leaf dry mass per unit area (Luo *et al.*, 1994). The C:N ratio usually increases in elevated [CO$_2$] (McGuire *et al.*, 1995).

Measurements of leaves at ambient [CO$_2$] from a variety of biomes show that the mass of N per unit dry mass normally decreases as leaf dry mass per unit area increases (Field & Mooney, 1986; Schulze *et al.*, 1994; Reich *et al.*, 1997). This pattern is consistent with the effect that elevated [CO$_2$] has on leaves. In that context, recent theoretical and empirical work (Roderick *et al.*, 1999a,b) has revealed a relatively straightforward interpretation of the well-known links between leaf composition and morphology. Those studies have shown that when the internal air spaces and liquids are also included, that leaf density is relatively conservative and that the leaf liquid content is positively correlated with the surface area:volume ratio. In this paper those relationships are used as the basis for interpreting the response of leaves to elevated [CO$_2$].

The aim of this paper is to establish whether the changes in leaf composition and morphology that

*Author for correspondence (tel +61 2 62494020; fax +61 2 62495095; e-mail Michael.Roderick@anu.edu.au).

have been observed at elevated $[CO_2]$ overlap those that have been observed at ambient $[CO_2]$. We begin by describing briefly a theoretical basis for the measurement of leaf properties based on those proposed by Roderick *et al.* (1999a). After that we use measurements of leaf properties that we have made at ambient $[CO_2]$ to develop an empirical model relating leaf thickness to the mass of liquid and dry matter, which are in turn related to the mass of C and N. Predictions from that model are then compared with a global database of leaf measurements that have been made at ambient $[CO_2]$. We then use data from several elevated $[CO_2]$ experiments reported in the literature to assess whether the changes in leaf composition and morphology follow the same pattern as that for leaves at ambient $[CO_2]$.

THE MEASUREMENT BASIS

Conceptual approach

Roderick *et al.* (1999a) developed two schemes by which to segment leaves. In the functional scheme, known as a–u–s–V (air space–solution–structure–volume basis), the solution is defined as the space enclosed by and inclusive of the outer membrane of all cells in the leaf; the remaining space is either structure (e.g. cell walls or cuticle) or internal air spaces. Although that scheme is convenient from a theoretical viewpoint (e.g. measures of solute concentration can be related to the solution), and the internal air spaces can be measured (Raskin, 1983), there are currently no simple techniques available to segment the solution from the structure. To address the practical need for measurement they developed a measurement scheme known as a–q–d–M (air space–liquid–dry matter–mass basis) in which the mass of liquid is given as computed as the difference between the fresh and dry mass. Although not as good from a biological viewpoint, the a–q–d–M scheme is a useful approximation that is convenient for measurement purposes because the liquid is largely confined to the solution. Both schemes are fully described by Roderick *et al.* (1999a).

Terminology

In the a–u–s–V scheme the volume of a leaf (V_1, m³; see Appendix A, Tables 4,5 for glossary of symbols) is:

$$V_1 = V_a + V_u + V_s \qquad \text{Eqn 1a}$$

where the subscripts refer to the air space (a), solution (u) and structure (s), respectively, and the mass of the leaf (m_1, kg) is:

$$m_1 = m_a + m_u + m_s \qquad \text{Eqn 1b}$$

The fractional air space (F_a) is defined as:

$$F_a = V_a / V_1 \qquad \text{Eqn 2}$$

In the a–q–d–M scheme, which is used for measurement, leaf mass is:

$$m_1 \approx m_q + m_d \qquad \text{Eqn 3}$$

where m_q is the mass of liquid, m_d is the mass of dry matter, and the mass of the internal air space (m_a) is ignored. The leaf liquid mass (L_q, kg m⁻²) and dry mass per unit leaf area (L_d, kg m⁻²) are defined as:

$$L_q = m_q / A_n, \quad L_d = m_d / A_n \qquad \text{Eqn 4}$$

where A_n (m²) is the area of the leaf projected normal to the leaf surface. The density of the leaf (ρ_1, kg m⁻³) is given by:

$$\rho_1 \approx \rho_{qd}(1 - F_a) \qquad \text{Eqn 5}$$

where ρ_{qd} is the density of the non-gaseous fraction. It follows that the thickness (z, m) of broad leaves is given by:

$$z = \rho_1^{-1}(L_q + L_d) \qquad \text{Eqn 6}$$

Similar expressions can also be developed for needle and cylindrical leaves as appropriate.

To describe the composition, the liquid content (Q) of leaves is defined as:

$$Q = m_q / m_1 \qquad \text{Eqn 7}$$

and the mass concentration of liquid ($[Q_1]$, kg m⁻³) and dry matter ($[D_{qd}]$, kg m⁻³) are given by:

$$[Q_1] = m_q / V_1, \quad [Q_{qd}] = m_q / [V_1(1 - F_a)] \qquad \text{Eqn 8a}$$
$$[D_1] = m_d / V_1, \quad [D_{qd}] = m_d / [V_1(1 - F_a)] \qquad \text{Eqn 8b}$$

where the subscripts 1 and qd refer to the whole leaf and the non-gaseous fraction, respectively. If the mass of N (N, kg) and C (C, kg) in the leaf are known, then the concentration of those substances can be expressed by using the above notation. The N and C content (mass basis) are given by:

$$N_q = N / m_q, \quad N_d = N / m_d \qquad \text{Eqn 9a}$$
$$C_q = C / m_q, \quad C_d = C / m_d \qquad \text{Eqn 9b}$$

where the subscripts q and d denote whether the content is expressed per unit mass of liquid or dry matter, respectively.

LEAF PROPERTIES AT AMBIENT $[CO_2]$

In this section we use measurements that we have made at ambient $[CO_2]$ to derive an empirical model of a leaf. That model is called the RSBS-model leaf because it is based on measurements of leaves collected outside the Research School of Biological

Table 1. *The morphology and composition of the RSBS-model leaf computed with Eqn 10 for a range of* z

z (m)	L_d (g m^{-2})	L_q (g m^{-2})	Q	N_d	N:C	$[D_l]$ (g cm^{-3})	$[Q_l]$ (g cm^{-3})	ρ_l (g cm^{-3})
0.000100	15.3	98.3	0.87	0.064	0.131	0.15	0.98	1.14
0.000200	72.0	140.9	0.66	0.020	0.040	0.36	0.71	1.07
0.000300	129.1	184.3	0.59	0.014	0.029	0.43	0.61	1.04
0.000400	186.3	227.8	0.55	0.012	0.025	0.46	0.57	1.03
0.000500	243.4	271.3	0.53	0.011	0.023	0.48	0.54	1.02
0.000600	300.6	314.8	0.51	0.010	0.021	0.50	0.52	1.02
0.000700	357.7	358.3	0.50	0.010	0.020	0.51	0.51	1.02

Note: $Q = L_q/(L_d+L_q)$ and $N_d = 0.01 L_q/L_d$.

Sciences. We then compare the predictions from that model with data collected from the literature.

The RSBS-model leaf

Measurements of the mass, dry mass, leaf volume, volume of internal air spaces, leaf area, and the C and N contents are shown in Appendix B for 27 leaves from 14 species collected in the immediate vicinity of our office on the campus of the Australian National University, Canberra, Australia. On the basis of those data, a clear relationship was found between F_a and ρ'_{qd} (ρ' is the specific gravity) (Fig. 4). Consequently, ρ'_l was relatively conservative (Fig. 5). Because the air space in leaves is rarely measured it is difficult to use that result for the purposes of this study. Consequently we have ignored variations in air space in the development of the RSBS-model leaf.

On the basis of the empirical relationships, the RSBS-model leaf (Appendix B) is:

$$z = 1.77 \times 10^{-6}L_d + 73 \times 10^{-6} \qquad \text{Eqn 10a}$$

$$z = 2.32 \times 10^{-6}L_q - 128 \times 10^{-6} \qquad \text{Eqn 10b}$$

$$C_d = 0.49 \qquad \text{Eqn 10c}$$

$$N_q = 0.01 \qquad \text{Eqn 10d}$$

where the respective units are: z (m), L_d (g m^{-2}), L_q (g m^{-2}). N_q and C_d are both mass fractions (computed as kg kg^{-1}) and are unitless. These relationships have been used to prepare Table 1.

Inspection of Table 1 indicates that Q declines as z increases. This linkage between composition and morphology can be understood most easily by reformulating Eqns 10a and 10b to derive expressions for the mass concentration of liquid and dry matter as:

$$[D_l] = (z-73 \times 10^{-6})/1.77 \times 10^{-6}z \qquad \text{Eqn 11a}$$

$$[Q_l] = (z+128 \times 10^{-6})/2.32 \times 10^{-6}z \qquad \text{Eqn 11b}$$

which are plotted in Fig. 1. This plot shows the overall trend such that thin leaves have a high $[Q_l]$ (and hence low $[D_l]$) and that, as z increases, both $[Q_l]$ and $[D_l]$ eventually converge at a value of approx. 0.5 g cm^{-3}. Note that because N is a constant

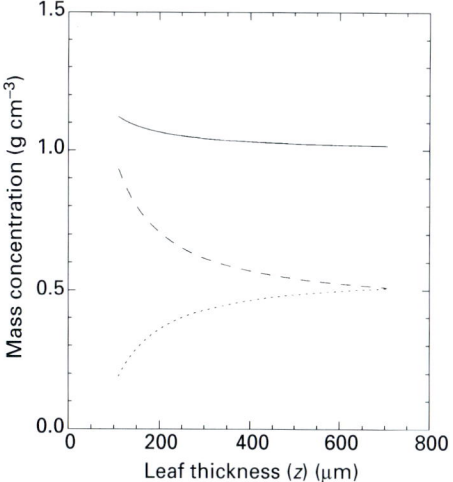

Fig. 1. The relationship between z and $[D_l]$ (dotted line), $[Q_l]$ (broken line) and ρ_l (full line) for the RSBS-model leaf (as in Eqn 11).

fraction of the mass of liquid (Eqn 10d) but C is a constant fraction of the mass of dry matter (Eqn 10c), it follows that:

$$\frac{N}{C} \propto \frac{m_q}{m_d} \qquad \text{Eqn 12a}$$

and, for the typical range in Q for most leaves, Eqn 12a can be reduced to:

$$N:C \propto Q + K_1 \qquad \text{Eqn 12b}$$

where K_1 is a constant. Similarly, Eqns 10a and 10b can be rearranged to show that (Roderick *et al.*, 1999b):

$$\Lambda_l \propto Q + K_2 \qquad \text{Eqn 12c}$$

where K_2 is a constant, Λ_l is the surface area:volume ratio of a leaf, which for broad leaves can be approximated as $2/z$ (Roderick *et al.*, 1999a).

Testing the RSBS-model leaf

Unfortunately, the liquid content of leaves is rarely measured or reported. Consequently we test the RSBS-model leaf by using measurements of L_d and N_d compiled from published data. That database

Table 2. *Details of the database used for testing the RSBS-model leaf at ambient* $[CO_2]$

Obs.	Locality	Description	Source
104	Global	Various spp.	Schulze *et al.* (1994)
6	Australia	*Eucalyptus* spp.	Mooney *et al.* (1978)
14	Australia	*Eucalyptus* spp.	Landsberg & Gillieson (1995)
25	Australia	Various spp.	Specht & Rundel (1990)
4	Australia	*Eucalyptus pauciflora*	Körner & Cochrane (1985)
32	Global	Various spp.	Specht (1988)
7	USA	Various tree spp.	Harley & Baldocchi (1995)
2	USA	*Quercus* spp.	Hollinger (1992)
9	South Africa	Various spp.	Midgley *et al.* (1995)
37	USA	Various tree spp.	Jose & Gillespie (1996)
7	Central America	Various *Piper* spp.	Chazdon & Field (1987)
40	Australia	*Eucalyptus blakelyi*	Landsberg (1990)

The database of Schulze *et al.* (1994) includes leaves from a variety of plant types (i.e. broadleaved crops, cereal crops, deciduous conifers, evergreen conifers, monsoonal forest, sclerophyllous scrub, temperate deciduous trees, temperate deciduous fruit trees, temperate grassland, temperate evergreen broadleaf tree, temperate evergreen crop, tropical deciduous forest, tropical fruit plantations, sugarcane and tropical rainforest) from different biomes around the world. All of the additional 187 leaves fall into one of the above categories. (Obs. is the number of observations from each study.)

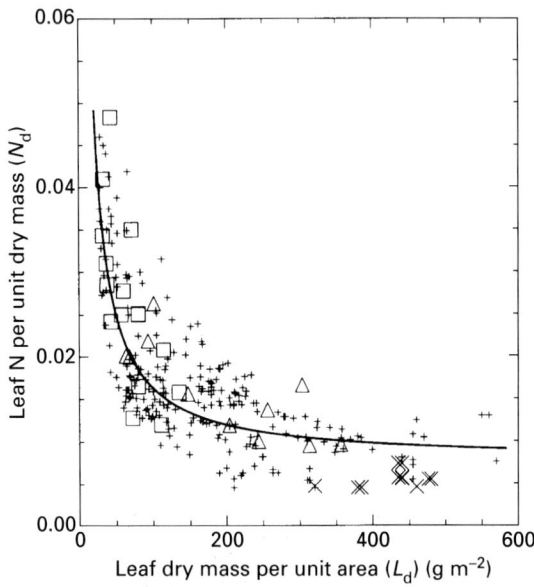

Fig. 2. Comparison between prediction with the RSBS-model leaf (full line) and measurements from the literature (see Table 2; $n = 287$). The species were classified as: +, dicotyledons; ×, monocotyledons excluding Poaceae; □, Poaceae; △, gymnosperms.

contains the 104 measurements of L_d and N_d compiled by Schulze *et al.* (1994) and a further 187 measurements from a variety of published sources and is summarized in Table 2. The database is available from the authors on request.

Our methodology implicitly assumes that the ambient $[CO_2]$ is the same for all leaves in Table 2. However, $[CO_2]$ is usually higher in the soil than the air (Corbett, 1969) owing to both autotrophic and heterotrophic respiration (Ellis, 1969). It follows that a gradient in $[CO_2]$ would usually exist between the soil and air and that plants growing close to the soil surface should experience higher levels of $[CO_2]$

depending on weather conditions (Buchmann *et al.*, 1997). Those gradients have also been observed to extend into the canopies of tall trees (Grace *et al.*, 1995). It is therefore unlikely that the leaves described in Table 2 (or in Table 4) all developed in the same $[CO_2]$. Because we have no control over $[CO_2]$ we assume that if it were significant then it would be apparent in the relationships.

Predictions from the RSBS-model leaf have been compared with the global database in Fig. 2. On the basis of the overall agreement evident in Fig. 2, we can interpret the global pattern in terms of the relationships described in Eqn 12. According to those equations, the thinnest leaves have the highest Q; as z increases (and Λ_l declines) then Q declines in direct proportion to Λ_l. Consequently, leaves that have the highest mass of N per unit dry mass will be leaves with the least fractional dry mass (and hence the highest Q).

LEAF PROPERTIES AT ELEVATED $[CO_2]$

To determine the effect of elevated $[CO_2]$ on the composition and morphology of leaves we initially compiled a large database from elevated $[CO_2]$ experiments reported in previous studies. We found that the trends were all essentially the same, so we simplified the database by retaining any data for which Q (or L_q) had been measured. We also retained data from experiments that imposed light and N treatments on both herbaceous and woody species. The final database is composed of 23 leaves from six separate studies (Table 3). Unfortunately we were unable to locate data showing the effect of elevated $[CO_2]$ on thick (high L_d) leaves. This is probably because most experiments use herbaceous species or the young leaves from tree seedlings (see the Discussion section).

Table 3. *Summary of experiments showing the impact of elevated [CO$_2$] on the composition and morphology of leaves*

Obs.	Ref. no.	[CO$_2$] (Pa)	L_d (g m^{-2})	N_c	L_q (g m^{-2})	Q (%)	[CO$_2$] (Pa)	L_d (g m^{-2})	N_d	L_q (g m^{-2})	Q (%)	Class	Species	Details
1	1	35	49	0.042	138	74	70	59	0.033	133	69	D	*Glycine max*	HL, HN, 21–42
2	1	35	43	0.023	102	70	70	64	0.012	110	63	D	*Glycine max*	HL, LN, 21–42
3	1	35	14	0.051	54	79	70	14	0.052	54	79	D	*Glycine max*	LL, HN, 21–42
4	1	35	14	0.051	54	79	70	14	0.052	54	79	D	*Glycine max*	LL, LN, 21–42
5	2	38	50	0.024		69	64	59	0.019		66	D	*Populus tremuloides*	60
6	2	38	56	0.022		54	64	61	0.023		51	D	*Quercus rubra*	60
7	2	38	59	0.036		60	64	58	0.030		61	D	*Acer saccharum*	60
8	3	32	49	0.052			64	88	0.025			D	*Gossypium hirsutum*	HN, 35
9	3	32	78	0.017			64	115	0.009			D	*Gossypium hirsutum*	LN, 35
10	4	30	55	0.033			95	95	0.023			D	*Chenopodium album*	21–56
11	4	30	40	0.037			95	64	0.026			D	*Solanum tuberosum*	21–56
12	4	30	55	0.027			95	84	0.024			D	*Solanum melongena*	21–56
13	4	30	80	0.020			95	138	0.008			D	*Brassica oleracea*	21–56
14	5	33	69	0.034			66	104	0.025			D	*Eucalyptus camaldulensis*	HN, 84
15	5	33	89	0.027			66	129	0.018			D	*Eucalyptus camaldulensis*	LN, 84
16	5	33	96	0.031			66	130	0.025			D	*Eucalyptus pauciflora*	HN, 100
17	5	33	79	0.025			66	116	0.020			D	*Eucalyptus pauciflora*	LN, 100
18	5	33	58	0.032			66	84	0.024			D	*Eucalyptus cypellocarpa*	HN, 84
19	5	33	56	0.027			66	83	0.019			D	*Eucalyptus cypellocarpa*	LN, 84
20	5	33	59	0.042			66	80	0.035			D	*Eucalyptus pulverulenta*	HN, 100
21	5	33	57	0.032			66	88	0.026			D	*Eucalyptus pulverulenta*	LN, 100
22	6	38	121	0.020			64	128	0.018			D	*Nothofagus fusca*	Approx. 365
23	6	38	105	0.014			64	116	0.017			G	*Pinus radiata*	Approx. 365

The partial pressure of CO$_2$ imposed during each experiment is shown. The class is as described in the legend to Fig. 2. The details describe multiple treatments (if applicable) and the length of time (in d) for which the treatments were applied. The treatments are signified with a two-character alphanumeric code: the first character signifies the level (H, high; L, low) of resource; the second character signifies the resource being modified (L, light; N, nitrogen). The references (Ref. no.) are: 1, Sims *et al.* (1998b); 2, Lindroth *et al.* (1993); 3, Wong (1990); 4, Sage *et al.* (1989); 5, Wong *et al.* (1992); 6, Hogan *et al.* (1996).

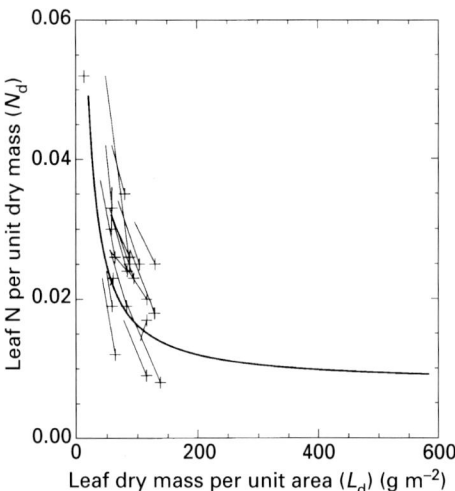

Fig. 3. The trajectories of leaves at elevated $[CO_2]$ (as in Table 3). Each line represents the change vector induced by elevated $[CO_2]$ where the + at the end of each line represents the leaf at elevated $[CO_2]$. The thick full line is the RSBS-model leaf. Note that there is no line for the + at the top left-hand corner of the plot because the leaf properties were almost identical in the low light treatment (Obs. 3 and 4 in Table 3).

According to the data in Fig. 3, elevated $[CO_2]$ induces changes in leaf composition and morphology that follow the same overall trend in leaf properties evident at ambient $[CO_2]$. On the basis of the RSBS-model leaf we interpret the major trends in Fig. 3 as a decrease in Q in elevated $[CO_2]$, which is generally consistent (excluding Obs. no. 7) with the few measurements of Q shown in Table 3. Note that in the low-light treatment (Obs. nos 3 and 4) neither Q nor N_d changed in response to $[CO_2]$ or the N supply.

Several previous studies have found that the decline in N_d under high $[CO_2]$ is largely due to the accumulation of total non-structural carbohydrates (TNCs) (Wong, 1990; Körner & Miglietta, 1994; Sims *et al.*, 1998a). Our interpretation is consistent with those data because an accumulation of TNCs within the solution would presumably displace water, which would result in a decrease in Q.

DISCUSSION

Ideally we should have used direct measurements of the liquid content and fractional air space as the basis for the comparisons in this paper. This was not possible because those quantities are currently not routinely measured. However, in view of the significance attributed to leaf density and fractional air space by Roderick *et al.* (1999b) and the obvious importance of the leaf liquids, this should be a priority area for future research.

Variations in fractional air space and leaf density will affect the relationships between z and both L_d

and L_q, and the parameters of those regressions (Eqns 10a and 10b) will not be globally applicable (Roderick *et al.*, 1999b). However, air space usually contributes to leaf thickness more than leaf area. Consequently, measurements of L_d should be relatively unaffected by variations in fractional air space, and the basic principles (i.e. Eqn 12) that follow from the RSBS-model leaf should be broadly applicable. Those principles are also consistent with theoretical predictions (Roderick *et al.*, 1999a).

The predictions from the RSBS-model leaf are in general agreement with leaf properties measured in studies around the world (Fig. 2), indicating that the basic relationships are globally applicable (excluding succulent leaves such as those of CAM plants). The underlying basis of those relationships have been described by Roderick *et al.* (1999b). At elevated $[CO_2]$ the same trends are also apparent and interpretations based on the RSBS-model leaf indicate that a decline in Q is the most likely reason for the changes that have been observed. Because the N:C ratio increases with Q (Eqn 12b), it follows that the N:C ratio should decline at elevated $[CO_2]$. This prediction is consistent with increases in the C:N ratio of leaves that are usually reported as a consequence of elevated $[CO_2]$ (McGuire *et al.*, 1995).

It follows that much of the variation in leaf properties must also be correlated with leaf age. For example, it has previously been found that specific leaf area (SLA) is negatively correlated with leaf longevity (Reich, 1993; Reich *et al.*, 1997). Because SLA is the inverse of L_d, which increases with z (Eqn 10a), it follows that SLA must increase with Λ_l (Roderick *et al.*, 1999b). Thus, leaf longevity must also be negatively correlated with Λ_l.

That proposition can be tested by noting that from a functional viewpoint the longevity of a leaf is determined by the length of time during which it can exchange gas and intercept light. The major limitation on gas exchange is the supply of water to the leaves. A gradient in water potential is necessary for water to flow from the roots to the leaves (Slatyer, 1967); this gradient, coupled with the link between composition and morphology (Eqn 12), is presumably one of the reasons for the numerous canopy gradients in leaf properties that have been observed. In contrast, the capacity for light interception is determined largely by the amount of shade cast by leaves and stems above the leaf in question. Ignoring the water limitation, the functional longevity of a leaf from a typical erect plant should be correlated with the time it takes for the leaf to become shaded. Leaves with a high liquid content would tend to produce leaf volume faster (in mesic conditions), and these leaves would also cast the most shade per unit volume because they have a high Λ_l. It follows that the (functional) longevity of a leaf should be negatively correlated with Λ_l (and hence Q), which is consistent with the observations noted above. Consequently, as leaves track around the curve (from top left to bottom right) in Figs 2 and 3, the longevity increases and this curve is a trajectory.

Under elevated [CO_2] the potential supply of carbohydrate is enhanced and over a given time interval those leaves that are exposed to elevated [CO_2] should travel farther around the curve than they would at lower levels of [CO_2] if all other factors (e.g. light, water and mineral supply) were kept constant. This is generally consistent with the data shown in Fig. 3. However, an enhanced rate of accumulation of dry matter does not imply that the final position of a leaf on that curve would change (Coleman *et al.*, 1993).

Unfortunately, we are unable to test from the available data whether the final position on that curve does vary. That should be a high priority for future research and all future comparisons should be made on a volumetric basis.

REFERENCES

Buchmann N, Guehl J-M, Barigah TS, Ehleringer JR. 1997. Interseasonal comparison of CO_2 concentrations, isotopic composition, and carbon dynamics in an Amazonian rainforest (French Guiana). *Oecologia* **110**: 120–131.

Chazdon RL, Field CB. 1987. Determinants of photosynthetic capacity in six rainforest *Piper* species. *Oecologia* **73**: 222–230.

Coleman JS, McConnaughay KDM, Bazzaz FA. 1993. Elevated CO_2 and plant nitrogen use: is reduced tissue nitrogen concentration size-dependent? *Oecologia* **93**: 195–200.

Corbett JR. 1969. *The living soil.* Sydney, Australia: Martindale Press.

Drake BG, Gonzalez-Meler MA, Long SP. 1997. More efficient plants: a consequence of rising atmospheric CO_2. *Annual Review of Plant Physiology and Plant Molecular Biology* **48**: 609–639.

Ellis RC. 1969. The respiration of the soil beneath some *Eucalyptus* forest stands as related to the productivity of the stands. *Australian Journal of Soil Research* **7**: 349–357.

Farquhar GD, von Caemmerer S, Berry J. 1980. A biochemical model of photosynthetic CO_2 assimilation in leaves of C_3 species. *Planta* **149**: 78–90.

Field CB, Mooney HA. 1986. The photosynthesis–nitrogen relationship in wild plants. In: Givnish TJ, ed. *On the economy of plant form and function.* Cambridge, UK: Cambridge University Press, 22–55.

Grace J, Lloyd J, McIntyre J, Miranda AC, Meir P, Miranda HS, Nobre C, Moncrief J, Massheder J, Malhi Y, Wright I, Gash J. 1995. Carbon dioxide uptake by an undisturbed tropical rain forest in southwest Amazonia, 1992 to 1993. *Science* **270**: 778–780.

Harley PC, Baldocchi DD. 1995. Scaling carbon dioxide and water vapour exchange from leaf to canopy in a deciduous forest. I. Leaf model parametrization. *Plant Cell and Environment* **18**: 1146–1156.

Hogan KP, Whitehead D, Kallarackal J, Buwalda JG, Meekings J, Rogers GND. 1996. Photosynthetic activity of leaves of *Pinus radiata* and *Nothofagus fusca* after 1 year of growth at elevated CO_2. *Australian Journal of Plant Physiology* **23**: 623–630.

Hollinger DY. 1992. Leaf and simulated whole-canopy photosynthesis in two co-occurring tree species. *Ecology* **73**: 1–14.

IPCC. 1996. *Climate change* 1995: *the science of climate change.* Cambridge, UK: Cambridge University Press.

Jose S, Gillespie AR. 1996. Aboveground production efficiency and canopy nutrient contents of mixed-hardwood forest communities along a moisture gradient in the central United States. *Canadian Journal of Forest Research* **26**: 2214–2223.

Körner C, Cochrane PM. 1985. Stomatal responses and water relations of *Eucalyptus pauciflora* in summer along an elevational gradient. *Oecologia (Berlin)* **66**: 443–455.

Körner C, Miglietta F. 1994. Long term effects of naturally elevated CO_2 on mediterranean grassland and forest. *Oecologia* **99**: 343–351.

Landsberg J. 1990. Dieback of rural eucalypts: does insect herbivory relate to dietary quality of tree foliage. *Australian Journal of Ecology* **15**: 73–87.

Landsberg J, Gillieson DS. 1995. Regional and local variations in insect herbivory, vegetation and soils of eucalypt associations in contrasted landscape positions along a climatic gradient. *Australian Journal of Ecology* **20**: 299–315.

Lindroth RL, Kinney KK, Platz CL. 1993. Responses of deciduous trees to elevated atmospheric CO_2: productivity, phytochemistry, and insect performance. *Ecology* **74**: 763–777.

Luo Y, Field CB, Mooney HA. 1994. Predicting responses of photosynthesis and root fraction to elevated [CO_2] – interactions among carbon, nitrogen, and growth. *Plant, Cell and Environment* **17**: 1195–1204.

McGuire AD, Melillo J, Joyce LA. 1995. The role of nitrogen in the response of forest net primary production to elevated atmospheric carbon dioxide. *Annual Review of Ecology and Systematics* **26**: 473–503.

Midgley JJ, Vanwyk GR, Everard DA. 1995. Leaf attributes of South African forest species. *African Journal of Ecology* **33**: 160–168.

Mooney HA, Ferrar PJ, Slatyer RO. 1978. Photosynthetic capacity and carbon allocation patterns in diverse growth forms of *Eucalyptus. Oecologia (Berlin)* **36**: 103–111.

Raskin I. 1983. A method for measuring leaf volume, density, thickness, and internal gas volume. *HortScience* **18**: 698–699.

Reich, PB. 1993. Reconciling apparent discrepancies among studies relating life span, structure and function of leaves in contrasting plant life forms and climates: 'The blind men and the elephant retold'. *Functional Ecology* **7**: 721–725.

Reich PB, Walters MB, Ellsworth DS. 1997. From tropics to tundra: global convergence in plant functioning. *Proceedings of the National Academy of Sciences, USA* **94**: 13730–13734.

Roderick ML, Berry SL, Noble IR, Farquhar GD. 1999a. A theoretical approach to linking the composition and morphology with the function of leaves. *Functional Ecology.* (In press.)

Roderick ML, Berry SL, Saunders AR, Noble IR. 1999b. On the relationship between the composition, morphology and function of leaves. *Functional Ecology.* (In press.)

Sage RF, Sharkey TD, Seemann JR. 1989. Acclimation of photosynthesis to elevated CO_2 in five C_3 species. *Plant Physiology* **89**: 590–596.

Schulze E-D, Kelliher FM, Körner C, Lloyd J, Leuning R. 1994. Relationships among maximum stomatal conductance, ecosystem surface conductance, carbon assimilation rate, and plant nutrition: a global ecology scaling exercise. *Annual Review of Ecology and Systematics* **25**: 629–660.

Sims DA, Seemann JR, Luo Y. 1998a. Elevated CO_2 concentration has independent effects on expansion rates and thickness of soybean leaves across light and nitrogen gradients. *Journal of Experimental Botany* **49**: 583–591.

Sims DA, Seemann JR, Luo Y. 1998b. The significance of differences in the mechanisms of photosynthetic acclimation to light, nitrogen and CO_2 for return on investment in leaves. *Functional Ecology* **12**: 185–194.

Slatyer RO. 1967. *Plant–water relationships.* London, UK: Academic Press.

Specht RL. 1988. *Mediterranean-type ecosystems – a data source book.* Dordrecht, The Netherlands: Kluwer Academic Publishers.

Specht RL, Rundel PW. 1990. Sclerophylly and foliar nutrient status of mediterranean-climate plant communities in southern Australia. *Australian Journal of Botany* **38**: 459–474.

Wong SC. 1990. Elevated atmospheric partial pressure of CO_2 and plant growth. II. Non-structural carbohydrate content in cotton plants and its effect on growth. *Photosynthesis Research* **23**: 171–180.

Wong SC, Kriedemann PE, Farquhar GD. 1992. CO_2 × nitrogen interaction on seedling growth of four species of Eucalypt. *Australian Journal of Botany* **40**: 457–472.

APPENDIX A. GLOSSARY OF SYMBOLS

Table 4. *List of subscripts used to describe leaf components*

Leaf subscripts	Definition
l	Leaf
a	Air space
q	Liquid
d	Dry matter
u	Solution
s	Structure
qd	Non-gaseous fraction

Table 5. *Main symbols used in this paper*

Symbol	Units	Description
m	kg	Mass
V	m^3	Volume
ρ	kg m^{-3}	Density
ρ'	–	Specific gravity
Q	–	Liquid content ($= m_q/m_l$)
$[Q_i]$	kg m^{-3}	Mass concentration of liquid; subscripts used are $_{qd, \; l}$; e.g. $[Q_l] = m_q/V_l$
$[D_i]$	kg m^{-3}	Mass concentration of dry matter; subscripts used are $_{qd, \; l}$; e.g. $[D_{qd}] = m_d/V_{qd}$
F_a	–	Fractional air space of leaf ($= V_a/V_l$)
C, N	kg	Mass of C and N
C_i, N_i	–	C and N content; subscripts used are $_{q, \; d, \; l}$. Computed as kg kg^{-1}, e.g. $C_d = C/m_d$, $N_l = N/m_l$
A_n	m^2	Projected leaf area measured normal to leaf surface
Λ	m^{-1}	Surface area:volume ratio
L	kg m^{-2}	Leaf mass per unit area; subscripts used are $_{q, \; d}$; e.g. $L_d = m_d/A_n$
z	m	Leaf thickness

APPENDIX B. LEAF MEASUREMENTS

Measurements of the mass, dry mass, leaf volume, volume of internal air spaces, leaf area, and the C and N contents are listed in Table 6 for 27 individual (broad) leaves from 14 different species over a large range of leaf thickness. The leaves were collected in the immediate vicinity of the Research School of Biological Sciences on the campus of the Australian National University during November and December 1997. The volume of each leaf and the associated internal air space were measured by using the Archimedes principle and followed the proce-

dures of Raskin (1983). The experiment is described fully by Roderick *et al.* (1999b).

Data from Table 6 have been used to prepare Figs 4 and 5. The error bars shown in each figure were computed by using the estimated standard deviations for each measurement and the law of propagation of variance (Roderick *et al.*, 1999b).

According to Fig. 4a, F_a increases with ρ'_{qd} (ρ is the specific gravity). Consequently, ρ'_l varies within a much smaller range (0.9–1.4; Fig. 4b). Because ρ'_l was relatively conservative, we ignored the variations in density and established regressions based on leaf thickness (z) for the purpose of this study (Fig. 5). Note that C_d and N_q were both independent of z.

Table 6. *Measurements of 27 leaves from 14 species growing on the campus of the Australian National University, Canberra, Australia*

ID	m_l (g)	m_d (g)	V_l (mm³)	V_a (mm³)	A_n (mm²)	C_d (%)	N_d (%)	Code	Species
SD	0.001	0.001	1.9	2.1	5	0.5	0.02		
13A	0.2626	0.1298	293	2	711	51.9	0.81	BET3	*Callistemon viminalis*
14A	0.3008	0.1463	333	3	928	52.1	1.01	BET3	*Callistemon viminalis*
15A	0.0415	0.0190	41	7	98	44.1	1.07	BET1	*Podocarpus macrophylla*
16A	0.1095	0.0523	110	17	291	41.1	0.91	BET3	*Podocarpus macrophylla*
19A	0.5084	0.1674	453	16	2378	43.3	2.86	BDT1	*Ulmus procera*
20A	0.3920	0.1112	327	11	2437	43.0	2.73	BDT3	*Ulmus procera*
21A	0.2864	0.1168	297	17	1142	47.8	2.80	BDT1	*Salix babylonica*
22A	0.1765	0.0401	168	5	1192	45.4	3.78	BDT4	*Salix babylonica*
25A	0.8908	0.2361	696	267	3707	49.0	2.56	BET3	*Ginkgo biloba*
26A	0.5587	0.1304	426	127	2676	48.0	2.33	BET3	*Ginkgo biloba*
27A	0.0370	0.0066	32	9	227	45.3	3.46	BRH3	*Convolvulus mauritanicus*
28A	0.0409	0.0080	35	10	272	45.5	4.01	BRH4	*Convolvulus mauritanicus*
31A	0.3845	0.1716	391	17	1284	53.8	1.35	BET1	*Eucalyptus radiata*
32A	0.5391	0.2576	567	14	1735	51.1	1.40	BET1	*Eucalyptus radiata*
35A	0.0525	0.0221	49	10	138	50.3	1.17	BET1	*Podocarpus macrophylla*
36A	0.1067	0.0486	105	19	313	43.3	0.89	BET3	*Podocarpus macrophylla*
37A	0.4670	0.2279	498	37	737	54.1	0.87	BET1	*Eucalyptus brockwayi*
38A	0.4704	0.2209	492	33	811	54.3	0.82	BET3	*Eucalyptus brockwayi*
41A	0.6132	0.3083	586	118	1031	53.5	1.67	BES1	*Grevillea johnsonii*
42A	0.5656	0.2599	490	150	983	52.0	1.24	BES3	*Grevillea johnsonii*
45A	0.5624	0.2146	406	206	1488	51.9	1.47	BES1	*Ilex heterophylla*
46A	0.4828	0.1923	388	145	1214	50.6	1.38	BES3	*Ilex heterophylla*
48A	1.2776	0.4623	1206	188	3356	57.6	2.00	BET2	*Eucalyptus pauciflora*
49A	1.2464	0.4458	1170	78	3305	54.6	1.80	BET1	*Eucalyptus kitsoniana*
50A	2.0011	0.9926	2090	36	4001	43.7	1.13	BET2	*Eucalyptus kitsoniana*
51A	2.4692	1.3070	2556	258	4012	54.0	1.52	BET1	*Eucalyptus verrucosa*
52A	2.2768	1.1681	2290	197	3610	46.7	1.21	BET2	*Eucalyptus verrucosa*

Measurements are m_l (mass), m_d (dry mass), V_l (leaf volume), V_a (air space volume), A_n (projected leaf area), C_d (mass of C per unit dry mass) and N_d (mass of N per unit dry mass). The four-digit alphanumeric code is based on (in the following order): leaf shape (B, broadleaf; C, cylindrical; N, needle), leaf habit (D, deciduous; E, evergreen; R, raingreen), plant form (G, grass; H, herb; S, shrub; T, tree; V, vine) and the estimated local light conditions (1, full sun; 2, part sun; 3, part shade; 4, full shade). The mass of liquid (m_q) is calculated as the difference between m_l and m_d. The estimated SD for each measurement is shown in the first row.

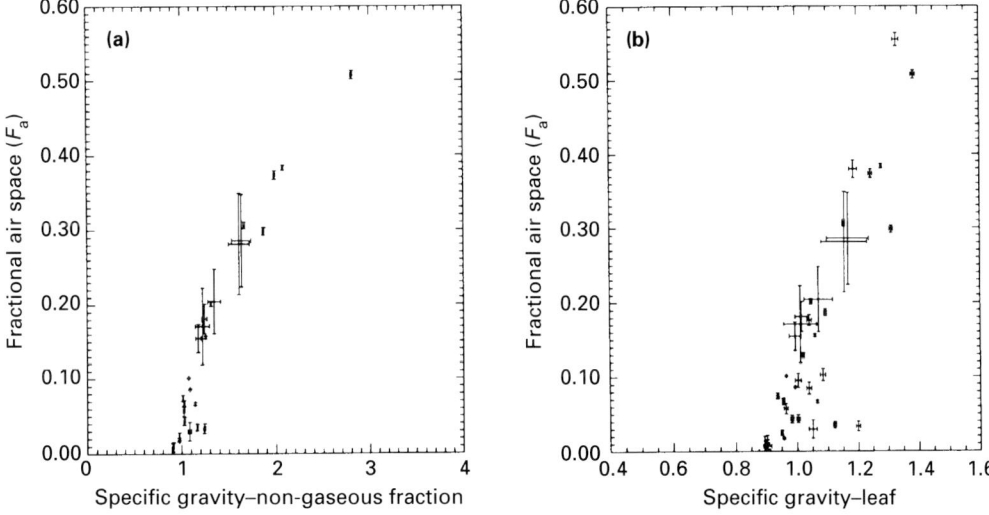

Fig. 4. Relationships between (a) ρ'_{qd} and F_a, and (b) ρ'_l and F_a, based on data in Table 6. (ρ' is the specific gravity.)

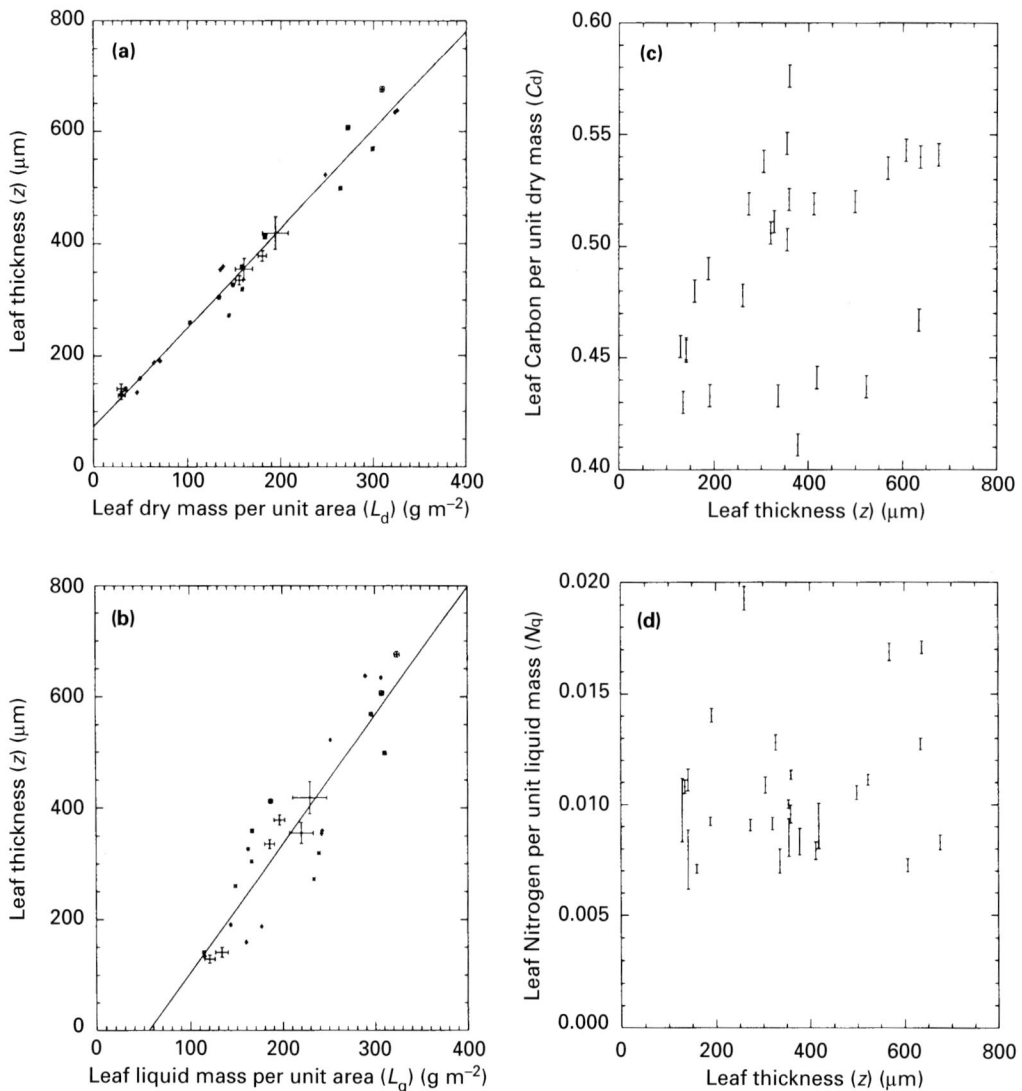

Fig. 5. Relationships between z (computed as V_l/A_n) and (a) L_d ($y = 1.77x + 73$, $r^2 = 0.97$), (b) L_q ($y = 2.32x - 128$, $r^2 = 0.83$), (c) C_d ($y = 0.49 \pm 0.05$) and (d) N_q ($y = 0.01 \pm 0.003$), based on the data ($n = 27$) in Table 6.

New Phytol. (1999), **143**, 73–81

Leaf structure and chemical composition as affected by elevated CO_2: genotypic responses of two perennial grasses

CATHERINE ROUMET*, GÉRARD LAURENT AND JACQUES ROY

Centre d'Ecologie Fonctionnelle et Evolutive, CNRS-UPR 9056, 1919 Route de Mende, 34293 Montpellier Cedex 5, France

Received 5 October 1998; accepted 31 March 1999

SUMMARY

Genotypic variability was studied in two Mediterranean grass species, *Bromus erectus* and *Dactylis glomerata*, with regard to the response to CO_2 of leaf total non-structural carbohydrate concentration ($[TNC]_{lf}$), specific leaf area (SLA), and leaf carbon and nitrogen concentrations ($[C]_{lf}$ and $[N]_{lf}$, respectively). Fourteen genotypes of each species were grown together on intact soil monoliths at ambient and elevated CO_2 concentrations (350 and 700 $\mu mol\ mol^{-1}$, respectively). In both species, the most consistent effect of elevated CO_2 was an increase in $[TNC]_{lf}$ and a decrease in leaf nitrogen concentration when expressed either as total dry mass $[N_m]_{lf}$, structural dry mass $[N_m st]_{lf}$ or leaf area $[N_a]_{lf}$. The SLA decreased only in *D. glomerata*, due to an accumulation of total non-structural carbohydrates and to an increase in leaf density. No genotypic variability was found for any variable in *B. erectus*, suggesting that genotypes responded in a similar way to elevated CO_2. In *D. glomerata*, a genotypic variability was found only for $[Cst]$, $[N_m]_{lf}$, $[N_m st]_{lf}$ and $[N_a]_{lf}$. Since $[N_m]_{lf}$ is related to plant growth and is a strong determinant of plant–herbivore interactions, our results suggest evolutionary consequences of elevated CO_2 through competitive interactions or herbivory.

Key words: elevated CO_2, intraspecific variability, nitrogen, specific leaf area, total non-structural carbohydrate.

INTRODUCTION

One of the most consistent long-term effects of elevated CO_2 on plants is on leaf structure and chemical composition (Luo *et al.*, 1994; Poorter *et al.*, 1997; Cotrufo *et al.*, 1998). In most cases, elevated CO_2 increases leaf total non-structural carbohydrate concentration ($[TNC]_{lf}$) and decreases specific leaf area (SLA; the ratio between leaf area and leaf biomass) and leaf N concentration expressed as dry mass ($[N_m]_{lf}$). Although these effects appear to be fairly general, there is considerable variation in their magnitude across species and environments (Körner & Miglietta, 1994; Luo *et al.*, 1994; Körner *et al.*, 1995; Roumet *et al.*, 1996; Poorter *et al.*, 1997). At least part of the variability in SLA and $[N_m]_{lf}$ responses to elevated CO_2 arises from interspecific differences in TNC accumulation (Stulen *et al.*, 1998). These differences can cause an increase in leaf thickness (Vu *et al.*, 1989; Radoglou & Jarvis, 1990), one of the components of SLA, and an N

dilution. Recalculation of SLA on a structural dry mass basis (SLAst) resulted in total elimination of the CO_2 effect in many species (Roumet *et al.*, 1996; Den Hertog *et al.*, 1998), whereas in others the decrease persisted, suggesting additional changes in leaf anatomy (Den Hertog *et al.*, 1996) or tissue composition. The same diversity of response was found for $[N_m]_{lf}$ when expressed on a structural dry mass basis ($[N_m st]_{lf}$) (Körner & Miglietta, 1994; Poorter *et al.*, 1997). It is not clear whether the remaining effect of CO_2 on $[N_m st]_{lf}$ results from: N limitation in the growing medium, which could cause an imbalance between uptake of C and N; a reduced demand for N in CO_2-enriched conditions (Stitt, 1991; Drake *et al.*, 1997); or an ontogenetic effect. Because CO_2 stimulates growth, and both SLA and $[N_m]_{lf}$ change with plant size, the decrease in $[N_m]_{lf}$ observed when plants grown in ambient and elevated CO_2 are compared at a given age could be the result of plants grown at elevated CO_2 being larger (Coleman *et al.*, 1993). In addition, $[N_m]_{lf}$ is often negatively correlated to SLA (Reich & Walters, 1994; Garnier *et al.*, 1997; Pyankov *et al.*, 1999); thus a further question is whether the decrease in

*Author for correspondence (fax +33 4 67 41 21 38; e-mail roumet@cefe.cnrs-mop.fr).

$[N_m]_{lf}$ is linked to the decrease in SLA, as predicted by the model of Luo *et al.* (1994).

Interspecific variability in the CO_2 response of leaf structure and chemical composition might explain differences in the growth response of species to elevated CO_2, as carbohydrate accumulation, N concentration and SLA are implicated in the responses of photosynthesis and growth to elevated CO_2 (Farrar & Williams, 1991; Stitt, 1991; Roumet & Roy, 1996). As an example, accumulation of TNC may induce feedback inhibition of photosynthesis (Stitt, 1991); species which accumulate large amounts of TNC in their leaves when grown at elevated CO_2 may be less responsive in term of biomass, although such a negative relationship has not been tested. There have been studies of interspecific differences in these mechanisms, but little attention has been paid to intraspecific differences. Genotypes may differ in their responses to environmental conditions such as N or light (Schlichting, 1986), and intraspecific variability in CO_2 response has been found for germination (Andalo *et al.*, 1996), growth (Wulff & Alexander, 1985), biomass (Ziska & Teramura, 1992; Leadley & Stöcklin, 1996), and reproductive effort (Curtis *et al.*, 1994). Thus there may be intraspecific variability in CO_2 responses for leaf structure and chemical composition. In addition, genetic variability provides raw material to gain an understanding of the potential evolutionary consequences of elevated CO_2, to select for optimal genotypes, or to analyse the range of variation in CO_2 response independently of phylogenetic differences. It is also expected to alter plant–herbivore interactions (Fajer *et al.*, 1989; Lawler *et al.*, 1997) and decomposition processes (Coûteaux *et al.*, 1991; Cotrufo *et al.*, 1994).

This study examines the range of variation in CO_2 responses of $[TNC]_{lf}$, SLA, $[C]_{lf}$, and $[N_m]_{lf}$ for different genotypes belonging to two perennial grass species, grown under conditions of competition for light and nutrients. The objectives were: to test whether genetic variability in $[TNC]_{lf}$, SLA, $[C]_{lf}$ and $[N_m]_{lf}$ responses to elevated CO_2 is present in natural populations; to identify the underlying causes of intraspecific differences in the response of SLA and $[N_m]_{lf}$ to elevated CO_2; and to test whether variation in $[TNC]_{lf}$ response scales negatively with the biomass response of the different genotypes to elevated CO_2.

MATERIALS AND METHODS

This study was conducted on two perennial grass species commonly found in Mediterranean rangelands: *Bromus erectus* Huds. and *Dactylis glomerata* L. ssp. *hispanica*. In April 1995, 14 individuals of each species were collected from calcareous grassland near Montpellier, France, excavated along a transect at 5 m intervals. As both species are caespitose grasses forming small clumps (5–30 cm in diameter), each individual can probably be considered a distinct genotype (R. Lumaret, pers. comm.). Each genotype was vegetatively propagated by separating its tillers. Ramets were then grown in pots under productive conditions and vegetatively propagated again in October 1995 before being transplanted into intact soil monoliths ($71 \times 71 \times 28$ cm) extracted from an old field on the CNRS experimental field in Montpellier. The soil was a clay loam with low organic matter (1.8–2.5%), pH 8.2, and total C and total N concentrations of 1.21% and 1.19 mg kg^{-1}, respectively. In each monolith, 12 ramets per genotype and per species were selected for uniformity, and planted in rows at a density of 700 plants m^{-2}, with 3.5 cm between rows and 4 cm between plants in a row. Ramets of each species were planted alternately along a row, and genotypes were randomly distributed within a species. Each plant had six equidistant neighbours, two of its own species and four of the other. Eight similar monoliths were planted. After planting, monoliths were watered with deionized water until soil saturation. Leaf water potential was measured at midday with a pressure chamber (PMS Instruments, Corvallis, USA), after watering and then twice weekly. When the water potential reached values 0.4 MPa below the postwatering value, 10 l of deionized water was applied slowly to avoid draining. From October to March, each monolith was watered approx. twice a month and received a total amount of 73 l (corresponding to 206 mm of rainfall). The eight monoliths were distributed in four naturally lit glasshouses. The environmental conditions (temperature, saturation vapour pressure deficit (VPD) and global radiation) did not differ significantly between the four glasshouses ($P < 0.05$, data not shown). Temperature and VPD in the glasshouses were controlled to follow the conditions outside, being maintained at an average $0.5 \pm 2.3°C$ and 210 ± 440 Pa, respectively, lower than outside. Two glasshouses were run at 350 µmol mol^{-1} and two at 700 µmol mol^{-1} atmospheric CO_2. The two monoliths in each glasshouse were rotated every 2 months.

In late March, during the peak of vegetative growth, the shoots of the 168 central plants (six plants per genotype and per species) in each monolith were harvested by clipping plants at the soil surface. Laminas were separated from sheaths, and their fresh weight determined immediately. Lamina area was measured on three plants per genotype and per monolith (12 plants per genotype and per CO_2 treatment) with a leaf area meter (Delta-T Devices, model MK2, Cambridge, UK) on a subsample representing approx. 50% of the total plant leaf mass. Leaves were oven-dried for 48 h at 60°C before weighing.

Leaf chemistry was determined on a subset of the 28 original genotypes. Nine genotypes of *B. erectus*

Table 1. *Significance levels for the effect of CO_2, genotype and their interaction on carbohydrate concentration, specific leaf area, and carbon and nitrogen concentrations*

	Bromus erectus			Dactylis glomerata		
	CO_2	Genotype	$CO_2 \times$ Genotype	CO_2	Genotype	$CO_2 \times$ Genotype
[TNC]	**	***	ns	***	*	ns
[WSC]	**	***	ns	***	*	ns
[ESC]	**	***	(a)	***	*	ns
SLA	ns	***	ns	***	***	ns
SLAst	ns	***	ns	ns	***	ns
[C]	**	**	ns	**	***	(a)
[Cst]	ns	***	***	***	***	***
[N_m]	***	***	ns	***	***	*
[N_mst]	***	***	ns	***	ns	**
[N_a]	***	***	ns	***	***	*

[TNC], total non-structural carbohydrate concentration; [WSC], water-soluble carbohydrate concentration; [ESC], ethanol-soluble carbohydrate concentration; SLA, specific leaf area expressed as dry mass; SLAst, specific leaf area expressed as structural dry mass; [C], leaf carbon concentration expressed as dry mass, or [Cst] as structural dry mass basis; [Nm] leaf nitrogen concentration expressed as dry mass, [Nmst] as structural dry mass; or [Na] as leaf area. Significance levels: ***, $P < 0.001$; **, $P < 0.01$; *, $P < 0.05$; (a), $0.01 < P < 0.05$.

and 10 of *D. glomerata*, covering the whole range of responses of shoot biomass to CO_2, were chosen. Concentrations of N and C were analysed on three plants per genotype and per monolith (12 plants per genotype and CO_2 treatment) with a carbon–hydrogen–nitrogen analyser (Carlo Erba Instruments, model EA 1108, Milan, Italy).

Total non-structural carbohydrate analysis was carried out on 20 mg of ground material following the method of Farrar (1993). Samples were extracted in 90% (v/v) ethanol at 80°C to yield ethanol-soluble sugars. Starch and fructans (water-soluble sugars) contained in the residue were extracted at 100°C for 1 h in water; starch was then hydrolysed using amyloglucosidase in acetate:acetic acid buffer, pH 4.5. The total carbohydrate in each fraction was determined using the phenol–sulphuric acid method of Dubois *et al.* (1956).

SLA, C and N concentrations were expressed as dry mass (SLA, $[C]_{lf}$, and $[N_m]_{lf}$, respectively) or as structural dry mass (SLAst, $[Cst]_{lf}$, and $[N_m st]_{lf}$, respectively) after subtraction of tissue non-structural carbohydrate content from the total dry mass.

For each species, data were analysed using ANOVA to test for significant effects of CO_2, genotype and $CO_2 \times$ genotype interaction. The relative effect of CO_2 on characteristics of the leaves was estimated by the ratio (parameter measured at elevated CO_2):(parameter measured at ambient CO_2). These ratios were $\log_e$ transformed for the statistical tests. Relationships between parameters were tested by linear regression (Statgraphics Plus, Manugistic, Rockville, MA, USA).

RESULTS

Effect of CO_2 on $[TNC]_{lf}$

On average, $[TNC]_{lf}$ was 16.8 and 29.4% higher at elevated than at ambient CO_2 for *B. erectus* and *D. glomerata*, respectively. In *D. glomerata* both water- and ethanol-soluble carbohydrate concentrations were similarly increased, whereas in *B. erectus* water-soluble carbohydrate concentration increased more than ethanol-soluble (24.9 vs 10.8%, respectively). Although no significant $CO_2 \times$ genotype interaction was found for $[TNC]_{lf}$ (Table 1; Fig. 1a,b), considerable variation existed between genotypes (Fig. 2), with responses ranging from highly negative to highly positive (from −25.6 to 48.3% for *B. erectus* and from −1.2 to 67.7% for *D. glomerata*; Fig. 2). There appeared to be no relationship between the CO_2 response of $[TNC]_{lf}$ and the changes in shoot biomass induced by elevated CO_2 (Fig. 3).

Effect of CO_2 on specific leaf area

The two species differed in their overall response to elevated CO_2. In *B. erectus*, neither SLA nor SLAst was significantly affected by CO_2. In *D. glomerata*, SLA decreased significantly by 6% (Table 1; Fig. 1c,d), while SLAst was not affected by CO_2 (Table 1; Fig. 2). The range of variation between genotypes for the SLA response to elevated CO_2 (from −10.5 to 4.9% for *B. erectus* and from −13 to −1.5% for *D. glomerata*) was low compared with the range of variation in $[TNC]_{lf}$ response (Fig. 2). No

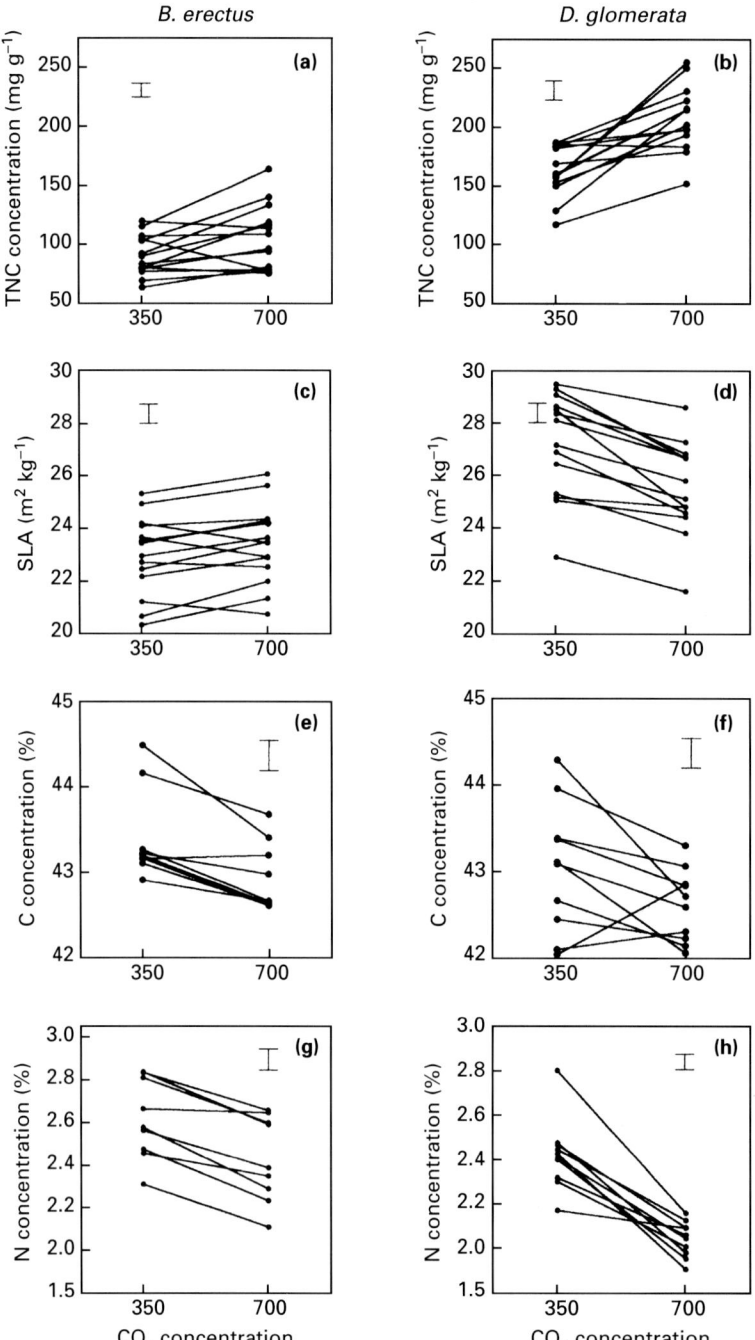

Fig. 1. Effect of elevated CO_2 on total non-structural carbohydrate concentration (TNC; a,b), specific leaf area (SLA; c,d), leaf carbon (C; e,f) and nitrogen (N; g,h) concentrations (expressed as dry mass) of different genotypes of *Bromus erectus* (a,c,e,g) and *Dactylis glomerata* (b,d,f,h) grown in competition in soil monoliths. Identical genotypes are connected by a solid line. For TNC concentration, each point is the mean of four replicates per genotype. For SLA, $[C]_{lf}$ and $[N_m]_{lf}$ concentrations, each point is the mean of 12 replicates per genotype. Bars indicate SE.

significant $CO_2 \times$ genotype interaction was found for SLA or SLAst in either species (Table 1).

The CO_2 response of SLA (expressed as dry mass) was plotted against the CO_2 response of its two components; leaf density estimated by leaf dry matter content (dry mass:fresh mass), and leaf thickness estimated by the ratio of fresh leaf mass:leaf area (Dijkstra, 1990; Garnier & Laurent, 1994). Fig. 4 shows that the CO_2 response of SLA

was negatively correlated with the CO_2 response of leaf density (Fig. 4a), whereas it was not correlated with the CO_2 response of leaf thickness (Fig. 4b).

Effect of CO_2 on leaf carbon and nitrogen concentrations

For both species, there was a slight (1%) but significant decrease in $[C]_{lf}$ (Table 1; Fig. 2). The

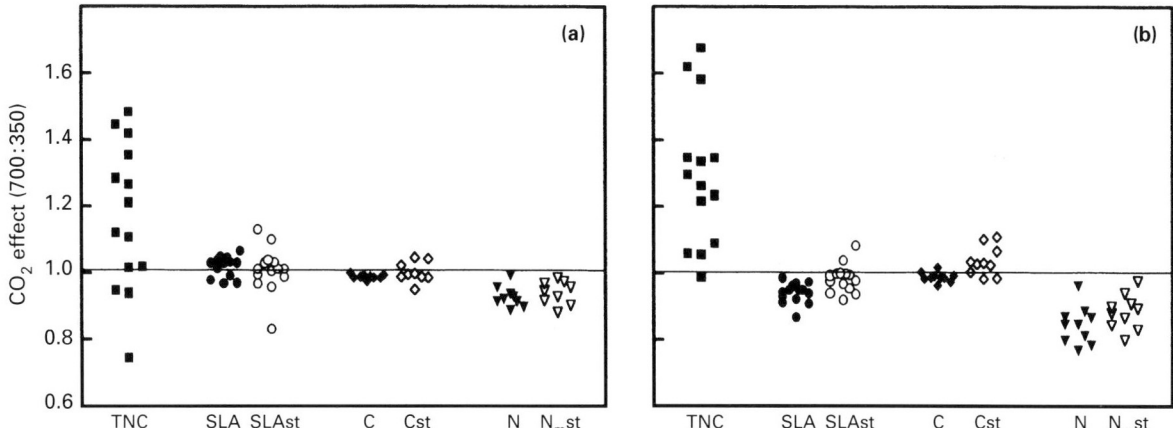

Fig. 2. Proportional change in leaf total non-structural carbohydrate concentration (TNC), specific leaf area (SLA), leaf carbon (C) and nitrogen (N) concentrations expressed as dry mass or structural dry mass (SLAst, Cst and N_mst, respectively) of different genotypes of *Bromus erectus* (a) and *Dactylis glomerata* (b) grown at ambient (350) or elevated (700 μmol mol^{-1}) CO₂. Each point represents a genotype.

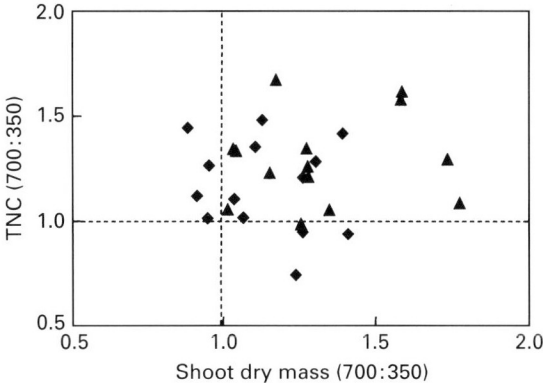

Fig. 3. Response of total non-structural carbohydrates (TNC 700/350) to elevated CO₂ at 700 and 350 μmol mol^{-1} as a function of the CO₂ response of shoot dry mass (ratio between dry mass at 700 and 350 μmol mol^{-1}) of different genotypes of *Bromus erectus* (diamonds) and *Dactylis glomerata* (triangles). Each point represents a genotype.

larger decrease was in genotypes with a high C concentration at ambient CO₂ (Fig. 1e,f). After recalculation of C concentration on a structural dry mass basis, on average the $[Cst]_{lf}$ of *B. erectus* was not affected by elevated CO₂, whereas that of *D. glomerata* was slightly increased (1%) (Table 1; Fig. 2). For both species the CO₂ × genotype interaction was significant (Table 1).

Growth at elevated CO₂ decreased $[N_m]_{lf}$ for all genotypes (Table 1; Fig. 1e,f). In *B. erectus* the response of $[N_m]_{lf}$ ranged from −11 to −1% (Fig. 2), with an average decrease of −7%; the CO₂ × genotype interaction was not significant (Table 1; Fig. 1g,h). In *D. glomerata*, the response ranged from −23 to −3.6% with an average of −15.5% (Fig. 2), and the CO₂ × genotype interaction was significant (Table 1; Fig. 1g,h). This interaction was due mainly to the genotype which had the highest $[N_m]_{lf}$ at ambient CO₂. After correction for TNC,

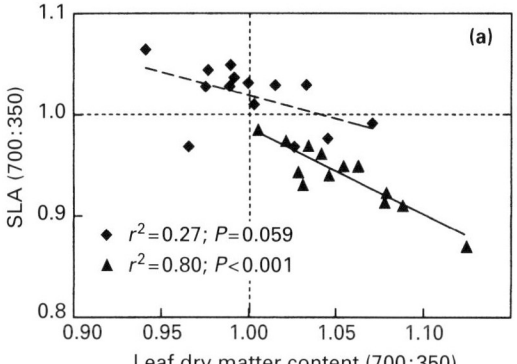

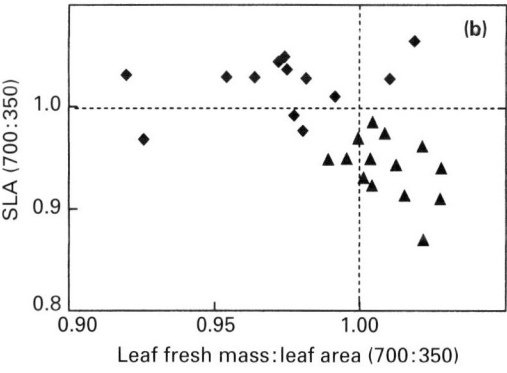

Fig. 4. Response of SLA to elevated CO₂ (ratio between SLA at 700 and 350 μmol mol^{-1}) as a function of the CO₂ response of (a) leaf dry matter content and (b) ratio of leaf fresh weight:leaf area, of different genotypes of *Bromus erectus* (diamonds) and *Dactylis glomerata* (triangles). Each point represents a genotype. Lines denote a relationship within each species; significance levels are indicated on (a), relationship was not significant for both species on (b).

the decrease of N concentration persisted; $[N_m st]_{lf}$ was decreased by 5.9% in *B. erectus* and by 11.4% in *D. glomerata* (Fig. 2). The N concentration expressed as leaf area ($[N_a]_{lf}$) was also decreased by elevated CO₂, by 8.4 and 9.1% in *B. erectus* and *D. glomerata*,

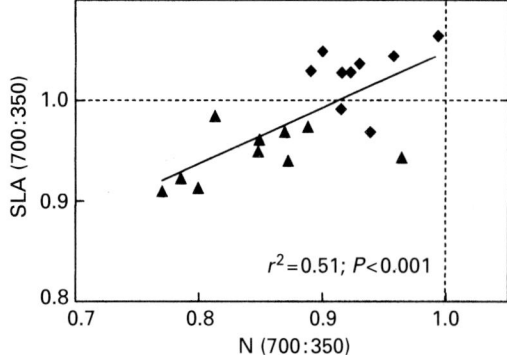

Fig. 5. Correlation between the SLA response to elevated CO_2 (ratio between SLA at 700 and 350 µmol mol^{-1}) and the effect of elevated CO_2 on leaf N concentration expressed as dry mass (N_m) of different genotypes of *Bromus erectus* (diamonds) and *Dactylis glomerata* (triangles). Each point represents a genotype. Lines denote a relationship across the two species; significance levels are indicated on the figure.

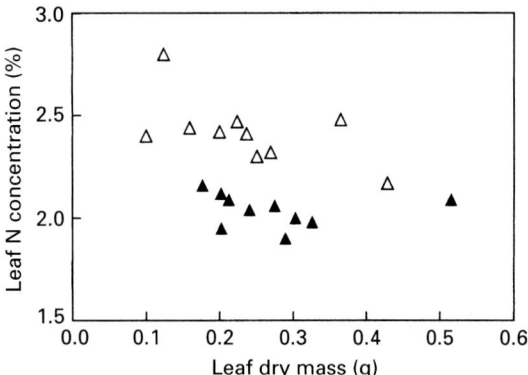

Fig. 6. Leaf N concentration as a function of leaf dry mass for different genotypes of *Dactylis glomerata* grown at ambient (open triangles) or elevated (filled triangles) CO_2. Each point represents a genotype.

respectively. The relationship between changes in $[N_m]_{lf}$ and changes in SLA was significant across, but not within, species (Fig. 5).

As elevated CO_2 increases growth, and $[N_m]_{lf}$ concentration changes with plant biomass (Coleman *et al.*, 1993), $[N_m]_{lf}$ was plotted against leaf biomass to test whether the reduction of $[N_m]_{lf}$ in elevated CO_2 was the consequence of higher biomass. In *B. erectus*, the range of variation for leaf biomass and $[N_m]_{lf}$ was low, and no relationship was found between leaf dry mass and $[N_m]_{lf}$ either at ambient or elevated CO_2. In *D. glomerata*, $[N_m]_{lf}$ declined with the leaf dry mass, and the regression lines were different for each CO_2 concentration (Fig. 6). For a given biomass, $[N_m]_{lf}$ was reduced by elevated CO_2.

DISCUSSION

The effect of elevated CO_2 on leaf structure and chemical composition has generally been based on mean responses estimated from a small number of individuals of a species. Variation within species was assumed to be negligible. The existence of genetic variation in the response to elevated CO_2 should be examined because it determines natural selection, and could act, for example, to increase the frequency of better adapted genotypes at elevated CO_2. In the present study, very few $CO_2 \times$ genotype interactions were statistically significant. None of the parameters measured in *B. erectus*, except [Cst], supported the existence of genetic variability in CO_2 responsiveness, suggesting that genotypes respond in a similar fashion to elevated CO_2. In *D. glomerata*, significant $CO_2 \times$ genotype interactions were found only for [Cst], $[N_m]_{lf}$, $[N_m st]_{lf}$ and $[N_a]_{lf}$. A large diversity of responses have been found in other intraspecific studies on the effect of elevated CO_2 on growth and reproductive traits. No significant $CO_2 \times$ genotype interaction was found for leaf secondary compounds in *Plantago lanceolata* (Fajer *et al.*, 1992), or for shoot dry mass in *Festuca ovina* (Leadley & Stöcklin, 1996), or in 12 perennial species including *D. glomerata* (Lüsher *et al.*, 1998). By contrast, there is genetic variability in the response to elevated CO_2 of lifetime fecundity in *Raphanus raphanistrum* (Curtis *et al.*, 1994); seed maturation and seedling development in *P. lanceolatus* (Wulff & Alexander, 1985); and biomass production in *Arabidopsis thaliana* (Norton *et al.*, 1995) and *B. erectus* (Leadley & Stöcklin, 1996). The expression of genetic variation also depends on the environment: for example, Schmid *et al.* (1996) found that the $CO_2 \times$ genotype interaction measured for the ramet length of *Prunella grandiflora* was not significant when plants were grown in the field with competition, but became significant when plants were grown without competition. This cautions against premature conclusions: CO_2 responses could be modified by other $CO_2 \times$ environment interactions. Nevertheless, in our study the significant $CO_2 \times$ genotype interaction for [Cst], $[N_m]_{lf}$, $[N_m st]_{lf}$ and $[N_a]_{lf}$ indicated that there is potential for selection due to increasing CO_2 concentration. This may have important consequences for the photosynthetic and growth responses of genotypes. Competitive interactions within plant communities (Leadley & Stöcklin, 1996) and between plants and herbivores might therefore be modified.

In both species, the most consistent effect of elevated CO_2 was an increase in $[TNC]_{lf}$ and a decrease in $[N_m]_{lf}$, whereas SLA was decreased only in *D. glomerata* and not in *B. erectus*. The mean increase in $[TNC]_{lf}$ (17% in *B. erectus* and 29% in *D. glomerata*) was low compared with that reported in other studies: 47% for 13 species grown near a natural CO_2 spring (Körner & Miglietta, 1994); 49% for 27 species grown individually in non-limiting conditions (Poorter *et al.*, 1997); and 42% for 17 tropical tree species (Würth *et al.*, 1998). Within a species, the range of variation in the response of

[TNC]$_\text{lf}$ to elevated CO_2 (from -30 to 50% for *B. erectus* and from -2 to 70% for *D. glomerata*) represents half the range observed between different species (3–171% reported by Körner & Miglietta, 1994; 10–150% by Poorter *et al.*, 1997). The causes of such inter- and intraspecific variability remain unclear. Differences in the phloem-loading system (apoplastic vs symplastic loaders) may explain part of the interspecific variability (Körner *et al.*, 1995), but this hypothesis cannot explain intraspecific variability. On the contrary, there may be differences in sink activity between genotypes, particularly in those perennial species that can store carbohydrates in their stubble. Insufficient sink demand in plants grown at elevated CO_2 can result in an accumulation of TNC and a down-regulation of photosynthesis (Stitt, 1991). Consistent with that view, species with small sinks (slow-growing species) have a lower growth response to CO_2 than other species (crops and fast-growing species) (Poorter *et al.*, 1996). Larger accumulation of TNC should therefore be expected in leaves of the less responsive genotypes. However, in the present study there was no relationship between biomass response and TNC accumulation (Fig. 3).

A reduction in SLA is a common response to CO_2 enrichment (Curtis, 1996; Poorter *et al.*, 1996; Bertin *et al.*, 1999; Gunn *et al.*, 1999). Surprisingly, in the present study SLA significantly decreased only for *D. glomerata* and not for *B. erectus*. Plotting the variation of SLA against that of each of its components (leaf thickness and leaf density; Fig. 4a,b) showed that the intraspecific variability in CO_2 response of SLA was attributable more to modifications in leaf density than to modifications of leaf thickness. Generally, the decrease in SLA in plants grown at elevated CO_2 is attributed to thicker leaves with more cell layers, or to increased cell size (Sims *et al.*, 1998). Changes in leaf density have been reported less often; they could result from anatomical changes, for example in the leaf cuticle and in the proportion of mesophyll protoplasm, since these are correlated with leaf density (Garnier & Laurent, 1994; Ü. Niinemets, unpublished). Decrease in SLA is often attributed to accumulation of TNC. In the present study, correction for TNC eliminated most of the effect of CO_2 on SLA. Therefore, as in *P. major* (Den Hertog *et al.*, 1998) and in many other species (Roumet *et al.*, 1996), the decrease in SLA observed in *D. glomerata* was caused mostly by TNC accumulation. In a plant community, the interaction of light and CO_2 could also be critical in determining SLA response, as decreasing irradiance has been reported to increase SLA (Niinemets, 1997). To test this hypothesis, the degree of shading of each plant was estimated by the ratio between its leaf area and the leaf area of its six neighbours. *B. erectus* showed a lower biomass than *D. glomerata* and probably experienced more shading

as a consequence. However, within each species no relationship was found between the degree of shading of each genotype and the response of SLA to elevated CO_2 (data not shown).

Increasing CO_2 concentration decreased the N concentration in all genotypes when expressed as dry mass. Similar results have been obtained in many experiments where species were grown either in non-limiting conditions (Luo *et al.*, 1994; Den Hertog *et al.*, 1996; Poorter *et al.*, 1997); in limiting conditions (Cotrufo *et al.*, 1998); or in natural ecosystems under conditions of competition for nutrients (Körner & Miglietta, 1994; Niklaus *et al.*, 1998). The magnitude of this decrease is highly variable among species. The present study demonstrates that this is also true within a species, as a significant $CO_2 \times$ genotype interaction was found for N concentration when expressed as total dry mass, structural dry mass or leaf area. The decrease in N concentration remained important when N was expressed as structural dry mass or leaf area, suggesting that [N$_\text{m}$]$_\text{lf}$ decreased independently of an accumulation of TNC or a change in SLA. Coleman *et al.* (1993) argued that the decrease in [N$_\text{m}$]$_\text{lf}$ at elevated CO_2 is the consequence of accelerated growth. This was not confirmed for genotypes of *D. glomerata*, which always had lower leaf N concentration at elevated than at ambient CO_2 when compared at a similar biomass (Fig. 6). In addition, these results showed that the efficiency with which genotypes use their N to gain biomass was increased at elevated CO_2. This result could be explained by a lower amount of N invested in Rubisco, as found for *B. erectus* grown in natural conditions (Sage *et al.*, 1997). A possible cause of the decrease in [N$_\text{m}$st]$_\text{lf}$ could be N limitation at elevated CO_2. In accordance with that hypothesis, we found that the amount of N captured by the community as a whole was similar at ambient or elevated CO_2 (106.6 and 105.6 g N, respectively). It can therefore be hypothesized that the genotypes showing the largest decrease in [N$_\text{m}$st]$_\text{lf}$ were the least competitive for N acquisition at elevated CO_2.

In the present study, *B. erectus* had no genotypic variability for any of the variables except [Cst], suggesting that genotypes responded in similar way to elevated CO_2. In *D. glomerata*, a genotypic variability was found only for [Cst], [N$_\text{m}$]$_\text{lf}$, [N$_\text{m}$st]$_\text{lf}$ and [N$_\text{a}$]$_\text{lf}$. The consequences of these results at the community and ecosystem levels are difficult to assess, first because they can change across species and environment, and second because leaf N concentration is not directly related to plant fitness. Nevertheless, N concentration could directly alter plant–herbivore interactions and, indirectly, other fitness traits because it is implicated in the control of photosynthesis and response of growth to elevated CO_2. *D. glomerata*, which has a higher genetic variability for N response to elevated CO_2, might have an additional long-term advantage in competition with *B. erectus*. Further analyses of a large set of fitness-related traits are needed.

ACKNOWLEDGEMENTS

We thank Nelly Garcia for the sugar analysis, Françoise Lafont for her precious advice on the use of the elemental analyser, and the technical personnel from the physiological ecology group and from the experimental field of the CEFE for extraction of the monoliths and management of the glasshouses. This research was funded by the French Foreign Affairs Office within a Swiss–French collaborative research programme.

REFERENCES

Andalo C, Godelle B, Lefranc M, Mousseau M, Till-Bottraud I. 1996. Elevated CO_2 decreases seed germination in *Arabidopsis thaliana*. *Annals of Botany* **2**: 129–135.

Bertin N, Tchamitchian M, Baldet P, Devaux C, Brunel B, Gary C. 1999. Contribution of carbohydrate pools to the variations in leaf mass per area within a tomato plant. *New Phytologist* **143**: 53–62.

Coleman JS, McConnaughay KDM, Bazzaz FA. 1993. Elevated CO_2 and plant nitrogen-use: is reduced tissue nitrogen concentration size-dependent? *Oecologia* **93**: 195–200.

Cotrufo MF, Ineson P, Rowland AP. 1994. Decomposition of tree litters grown under elevated CO_2: effect of litter quality. *Plant and Soil* **163**: 121–130.

Cotrufo MF, Ineson P, Scott A. 1998. Elevated CO_2 reduces the nitrogen concentration of plant tissues. *Global Change Biology* **4**: 43–54.

Coûteaux MM, Mousseau M, Celerier ML, Bottner P. 1991. Increased atmospheric CO_2 and litter quality: decomposition of sweet chestnut leaf litter with animal food webs of different complexities. *Oikos* **61**: 54–64.

Curtis PS. 1996. A meta-analysis of leaf gas exchange and nitrogen in trees grown under elevated carbon dioxide. *Plant, Cell and Environment* **19**: 127–137.

Curtis PS, Snow AA, Miller AS. 1994. Genotype-specific effects of elevated CO_2 on fecundity in wild radish (*Raphanus raphanistrum*). *Oecologia* **97**: 100–105.

Den Hertog J, Stulen I, Fonseca F, Delea P. 1996. Modulation of carbon and nitrogen allocation in *Urtica dioica* and *Plantago major* by elevated CO_2: impact of accumulation of nonstructural carbohydrates and ontogenetic drift. *Physiologia Plantarum* **98**: 77–88.

Den Hertog J, Stulen I, Posthumus F, Poorter H. 1998. Interactive effects of growth-limiting N supply and elevated atmospheric CO_2 concentration on growth and carbon balance of *Plantago major*. *Physiologia Plantarum* **103**: 451–460.

Dijkstra P. 1990. Cause and effect of differences in specific leaf area. In: Lambers H, Cambridge ML, Konings H, Pons TL, eds. *Causes and consequences of variation in growth rate and productivity of higher plants*. The Hague, The Netherlands: Academic Press, 125–140.

Drake BG, Gonzàlez-Meyer MA, Long SP. 1997. More efficient plants: a consequence of rising atmospheric CO_2? *Annual Review of Plant Physiology and Molecular Biology* **48**: 607–637.

Dubois M, Gilles KA, Hamilton JK, Rebers PA, Smith F. 1956. Colorimetric method for determination of sugars and related substances. *Analytical Chemistry* **28**: 350–356.

Fajer ED, Bowers MD, Bazzaz FA, 1989. The effects of enriched carbon dioxide atmospheres on plant–insect–herbivore interactions. *Science* **2431**: 1198–1200.

Fajer ED, Bowers MD, Bazzaz FA. 1992. The effect of nutrients and enriched CO_2 environments on production of carbon-based allelochemicals in *Plantago*: a test of the carbon/nutrient balance hypothesis. *American Naturalist* **140**: 707–723.

Farrar JF. 1993. Carbon partitioning. In: Hall DO, Scurlock JMO, Bolhar-Nordenkampf HR, Leegood SC, Long SP, eds. *Photosynthesis and production in a changing environment. A field and laboratory manual*. London, UK: Chapman & Hall, 232–246.

Farrar JF, Williams ML. 1991. The effects of increased atmospheric carbon dioxide and temperature on carbon partitioning, source–sink relations and respiration. *Plant, Cell and Environment* **14**: 819–830.

Garnier E, Cordonnier P, Guillerm J-L, Sonie L. 1997. Specific leaf area and leaf nitrogen concentration in annual and perennial grass species growing in Mediterranean old-field. *Oecologia* **111**: 490–198.

Garnier E, Laurent G. 1994. Leaf anatomy, specific mass and water content in congeneric annual and perennial grass species. *Journal of Ecology* **80**: 725–736.

Gunn S, Farrar JF, Collis BE, Nason M. 1999. Specific leaf area in barley: individual leaves versus whole plants. *New Phytologist* **143**: 45–51.

Körner Ch, Miglietta F. 1994. Long term effects of naturally elevated CO_2 on mediterranean grassland and forest trees. *Oecologia* **99**: 343–351.

Körner Ch, Pelaez-Riedl S, van Bel AJE. 1995. CO_2 responsiveness of plants: a possible link to phloem loading. *Plant Cell and Environment* **18**: 595–600.

Lawler IR, Foley WJ, Woodrow IE, Cork SJ. 1997. The effects of elevated CO_2 atmospheres on the nutritional quality of *Eucalyptus* foliage and its interaction with soil nutrient and light availability. *Oecologia* **109**: 59–68.

Leadley PW, Stöcklin J. 1996. Effects of elevated CO_2 on model calcareous grasslands: community, species, and genotype level responses. *Global Change Biology* **2**: 389–397.

Luo Y, Field CB, Mooney HA. 1994. Predicting responses of photosynthesis and root fraction to elevated $[CO_2]$: interactions among carbon, nitrogen and growth. *Plant, Cell and Environment* **17**: 1195–1204.

Lüsher A, Hendrey GR, Nösberger J. 1998. Long-term responsiveness to free air CO_2 enrichment of functional types, species and genotypes of plants from fertile permanent grassland. *Oecologia* **113**: 37–45.

Niinemets Ü. 1997. Role of foliar nitrogen in light harvesting and shade tolerance of four temperate deciduous woody species. *Functional Ecology* **11**: 518–531.

Niklaus PA, Leadley PW, Stöcklin J, Körner Ch. 1998. Nutrient relations in calcareous grassland under elevated CO_2. *Oecologia* **116**: 67–75.

Norton LR, Firbank LG, Watkinson AR. 1995. Ecotypic differentiation of response to enhanced CO_2 and temperature levels in *Arabidopsis thaliana*. *Oecologia* **104**: 394–396.

Poorter H, Roumet C, Campbell BD. 1996. Interspecific variation in the growth response of plants to elevated CO_2: a search for functional types. In: Körner Ch, Bazzaz F, eds. *Carbon dioxide, populations and communities*. San Diego, CA, USA: Academic Press, 375–412.

Poorter H, van Berkel Y, Baxter R, den Hertog J, Dijkstra P, Gifford RM, Griffin KL, Roumet C, Roy J, Wong SC. 1997. The effect of elevated CO_2 on the chemical composition and construction costs of leaves of 27 C_3 species. *Plant, Cell and Environment* **20**: 472–482.

Pyankov VP, Kondratchuk AV, Shipley B. 1999. Leaf structure and specific leaf mass: the alpine desert plants of the Eastern Pamirs, Tadjikistan. *New Phytologist* **143**: 131–142.

Radoglou KM, Jarvis PG. 1990. Effects of CO_2 enrichment on four poplar clones. I. Growth and leaf anatomy. *Annals of Botany* **65**: 617–626.

Reich PB, Walters MB. 1994. Photosynthesis–nitrogen relations in Amazonian tree species. II. Variation in nitrogen *vis-a-vis* specific leaf area influences mass- and area-based expressions. *Oecologia* **97**: 73–81.

Roumet C, Bel M-P, Sonié L, Jardon F, Roy J. 1996. Growth response of grasses to elevated CO_2: a physiological plurispecific analysis. *New Phytologist* **133**: 595–603.

Roumet C, Roy J. 1996. Prediction of the growth response to elevated CO_2: a search for physiological criteria in closely related grass species. *New Phytologist* **134**: 615–621.

Sage RF, Schäppi B, Körner Ch. 1997. Effect of atmospheric CO_2 enrichment on Rubisco content in herbaceous species from high and low altitude. *Acta Oecologica* **18**: 183–192.

Schlichting CD. 1986. The evolution of phenotypic plasticity in plants. *Annual Review of Ecology and Systematics* **17**: 667–693.

Schmid B, Birrer A, Lavigne C. 1996. Genetic variation in the response of plant populations to elevated CO_2 in a nutrient-poor, calcareous grassland. In: Körner Ch, Bazzaz F, eds. *Carbon dioxide, populations and communities.* San Diego, CA, USA: Academic Press, 31–50.

Sims DA, Seemann JR, Luo Y. 1998. Elevated CO_2 concentration has independent effects on expansion rates and thickness of soybean leaves across light and nitrogen gradients. *Journal of Experimental Botany* **49**: 583–591.

Stitt M. 1991. Rising CO_2 levels and their potential significance for carbon flux in photosynthetic cells. *Plant, Cell and Environment* **14**: 741–762.

Stulen I, Den Hertog J, Fonseca F, Steg K, Posthumus F, Van der Kooij TAW, De Kok LJ. 1998. Impact of elevated atmospheric CO_2 on plants. In: De Kok LJ, Stulen I, eds. *Responses of plant metabolism to air pollution and global change.* Leiden, The Netherlands: Backhuys Publishers, 167–179.

Vu JCV, Allen LH, Bowes G. 1989. Leaf ultrastructure, carbohydrates and protein of soybeans grown under CO_2 enrichment. *Environmental and Experimental Botany* **29**: 141–147.

Wulff RD, Alexander HM. 1985. Intraspecific variation in the response to CO_2 enrichment in seeds and seedlings of *Plantago lanceolatus* L. *Oecologia* **66**: 458–460.

Würth MKR, Winter K, Körner Ch. 1998. Leaf carbohydrate responses to CO_2 at the top of a tropical forest. *Oecologia* **116**: 18–25.

Ziska LH, Teramura AH. 1992. Intraspecific variation in the response of rice (*Oryza sativa*) to increased CO_2 – photosynthetic, biomass and reproductive characteristics. *Physiologia Plantarum* **84**: 269–276.

Profiles of photosynthetic oxygen-evolution within leaves of *Spinacia oleracea*

T. HAN, T. VOGELMANN* AND J. NISHIO

Botany Department, University of Wyoming, Laramie, WY 82071, USA

Received 22 October 1998 ; accepted 16 April 1999

SUMMARY

Oxygen evolution was measured from mesophyll tissues in spinach leaves using a photoacoustic technique. The photosynthetic capacity of individual cell layers was measured by directing microscopic beams of light, 40 μm wide, to cells exposed within a leaf cross section. The resulting profile for oxygen-evolution potential was relatively flat, indicating a uniform capacity for photosynthesis in leaf mesophyll tissues. Two experimental approaches were used to estimate the photosynthetic performance of individual mesophyll cell layers when white light was applied to the adaxial leaf surface. These experiments indicated that oxygen was produced relatively uniformly across the mesophyll and that oxygen evolution increased with irradiance of the white light applied to the leaf surface. The measured profiles for oxygen evolution and capacity are flatter than previous measurements of profiles of fixed carbon and estimates of profiles for absorbed light within spinach leaves.

Key words: leaf anatomy, palisade mesophyll, oxygen evolution, photoacoustics, photosynthesis, spinach (*Spinacia oleracea*), structure–function.

INTRODUCTION

Photosynthesis at the level of the whole leaf is determined by the combined activity of the mesophyll layers. Within each layer, photosynthetic performance is determined by the amount of light and CO_2 within that layer and the photosynthetic capacity of the chloroplasts. In order to examine relationships between leaf structure and whole-leaf photosynthetic performance, it is necessary to be able to measure light, CO_2 and photosynthesis at the tissue level. Using microscopic light sensors inserted into leaves, it has been shown that light declines rapidly and exponentially within the leaf (Vogelmann *et al.*, 1989; Cui *et al.*, 1991; Vogelmann & Martin, 1993; Vogelmann *et al.*, 1996). In spinach, light within the blue and red parts of the spectrum is absorbed almost completely by the initial palisade cell layers and it is primarily green light that remains for photosynthesis in tissues located deeper within the leaf (Cui *et al.*, 1991; Sun *et al.*, 1998). Correspondingly, the distribution of absorbed quanta across the mesophyll tissues is very steep, implying that there could be steep gradients in photosynthetic performance within the leaf (Evans, 1995). Measurements of the amount of C fixed by mesophyll layers of spinach (Nishio *et al.*, 1993; Sun *et al.*, 1996, 1998) and *Vicia* (Jeje &

Zimmermann, 1983) have shown that the rate of photosynthesis has a Gaussian profile across the leaf. In *Vicia*, the profile of C fixation was Gaussian, with maximum activity in the middle of the leaf. Spinach has a similar Gaussian profile where, progressing from the adaxial (upper) surface into the mesophyll, the amount of fixed C increases to a maximum in tissues located 30–40% midway through the leaf (Nishio *et al.*, 1993; Sun *et al.*, 1996). C fixation declined rapidly in the spongy mesophyll to a minimum near the abaxial (lower) leaf surface. In spinach, profiles for Rubisco and chlorophyll contents were also Gaussian, but C fixation was correlated more closely to Rubisco than chlorophyll (Nishio *et al.*, 1993). It has been proposed that the Gaussian profiles in C fixation in spinach are a function of the light gradient and the chlorophyll content of the tissues, and that they reflect the distribution of absorbed quanta (Evans, 1995). Thus far, profiles of absorbed light in leaves have been calculated, but it may be possible to measure them by using laser-induced chlorophyll fluorescence (Takahashi *et al.*, 1994; Koizumi *et al.*, 1998). This experimental approach consists of irradiating the adaxial surface of a leaf cross section with monochromatic light and, using image analysis, measuring the amount of chlorophyll fluorescence that is emitted from mesophyll tissues in that cross section. Presumably the amount of chlorophyll fluorescence emitted by each mesophyll layer is proportional to

*Author for correspondence (tel +1 307 766 6293; fax +1 307 766 2851; e-mail tvogel@uwyo.edu).

the amount of light absorbed by that layer. Thus, the profile for chlorophyll fluorescence across the tissues in the cross section should coincide with the profile of absorbed quanta.

In order more completely to evaluate the functional significance of leaf structure with respect to photosynthesis, it is necessary to be able to measure photosynthesis at the tissue level in real time and under a variety of experimental conditions. We have recently described a photoacoustic instrument that makes such measurements possible (Han & Vogelmann, 1999). With this device, microbeams of light are directed to individual cell layers within leaf cross sections and O_2 evolution is measured from these layers. Here, we use this photoacoustic device to measure the O_2-evolution potential of mesophyll cells within spinach leaves and the apparent profile of O_2 evolution across the leaf when light enters through the adaxial surface.

MATERIALS AND METHODS

Plants and growth conditions

Spinach (*Spinacia oleracea* cv. hybrid 424; Ferry-Morse Seed Company, Modesto, CA, USA) was grown in environmental chambers as previously described (Nishio *et al.*, 1993). Plants were grown under simulated light conditions with fluorescent and incandescent bulbs. High-light plants were grown under an irradiance of 600–800 μmol m^{-2}s^{-1}, measured by an LI-185B quantum meter (LiCor, Lincoln, NE, USA). Temperature was $23 \pm 2°C$ during the light period (12 h) and $17 \pm 2°C$ during the dark period. Leaves, 600–800 μm thick, were used for experiments from plants 5–6 wk old.

Photoacoustic measurements of leaf disks

Photoacoustic measurements were made with an instrument previously described (Han & Vogelmann, 1999). The photoacoustic cell was gas permeable and was contained within a gas-tight housing that was flooded with gas consisting of 21% O_2, 79% N and 360 ppm CO_2 (Scottys Gas, Denver, CO, USA). Gas exchange between the housing and photoacoustic cell was sufficient to maintain CO_2 concentration around the leaf sample well above the CO_2 compensation point. Experiments were done on leaf cross sections and leaf disks, with these irradiated in different ways.

For experiments with leaf disks (1-cm diameter), modulated and nonmodulated white light was provided by a quartz–halogen lamp (21 V, 150 W; EKE, Ushio, Japan) coupled into a bifurcated fiber-optic cable. In one arm of the cable the light was modulated simultaneously at 3 and 470 Hz by two mechanical choppers and drivers (models 5R541 and 5R540; Stanford Research Systems, Sunnyvale, CA,

USA). Continuous light (2400 μmol m^{-2}s^{-1}) was introduced into the other arm of the fiber-optic cable, controlled by a manually operated shutter. Irradiance in each arm of the cable was controlled by changing the distance between the arm of the cable and the lamp.

Modulated photoacoustic signals were detected using a random-incidence and pressure-response microphone (model 2560; Larson Davis, Provo, UT, USA) that had a response range of 1–10 000 Hz and a sensitivity of 47.5 mV Pa^{-1}. Signals were routed to a second preamplifier (model 5R560; Stanford Research Systems), then to two lock-in amplifiers (model 5R830 DSP; Stanford Research Systems) and finally to a dual-channel, strip-chart recorder (model BD112; Kipp and Zonen, Delft, The Netherlands).

Leaf disks were positioned within the chamber such that light was directed to the adaxial or abaxial leaf surface. Photoacoustic signals were measured as the irradiance of the modulated light was changed from 60–1200 μmol m^{-2}s^{-1} and values for energy storage, photochemical loss and O_2 evolution calculated as described previously (Poulet *et al.*, 1983).

Measurement of oxygen-evolution potential of cell layers within leaf cross sections

To estimate the O_2-evolution potential of individual mesophyll layers, microscopic beams of light were directed to leaf cross sections (Fig. 1a,b). A leaf sample (1 mm wide, 10 mm long and 600–800 μm thick) was excised with a razor blade, placed in the photoacoustic cell and individual cell layers were irradiated with microscopic beams of light 40 μm wide (Fig. 1a). Actinic light at 635 nm was provided by a 10 mW laser diode (HL6320G; Hitachi, Tokyo, Japan) and diode laser driver (Model O6DLD2O1; Melles Griot, Irvine, CA, USA). This light was modulated with a mechanical chopper. The light was focused with a cylindrical lens and directed into the photoacoustic cell by a surface-silvered mirror. The distance between the cylindrical lens and leaf sample was adjusted such that the modulated light formed a narrow linear beam (40 μm × 6 mm) that could be placed on individual mesophyll cell layers. The position of the light on the sample was viewed with a long-working-distance microscope (Gaertner Scientific, Chicago, IL, USA).

A nonmodulated saturating beam of light (690 nm) was provided by a 30 mW laser diode (model TOLD9215; Toshiba, Tokyo, Japan). This light was focused by a second cylindrical lens and directed into the photoacoustic cell such that it formed a linear beam (70 μm × 7 mm) that covered the region of the sample irradiated by the modulated light (Fig. 1b). The modulated laser line was then moved 40 μm to a new position on the leaf and the procedure repeated.

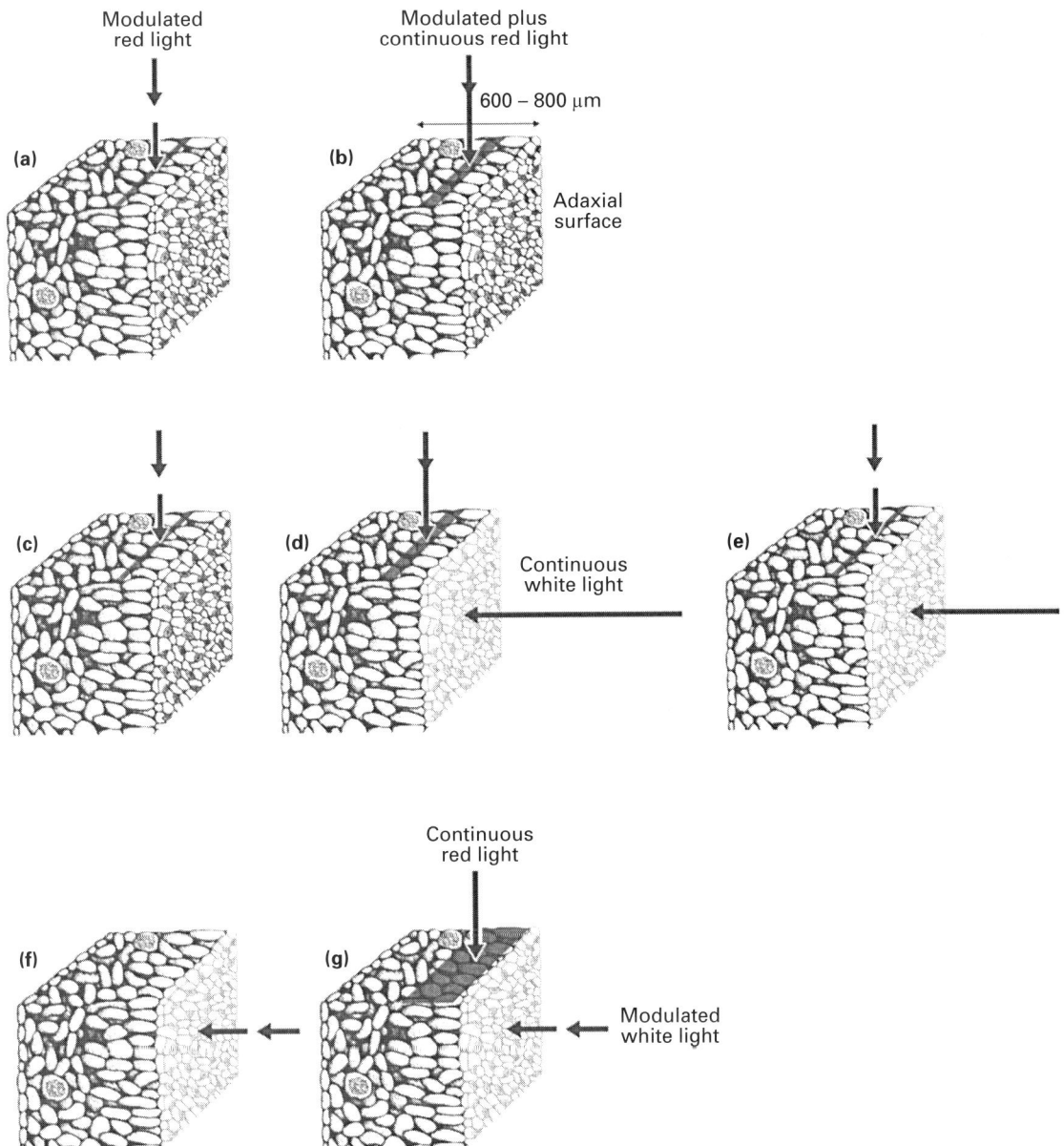

Fig. 1. Experimental designs for measurement of O_2 evolution from spinach mesophyll cells. To provide corroboration between experiments, O_2 evolution was measured by applying light to the leaf sample in three different ways (a and b; c–e; f and g). The spinach leaf sample consisted of a linear section, 600–800 μm thick, 10 mm long and 1 mm wide. Microscopic bands of light, 40 μm wide and 10 mm long, were applied to the cross-sectional surface of the leaf sample.

Measurements of photoacoustic signals from leaf cross sections are complicated by the presence of damaged cells at the cut surface. These cells elevate the photothermal component of the photoacoustic signal and prevent the measurement of photochemical loss at high frequency. Thus, O_2-evolution signals were approximated by a simplified procedure as described previously (Han & Vogelmann, 1999). Briefly, application of modulated light to a cell layer results in a modulated photoacoustic signal that has the following components:

$$A = A_{th} + A_{ox}$$ Eqn 1

(A, photoacoustic signal; A_{th}, thermal component; A_{ox}, O_2-evolution component). In healthy leaves, approximately half of the photoacoustic signal consists of the O_2-evolution component and the other half (heat) represents the energy that is not used by the photosynthetic system. With the application of nonmodulated saturating light, photosynthetic electron transport is saturated and the modulated light no longer elicits O_2 evolution. In this situation $A_{ox} = 0$, and the photoacoustic signal (A^*) consists of the thermal component only:

$$A^* = A_{th}^*$$ Eqn 2

We estimated relative rates of O_2 evolution between different tissues in the leaf using the following equation:

$$A_{ox} = (A - A^*)$$ Eqn 3

This procedure is valid only if several conditions are met. In reality, A_{th}^* is larger than A_{th} because with the application of nonmodulated saturating light the photochemical energy that was devoted to O_2 evolution (modulated light) is now released as heat. The difference between A_{th}^* and A_{th} is referred to as photochemical loss. The ratio of photochemical loss to A_{th}^* is referred to as energy storage. Estimation of O_2 evolution using Eqn 3 relies on the assumption that energy storage is a relatively small quantity compared to O_2 evolution. Typical values of energy storage at low irradiance are in the range of 20% (Herbert *et al.*, 1990; Havaux, 1992). Two other conditions are that the ratio of photochemical loss to O_2 evolution should remain relatively constant with irradiance and that palisade and spongy tissues have similar ratios. As shown here by experiments with leaf disks, this appears to be the case for spinach. A final consideration is that the light gradient should be similar in different tissues such that equivalent depth profiles are measured photoacoustically. Measurements described here were done to ascertain whether this was the case.

Measurement of oxygen evolution from individual cell layers when the upper leaf surface is irradiated with continuous white light

Experiments were done to determine the manner in which the profiles for O_2 evolution potential, measured from the exposed surface of a leaf cross section, were altered by the application of nonmodulated white light to the adaxial leaf surface (Fig. 1c–e). This experiment was designed to determine how much O_2 was evolved from individual cell layers when there was a light gradient across the leaf and in the presence of various metabolic feedback controls. The experimental procedure consisted of irradiating an individual cell layer within a cross section with modulated 635-nm light (4200 μmol m^{-2} s^{-1}), 40 μm wide and 6 mm long (Fig. 1c; Eqn 1) and measuring the resulting photoacoustic signal. The maximum photothermal signal (A^*; Eqn 2) was determined by applying continuous white light (1050 μmol m^{-2} s^{-1}) – obtained from a quartz–halogen bulb (150 W; EKE) and directed to the leaf sample through a fiber-optic cable – to the adaxial leaf surface (Fig. 1d). At the same time, continuous red light (690 nm, 700 μm wide by 8 mm long, 5500 μmol m^{-2} s^{-1}) was superimposed over the band of modulated light on the surface of a cross section. Photoacoustic signals (A^w; Eqn 4) from a cell layer in the presence of white light applied to the adaxial surface, was measured by turning off the continuous 690-nm red light and applying white light to the adaxial leaf surface in sequence at 1050, 570 and 220 μmol m^{-2} s^{-1} (Fig. 1e).

$$A^w = A_{th} + A_{ox}^w \qquad \text{Eqn 4}$$

The modulated 40-μm band of 635-nm light was then moved to a new location on the leaf cross section and the procedure repeated. In the presence of nonmodulated white light, O_2 evolution (P, %), was determined at 40 μm increments across the cross section using the following equation:

$$P = (A_{ox} - A_{ox}^w)/A_{ox} \times 100 \qquad \text{Eqn 5}$$

Measurement of oxygen evolution from individual cell layers when the upper leaf surface is irradiated with modulated white light

The photosynthetic performance of individual cell layers, in the presence of white light applied to the adaxial leaf surface, was examined further (Fig. 1f,g). Experiments were performed in which modulated white light (690 μmol m^{-2} s^{-1}) was directed to the leaf surface (Fig. 1f). This resulted in a modulated release of O_2 from all of the photosynthetic cells in the sample. The contribution of mesophyll tissues to the total modulated O_2 release was determined by placing a 200 μm wide beam of continuous and saturating 635-nm light (10700 μmol m^{-2} s^{-1}) upon broad regions within the cross section (Fig. 1g) and measuring the decrease in the photoacoustic signal elicited by the modulated white light. The 200-μm-wide band of light was required to produce a measurable signal against the relatively large O_2-evolution signal of the cross section.

Measurement of penetration of 635-nm light into palisade and spongy mesophyll tissues

In experiments with leaf cross sections, palisade and spongy mesophyll tissues were irradiated with microbeams of light. The size of the photoacoustic signal from these tissues depends on several factors, including the depth to which light penetrates. In order to compare light propagation in palisade and spongy tissues, attenuation profiles of 635-nm light were measured in leaf cross sections using image analysis. A leaf cross section, similar to that used for photoacoustic measurements, was cut such that it formed a 30° wedge of tissue (Fig. 2) and placed on a microscope slide on the stage of an IMT-2 inverted microscope (Olympus, Tokyo, Japan). A linear band of 635-nm light, 40 μm wide, was positioned with a micromanipulator at different positions on the cross section. The light that was transmitted through the wedge of tissue was passed through a 633 nm-narrow-band interference filter (Corion P1–633-F-H636; Corion, Holliston, MA, USA) to eliminate chlorophyll fluorescence, and captured as an image using a liquid nitrogen-cooled CCD camera (CH270 camera head, CF200A 16/40 camera electronics unit, AT200 controller board, 35-mm shutter; all Photometrics, Tucson, AZ, USA) with a 16-bit resolution of dynamic range and PMIS image-processing software (Photometrics). The 40-μm band of 635-nm

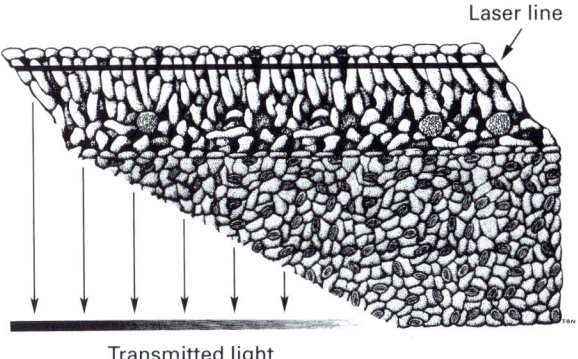

Fig. 2. Measurement of attenuation of 635-nm light in palisade and spongy mesophyll tissues in leaf cross sections. A spinach leaf cross section was cut at 30° to form a wedge of variable tissue thickness. A linear beam of 635-nm light, 40 μm wide, was positioned on either the palisade or spongy mesophyll and the amount of light transmitted through the wedge was captured as an image with a CCD camera. Light attenuation profiles were determined by plotting gray-scale values of pixels from the image against the corresponding thickness of the tissue in the wedge.

light was positioned on the palisade and spongy mesophyll tissues and light-attenuation profiles within the two tissues were constructed from gray-scale values for pixels, analysed from the wedge-shaped region of the cross section, on the captured images.

RESULTS

The ratio of photochemical loss to oxygen evolution is relatively constant with irradiance in leaf disks

Light-response curves for O_2 evolution were measured when white light was directed to the adaxial or abaxial surface of leaf disks (Fig. 3). O_2-evolution curves were plotted with (A_{ox}) and without (a_{ox}) corrections for photochemical loss. The corrected and uncorrected curves were similar in shape. Expressing the data as a_{ox}/A_{ox} (Fig. 4) showed a relatively constant relationship with irradiance, indicating a constant ratio of photochemical loss to O_2 evolution irrespective of irradiance. Also, this ratio

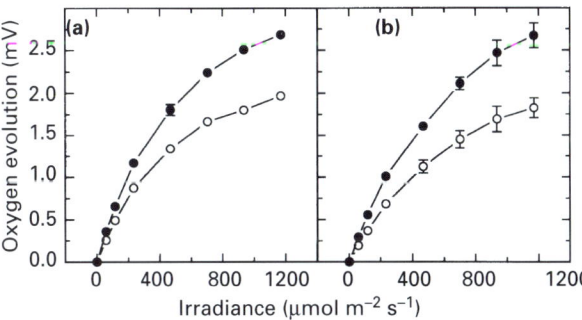

Fig. 3. Light-response curves for O_2 evolution in spinach leaf disks. (a) Leaf disk irradiated on its adaxial surface. (b) Leaf disk irradiated on its abaxial surface. O_2 evolution is shown with (filled circles) and without (open circles) corrections for photochemical loss. Bars: SE ($n = 3$).

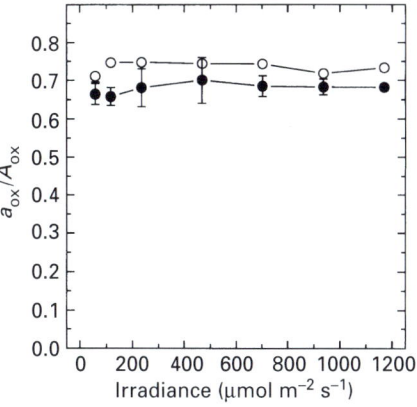

Fig. 4. The ratio of uncorrected (a_{ox}) to corrected (A_{ox}) values of O_2 evolution vs irradiance in spinach leaf disks. Data are from Fig. 3. Measurements are shown when the leaf was irradiated on its upper (open circles) and lower (filled circles) leaf surfaces.

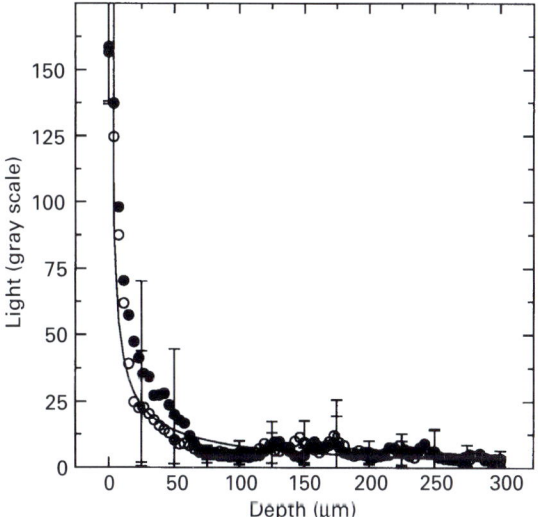

Fig. 5. Attenuation profiles of 635-nm light in spinach leaf tissues. Palisade (open circles), spongy mesophyll (filled circles). Measurements were made as described in Fig. 2. Bars: SE ($n = 12$).

was similar when the leaf was irradiated adaxially or abaxially, suggesting that it was similar for spongy and palisade mesophyll tissues.

Penetration of 635-nm light is similar in mesophyll tissues in cross sections

Positioning a 40 μm wide line of 635-nm light on the palisade or spongy mesophyll in leaf cross sections showed that both tissues rapidly attenuated the light (Fig. 5). Light attenuation was similar in both palisade and spongy mesophyll and light fell to 10% of initial values after passage through 50 μm of tissue. The width of the laser line did not increase as it passed through the tissues, suggesting that attenuation by light scattering was minimal. The similar light-attenuation profiles in palisade and spongy mesophyll simplifies interpretation of the subsequent O_2-evolution measurements.

*Oxygen-evolution potential of spinach mesophyll cells
is relatively uniform in leaf cross sections*

The ability to direct microscopic beams of light to
individual cell layers, exposed in a leaf cross section,
made it possible to measure the capacity of these
cells to evolve O_2. The resulting profile for O_2-
evolution potential was relatively uniform across the
leaf tissues (Fig. 6). A particular profile for a spinach
leaf was consistent and reproducible for that leaf, but
there was some variation between leaves. Sometimes
there was a depression in O_2 evolution (200–300 μm
within the leaf); other times it was relatively uniform
across the mesophyll, as shown in Fig. 6.

*Irradiation of the upper leaf surface with continuous
or modulated white light reveals profiles of oxygen
evolution across mesophyll tissues*

The application of continuous white light to the
adaxial leaf surface resulted in a relatively uniform
decrease in O_2 production across the leaf (Fig. 7),
whereas it was less uniform when modulated light

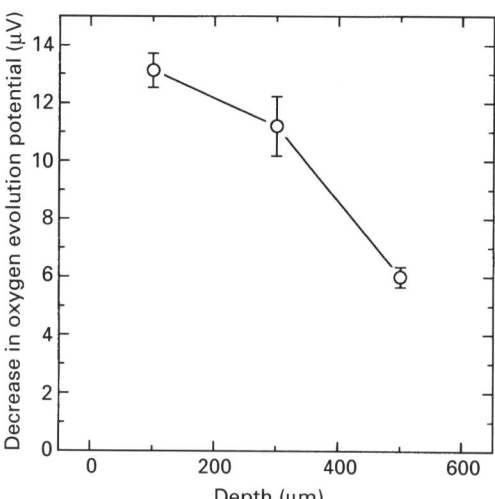

Fig. 8. Profiles of O_2 evolution across the leaf in the
presence of modulated white light incident upon the
adaxial leaf surface. Measurements were made as described
in Fig. 1f,g. Bars: SE ($n = 3$).

was applied to the adaxial surface (Fig. 8). These
O_2-evolution profiles probably result from the way
in which light enters the leaf. In one experiment
(Fig. 7), a modulated 40-μm band of light was used
to measure how much O_2 was evolved from exposed
mesophyll cells and a profile for potential O_2
evolution was constructed for the cross section. With
the application of continuous white light to the
adaxial leaf surface, the decline in O_2 evolution was
measured at 40-μm increments across the cross
section. The resulting data in Fig. 7 shows how the
application of continuous white light to the adaxial
leaf surface affects the potential to evolve O_2 by the
mesophyll cell layers. Adding white light to the leaf
surface caused a uniform decrease in O_2-evolution
potential across the mesophyll tissues. The decreases
in O_2 evolution were relatively uniform irrespective
of the irradiance of the applied white light, which
was varied between 220 and 1050 μmol m^{-2} s^{-1} (Fig.
7b).

A different procedure was used to measure O_2-
evolution profiles in the experiment shown in Fig. 8.

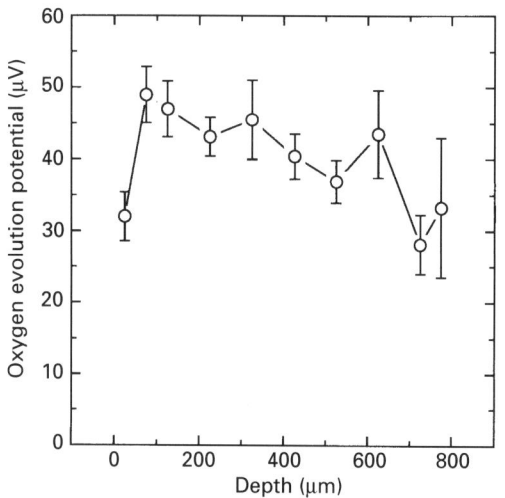

Fig. 6. Profile for O_2-evolution potential (A_{ox}) across the
leaf. Measurements were made as described in Fig. 1a,b.
Bars: SE ($n = 4$).

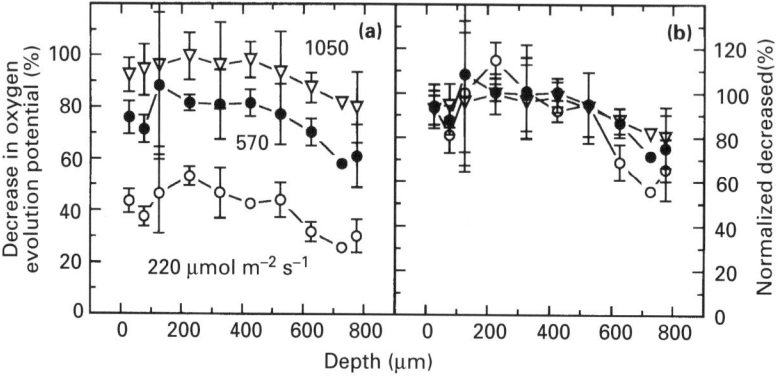

Fig. 7. Profiles of O_2 evolution across the leaf in the presence of continuous white light incident upon the
adaxial leaf surface. Measurements were made as described in Fig. 1c–e. (a) O_2-evolution profiles in the
presence of 220, 570 and 1050 μmol m^{-2} s^{-1} white light. (b) O_2-evolution profiles normalized to 100%. Curves
were aligned by setting the mean of three data points (200–400 μm) to 100. Bars: SE ($n = 3$).

Here, a modulated release of O_2 from all the mesophyll cells was generated by applying modulated white light to the upper leaf surface. By applying a 200-μm band of saturating red light to the surface of a cross section, it was possible to eliminate and quantify modulated O_2 evolution from that region of the cross section. The rationale for this sampling procedure is that it may reveal the O_2-evolution profile that occurs in response to the light gradient within the leaf. Data in Fig. 8 are consistent with this idea and show a less-uniform profile for O_2 evolution than in Fig. 7.

DISCUSSION

Origin of photoacoustic signals within tissues and rationale for measurement of A_{ox}, the oxygen-evolution component

As an indicator of O_2-evolution capacity, comparison of A_{ox} from palisade and mesophyll tissues is only valid if the origin and propagation of photoacoustic signals is similar between the two tissues. The origin and characteristics of thermal and O_2-evolution signals have been described in several reviews (Bults et al., 1982; Buschmann & Prehn, 1990; Fork & Herbert, 1993; Malkin & Canaani, 1994; Malkin, 1996) and a complete description of the origin and propagation of these signals within leaves is not simple. Photoacoustic detection depends on the migration of heat and O_2 from the chloroplasts through the cell walls and into the intercellular air spaces. From there, the photoacoustic signals must propagate out of the leaf and into the gas phase of the photoacoustic cell.

Propagation of the thermal signal, O_2-evolution signal or both may be affected by internal leaf anatomy as well as by the epidermis. For example, in experiments with leaf disks of normal and herbicide-resistant biotypes of *Chenopodium album*, it was found the $A_{ox}:A_{th}$ ratio, a measure of quantum yield, had different relationships with modulation frequency, indicating that the photoacoustic signals exited the leaves differently (Havaux, 1989). Based on frequency analysis, this difference was attributed to different O_2-diffusion characteristics of leaves. In another study of leaf disks, it was found that in some species, such as *Passiflora edulis*, water stress appeared to alter the O_2-diffusion pathway, whereas in others, such as *Casimiroa edulis*, it did not (Havaux et al., 1986). Moreover, it was found that there were two thermal signals that could be separated by modulation frequency. One thermal signal was attributed to the heat that migrated out of the cells into the intercellular air space and out of the leaf. The other signal was attributed to the heat that migrated directly out of the cells and into the gas phase of the photoacoustic cell. In other studies with leaf disks, it was found that it was very difficult to

prevent the O_2-evolution signal from exiting the leaf, even when it was covered with vacuum grease, leading to the suggestion that the O_2-evolution signal may be propagated, at least in part, by a 'drum-like' mechanism (Fork & Herbert, 1993). Given the complexity of the potential interactions between these signals and leaf structure, the details associated with signal origin and propagation should probably be studied further.

With respect to our measurements of the O_2-evolution capacity of different tissues, the use of leaf cross sections may avoid some of the complications associated with using leaf disks. With cross sections, the intercellular spaces are in direct contact with the gas phase of the photoacoustic cell, thus eliminating potential problems associated with stomatal closure, or migration of photoacoustic signals across the epidermis. In addition, the use of a low chopping frequency (3 Hz) to measure O_2 evolution has distinct advantages over higher chopping frequencies. For example, a frequency analysis of *Philodendron sp.* showed a maximum amplitude for O_2 evolution at 2 Hz (Frandas et al., 1997). This may be attributable in part to maximizing the O_2 diffusion length within cells, another advantage when comparing tissues with different anatomy.

Other potential complications in comparing O_2-evolution signals from palisade and spongy mesophyll tissues include the possibility that the ratio of photochemical loss to O_2 evolution varies between tissues. Unfortunately, it is not currently possible directly to measure photochemical loss from tissues exposed in leaf cross sections. Because the cells near the surface are damaged, the chloroplasts have a reduced capacity for photosynthesis. Upon absorption of light, most of the energy is released as heat, thereby greatly increasing the photothermal component of the photoacoustic signal. At the higher frequencies used to measure photochemical loss, the sampling depth is limited to the initial 10 μm of the cross section. For these reasons, we have not been successful in measuring photochemical loss from these cells. To approach the problem another way, we used spinach leaf disks. Presumably, when leaf disks are irradiated adaxially, most of the photoacoustic signals come from the palisade. Similarly, when irradiated abaxially, most of the photoacoustic signals should come from the spongy mesophyll. Results from these experiments showed that photochemical loss in spinach is proportional to O_2 evolution (Fig. 4), a finding that agrees with previous results from wheat and pea (Kanstad et al., 1983). In addition, palisade and spongy tissues in spinach appear to have similar ratios of photochemical loss to O_2 evolution (Fig. 4), eliminating a potential complication with our simplified measurement of A_{ox}.

A final consideration is that dissimilar light gradients within palisade and spongy mesophyll tissues could result in different photoacoustic sampling profiles. However, this does not appear to be a serious issue in spinach-leaf cross sections because profiles for light attenuation (635 nm) in palisade and spongy mesophyll appear similar (Fig. 5). This is not surprising, because palisade and spongy mesophyll have uniformly high proportions of intercellular air space (Warmbrodt & Van Der Woude, 1990; James *et al.*, 1999), which would tend to scatter the light uniformly. In addition, the light-guiding capacity of the palisade mesophyll (Vogelmann & Martin, 1993) is lost when light enters the tissue laterally instead of adaxially. Further measurements will help resolve some of the complications associated with measuring photoacoustic signals from discrete cell layers within leaves.

A close match between capacity for, and relative rates of, oxygen evolution

By moving a modulated beam of light across the mesophyll cells exposed in a leaf cross section, we attempted to measure the relative capacity of each cell layer to evolve O_2. For the modulated light, a relatively high irradiance was necessary (3700 µmol m^{-2} s^{-1}) to provide a strong O_2-evolution signal. This is not unexpected because the light must pass through damaged cells near the cut surface to underlying, more photosynthetically active cells. Approx. 80% of the modulated light was absorbed by the cells at the surface of the cross section, increasing the photothermal signal relative to the O_2-evolution signal. This resulted in a lower relative quantum yield for O_2 evolution, and A_{ox}/A_{th} for a cross section was typically 0.5 as opposed to 1.5 for similar measurements made from the adaxial surface of a leaf disk. Cells near the cut surface did not lose all their ability to evolve O_2, rather there was a gradient in O_2-evolution capacity, extending from a minimum of 40% near the surface of the cross section to full capacity in cells located 100 µm from the cut surface (Han & Vogelmann, 1999). The zone of damage varied and, depending on the experiment, extended from one to three cell layers thick. Despite these unavoidable experimental complications, it seems reasonable to assume that damage caused by cutting should be uniform across the cross section. This would not alter the shape of the profile for potential O_2 evolution. Although there was some variability between measurements, the profile for potential O_2 evolution (Fig. 6) was remarkably flat.

Measurements of photosynthetic performance of leaf mesophyll cells were made in two ways to ascertain the shape of the O_2-evolution profile when the leaf sample was irradiated with white light on its adaxial surface (Figs 7, 8). Taken together with the profile for O_2-evolution potential (Fig. 6), it appears that O_2 evolution decreased uniformly across the mesophyll tissues when white light was applied to

the leaf surface. This suggests that there may be a remarkable amount of integration of photosynthetic performance between the cell layers. The data in Fig. 8 show a less-uniform O_2-evolution profile, which may be a result of the light gradient that is created when white light enters through the adaxial leaf surface.

Comparison of profiles for oxygen evolution with absorbed light and CO_2 fixation

It has been assumed that photosynthesis within leaves is directly related to the amount of light absorbed by the mesophyll layers (Evans *et al.*, 1993; Evans, 1995; Richter & Fukshansky, 1996a, b, 1998). However, the relatively flat profiles for O_2 evolution measured so far do not coincide very closely with estimates of profiles of absorbed light within spinach or other leaves. In spinach, profiles of absorbed light were calculated based upon a modified Beer–Lambert law and the amount of C fixed within the leaf (Evans *et al.*, 1993). The calculated absorption profiles were biphasic, showing a maximum in absorbed quanta 150 µm within the mesophyll. Profiles of absorbed light were also calculated under a variety of experimental conditions for *Catalpa* leaves (Richter & Fukshansky, 1996a, b, 1998). Despite the fact that *Catalpa* has a more strongly developed palisade than spinach, calculated absorption profiles were biphasic. Maximum absorption rates were located at different positions within the mesophyll depending upon wavelength, collimation and leaf orientation.

Profiles of absorbed light have been estimated experimentally from the distribution of chlorophyll fluorescence in mesophyll tissues in leaf cross sections irradiated on their adaxial surface and observed through a microscope (Takahashi *et al.*, 1994; Koizumi *et al.*, 1998). In these experiments, it was assumed that the amount of chlorophyll fluorescence from mesophyll tissues was proportional to the amount of light absorbed. In *Camellia* leaves irradiated with green light (515 nm), chlorophyll fluorescence profiles had a fluorescence maximum 150 µm beneath the irradiated surface. This maximum shifted to 100 µm when the leaf sample was irradiated with blue light (488 nm). Chlorophyll fluorescence profiles were also measured in mesophyll tissues of mangrove and were similar to those of *Camellia* except that the fluorescence profiles were shifted deeper into the leaf by the presence of a largely chlorophyll-free hyaline layer near the adaxial surface.

Similar to estimates of absorbed light, measured profiles of C fixation in spinach (Nishio *et al.*, 1993) and *Vicia* (Jeje & Zimmermann, 1983) were Gaussian. How can the relatively flat profiles for O_2 evolution within spinach be explained? Photosynthetic performance of mesophyll cell layers within spinach can be altered by growth conditions (e.g. sun vs shade leaves, (Nishio *et al.*, 1993)), and there is some variability between leaves. Nonetheless, the plants used in our experiments were grown similarly to those used in previous work and it would be difficult to reconcile the dissimilarity in O_2 evolution with other profiles based upon differences between plants and growth conditions.

Photoacoustic measurements of O_2 evolution were made on leaf pieces as opposed to previous measurements of C fixation, which were made on leaves attached to the plant. However, in both cases, profiles for absorbed light across the leaf tissues should be similar, and if the assumption is made that C fixation follows absorbed light, then it is difficult to reconcile the differences between profiles for O_2 and fixed C based upon differences in experimental methods. Similarly, there may be a difference in permeability to CO_2 where, in leaf pieces, CO_2 can gain direct entry into the mesophyll through the cut surfaces as opposed to intact leaves where it enters via stomata. Although opinions vary (Parkhurst & Mott, 1990; Parkhurst, 1994; Evans & von Caemmerer, 1996), a number of studies have concluded that, in most leaves, the mesophyll is highly permeable to gas and that CO_2 concentration gradients within the leaf are rather small (Evans & von Caemmerer, 1996). Again, the differences between profiles of O_2 evolution and CO_2 fixation are difficult to explain based upon the assumption that photosynthetic performance is directly related to the amount of light absorbed by mesophyll tissues.

It may be that the differences between O_2 evolution and C fixation profiles indicate electron-transport capacity in excess of what is needed to fix C. Although C is a major sink for electrons, dissipatory pathways can consume many of the electrons released in photosynthesis. Under saturating light, it was pointed out that only one out of three electrons may be used to reduce C (Osmond & Grace, 1995). The remaining electrons can be channeled, in roughly equal amounts, to photo-respiration and the Mehler reaction. Since the photoacoustic method detects the release of O_2 from photosystem II, it provides a measure of the total electrons available. Under our experimental conditions, where the sample was exposed to an atmosphere consisting of 20% O_2 and 360 ppm CO_2, a significant part of the electron flow could be directed to photorespiration and the Mehler reaction, and it may be that different fractions of electrons are allocated to C fixation within the leaf mesophyll, thus explaining the disparity between profiles of absorbed light, O_2 evolution and C fixation. A more-detailed analysis of the light regime and the photosynthetic characteristics of the individual mesophyll layers within spinach leaves is currently under way to evaluate this and related hypotheses.

ACKNOWLEDGEMENTS

This research was funded by grants from the National Science Foundation (DBI-9724499) and USDA (93–37100–8855 and 96–35100–3167).

REFERENCES

Bults G, Horwitz BA, Malkin S, Cahen D. 1982. Photoacoustic measurements of photosynthetic activities in whole leaves – photochemistry and gas exchange. *Biochimica et Biophysica Acta* **679**: 452–465.

Buschmann C, Prehn H. 1990. Photoacoustic spectroscopy – photoacoustic and photothermal methods. In: Linskens HF, Jackson JF, eds. *Modern methods in plant analysis*. Berlin, Germany: Springer-Verlag, 148–180.

Cui M, Vogelmann TC, Smith WK. 1991. Chlorophyll and light gradients in sun and shade leaves of *Spinacia oleracea*. *Plant, Cell and Environment* **14**: 493–500.

Evans JR. 1995. Carbon fixation profiles do reflect light absorption profiles in leaves. *Australian Journal of Plant Physiology* **22**: 865–873.

Evans JR, Jakobsen I, Ögren E. 1993. Photosynthetic light-response curves. 2. Gradients of light absorption and photosynthetic capacity. *Planta* **189**: 191–200.

Evans JR, von Caemmerer S. 1996. Carbon dioxide diffusion inside leaves. *Plant Physiology* **110**: 339–346.

Fork DC, Herbert SK. 1993. The application of photoacoustic techniques to studies of photosynthesis. *Photochemistry and Photobiology* **57**: 207–220.

Frandas A, Jalink H, van der Schoor R. 1997. Low frequency photoacoustics for monitoring the photobaric component *in vivo* of green leaves. *Photosynthesis Research* **52**: 65–67.

Han T, Vogelmann TC. 1999. A photoacoustic spectrometer for measuring heat dissipation and oxygen quantum yield at the microscopic level within leaf tissues. *Journal of Photochemistry and Photobiology* **48**: 158–165.

Havaux M. 1989. Photoacoustic characteristics of leaves of atrazine-resistant weed mutants. *Photosynthesis Research* **21**: 51–59.

Havaux M. 1992. Photoacoustic measurements of cyclic electron flow around photosystem I in leaves adapted to light-states 1 and 2. *Plant and Cell Physiology* **33**: 799–803.

Havaux M, Canaani O, Malkin S. 1986. Photosynthetic responses of leaves to water stress, expressed by photoacoustics and related methods. *Plant Physiology* **82**: 827–833.

Herbert SK, Fork DC, Malkin S. 1990. Photoacoustic measurement *in vivo* of energy storage by cyclic electron flow in algae and higher plants. *Plant Physiology* **94**: 926–934.

James SA, Smith WK, Vogelmann TC. 1999. Ontogenetic differences in mesophyll structure and chlorophyll distribution in *Eucalyptus globulus* ssp. globulus (Myrtacease). *American Journal of Botany* **86**: 198–207.

Jeje A, Zimmermann M. 1983. The anisotropy of the mesophyll and CO_2 capture sites in *Vicia faba* L. leaves at low light intensities. *Journal of Experimental Botany* **34**: 1676–1694.

Kanstad SO, Cahen D, Malkin S. 1983. Simultaneous detection of photosynthetic energy storage and oxygen evolution in leaves by photothermal radiometry and photoacoustics. *Biochimica et Biophysica Acta* **722**: 182–189.

Koizumi M, Takahashi K, Mineuchi K, Nakamura T, Kano H. 1998. Light gradients and the transverse distribution of chlorophyll fluorescence in mangrove and *Camellia* leaves. *Annals of Botany* **81**: 527–533.

Malkin S. 1996. Biophysical techniques in photosynthesis. In: Amesz J, Hoff AJ, eds. *Advances in photosynthesis*. Boston, USA: Kluwer Academic Publishers, 191–206.

Malkin S, Canaani O. 1994. The use and characteristics of the photoacoustic method in the study of photosynthesis. *Annual Review of Plant Physiology and Plant Molecular Biology* **45**: 493–526.

Nishio J, Sun J, Vogelmann TC. 1993. Gradients of carbon fixation do not follow light gradients within leaves. *The Plant Cell* **5**: 953–961.

Osmond CB, Grace SC. 1995. Perspectives on photoinhibition and photorespiration in the field: Quintessential inefficiencies of the light and dark reactions of photosynthesis? *Journal of Experimental Botany* **46**: 1351–1362.

Parkhurst DF. 1994. Diffusion of CO_2 and other gases in leaves. *New Phytologist* **126**: 449–479.

Parkhurst DF, Mott K. 1990. Intercellular diffusion limits to CO_2 uptake in leaves. *Plant Physiology* **94**: 1024–1032.

Poulet P, Cahen D, Malkin S. 1983. Photoacoustic detection of photosynthetic oxygen evolution from leaves – quantitative analysis by phase and amplitude measurements. *Biochimica et Biophysica Acta* **724**: 433–446.

Richter T, Fukshansky L. 1996a. Optics of a bifacial leaf. 1. A novel combined procedure for deriving the optical parameters. *Photochemistry and Photobiology* **63**: 507–516.

Richter T, Fukshansky L. 1996b. Optics of a bifacial leaf. 2. Light regime as affected by leaf structure and the light source. *Photochemistry and Photobiology* **63**: 517–527.

Richter T, Fukshansky L. 1998. Optics of a bifacial leaf. 3. Implications for the photosynthetic performance. *Photochemistry and Photobiology* **68**: 337–352.

Sun J, Nishio JN, Vogelmann TC. 1996. High light alters photosynthetic carbon fixation gradients across sun and shade leaves. *Plant Cell & Environment* **19**: 1261–1271.

Sun J, Nishio JN, Vogelmann TC. 1998. Green light drives CO_2 fixation deep within leaves. *Plant & Cell Physiology*. (In press.)

Takahashi K, Mineuchi K, Nakamura T, Koizumi M, Kano H. 1994. A system for imaging transverse distribution of scattered light and chlorophyll fluorescence in intact rice leaves. *Plant, Cell and Environment* **17**: 105–110.

Vogelmann TC, Bornman JF, Josserand S. 1989. Photosynthetic light gradients and spectral regime within leaves of *Medicago* sativa. *Proceedings of the Philosophical Transactions of the Royal Society* (*London*) **323**: 411–421.

Vogelmann TC, Martin G. 1993. The functional significance of palisade tissue: penetration of directional vs diffuse light. *Plant, Cell and Environment* **16**: 65–72.

Vogelmann TC, Nishio JN, Smith WK. 1996. Leaves and light capture: Light propagation and gradients of carbon fixation within leaves. *Trends in Plant Science* **1**: 65–70.

Warmbrodt RD, Van Der Woude WJ. 1990. Leaf of *Spinacia oleracea* (spinach): Ultrastructure, and plasmodesmatal distribution and frequency, in relation to sieve-tube loading. *American Journal of Botany* **77**: 1361–1377.

Leaf anatomy enables more equal access to light and CO_2 between chloroplasts

JOHN R. EVANS

Environmental Biology Group, Research School of Biological Sciences, Australian National University, GPO Box 475, Canberra, ACT 2601, Australia
(fax +61 2 6249 4919; e-mail evans@rsbs.anu.edu.au)

Received 5 October 1998; accepted 11 February 1999

SUMMARY

The function of a leaf is photosynthesis, which requires the interception of light and access to atmospheric CO_2 while controlling water loss. This paper examines the influence of leaf anatomy on both light capture and CO_2 diffusion. As photosynthetic metabolism is spread between many chloroplasts, a leaf faces the challenge of matching light capture by a given chloroplast with the metabolic capacity of that chloroplast. Chloroplasts nearest the leaf surface receive the greatest irradiance and therefore absorb more light per unit chlorophyll than chloroplasts in the centre of a leaf. Electron transport and carbon fixation capacities per unit of chlorophyll decline with increasing depth in the leaf, to compensate for the decline in light absorbed per unit chlorophyll. Many key photosynthetic protein complexes in chloroplasts have nuclear encoded genetic information. Consequently, all chloroplasts within a given cell have a similar metabolic complement, which limits the potential gradient of photosynthetic capacity per unit chlorophyll across the leaf. A simple model couples light absorption through the leaf (based on the Beer–Lambert law) with the profile of chlorophyll through a leaf and the gradient in photosynthetic capacity. It is validated by comparison with $^{14}CO_2$ fixation profiles through spinach leaves obtained in various studies. The model can account for published ^{14}C fixation profiles obtained with blue, red and green light of different irradiances and white light applied in different combinations to the adaxial and abaxial surfaces of spinach leaves. The model confirms that spongy mesophyll increases the apparent extinction coefficient of chlorophyll compared to palisade tissue. The palisade tissue nearest the surface which receives light facilitates the penetration of light to a greater depth, while spongy mesophyll promotes scattering to enhance light absorption, thus reducing the gradient in light absorbed per unit chlorophyll through a leaf. CO_2 fixation faces a diffusional limitation, which necessitates Rubisco to be spread evenly across the cell walls exposed to intercellular airspace. Mesophyll cell structure reflects the need to have a large cell surface per unit volume exposed to airspaces. The regular array of columnar cells in palisade tissue, or cell lobing in monocot leaves, results in greater exposed surface per unit tissue volume than spongy mesophyll. The exposed surface area per unit leaf area scales with photosynthetic capacity such that the difference in CO_2 partial pressure between substomatal cavities and the sites of carboxylation within chloroplasts is, on average, independent of photosynthetic capacity of the leaf. However, Rubisco specific activity declines as the Rubisco content per unit leaf area increases due to greater internal diffusional limitations.

Key words: light absorption profiles, internal conductance, chloroplast surface area, Rubisco, acclimation to light, CO_2 diffusion.

INTRODUCTION

The function of a leaf is photosynthesis, the capture of light and conversion of that energy into chemical bonds during CO_2 assimilation. While all C_3 leaves share an identical photosynthetic mechanism, an amazing diversity of leaf structures exist in order to achieve the same end. This presumably reflects the conflicting biochemical requirements of light capture and CO_2 uptake versus mechanical requirements for strength and durability. Different environmental niches emphasize different characteristics, so suites of traits may tend to prevail in a particular environment. This paper focuses on the interplay between leaf structure and both light capture and CO_2 fixation.

Chloroplasts are the fundamental units for photosynthesis. Typically, there are about 10 million chloroplasts in each square centimetre of leaf. The composition of chloroplasts is flexible, being particularly responsive to the light environment which alters the relative abundance of many of the protein complexes (Anderson, 1986). Nuclear gene components are present in all of the major complexes, resulting in coordinated regulation of all chloroplasts in a given cell. However, each cell may contain its own unique complement of chloroplasts. This enables a leaf to fine-tune the deployment of resources, resulting in greater daily photosynthesis per unit of protein than would be achieved in the absence of specialization. Considerable evidence for this specialization has been obtained from studies of chloroplast ultrastructure or biochemical measurements of chloroplasts isolated from paradermal leaf sections (Terashima, 1989).

Light capture by a leaf ought to be a complicated process to describe, compared with the absorption of light by pigments in solution. Firstly, pigments form complexes with protein in regular arrays in thylakoid membranes. Secondly, the density of thylakoid membranes varies through a leaf. Thirdly, due to the difference in refractive index between air and water, light is scattered within the leaf by air–cell-wall interfaces. While complicated optical treatments have been attempted (Fukshansky & Remisowski, 1992; Richter & Fukshansky, 1996), an alternative approach has been to measure light directly within intact leaves by inserting fibre-optic microprobes (Vogelmann & Björn, 1984; Vogelmann *et al.*, 1988, 1989; Cui *et al.*, 1991). This method yields profiles of space irradiance that decline rapidly through the initial palisade tissue.

CO_2 assimilation within the leaf has been directly measured by paradermal sectioning after ^{14}C labelling (Nishio *et al.*, 1993). This painstaking work revealed that CO_2 fixation peaked about one-third of the way into a spinach leaf. Nishio *et al.* concluded that 'the carbon fixation gradient did not follow the leaf internal light gradient' and that 'somehow, the light gradient is disconnected from CO_2 fixation'. Challenged by these statements, I re-analysed their data (Evans, 1995) using a model based on the work of Terashima & Saeki (1983, 1985), and concluded that the observed ^{14}C fixation profiles were consistent with light absorption profiles, based on the distribution of Chl through the leaf, and with photosynthetic capacity, based on the profile of Rubisco through the leaf. Subsequently, new ^{14}C fixation data have become available that enable a more rigorous test of this approach. Thus the first objective of this paper is to examine the profile of light absorption through leaves. Several aspects are considered: absorption of monochromatic light; the influence of mesophyll cell structure on the profile of light absorption in spinach; and absorptance of leaves in general.

To avoid dehydration, leaves use stomata to regulate gaseous diffusion across the epidermis. Traditional gas-exchange techniques routinely calculate the CO_2 partial pressure inside the leaf, which is reduced below ambient partial pressure due to uptake in photosynthesis and the restriction to diffusion imposed by stomata. The model of leaf photosynthesis of Farquhar *et al.* (1980) centres around the biochemical properties of Rubisco. The CO_2 assimilation rate depends principally on Rubisco content and CO_2 partial pressure. While the CO_2 partial pressure inside the leaf calculated from gas exchange has been used for this, in reality the CO_2 partial pressure at the sites of carboxylation is lower (Evans & von Caemmerer, 1996). The extensive review of diffusion within leaves by Parkhurst (1994) emphasized the gas phase. Subsequent work combining gas exchange with other techniques has confirmed that diffusion through the liquid phase is generally more limiting than the gaseous phase within the leaf, with a crucial factor being the surface area of chloroplasts exposed to intercellular airspace per unit leaf area (Evans *et al.*, 1994). The second objective of this paper is to examine the recent evidence concerning CO_2 diffusion within leaves with respect to leaf structure, and its impact on the performance of Rubisco.

THE SPINACH LEAF

The use of a thick leaf for paradermal sectioning enables finer resolution of the profiles of Chl, Rubisco and light than could be achieved with thin leaves. Spinach leaves are ideal as they are around 600 μm thick, allowing 15 layers to be resolved using 40-μm sections. In addition, there exists a wealth of biochemical knowledge on this species. Spinach leaves are bifacial, having several layers of palisade mesophyll cells beneath the adaxial (upper) surface and spongy mesophyll adjacent to the abaxial (lower) surface. A freeze-fractured view of market spinach obtained by scanning electron microscopy is shown in Fig. 1. By using an uncoated specimen it is possible to visualize the chloroplasts appressed to the cell walls. Ice forms on the frozen specimen during evacuation in the microscope and obscures the view of chloroplasts on some cells. At this magnification, the view is less dramatic than for that of a tobacco leaf (Evans & von Caemmerer, 1996) because the spinach leaf is so much thicker. The profiles of Chl and Rubisco obtained by Nishio *et al.* (1993) are aligned against the micrograph. Chl content increases steadily, reaching a maximum in the centre of the leaf near the transition from palisade to spongy mesophyll, before declining slightly towards the abaxial surface.

For the modelling that follows, a polynomial function was fitted to the Chl profile which was used to predict Chl content at any depth. Rubisco content shows a narrower peak, reached midway through the palisade tissue before declining to 40% of the maximum near the abaxial surface. In the right-hand panel of Fig. 1, the profile of ^{14}C fixation following adaxial illumination with white light is shown, along with predicted profiles of light and light absorption.

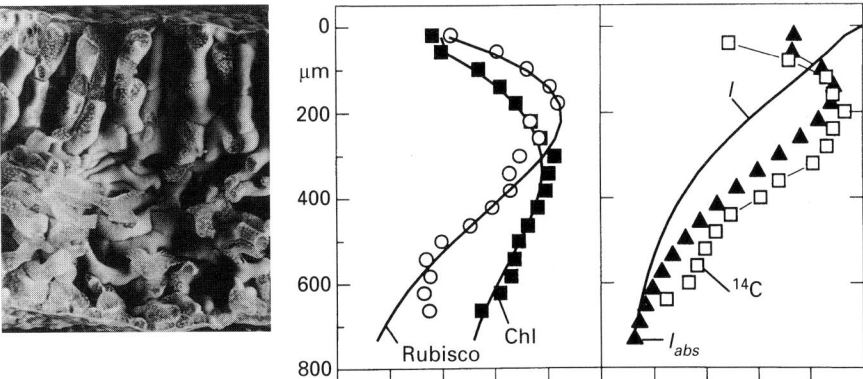

Fig. 1. Spinach leaf transverse fracture (left) and profiles of Chl, Rubisco (centre), ^{14}C fixation, white light, I and absorbed light, I_{abs} (right). Market spinach was snap-frozen in liquid nitrogen and fractured before being placed in the scanning electron microscope and viewed as an uncoated specimen. The Chl (closed square), Rubisco (open circle) and ^{14}C fixation (open square) data are from Nishio *et al.* (1993). The function used to model the cumulative Chl profile with depth (µm) from the adaxial surface was Chl $(\times 10^{-2}$ g m$^{-2}) = 1.057 + 0.0352D + 2.382E - 4D^2 - 3.601E - 7D^3 + 1.655E - 10D^4$, with the scale ranging from 0 to 0.05 g m^{-2}. The Rubisco data are relative units, scaling from 0 to 1. The profile of light absorbed was calculated using the Beer–Lambert law, the Chl profile and an extinction coefficient of 1350 m^2 mol^{-1} (see Evans, 1995 for details). The light absorbed and ^{14}C fixation scale ranges from $0-0.1$, while the white light profile scales from $0-1$.

THE MODEL

A model linking light absorption with CO_2 fixation

Light absorption was calculated assuming the Beer–Lambert law, which states that absorptance is given by the product of the extinction coefficient, pigment concentration and path length. The Beer–Lambert law assumes that the system has parallel, monochromatic light and that the pigments are isotropically dispersed. All three of these assumptions are violated in the leaf, yet despite this, calculations based on the Beer–Lambert law can explain much of the detail that can be resolved from work with paradermal sections of both 'sun' and 'shade' leaves of spinach (Evans, 1995). Predicted light absorption is maximal about 150 µm into the leaf, despite irradiance declining nearly exponentially through the leaf (Fig. 1). This is because pigment content per layer increases more rapidly than the decline in irradiance.

In order to test the modelling of light absorption through a leaf with profiles of ^{14}C fixation, it is necessary to link the two. This is achieved by incorporating the profile of Rubisco through the leaf which is used to define the light-saturated rate of photosynthesis for each layer. Each layer is assumed to operate along a non-rectangular hyperbolic response that can be described by three parameters: the maximum quantum yield; a curvature factor; and a maximum rate (Evans *et al.*, 1993; Ögren & Evans, 1993). It is assumed that all layers have the same maximum quantum yield and curvature factor. Consequently, it is necessary to specify only the profile of maximum rate, which is assumed to be directly proportional to Rubisco content (see Evans, 1995 for more detailed justification). Three profiles of Rubisco content through spinach leaves have been

measured, and show that Rubisco per unit Chl declines linearly with cumulative Chl (Fig. 2). Remarkably, the three data sets show the same relative change, with only minor deviations in the first and last layers sampled. The model therefore calculates the profile of Rubisco with depth from the profile of Chl given in Fig. 1 and the relationship in Fig. 2 (see Rubisco curve, centre panel, Fig. 1). The photosynthetic capacity of a leaf is then simply adjusted by multiplying each layer by a constant.

The linear decline in Rubisco per unit Chl does not quite match the curve predicted for the absorption of white light per unit Chl (Fig. 2). However, it illustrates the changing composition of chloroplasts required to try and match light capture with carbon metabolism. The mismatch is also evident in Fig. 1 where the ^{14}C fixation profile is offset to slightly greater depth relative to the profile of absorbed light. For Rubisco to function efficiently, it requires ready access to intercellular airspace, as discussed below. The large changes in Rubisco per unit Chl mean that chloroplasts near the abaxial surface have to fit in much more Chl per unit of Rubisco than chloroplasts near the adaxial surface. The structural manifestation of this is the change in number of thylakoid membranes per grana stack, which increase from four near the adaxial surface, to seven in the spongy mesophyll (Terashima *et al.*, 1986; Terashima & Evans, 1988). At the same time, thylakoid membranes occupy a greater proportion of the chloroplast volume (Terashima & Evans, 1988, Makovetskii & Manzhulin, 1990).

A test of light capture with monochromatic light

Since this model was published, additional ^{14}C fixation profiles using monochromatic light of various irradiances to the adaxial leaf surface have

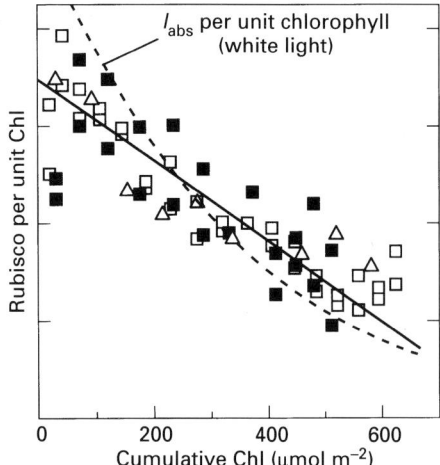

Fig. 2. Function relating Rubisco per unit Chl with cumulative Chl from the adaxial surface. Data are from paradermally sectioned spinach leaves obtained by Terashima & Inoue, 1985 (open triangles) and Nishio *et al.*, 1993 (open squares, sun leaves; closed squares, shade leaves). The regression function is Rubisco = $(34.8 - 0.461 \times \Sigma\text{Chl})$, where ΣChl is given by the Chl versus depth function given in Fig. 1. It is converted to photosynthetic capacity by multiplying by the Chl content of the layer and dividing by a scaling factor (5.8). The profile of white light absorption per unit Chl (I_{abs}) is shown for comparison, calculated as in Fig. 1 (broken curve).

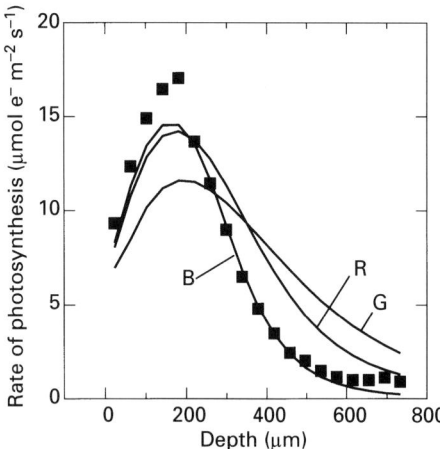

Fig. 3. Profile of photosynthesis with depth (closed square) for a spinach leaf given 500 μmol blue quanta m^{-2} s^{-1} to the adaxial surface. ^{14}C fixation data are from Sun *et al.* (1998) and the curves are calculated for blue (B), red (R) and green (G) light using the following values for the extinction coefficient: 3636, 2106, 900 m^2 (mol Chl)$^{-1}$, respectively.

become available (Sun, 1996; Sun *et al.*, 1998). They provide a robust test of the model, because the absolute response to a variety of lights given to the adaxial surface can be examined for a given leaf. Seven profiles were obtained using sun-type spinach leaves: 500 μmol quanta m^{-2} s^{-1} of blue, red or green light; 200 μmol quanta m^{-2} s^{-1} of red or green light; and 50 μmol quanta m^{-2} s^{-1} of blue or green light. Data were reported as c.p.m. per section, and the mean values used were obtained from four to ten replicates, omitting the error bars which were approx. 30% of the mean. The model has two

variables: a scaling factor to convert Rubisco content to maximum photosynthetic rate, and an extinction coefficient. Initially, the scaling factor was varied until the photosynthetic rate of the leaf matched the observed irradiance response curve. The photosynthetic capacity of the sun-type spinach leaf was 207 μmol e$^-$ m^{-2} s^{-1}, which is comparable to the rate of 240 for spinach measured under slightly higher CO_2 conditions (Evans & Terashima, 1988). Having defined the value for the scaling factor, an extinction coefficient was obtained for each colour that best explained the ^{14}C profile measured under 500 μmol quanta m^{-2} s^{-1}. Different wavelengths penetrate to different depths in a leaf because both absorption and scattering are wavelength-dependent. Blue light is the most strongly absorbed and green light the most weakly absorbed, with red light intermediate, consistent with the absorption spectrum of chloroplasts.

For a given irradiance, the profile of ^{14}C fixation depends on the extinction coefficient (Fig. 3). Fixation peaks in blue light just above 200 μm and at slightly greater depth in red or green light. More importantly, the fraction of photosynthesis that occurs in the first 300 μm declines from 81% for blue to 67% for red and 57% for green light. The curve predicted for blue light matches the ^{14}C fixation data very closely from 200 μm onwards, but underestimates the peak rate around 180 μm. The peak rate could be better matched by increasing the photosynthetic capacity of those layers in the leaf. However, the rate predicted for red and green light would then be poorer. It should be remembered that the error associated with each mean data point was about 30%, so the predicted profile for blue light falls well within the envelope of uncertainty.

The predicted profiles for all three colours, each at different irradiances, are shown in Fig. 4 along with the light profiles. The model predictions match the ^{14}C profiles very well, accounting for changes due to colour as well as irradiance. Due to the strong absorption of blue light, only 10% penetrates deeper than 300 μm, yet it results in 19% of the photosynthesis. This is because light captured by the first few layers exceeds the capacity of the chloroplasts to convert it into carbohydrates, and is lost as heat. By contrast, 23% of red light is absorbed in the deeper layers, contributing 33% of photosynthesis, while for green light 43% of the absorbed light occurs in the deeper layers, accounting for 43% of photosynthesis. The match between green light absorption and CO_2 fixation occurs because the profile of green light absorption is similar to the profile of Rubisco.

The only noticable deviation of the model from the ^{14}C fixation data occurs in the first few layers at intermediate irradiance, where the predicted photosynthetic rate exceeds the observed rate. This has two possible causes. Firstly, because chloroplasts only line the periphery of a cell (e.g. Evans *et al.*,

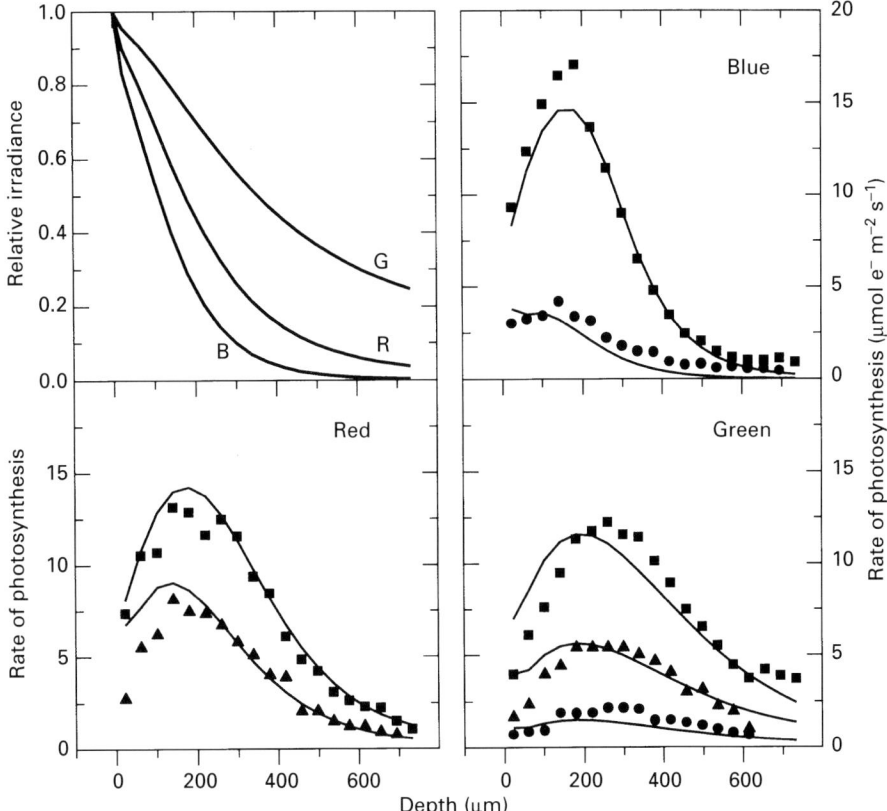

Fig. 4. Profile of relative irradiance and photosynthesis with depth for a spinach leaf given various adaxial light treatments. Irradiance profiles were calculated for blue, red and green light using the following values for the extinction coefficient: 3636, 2106, 900 m² (mol Chl)⁻¹, respectively, and the profile of Chl shown in Fig. 1. Irradiances were: squares, 500; triangles, 200; circles, 50 μmol quanta m⁻² s⁻¹; data are from Sun *et al.* (1998), with curves calculated from the model.

1994), collimated light entering the palisade tissue will generally travel some distance before encountering a chloroplast. From paradermal sections, it can be seen that chloroplasts occupy only about 20% of the plane. Consequently, the model is likely to overestimate the amount of monochromatic light absorbed in this region. While the model could be modified to incorporate a variable to deal with this, it would detract from the present simplicity, and in white light there is little evidence of any consistent error. Alternatively, the deviation could be used to estimate the fraction of light that bypasses chloroplasts in the initial layers. The addition of another variable improves the fit of the model for the first 200 μm under monochromatic light but not with white light. Secondly, measurement errors for the first section are greatest as it contains a variable proportion of epidermal tissue. This leads to the greatest uncertainty in the Chl and Rubisco contents where they have the largest impact.

DISCUSSION

Effect of mesophyll structure on light capture

The values for the extinction coefficients cannot be directly equated to those obtained for pigments in solution, because the pigments are not distributed uniformly through the tissue, and scattering of light increases with depth leading to path lengthening. The consequence of path lengthening is evident in the work of Terashima & Saeki (1983), who showed that the apparent extinction coefficient for red and green light was greater in spongy tissue.

Providing light to the abaxial surface of the leaf during ¹⁴C labelling provides two useful tests for the model. Firstly, it uncouples the light absorption profile from the Rubisco profile; and secondly, it shows whether mesophyll structure alters the apparent extinction coefficient. In a separate experiment from that described above, Sun (1996) carried out ¹⁴C labelling of sun leaves of spinach that were given white light to either or both surfaces. In modelling this data set, the same leaf parameters were used apart from one change: as the layers were aligned slightly differently, it became evident that the photosynthetic rate for the first adaxial layer was overestimated, and consequently the Rubisco content for this layer was halved.

As white light was used, an extinction coefficient was fitted to the data with adaxial light of 800 μmol quanta m⁻² s⁻¹. The value 1350 m² mol⁻¹ is slightly less than that derived in the previous paper for white light (1500 m² mol⁻¹; Evans, 1995). A poor fit to data

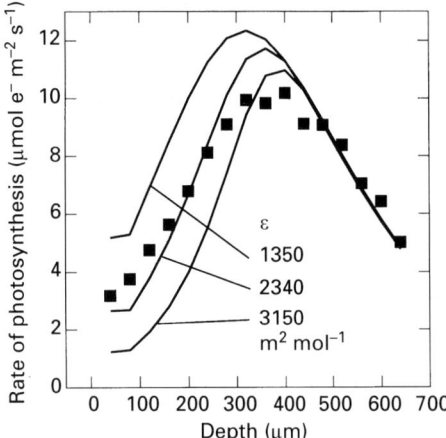

Fig. 5. Profile of photosynthesis with depth for a spinach leaf given white light to the abaxial surface. Data are from Sun (1996) with curves calculated from the model using different extinction coefficients. Abaxial irradiance was 800 µmol quanta m^{-2} s^{-1}.

obtained with abaxial light was observed using 1350 m^2 mol^{-1} (Fig. 5). The first six layers in from the abaxial surface fitted well because they reflect light-saturated capacity, i.e. Rubisco content. At deeper layers, the model overestimated the photosynthetic rate because too much light was reaching these layers. Good agreement between the model and ^{14}C fixation data could be restored if a larger extinction coefficient of 2340 m^2 mol^{-1} was used. Choosing an even higher extinction coefficient (3150 m^2 mol^{-1}) resulted in too much light absorption in the first eight layers, leaving too little for photosynthesis in the adaxial half of the leaf.

The requirement by the model for a greater extinction coefficient for abaxial light is consistent with expectations. Spongy tissue is known to scatter light more than the regular array of palisade cells. By increasing the path length of light, absorption per unit Chl increases in spongy tissue, which is equivalent to a greater apparent extinction coefficient. Terashima & Saeki (1983) observed a 47% increase in extinction coefficient for 680 nm light, and an 88% increase for 550 nm light, for spongy versus palisade tissue in a *Camellia* leaf. Infiltrating the leaf with oil, which had a refractive index comparable to that of the cells, reduced the apparent extinction coefficient for spongy and palisade tissue by 39 and 26%, respectively, for 680 nm light. Bornman *et al.* (1991) also showed that light does not penetrate as far through spongy tissue compared to palisade tissue in a *Medicago sativa* leaf. Spongy tissue therefore enhances the light capture per unit of Chl by scattering light. By contrast, palisade tissue minimizes light scattering and allows much of the light to bypass chloroplasts by guiding it down the centre of palisade cells. This enables light to penetrate further into the leaf, thereby spreading light capture more evenly between chloroplasts. This was shown by Vogelmann & Martin (1993),

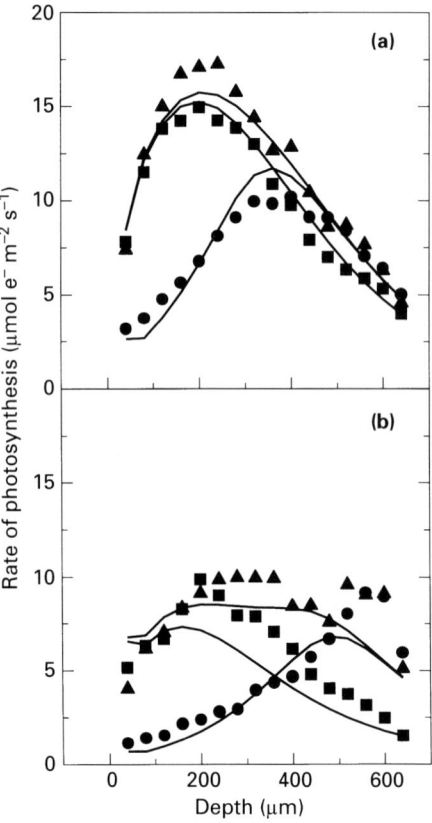

Fig. 6. Profiles of photosynthesis with depth for spinach leaves given white light to one or both surfaces. Data are from Sun (1996) with curves calculated from the model using an extinction coefficient of 1350 and 2340 m^2 (mol Chl)$^{-1}$ for adaxial and abaxial light, respectively. Squares, adaxial; circles, abaxial; triangles, both. (a) 800 µmol quanta m^{-2} s^{-1}; (b) 200 µmol quanta m^{-2} s^{-1}.

who compared *Thermopsis montana*, a legume with columnar palisade tissue, with that of the monocot *Smilacina stellata*, without palisade. Collimated green light was able to penetrate considerably further through the leaf with palisade mesophyll compared to the monocot leaf, whereas there was little difference when diffuse light was used. It is the combination of better light penetration with a greater cell surface area per unit of mesophyll volume that makes palisade tissue a more efficient structure in terms of photosynthesis than spongy mesophyll immediately adjacent to the adaxial surface. Conversely, spongy tissue adjacent to the abaxial surface increases the efficiency of capturing the small amount of remaining light for a given investment in pigment–protein complexes.

The ^{14}C fixation profiles with adaxial or abaxial white light also included a case where light was given to both surfaces (Fig. 6). The model predicts that receiving light simultaneously on both surfaces would increase photosynthesis only in the central part of the leaf. The initial profiles from either surface change little, either because such a small amount of additional light reaches the opposite side (when 200 µmol quanta m^{-2} s^{-1} are given), or because these layers are already light-saturated (when 800 μmol quanta m^{-2} s^{-1} are given). The model predicts lower photosynthetic rates under 200 μmol quanta m^{-2} s^{-1} at some depths than the observed ^{14}C fixation data. However, this is probably a reflection on the uncertainty of the ^{14}C profiles. Despite relatively small error bars, only three to four replicate leaves were used to obtain the low light data. At 600 μm depth, 200 μmol quanta m^{-2} s^{-1} to the abaxial surface resulted in 40% more ^{14}C fixation than 800 μmol quanta m^{-2} s^{-1}, which is highly unlikely.

Effect of specific leaf area on light absorption

The model describes light capture based on the Beer–Lambert law. It is sensitive to wavelength and mesophyll structure. Unfortunately, the only data available for testing the model are from spinach leaves, which have the classical dicotyledonous bifacial anatomy. In the work of Nishio et al. (1993), ^{14}C fixation profiles were obtained with spinach leaves that had been grown under sun conditions (800 μmol quanta m^{-2} s^{-1}) or shade conditions (200 μmol quanta m^{-2} s^{-1}, with an R:FR ratio of 0.25). This resulted in sun-type leaves having 15% more Chl per unit leaf area and being one third thicker than shade-type spinach leaves. The increased leaf thickness was mainly due to more palisade tissue, which had five rather than three cell layers. The palisade tissue of sun leaves also had narrower cells than that of shade leaves (26 versus 40 μm; Cui et al., 1991), resulting in a 40% increase in palisade mesophyll surface area per unit cell volume. Despite these differences, the model could describe ^{14}C fixation profiles in white light supplied to the adaxial surface, using the same apparent extinction coefficient of 1500 m^2 mol^{-1} for both leaf types (Evans, 1995). This suggests that the distribution of Chl through the leaf tissue is of the greatest importance in defining light capture.

While spongy mesophyll tissue does alter the apparent extinction coefficient, it is not yet possible to assess the impact of other mesophyll structures, apart from their impact on absorptance of the intact leaf. Absorptance is the fraction of light falling onto a leaf that is not reflected or transmitted. Absorptance is strongly related to Chl content by a hyperbolic function. Does the underlying mesophyll structure contribute directly to variation in leaf absorptance? This was addressed by examining leaves obtained from a diverse range of species (Evans, 1998a). The deviation of leaf absorptance from that predicted from Chl content was examined as a function of specific leaf area, which varied for a given species as a result of growth light environment. Changing specific leaf area had no effect on absorptance after accounting for any change in Chl content.

How much change in absorptance would we expect if a leaf had only palisade tissue? If we examine a spinach leaf with an apparent extinction coefficient of 1350 m^2 mol^{-1}, 15% of light that enters the leaf reaches the lower surface. Adding a different extinction coefficient for spongy tissue of 2340 reduces the light reaching the lower surface to 7%. The 8% difference ought to be detectable if a bifacial leaf were compared to a leaf which had only palisade tissue. Undoubtedly this analysis is too crude because, in reality, light becomes progressively more scattered, even by palisade tissue which would reduce the immediate impact of the transition to spongy tissue. No indication of an abrupt increase in light absorption was evident in the ^{14}C fixation profiles at the palisade–spongy boundary in spinach leaves (Fig. 4). However, an increase was evident in direct optical measurements of *Camellia* leaves (Terashima & Saeki, 1983). The fact that leaf absorptance can be well described simply by the Chl content, regardless of specific leaf area, suggests that internal leaf structure may play a role in altering the profile of light capture through a leaf, but it does not alter the absolute amount captured by a leaf.

Matching light capture to photosynthetic capacity

The patterns of ^{14}C fixation through the leaf are the consequence of interactions between light capture and photosynthetic capacity. Ideally, the profile of light capture matches that of photosynthetic capacity, as this results in the best use of protein invested in photosynthesis: it corresponds to each chloroplast operating at the point of its light response function, where an increase in incident light results in the same increase in photosynthesis for any chloroplast. This cannot be achieved exactly, for several reasons. Firstly, all chloroplasts in a given cell share a common nucleus which dictates the composition of all the chloroplasts in that cell. A considerable light gradient could exist along the length of a palisade cell which therefore could not be matched by different chloroplast properties along the cell. Secondly, light of different wavelengths will be absorbed in different profiles and it is only possible to match one profile.

Giving different monochromatic lights to a leaf leads to a dramatic mismatch between light absorption and photosynthetic capacity. This is shown in Fig. 7, where the rate of electron transport per absorbed quanta is shown as a function of depth for red, blue and green light of 500 μmol quanta m^{-2} s^{-1}. The assumption of a constant extinction coefficient which overestimates monochromatic light absorption in the first 200 μm results in uncertainty for the initial layers, but including this would still result in differences between profiles remaining for the different wavelengths. Blue light is absorbed most strongly, leading to light saturation near the adaxial surface which lowers the quantum yield. Since little blue light penetrates to the abaxial surface, these

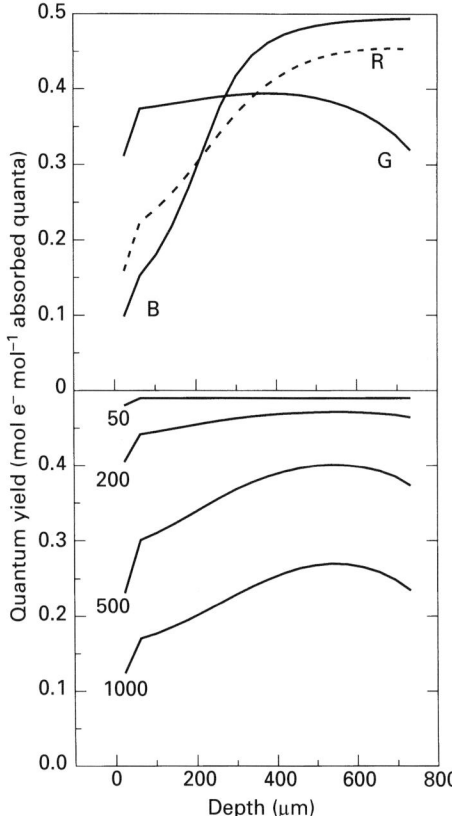

Fig. 7. Profiles of quantum yield of electron transport with depth through a sun spinach leaf calculated from the model with adaxial illumination. Upper panel calculated for blue, red and green light of 500 µmol quanta m^{-2} s^{-1}; lower panel calculated for white light of different irradiances as shown.

layers have maximal quantum yield. On the other hand, the absorption profile of green light nearly matches the Rubisco profile, which results in the quantum yield being stable through the leaf. Red light is intermediate between blue and green. It would be intriguing to grow leaves under different monochromatic lights and observe chloroplast ultrastructure to see if there were more dramatic gradients in thylakoid number per grana in leaves from blue versus green light.

Quantum yield profiles in white light of different irradiances are shown in the lower panel of Fig. 7. Quantum yield declines as irradiance increases, and the adaxial layers are relatively more light saturated than the abaxial layers. This should mean that estimates of photosynthetic electron transport based on Chl fluorescence, elicited by a measuring beam that does not penetrate very deeply into a leaf, underestimates the whole leaf rate (Evans *et al.*, 1993). The kink in all the profiles between the first and second layers reflects the uncertainty in the model associated with the Rubisco content and light absorption in the first layer. The fact that the profiles are not flat indicates that photosynthetic capacity and light absorption do not covary exactly. The resulting light–response curve of leaf photosynthesis

thus has a curvature factor of 0.69 compared to that defined for the chloroplast of 0.86, and the integral of the curve achieves 95.6% of that of an ideal leaf.

It should be remembered that the underlying photosynthetic biochemistry is flexible and able to acclimate to changing circumstances. When eucalypt leaves were restrained to a particular orientation and subsequently re-oriented, light–response curves provided evidence for a re-alignment of the biochemistry (Ögren & Evans, 1993). The change took place over about 7 days for both bifacial and isolateral leaf types.

Restrictions to CO$_2$ diffusion

To survive on land with a limited water supply requires that plants regulate the loss of water in exchange for CO$_2$ uptake. This is achieved by restricting gaseous diffusion into and out of the leaf through stomata. Stomatal morphology and density vary considerably between species, and stomatal responses to light, humidity and CO$_2$ have been widely studied by conventional gas-exchange techniques. The CO$_2$ partial pressure inside the leaf has routinely been calculated in the process. Leaf photosynthesis can be predicted if Rubisco activity and CO$_2$ partial pressure at the sites of carboxylation are known (Farquhar & von Caemmerer, 1982). It is frequently assumed that the CO$_2$ partial pressure at the sites of carboxylation is sufficiently close to that calculated in the intercellular airspaces for the two to be equated. However, several techniques are now available that have enabled the CO$_2$ partial pressure at the sites of carboxylation to be calculated (Evans & von Caemmerer, 1996). They reveal that the drawdown in CO$_2$ partial pressure inside actively photosynthesizing leaves is nearly as great as the drawdown through stomata. For ease of argument, standard atmospheric pressure is assumed to enable the pressure unit to be omitted from internal conductance. Otherwise, because dissolution of CO$_2$ into liquid is pressure-dependent, the units of atmospheric pressure are usually needed (see Harley *et al.*, 1992).

Assembling the growing number of published measurements is quite revealing. It is well known that photosynthetic capacity is strongly correlated with stomatal conductance between species (Körner *et al.*, 1979; Wong *et al.*, 1979; Yoshie, 1986). A similar relationship exists between photosynthetic capacity and internal conductance (from substomatal cavities to the sites of carboxylation; Fig. 8). The data have been separated into two groups: mesophytic leaves which are short-lived, herbaceous or deciduous species; and sclerophytic leaves which either are evergreen or have a low specific leaf area. As the fluorescence method becomes unreliable when internal conductances exceed 0.3 mol m^{-2} s^{-1}, only data collected using the isotopic method were used

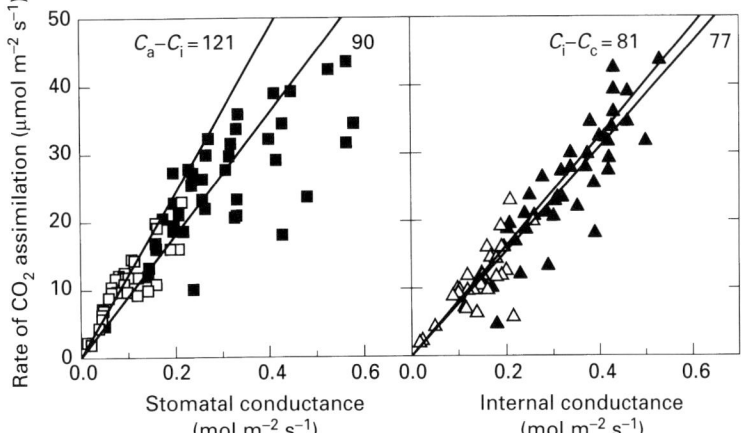

Fig. 8. Relationships between stomatal or internal conductance and the rate of CO_2 assimilation measured under high irradiance and ambient CO_2. Data have been placed in two leaf groups: solid symbols, mesophytic; open symbols, sclerophytic. Data are from von Caemmerer & Evans, 1991; Lloyd *et al.*, 1992; Loreto *et al.*, 1992; Evans *et al.*, 1994; Epron *et al.*, 1995; Syvertsen *et al.*, 1995; Evans & Vellen, 1996; Roupsard *et al.*, 1996; Lauteri *et al.*, 1997. Solid lines from the origin indicate mean drawdown in CO_2 mole fraction (μmol mol^{-1}) across stomata (C_a–C_i), or from substomatal cavities to the sites of carboxylation (C_i–C_c).

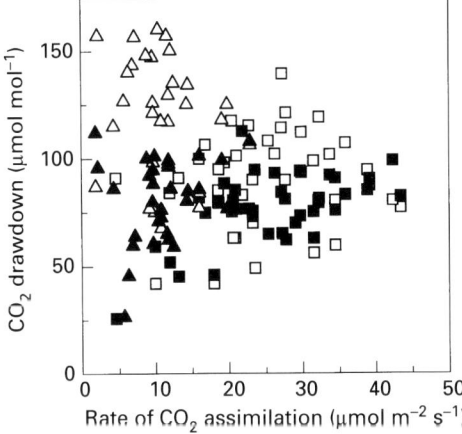

Fig. 9. CO_2 drawdown versus the rate of CO_2 assimilation measured under high irradiance and ambient CO_2. Solid symbols, C_i–C_c; open symbols, C_a–C_i; triangles, sclerophytic leaves; squares, mesophytic leaves.

for the mesophytic species. For the sclerophytic species, data collected by either method were used, as direct comparisons between the two methods agreed closely when internal conductance was less than 0.3 mol m^{-2} s^{-1} (Loreto *et al.*, 1992). Overall, the two groups share a common relationship between photosynthetic capacity and internal conductance.

Dividing CO_2 assimilation rate by conductance yields the drawdown in CO_2 mole fraction from ambient air to intercellular airspaces, C_a–C_i, or from intercellular airspaces to the sites of carboxylation, C_i–C_c, depending on whether stomatal or internal conductance is used (Fig. 9). While this calculation reveals greater scatter in the data, it is clear that no trend exists between C_i–C_c and rate of CO_2 assimilation measured under high irradiance. The mean values for C_i–C_c were 77 ± 3 and 81 ± 4 μmol mol^{-1} for the mesophytic and sclerophytic groups, respectively. While leaves of woody species tend to

have lower internal conductances relative to herbaceous species, because they also have lower photosynthetic capacities, the internal drawdown is similar for both groups. On average, the drawdown across stomata, C_a–C_i, was greater for sclerophyllous leaves compared to the mesophytic leaf group (121 ± 5 c.f. 90 ± 3 μmol mol^{-1}). Thus a slightly lower partial pressure at the site of carboxylation in sclerophyllous leaves is attributable to relatively greater limitations imposed by stomata rather than restrictions to diffusion within the leaf.

Relationship between internal conductance and exposed chloroplast surface area

Internal conductance consists of two components: gaseous diffusion through the intercellular airspaces, and the liquid diffusion pathway from the cell wall to the sites of carboxylation. The techniques available for calculating the internal drawdown, C_i–C_c, cannot separate these two components without the additional effort of making measurements in two different gas compositions which alter the diffusivity of CO_2. For example, comparing air with helox (80% helium, 20% oxygen) provides extra information that can be used to assess the magnitude of the drawdown in the intercellular airspaces, because the diffusivity of CO_2 is 2.3 times greater in helox than in air. Parkhurst & Mott (1990) found that helox increased photosynthetic rates of amphistomatous and hypostomatous leaves by 2 and 12%, respectively. Genty *et al.* (1998) combined helox with Chl fluorescence imaging to separate the gaseous and liquid diffusion pathways. They measured both amphistomatous *Populus koreana* × *trichocarpa* cv. Peace and heterobaric hypostomatous *Rosa rubiginosa* leaves. Switching from air to helox made no measurable difference to the CO_2-response curves,

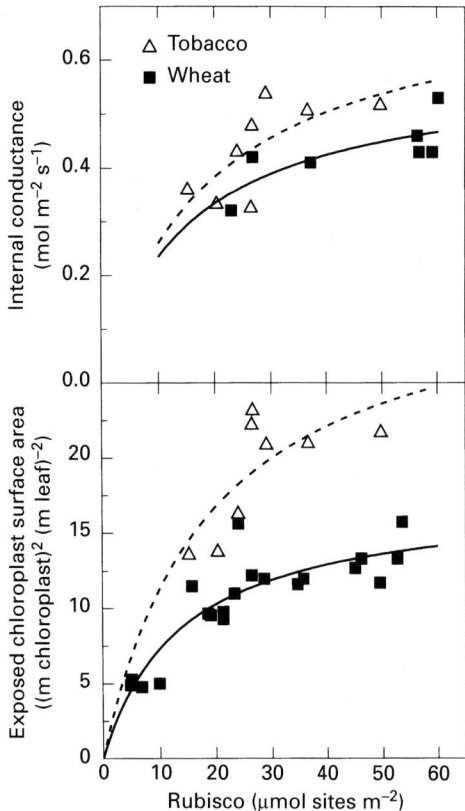

Fig. 10. Relationships between Rubisco content per unit leaf area and internal conductance or chloroplast surface area exposed to intercellular airspace per unit leaf area. Data for wheat are from Evans (1983b) and von Caemmerer & Evans (1991), and for wild-type tobacco from Evans *et al.* (1994). Internal conductance $= (0.736R)/(18.4+R)$ for tobacco and $(0.584R)/(15.0+R)$ for wheat; exposed chloroplast surface area $= (32.5R)/(18.8+R)$ for tobacco and $(17.4R)/(13.8+R)$ for wheat.

suggesting that for these two contrasting leaf types the limitation imposed by gaseous diffusion inside leaves was substantially less than that in the liquid phase. Sun (1996), working with spinach leaves, also found that the pattern of ^{14}C fixation was independent of which leaf surface the label was supplied to, consistent with a negligible gaseous diffusion limitation inside these leaves.

Diffusion in the liquid phase is restricted by the permeability of membranes and the thickness of cell wall, cytoplasm and chloroplasts. These restrictions are likely to be shared by all chloroplasts in a given leaf, so that the liquid phase conductance will be proportional to the surface area of chloroplasts exposed to intercellular airspace per unit leaf area. The maximum value that can be reached for a given leaf is set by the mesophyll anatomy, which defines the surface area of mesophyll cells exposed to intercellular airspace per unit leaf area. A causal correlation between internal conductance and exposed chloroplast surface area was put forward by von Caemmerer & Evans (1991). To test this, wild-type and transgenic tobacco having reduced amounts

of Rubisco were examined in order to separate the effects of photosynthetic capacity from chloroplast surface area. The results clearly showed that internal conductance was related to exposed chloroplast surface area (Evans *et al.*, 1994). Remarkably, the relationship between internal conductance and exposed chloroplast surface area found for tobacco also fits the data available for three other species: wheat, peach and citrus (Syvertsen *et al.*, 1995; Evans & Vellen, 1996; Evans, 1998b; Evans & Loreto, 1999). The slope yields a value of 24 mmol CO_2 (m chloroplast)$^{-2}$ s^{-1} for the liquid phase conductance per unit of chloroplast surface area exposed to intercellular airspace.

Relationship between internal conductance and Rubisco

To link CO_2 fixation to diffusion restrictions, we need to know how Rubisco content is related to chloroplast surface area. Data are available for both wheat and tobacco, although only in tobacco were all the measurements made on the same material. Both internal conductance and exposed chloroplast surface area show hyperbolic relationships with Rubisco content (Fig. 10). For a given Rubisco content, tobacco has greater internal conductance and more chloroplast surface area than wheat. The concentration of Rubisco in chloroplasts is also lower in tobacco (0.9–1.2 mM sites; Evans *et al.*, 1994) compared to wheat (1.5–2.2 mM sites; Evans, 1983b). Fig. 10 implies that internal conductance per unit exposed chloroplast surface area is greater in wheat than tobacco. However, the wheat data are from two different experiments. Internal conductance per unit of exposed chloroplast surface area in young, fully expanded wheat leaves (Evans & Vellen, 1996) was similar to that of tobacco (Evans *et al.*, 1994; see above). The hyperbolic relationship with chloroplast surface area indicates that increasing Rubisco content per unit leaf area is initially achieved by chloroplasts progressively covering more of the available exposed cell surface, but eventually when all the exposed surface is occupied, either the chloroplasts have to become thicker and/or the Rubisco concentration must increase. Because internal conductance cannot increase in direct proportion to Rubisco content, leaves with greater Rubisco content face greater internal diffusion resistances. This can be seen for wheat leaves where Rubisco activity calculated from the slope of CO_2 response curves near the CO_2 compensation point is related to the extractable Rubisco content (Fig. 11). Plants were grown either in summer or winter, but the curvature is apparent only when very high Rubisco contents were achieved. Two model lines are shown. The straight broken line represents the relationship expected for Rubisco with a catalytic rate constant of 5 mol CO_2 (mol sites)$^{-1}$ s^{-1}. The solid

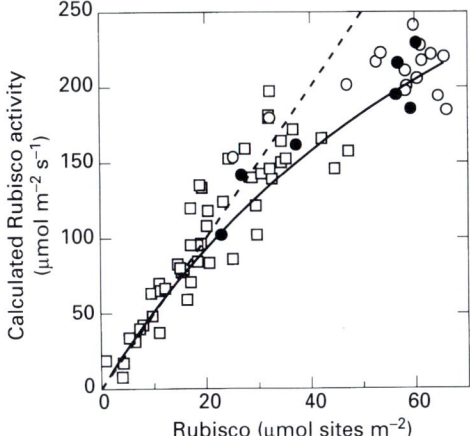

Fig. 11. Relationship between *in vivo* Rubisco activity and Rubisco content for wheat leaves. Data are from Evans (1983a) (open symbols) and von Caemmerer & Evans (1991) (solid symbols). Open square, winter; open and closed circles, summer. Broken line calculated using a catalytic turnover number of 5 mol CO_2 (mol sites)$^{-1}$ s^{-1} and kinetic constants from von Caemmerer *et al.* (1994). Curved line also includes the internal conductance shown in Fig. 10 (see Poorter & Evans, 1998).

curve also includes the hyperbolic relationship with internal conductance from Fig. 10, which accords with the observed data. Such curvilinear dependencies have been observed in *Phaseolus vulgaris* (von Caemmerer & Farquhar, 1981), sunflower (Eichelmann & Laisk, 1990), *Ipomoea* (Hikosaka, 1996), *Chenopodium album* and *Alocasia* (Hikosaka & Terashima, 1996) and between a range of woody and herbaceous species (Poorter & Evans, 1998).

Clearly, the best performance of Rubisco is achieved with a thin chloroplast that maximizes diffusive entry of CO_2. This needs to be offset by the structural cost of providing mesophyll cell surface that may become progressively larger. Mesophyll structure contributes to maximizing cell surface exposed to airspace, on average because palisade tissue has 2.6 times more surface per unit tissue volume than spongy tissue (Turrell, 1936). This enables more Rubisco to be deployed where light absorption is greatest.

Detailed stereological quantification of the volume fractions occupied by the vacuole, chloroplast and other organelles has been carried out on mesophyll cells from *Hordeum* (Winter *et al.*, 1993), *Spinacia* (Winter *et al.*, 1994) and *Solanum* (Leidreiter *et al.*, 1995). The fraction of an average mesophyll cell occupied by chloroplasts varied little, ranging between 16.4 and 19% for these species. A completely different approach has been taken by McCain *et al.* (1988) using NMR imaging of sun and shade leaves of *Acer platanoides*. They found that chloroplasts of sun leaves contained 17% of leaf water, which compares favourably with the stereological method for herbaceous leaves. For shade leaves, the proportion increased to 47% (McCain *et al.*, 1988).

Subsequently, better resolution enabled the profile of chloroplasts and total water versus depth to be obtained (McCain *et al.*, 1993). This suggests that the spatial picture within the leaf is more complex than that revealed by the stereological average. Data are needed on Rubisco profiles in relation to Chl for species other than spinach, in order to test the generality of predictions based on the model spinach leaf in terms of both light capture and CO_2 diffusion limitations.

ACKNOWLEDGEMENTS

My thanks to Eric Garnier for the invitation to talk and the opportunity to experience Montpellier, and to Tom Vogelmann for providing me with the data of Sun, without which I could not have prepared this paper.

REFERENCES

Anderson JM. 1986. Photoregulation of the composition, function and structure of thylakoid membranes. *Annual Review of Plant Physiology* **37**: 93–136

Bornman JF, Vogelmann TC, Martin G. 1991. Measurement of chlorophyll fluorescence within leaves using a fibre-optic microprobe. *Plant, Cell and Environment* **14**: 719–725

Cui M, Vogelmann TC, Smith WK. 1991. Chlorophyll and light gradients in sun and shade leaves of *Spinacia oleracea*. *Plant, Cell and Environment* **14**: 493–500

Eichelmann H, Laisk A. 1990. Content of ribulose-1,5-bisphosphate carboxylase and kinetic characteristics of photosynthesis of leaves. *Soviet Journal of Plant Physiology* **37**: 798–809.

Epron D, Godard D, Cornic G, Genty B. 1995. Limitation of net CO_2 assimilation rate by internal resistances to CO_2 transfer in the leaves of two tree species (*Fagus sylvatica* L. and *Castanea sativa* Mill.) *Plant, Cell and Environment* **18**: 43–51.

Evans JR. 1983a. Nitrogen and photosynthesis in the flag leaf of wheat (*Triticum aestivum* L.). *Plant Physiology* **72**: 297–302.

Evans JR. 1983b. Photosynthesis and nitrogen partitioning in leaves of *Triticum aestivum* and related species. PhD thesis, Australian National University, Canberra, Australia.

Evans JR. 1995. Carbon fixation profiles do reflect light absorption profiles in leaves. *Australian Journal of Plant Physiology* **22**: 865–873.

Evans JR. 1998a. Photosynthetic characteristics of fast and slow growing species. In: Lambers H, Poorter H, van Vuuren MMI, eds. *Inherent variation in plant growth. Physiological mechanisms and ecological consequences.* Leiden, Netherlands: Backhuys Publishers, 101–119.

Evans JR. 1998b. Carbon dioxide diffusion inside C₃ leaves. In *Proceeding of the XIth International Congress on Photosynthesis.* Dordrecht, Netherlands: Kluwer, in press.

Evans JR, von Caemmerer S. 1996. CO_2 diffusion inside leaves. *Plant Physiology* **110**: 339–346.

Evans JR, Loreto F. 1999. Acquisition and diffusion of CO_2 in higher plant leaves. In: Leegood RC, Sharkey TD, von Caemmerer S, eds. *Photosynthesis: physiology and metabolism.* Dordrecht, Netherlands: Kluwer, in press.

Evans JR, Terashima I. 1988. Photosynthetic characteristics of spinach leaves grown with different nitrogen treatments. *Plant Cell Physiology* **29**: 157–165

Evans JR, Vellen L. 1996. Wheat cultivars differ in transpiration efficiency and CO_2 diffusion inside their leaves. In: Ishii R, Horie T, eds. *Crop research in Asia: achievements and perspective.* Tokyo, Japan: Asian Crop Science Association, 326–329.

Evans JR, von Caemmerer S, Setchell BA, Hudson GS. 1994. The relationship between CO_2 transfer conductance and leaf anatomy in transgenic tobacco with a reduced content of Rubisco. *Australian Journal of Plant Physiology* **21**: 475–495.

Evans JR, Jakobsen I, Ögren E. 1993. Photosynthetic light response curves. 2. Gradients of light absorption and photosynthetic capacity. *Planta* **189**: 191–200.

Farquhar GD, von Caemmerer S, Berry JA. 1980. A biochemical model of photosynthetic CO_2 assimilation in leaves of C_3 species. *Planta* **149**: 78–80.

Farquhar GD, von Caemmerer S. 1982. Modelling of photosynthetic responses to environmental conditions. In: Lange OL, Nobel PS, Osmond CB, Ziegler H, eds. *Physiological plant ecology II. Water relations and carbon assimilation. Encyclopaedia of plant physiology* (New Series), Vol. 12B. Berlin, Germany: Springer Verlag, 549–587.

Fukshansky L, Remisowsky AMv. 1992. A theoretical study of the light microenvironment in a leaf in relation to photosynthesis. *Plant Science* **86**: 167–182.

Genty B, Meyer S, Piel C, Badeck F, Liozon R. 1998. CO_2 diffusion in leaves. In *Proceedings of the XIth International Congress on Photosynthesis*. Dordrecht, Netherlands: Kluwer, in press.

Harley PC, Loreto F, Di Marco G, Sharkey TD. 1992. Theoretical considerations when estimating the mesophyll conductance to CO_2 flux by analysis of the response of photosynthesis to CO_2. *Plant Physiology* **98**: 1429–1436.

Hikosaka K. 1996. Effects of leaf age, nitrogen nutrition and photon flux density on the organisation of the photosynthetic apparatus in leaves of a vine (*Ipomoea tricolor* Cav.) grown horizontally to avoid mutual shading of leaves. *Planta* **198**: 144–150.

Hikosaka K, Terashima I. 1996. Nitrogen partitioning among photosynthetic compounds and its consequence in sun and shade plants. *Functional Ecology* **10**: 335–343.

Körner C, Scheel JA, Bauer H. 1979. Maximum leaf diffusive conductance in vascular plants. *Photosynthetica* **13**: 45–82.

Lauteri M, Scartazza A, Guido MC, Brugnoli E. 1997. Genetic variation in photosynthetic capacity, carbon isotope discrimination and mesophyll conductance in provenances of *Castanea sativa* adapted to different environments. *Functional Ecology* **11**: 675–683.

Leidreiter K, Kruse A, Heineke D, Robinson DG, Heldt HW. 1995. Subcellular volumes and metabolite concentrations in potato (*Solanum tuberosum* cv. Désirée) leaves. *Botanica Acta* **108**: 439–444.

Lloyd J, Syvertsen JP, Kriedemann PE, Farquhar GD. 1992. Low conductances for CO_2 diffusion from stomata to the sites of carboxylation in leaves of woody species. *Plant, Cell and Environment* **15**: 873–899.

Loreto F, Harley PC, Di Marco G, Sharkey TD. 1992. Estimation of mesophyll conductance to CO_2 flux by three different methods. *Plant Physiology* **98**: 1437–1443.

Makovetskii AF, Manzhulin AV. 1990. Effect of illumination intensity on chloroplast ultrastructure in palisade and spongy tissue of the lupine leaf. *Soviet Journal of Plant Physiology* **37**: 937–945.

McCain DC, Croxdale J, Markley JL. 1988. Water is allocated differently to chloroplasts in sun and shade leaves. *Plant Physiology* **86**: 16–18.

McCain DC, Croxdale J, Markley JL. 1993. The spatial distribution of chloroplast water in *Acer platanoides* sun and shade leaves. *Plant, Cell and Environment* **16**: 727–733.

Nishio JN, Sun J, Vogelmann TC. 1993. Carbon fixation gradients across spinach leaves do not follow internal light gradients. *Plant Cell* **5**: 953–961.

Ögren E, Evans JR. 1993. Photosynthetic light response curves. 1. The influence of CO_2 partial pressure and leaf inversion. *Planta* **189**: 182–190.

Parkhurst DF. 1994. Diffusion of CO_2 and other gases in leaves. *New Phytologist* **126**: 449–479.

Parkhurst DF, Mott K. 1990. Intercellular diffusion limits to CO_2 uptake in leaves. *Plant Physiology* **94**: 1024–1032.

Poorter H, Evans JR. 1998. Photosynthetic nitrogen-use efficiency of species that differ inherently in specific leaf area. *Oecologia* **116**: 26–37.

Richter T, Fukshansky L. 1996. Optics of a bifacial leaf. 1. A novel combined procedure for deriving the optical parameters. *Photochemistry and Photobiology* **63**: 507–516.

Roupsard O, Gross P, Dreyer E. 1996. Limitation of photosynthetic activity by CO_2 availability in the chloroplasts of oak leaves from different species and during drought. *Annales des Sciences Forrestières* **53**: 243–254.

Rühle W, Wild A. 1979. The intensification of absorbance changes in leaves by light dispersion. *Planta* **146**: 551–557.

Sun J. 1996. Photosynthetic gradients across spinach leaves. PhD thesis, University of Wyoming, Laramie, USA.

Sun J, Nishio JN, Vogelmann TC. 1998. Green light drives CO_2 fixation deep within leaves. *Plant and Cell Physiology* **39**: 1020–1026.

Syvertsen JP, Lloyd J, McConchie C, Kriedemann PE, Farquhar GD. 1995. On the site of biophysical constraints to CO_2 diffusion through the mesophyll of hypostomatous leaves. *Plant, Cell and Environment* **18**: 149–157.

Terashima I. 1989. Productive structure of a leaf. In: Briggs WR, ed. *Photosynthesis*. New York, USA: Alan R Liss, 207–226.

Terashima I, Evans JR. 1988. Effects of light and nitrogen nutrition on the organization of the photosynthetic apparatus in spinach. *Plant and Cell Physiology* **29**: 143–155.

Terashima I, Inoue Y. 1985. Vertical gradient in photosynthetic properties of spinach chloroplasts dependent on intra-leaf light environment. *Plant and Cell Physiology* **26**: 781–785.

Terashima I, Saeki T. 1983. Light environment within a leaf. I. Optical properties of paradermal sections of *Camellia* leaves with special reference to differences in the optical properties of palisade and spongy tissues. *Plant and Cell Physiology* **24**: 1493–1501.

Terashima I, Saeki T. 1985. A new model for leaf photosynthesis incorporating the gradients of light environment and of photosynthetic properties of chloroplasts within a leaf. *Annals of Botany* **56**: 489–499.

Terashima I, Sakaguchi S, Hara N. 1986. Intra-leaf and intracellular gradients in chloroplast ultrastructure of dorsiventral leaves illuminated from the adaxial or abaxial side during their development. *Plant and Cell Physiology* **27**: 1023–1031.

Turrell FM. 1936. The area of the internal exposed surface of dicotyledon leaves. *American Journal of Botany* **23**: 255–263.

Vogelmann TC, Björn LO. 1984. Measurement of light gradients and spectral regime in plant tissue with a fibre optic probe. *Physiologia Plantarum* **60**: 361–368.

Vogelmann TC, Martin G. 1993. The functional significance of palisade tissue: penetration of directional versus diffuse light. *Plant, Cell and Environment* **16**: 65–72.

Vogelmann TC, Bornman JF, Josserand S. 1989. Photosynthetic light gradients and spectral regime within leaves of *Medicago sativa*. *Philosophical Transactions of the Royal Society of London, Series B* **323**: 411–421.

Vogelmann TC, Knapp AK, McClean TM, Smith WK. 1988. Measurement of light within plant tissues with fiber optic microprobes. *Physiologia Plantarum* **72**: 623–630.

von Caemmerer S, Evans JR. 1991. Determination of the average partial pressure of CO_2 in chloroplasts from leaves of several C_3 plants. *Australian Journal of Plant Physiology* **18**: 287–305.

von Caemmerer S, Farquhar GD. 1981. Some relationships between the biochemistry of photosynthesis and the gas exchange of leaves. *Planta* **153**: 376–387.

von Caemmerer S, Evans JR, Hudson GS, Andrews TJ. 1994. The kinetics of ribulose-1,5-bisphosphate carboxylase/oxygenase *in vivo* inferred from measurements of photosynthesis in leaves of transgenic tobacco. *Planta* **195**: 88–97.

Winter H, Robinson DG, Heldt HW. 1993. Subcellular volumes and metabolite concentrations in barley leaves. *Planta* **191**: 180–190.

Winter H, Robinson DG, Heldt HW. 1994. Subcellular volumes and metabolite concentrations in spinach leaves. *Planta* **193**: 530–535.

Wong SC, Cowan IR, Farquhar GD. 1979. Stomatal conductance correlates with photosynthetic capacity. *Nature* **282**: 424–426.

Yoshie F. 1986. Intercellular CO_2 concentration and water-use efficiency of temperate plants with different life-forms and from different microhabitats. *Oecologia* **68**: 370–374.

New Phytol. (1999), **143**, 105–117

Assessing leaf pigment content and activity with a reflectometer

J. A. GAMON* AND J. S. SURFUS

Department of Biology and Microbiology, California State University Los Angeles, 5151 State University Drive, Los Angeles, CA 90032, USA

Received 29 October 1998; accepted 3 March 1999

SUMMARY

This study explored reflectance indices sampled with a 'leaf reflectometer' as measures of pigment content for leaves of contrasting light history, developmental stage and functional type (herbaceous annual versus sclerophyllous evergreen). We employed three reflectance indices: a modified normalized difference vegetation index (NDVI), an index of chlorophyll content; the red/green reflectance ratio ($R_{RED}:R_{GREEN}$), an index of anthocyanin content; and the change in photochemical reflectance index upon dark–light conversions (ΔPRI), an index of xanthophyll cycle pigment activity. In *Helianthus annuus* (sunflower), xanthophyll cycle pigment amounts were linearly related to growth light environment; leaves in full sun contained approximately twice the amount of xanthophyll cycle pigments as leaves in deep shade, and at midday a larger proportion of these pigments were in the photoprotective, de-epoxidized forms relative to shade leaves. Reflectance indices also revealed contrasting patterns of pigment development in leaves of contrasting structural types (annual versus evergreen). In *H. annuus* sun leaves, there was a remarkably rapid increase in amounts of both chlorophyll and xanthophyll cycle pigments along a leaf developmental sequence. This pattern contrasted with that of *Quercus agrifolia* (coast live oak, a sclerophyllous evergreen), which exhibited a gradual development of both chlorophyll and xanthophyll cycle pigments along with a pronounced peak of anthocyanin pigment content in newly expanding leaves. These temporal patterns of pigment development in *Q. agrifolia* leaves suggest that anthocyanins and xanthophyll cycle pigments serve complementary photoprotective roles during early leaf development. The results illustrate the use of reflectance indices for distinguishing divergent patterns of pigment activity in leaves of contrasting light history and functional type.

Key words: leaf development, leaf pigments, anthocyanins, chlorophyll, xanthophyll cycle, leaf reflectometer, reflectance indices, photoprotection.

INTRODUCTION

Leaf pigments, including chlorophylls, carotenoids and anthocyanins, are well positioned for light absorption in particular wavebands and can readily be assessed with spectral reflectance. Because these pigments are optically detectable and serve either photosynthetic or photoprotective functions, they also provide an accessible 'handle' for evaluating relative photosynthetic activity, which can vary with leaf type (Gamon *et al.*, 1997). Because pigment content and photosynthesis are often linked to other leaf physiological or structural properties, it is possible to infer a number of critical leaf properties from leaf reflectance. For example, pigment content and related physiological properties vary with ontogeny (Sestak, 1985), senescence (Gausman *et al.*, 1971; Adams *et al.*, 1990; Merzlyak & Gitelson,

1995), nitrogen content (Khamis *et al.*, 1990), air pollution exposure (Heath, 1994; Wellburn, 1994), light history (Thayer & Björkman, 1990) and functional type (Gamon *et al.*, 1997). Consequently, reflectance assessment of leaf pigments can potentially provide useful indicators of integrated leaf physiology under a wide range of conditions.

Historically, a number of methods have been developed for assessing leaf pigment content. Common techniques involve pigment extraction followed by spectrophotometric determination (e.g. Arnon, 1949; Mancinelli *et al.*, 1975; Lichtenthaler, 1987); however, many spectrophotometric methods do not allow simultaneous determination of different pigment classes, which include chlorophylls, carotenoids (carotenes and xanthophylls) and anthocyanins. More involved methods are now available, including paper chromatography, thin-layer chromatography and high-performance liquid chromatography (HPLC), involving separation be-

*Author for correspondence (fax +1 323 343 6451; e-mail jgamon@calstatela.edu).

fore spectrophotometric determination (Harborne, 1984; Lichtenthaler, 1987; Britton & Young, 1993). While HPLC is often the method of choice for assessing leaf pigment concentration for physiological studies, it has many additional drawbacks that severely restrict its use. These limitations include high cost (both instrumental and operational costs), and the long time required for extraction and quantification (a single sample can take 30–60 min). Like all wet chemical methods, HPLC requires sample destruction. Clearly, HPLC quantification is not possible or desirable under many situations.

Spectral reflectance offers one alternative to destructive and time-consuming wet chemical extraction. Traditionally, accurate reflectance measurements require the leaf to be removed and placed in an illuminated integrating sphere attached to a spectroradiometer. The sphere provides a uniform, repeatable sampling geometry, enabling accurate and precise reflectance determination. Additionally, the sphere allows ready determination of leaf transmittance and absorptance. However, this sampling approach has several drawbacks, including expense (many spectroradiometers and illuminated integrating spheres are quite costly); awkwardness (most spheres are too bulky and delicate for easy use in field conditions); and the common need for destructive sampling (measurements are most easily made by bringing the leaf sample to the sphere). Furthermore, while a variety of sphere sizes and geometries are commercially available, most require that a sizeable, flat leaf region (typically $\geqslant 1$ cm in diameter) be inserted in a sample port. Much of the world's vegetation, including conifers, scrub vegetation and many desert plants, do not meet this size or shape requirement and thus cannot be sampled with the restrictive geometry of most integrating spheres. Recent advances in optical instrumentation, including stable, solid-state array detectors, and miniaturized fibre optics, now allow a re-assessment of field optical methods for leaf reflectance sampling. Additionally, the ability to sample tiny leaf regions non-destructively allows the exploration of changing leaf properties during development.

In the past, leaf reflectance has not been regarded as a particularly accurate method for assessing leaf pigment content. The main reason is that, unlike properly prepared leaf extracts which obey Beer's law, pigments in the intact leaf are integrally bound with leaf structures. This influence of structure leads to complex scattering patterns that can confound a simple, quantitative interpretation of pigment content (Butler, 1964; Fukshansky, 1981). However, in recent years a number of studies have reported quantitative or semi-quantitative estimates of chemical content in intact leaves or canopies, using indices derived from spectral reflectance at specific wavelengths. Because of the links between chemical content and leaf structure and function, a number of

important ecophysiological properties can be inferred from these reflectance indices, and some of these applications at leaf and canopy scales have recently been reviewed (Peñuelas & Filella, 1998; Gamon & Qiu, 1999). These applications include various expressions of chlorophyll (Gitelson & Merzlyak, 1994a), carotenoid: chlorophyll ratios (Peñuelas *et al.*, 1995a), xanthophyll cycle pigments and related photosynthetic performance (Gamon *et al.*, 1992, 1997; Peñuelas *et al.*, 1994, 1995b), nitrogen (Peñuelas *et al.*, 1994; Filella *et al.*, 1995), water content (Peñuelas *et al.*, 1993, 1994; Gao, 1996; Pinol *et al.*, 1998) and other measures of integrated leaf stress (Carter, 1994). It is becoming increasingly clear that reflectance indices offer convenient and non-intrusive tools for rapidly inferring a number of functionally important leaf and canopy properties. Additionally, knowledge of leaf structure and optical properties has advanced to a point where leaf biochemical content and structure can now be reliably modelled (Jacquemoud & Baret, 1990; Govaerts *et al.*, 1996; Jacquemoud *et al.*, 1996). The combination of improved models with a growing body of empirical results provides an increasingly solid foundation for quantitative assessment of critical leaf properties from optical sampling of intact leaves.

In this paper we examine a novel optical instrument for assessing leaf pigment content and, by proxy, the changing physiology and structure of intact leaves. In particular, three indices of chlorophyll, xanthophyll and anthocyanin pigment content are explored, as these pigments serve important photosynthetic and protective roles. We illustrate how these indices can be used to assess changing pigment content during development in leaves of contrasting structure and functional type, and how this status is influenced by growth light environment.

MATERIALS AND METHODS

Plant materials and culture

Plant materials and cultural conditions are summarized in Table 1. All plants were pot-grown seedlings and ranged from several weeks old (*Helianthus annuus* L.), to 1–2 yr old (*Pseudotsuga menziesii* (Mirb.) Franco and *Quercus agrifolia* Nee). Varying growth light levels were achieved through the use of neutral-density shadecloth applied to glasshouse or outdoor environments. Additional relevant details regarding species, culture and sampling conditions are provided in the Results section and in the figure legends.

Leaf development and structure

Changing leaf-blade thickness during development was estimated with a spring-loaded micrometer. In

Table 1. *Plant material and culture conditions*

Species	Common name	Leaf type	Culture conditions
Pseudotsuga menziesii	Douglas fir	Sclerophyllous, needle-leaf evergreen	Pot culture (outdoors in sun and shade), Los Angeles, CA, USA
Quercus agrifolia	Coast live oak	Sclerophyllous, broadleaf evergreen	Pot culture (outdoors in sun and shade), Los Angeles, CA, USA
Helianthus annuus	Sunflower	Herbaceous, broadleaf annual	Pot culture (glasshouse and outdoors in sun and shade), Los Angeles, CA, USA

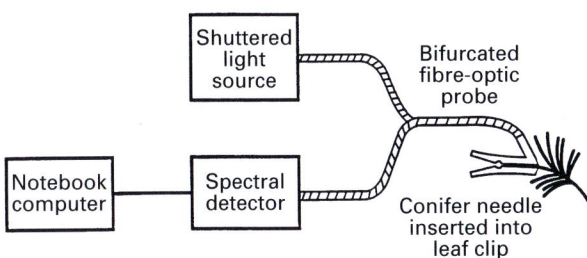

Fig. 1. Schematic diagram of leaf reflectometer, showing halogen light source with shutter, bifurcated fibre optic and spectral detector. The instrument can be controlled by a notebook or palmtop computer for field operation. A key to repeatable sampling is the leaf clip, which holds the common end of the bifurcated fibre optic at a fixed angle and distance relative to the leaf surface (see the Materials and Methods section). The modular design of the reflectometer allows a wide variety of alternate foreoptics to be attached to the detector, e.g. spectrophotometric cuvette holders, integrating spheres or fibre-optic probes for passive sampling.

general, three replicates were made on each leaf, and these replicates exhibited a high coefficient of variation (not shown). Although major leaf veins were avoided, small veins, leaf hairs and curving blade surfaces could not be avoided, contributing to this variability. Consequently, this method provided only a crude measure of thickness, and is presented as an index of qualitative trends rather than an accurate measure of thickness. Leaf area was estimated by weighing leaf tracings made on clear acetate sheets.

Reflectance instrumentation

Unless otherwise stated, all leaf reflectance measurements in this paper were conducted with a leaf reflectometer fitted with a leaf clip and bifurcated fibre optic (Fig. 1). The basic components of earlier prototypes have been defined (Gamon *et al.*, 1993, 1997; Peñuelas *et al.*, 1995b), but several advances have been made that warrant further description. Current versions of this instrument employ a solid-state silicon array detector (MMS-1, 300–1100 nm, NIR enhanced with visible blaze, Hellma Cells, Forest Hills, NY, USA) capable of a faster response

and wider dynamic range (16-bit) than previous prototypes. This detector has a bandwidth (full width half maximum) of approx. 10 nm, with 256 bands spaced at roughly even intervals between approx. 300 and 1150 nm (average band-to-band spacing 3.3 nm). The useful sampling range is further affected by the transmittance properties of the foreoptics which, in the case of this study, limited the effective spectral range to c. 400–1100 nm.

A key feature of the reflectometer that distinguishes it from other spectrometers is the unique foreoptics and leaf clip. New leaf-clip designs (e.g. UNI501, PP Systems, Haverhill, MA, USA) and optional, miniaturized fibre optics (e.g. part no. UNI410, PP Systems, with an optical diameter of approx. 0.6 mm) now allow sampling of leaf regions <1 mm in diameter. This small optical cross-section enables reflectance sampling on almost any type of leaf material, including needle leaves of many conifer, scrubland and desert species. The leaf clip enables the fibre tip to be held in a fixed geometry (60°, or 30° from normal) relative to the leaf surface, allowing accurate and repeatable leaf sampling. By coupling the clip to a flexible fibre, it is possible to make *in situ*, non-invasive measurements on most leaf types. Additionally, software improvements now allow simultaneous analysis of reflectance spectra, kinetic responses and mathematical expressions of reflectance (e.g. spectral indices, further defined in 'Reflectance measurement' in the Materials and Methods section) with a series of programmable, user-defined functions.

These improvements were driven by the need for a simple and portable instrument for remote sensing in the field, and by the fact that much of the world's vegetation cannot be sampled with conventional leaf reflectance instrumentation (e.g. integrating spheres), particularly under field conditions. In its current configuration, all optical and electronic components of our prototype reflectometer fit into a single package measuring $25 \times 31 \times 9$ cm. These dimensions were primarily determined by the footprint requirements of a typical laptop computer, used for data acquisition and storage in the field. Recently an improved, commercial version based in part on this prototype has been developed (UniSpec, part no. UNI003 with custom visible blaze, PP Systems), and offers a much smaller, more portable package for field or laboratory reflectance sampling. Results are incorporated here from both our prototype instrument and the newer commercial version, which can be used interchangeably with identical results.

Reflectance measurement

To calculate reflectance, leaf spectral radiance was divided by the radiance of a 99% reflective white reference standard (Spectralon, Labsphere, North Dutton, NH, USA). This normalization step avoided errors due to drift in the lamp or system electronics, and avoided the need for radiometric calibration. Periodic radiance scans of a mercury argon line source (model HG-1, Ocean Optics, Dunedin, FL, USA) ensured that the detector was in proper spectral (wavelength) calibration. In all cases, three to five reflectance spectra were determined for each leaf, and expressed as a mean.

To express pigment content, we calculated reflectance indices, which are simple mathematical expressions derived from reflectance spectra. These indices serve two general functions; they capture essential features of the spectra that scale with content of specific pigment classes (chlorophylls, xanthophylls and anthocyanins), and they reduce a large volume of data to a single, workable value to facilitate plotting and statistical analyses. Three indices were used in this study: (1) a modified normalized difference vegetation index (NDVI), an index of chlorophyll content; (2) the red:green reflectance ratio (R_{RED}: R_{GREEN}), an index of anthocyanin content; and (3) the photochemical reflectance index (PRI), an index of xanthophyll cycle pigment activity.

$$\text{NDVI (chlorophyll)} = (R_{750} - R_{705}) : (R_{750} + R_{705})$$
$$\text{Eqn 1}$$

$$\text{Red:green ratio (anthocyanin)} = R_{RED} : R_{GREEN}$$
$$\text{Eqn 2}$$

$$\text{PRI (xanthophyll)} = (R_{531} - R_{570}) : (R_{531} + R_{570})$$
$$\text{Eqn 3}$$

(R refers to reflectance, and the subscripts refer to a specific spectral band or wavelength. For example, RED refers to a broad red band (600–699 nm), and GREEN to a broad green band (500–599 nm). The numbers 531, 570, 705 and 750 refer to narrow wavebands at 531, 570, 705 and 750 nm, respectively. For all narrow wavebands, the effective bandwidth matched the detector bandwidth (full-width half-maximum) of approx. 10 nm).

In remote sensing, NDVI and related vegetation indices are normally derived from broad red and near-infrared (NIR) wavebands. These vegetation indices have been widely used since the 1970s as measures of fractional absorbed photosynthetically active radiation and green vegetation cover (e.g. Kumar & Monteith, 1981; Hatfield *et al.*, 1984; Bartlett *et al.*, 1990; Gamon *et al.*, 1995). While NDVI is most widely applied at the canopy to global scales using remotely positioned spectroradiometers, a number of studies have noted a correlation between NDVI and chlorophyll content at the leaf scale (e.g. Gitelson & Merzlyak, 1994a; Yoder & Pettigrew-Crosby, 1995). However, Gitelson & Merzlyak (1994a) noted that for leaf-scale assessment of chlorophyll content, this index can be improved by using a narrow waveband at the edge of the chlorophyll absorption feature (e.g. 705 nm) rather than at the middle, which results in a more linear relationship between this index and chlorophyll content. Our preliminary studies (not shown) generally supported this observation, and consequently we adopted the modified NDVI formulation of Gitelson & Merzlyak (1994a) and replaced the RED waveband with 705 nm and the NIR waveband with 750 nm.

For PRI, we followed the previous convention (Gamon *et al.*, 1997) of using R_{531} as an indicator of xanthophyll cycle activity and 570 nm as a reference wavelength. However, unlike previous studies that focused on diurnal changes of PRI (Gamon *et al.*, 1992; Peñuelas *et al.*, 1994, 1995b) or midday values of PRI (Gamon *et al.*, 1997), we explored a new sampling protocol that takes advantage of the built-in light source and kinetic capabilities of the reflectometer to force a rapid transition from the dark-adapted to the light-adapted state. This sampling approach, which we call dark-to-light transitions, rapidly causes the conversion of xanthophyll pigments from the epoxidized (violaxanthin) state to the de-epoxidized (antheraxanthin and zeaxanthin) state (Yamamoto, 1979; Gamon *et al.*, 1990). This method is analogous to many spectrophotometric assays that follow the kinetic response of a chemical reaction in solution. In this case, the goal is to use reflectance kinetics to determine pool sizes of photochemically active xanthophyll cycle pigments in the intact leaf. This sampling approach is based on preliminary observations (not shown) that dark-adapted leaves suddenly exposed to the equivalent of full sun (saturating light) with the reflectometer experienced near-complete de-epoxidation of the xanthophyll cycle pigment: i.e. zeaxanthin was the predominant form of xanthophyll cycle pigments following this treatment. From these observations we predicted that the change in PRI (ΔPRI) caused by saturating light could be used to estimate total xanthophyll cycle pool sizes (violaxanthin + antheraxanthin + zeaxanthin) in intact leaves. The basic method is as follows.

(1) Begin with a fully dark-adapted leaf (e.g. a leaf in its pre-dawn state, which can be extended either by shading the leaf or bringing the plant to a darkened laboratory the previous evening).

(2) Clamp a leaf clip on the leaf, with the instrument shutter closed.

(3) Open the shutter, immediately exposing the leaf to a pre-determined light intensity and forcing the leaf to begin conversion to the light state (see Gamon *et al.*, 1990, 1992). Upon shutter opening, begin automatic sampling of spectral reflectance at a pre-programmed time interval (typically every 30 s). (To attain full pigment conversion, we used a light intensity of 2000 µmol photons $m^{-2} s^{-1}$, i.e. saturating light. However, to explore the effect of light intensity on this conversion (light–response curves), intensity was altered by adjusting the lamp voltage. Although this caused slight spectral shifts due to changing lamp temperature, these shifts were normalized by reflectance calculation and did not cause any obvious experimental artefacts.)

(4) Using the index function of the reflectometer, monitor PRI kinetics during the dark-to-light transition to ensure attainment of a steady state. (Because xanthophyll cycle conversion is enzyme-dependent (Yamamoto, 1979) this transition is temperature-dependent. Under typical room temperatures (25–30°C), a complete transition usually takes *c.* 5 min, but this can vary depending on the physiological state of the leaf. In this study, all dark–light transitions were conducted in the 25–30°C range, and followed for 8–10 min to ensure attainment of steady state. Because of the rapid response of the instrument (scan time adjustable from several milliseconds to a few seconds) it is possible to capture reflectance in the dark state before enzymatic conversion of xanthophyll cycle pigments becomes apparent.)

(5) Once a steady light state has been attained, calculate ΔPRI from the difference between the initial (dark) and final (light) value. This ΔPRI can then be explored as a measure of active xanthophyll pigment pool sizes, or can be used to infer environmental or physiological conditions (see the Results and Discussion sections).

An alternative method to estimate xanthophyll pigment activity is to sample PRI under both pre-dawn and midday light conditions on the same leaf to derive a ΔPRI (expressed as the dawn PRI minus the midday PRI values). Unlike the method already outlined, which uses the built-in lamp as an active probe of xanthophyll cycle pigment pool sizes, this passive method detects the actual conversion of xanthophyll cycle pigments under ambient light conditions, and thus detects the portion of the total pigment pool used in photoprotection (see Frank *et al.*, 1994 for a discussion of this photoprotective mechanism). Results from both methods are compared.

Wet chemical assays

All chlorophyll and xanthophyll pigment samples were confirmed by extraction and subsequent HPLC analysis (model LC-10AS with detector SPD-10AV, Shimadzu, Kyoto, Japan) using the method of Thayer & Björkman (1990). Leaf anthocyanin content was assessed by extraction in hot $MeOH : HCl : H_2O$ (90:1:1, v:v:v) and subsequent absorbance measurements at 530 nm, corrected for chlorophyll absorbance at 657 nm (Mancinelli *et al.*, 1975). To estimate molar pigment concentrations from this corrected absorbance, we applied an extinction coefficient of $30\,000\ l\ mol^{-1}\ cm^{-1}$ (Murray & Hackett, 1991). Absorbance measurements were made with the reflectometer reconfigured as a spectrophotometer by replacing the bifurcated fibre and leaf clip (Fig. 1) with a spectrophotometric cuvette (CUV-UV, Ocean Optics). All pigments were expressed for a one-sided, projected leaf area.

To calibrate reflectance indices against pigment content, similar leaf regions sampled for reflectance were subsequently harvested and assayed for pigment content. Due to the varying leaf structures (needle versus broad leaves), and the much larger sample areas needed for pigment extraction (typically 1.5–2.5 cm^2 per sample), in most cases the exact leaf regions sampled with reflectance differed from those sampled for pigment content. However, to calibrate ΔPRI against HPLC estimates of xanthophyll pigments (Fig. 5c), a broad fibre (no. 2555.04, Fostec, Auburn, NY, USA) was coupled to the instrument to allow similar leaf areas to be used for reflectance and pigment extraction. Each reflectance sample consisted of three to five replicates expressed as a mean ($\pm SE$), and harvested leaf regions corresponding to these three to five samples were then pooled to allow sufficient material for pigment analysis. In the case of xanthophyll pigments, extracts were taken immediately following reflectance samples to ensure the same light conditions as the final reflectance readings. Chlorophyll and anthocyanin samples were collected on the same day as leaf reflectance.

RESULTS

Leaves of different ages and pigment content exhibited contrasting patterns of spectral reflectance, and some representative contrasts are illustrated in Fig. 2. For example, the loss of chlorophyll in a senescing *H. annuus* (sunflower) leaf caused an increase in yellow colour due to an increased reflectance between approx. 500 and 700 nm (Fig. 2a). By selecting a narrow waveband at the edge of this chlorophyll absorbance region (705 nm), and referencing this against a waveband not influenced by chlorophyll content (750 nm), it is possible to capture the effect of varying chlorophyll contents

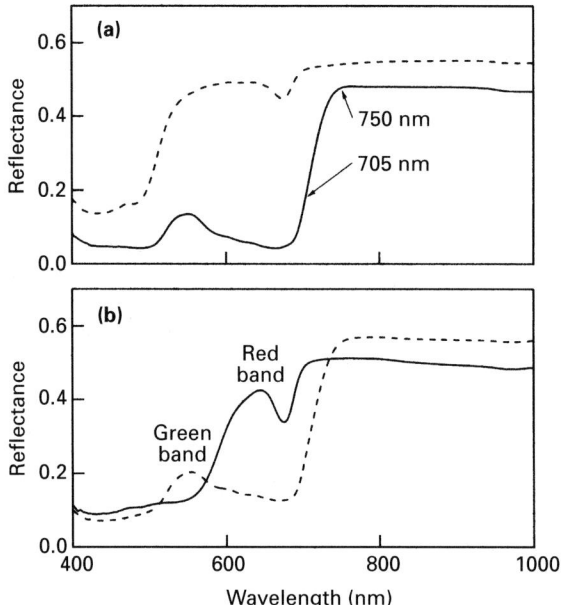

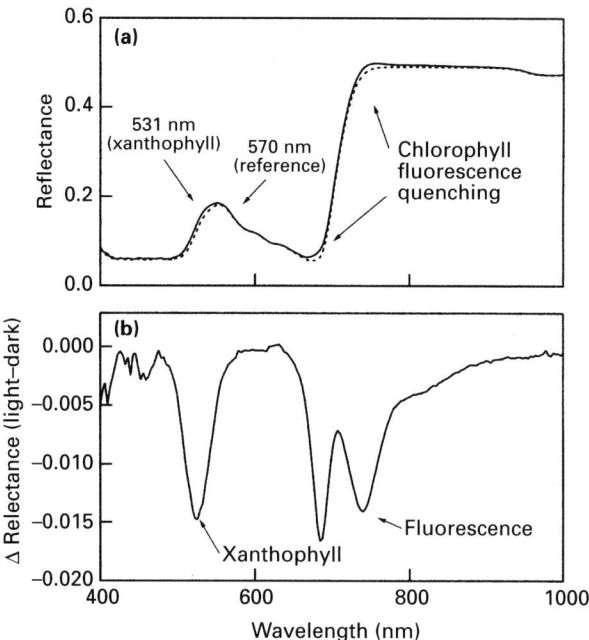

Fig. 2. (a) Representative reflectance spectra of a young green (solid line) and a senescing, yellow(dotted line) *Helianthus annuus* (sunflower) leaf. The labels at 705 and 750 nm indicate the wavebands used to assess chlorophyll content according to Eqn 1 (see 'Reflectance measurement in the "Materials and Methods" section'). (b) Representative reflectance spectra of a young, expanding *Quercus agrifolia* (oak) leaf exhibiting red pigmentation due to a high anthocyanin content (solid line) and a mature, green oak leaf with low anthocyanin content (dotted line). Red (600–699 nm) and green (500–599 nm) wavebands can be combined (Eqn 2) to form an index of anthocyanin content.

(Eqn 1). Additionally, temporary increases in anthocyanin pigmentation during early leaf expansion in *Q. agrifolia* (oak) caused a pronounced increase in red reflectance relative to a mature, green leaf (Fig. 2b). This change in anthocyanin pigmentation is estimated by the ratio $R_{RED}:R_{GREEN}$ (Eqn 2). As previously reported (Gausman, 1985), leaf structure also influenced leaf reflectance spectra; NIR reflectance increased slightly with increasing mesophyll thickness. Consequently, mature or senescing older leaves tended to have a higher NIR reflectance than younger, developing leaves (Fig. 2).

In contrast to the passive sampling approach used for anthocyanins and chlorophylls, xanthophyll pigment content was explored using an active technique of stimulating the xanthophyll cycle with light provided by the reflectometer. When dark-adapted leaves were exposed to bright light, subtle changes in apparent reflectance were visible at approx. 531, 685 and 738 nm (Fig. 3a). These changes were more apparent in the difference spectrum (light minus dark; Fig. 3b), and have been attributed to xanthophyll cycle pigment conversion and associated thylakoid energization (531 nm) and chlorophyll fluorescence quenching (685 and 738 nm), respectively (Gamon *et al.*, 1990, 1997). To test the use of the 531 nm Δ reflectance feature

Fig. 3. (a) Reflectance spectrum of a *Helianthus annuus* (sunflower) leaf in the dark state (solid line) and 10 min after exposure to white light (2000 µmol photons m^{-2} s^{-1}) using the halogen source (dotted line). (b) Difference spectrum (reflectance in dark state minus reflectance in light state) derived from the spectra in (a). Sampling in the dark state is possible because the rapid response of the reflectometer allows completion of reflectance sampling upon initial light exposure, before measurable xanthophyll pigment de-epoxidation occurs. The 531-nm Δ reflectance feature is due to xanthophyll pigment conversion, and the double feature at 685 and 738 nm is due to chlorophyll fluorescence quenching. Normalizing 531 nm reflectance against reflectance at a reference wavelength (570 nm) provides an expression of xanthophyll cycle pigment activity (see Eqn 3).

as an index of xanthophyll cycle pigment content, we exposed a series of dark-adapted sunflower leaves to a range of actinic light intensities in the range 14–2000 µmol photons m^{-2} s^{-1} by adjusting the reflectometer lamp voltage. Exposure to these varying light levels resulted in a corresponding variation in PRI responses over an 8-min period, with higher light intensities inducing greater declines in PRI (Figs 4, 5c).

To evaluate the ability of reflectance to determine actual pigment content, we compared each reflectance index (Eqns 1–3) to pigment amounts estimated by conventional, destructive, wet chemical methods. As expected, the chlorophyll index (Eqn 1) was strongly correlated with area-based total chlorophyll content, estimated by HPLC determination (Fig. 5a). Similarly, the anthocyanin index (Eqn 2) was strongly related to area-based pigment content estimated by destructive sampling and spectro-photometric quantification (Fig. 5b). The exact calibration and correlation of the chlorophyll index with extracted chlorophyll varied between species of contrasting leaf structure; for a given index value,

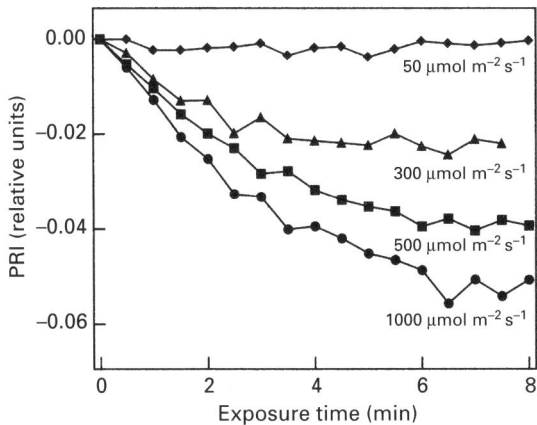

Fig. 4. Representative kinetic plots of photochemical reflectance index (PRI) during the dark-to-light transition (see Fig. 3) for *Helianthus annuus* (sunflower) leaves exposed to various PPFRs, as indicated. The PRI values at time zero (the point of initial light exposure) have been normalized to better illustrate the influence of actinic light intensity on PRI kinetics. The change in PRI (ΔPRI) between dark and light states provides an index of xanthophyll pigment de-epoxidation (see Fig. 5c).

the thicker leaves of *P. menziesii* (Douglas fir) tended to have a higher extracted chlorophyll value for a given surface area than the thinner leaves of *H. annuus* (Fig. 5a). Because the calibration of reflectance indices can vary with species, and because the goals of this study precluded further destructive sampling, pigment amounts were subsequently expressed as relative index values using Eqns 1–3 rather than as absolute molar units.

The change in PRI (ΔPRI) during the dark–light transition (Figs 3, 4; Eqn 3) was then compared with xanthophyll cycle pigment content as estimated by HPLC. In this case, ΔPRI values were tightly correlated with the amount of de-epoxidized xanthophyll cycle pigments produced by the light treatments, expressed as zeaxanthin plus half the amount of antheraxanthin (Fig. 5c).

Effect of growth light level

To explore the influence of growth light level on xanthophyll pigments, we then conducted a series of studies comparing ΔPRI of leaves from contrasting growth light environments. First, it was confirmed that both xanthophyll cycle pigment pool sizes and ΔPRI determined with saturating light did indeed vary with growth light environment (Fig. 6a). In this test, a close correlation ($r^2 = 0.93$) was also found between ΔPRI and total xanthophyll cycle pigment content (not shown).

Using a second set of sunflower leaves, ΔPRI measured with saturating light was then compared with ΔPRI determined in ambient light, by sampling dawn and midday PRI values for leaves in a range of growth light levels. This compared total xanthophyll cycle pigment pool sizes (the saturating light

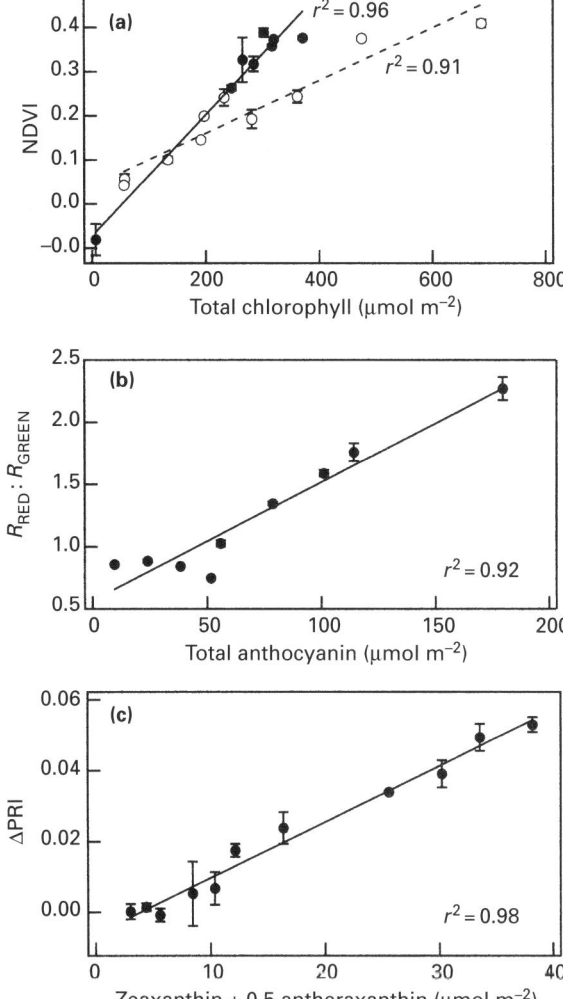

Fig. 5. (a) Comparison of a chlorophyll index (normalized difference vegetation index (NDVI) Eqn 1; see 'Reflectance measurement in the "Materials and Methods" section') to leaf chlorophyll content (determined by HPLC) for *Pseudotsuga menziesii* (closed symbols) and *Helianthus annuus* (open symbols). (b) Comparison of an anthocyanin index (Eqn 2) to anthocyanin content (determined spectrophotometrically) for *Quercus agrifolia*. (c) Photochemical reflectance index (ΔPRI) versus zeaxanthin plus one-half antheraxanthin content for previously dark-adapted sunflower leaves exposed to different actinic light levels, ranging from 14–2000 µmol photons $m^{-2} s^{-1}$ (see Fig. 4 for representative kinetics). Each point represents the mean (± 1 SE) for three to five reflectance samples from several leaves; these leaves were subsequently pooled for pigment quantification by spectrophotometry or HPLC. Leaves in (a) and (b) were sampled with the miniature bifurcated probe and leaf clip (see the Materials and Methods section); leaves in (c) were sampled with a broader fibre optic (no. 2555.04, Fostec). This fibre optic matched the diameter of our leaf punch, ensuring that reflectance measured a similar leaf area to that subsequently sampled for xanthophyll pigment content. (c) A summary of the results of 11 separate experimental treatments (11 points), each replicated three times (three leaves from three separate plants).

method) with the amount of xanthophyll pigments actually engaged in photoprotection (the ambient light method). The ΔPRI determined by both

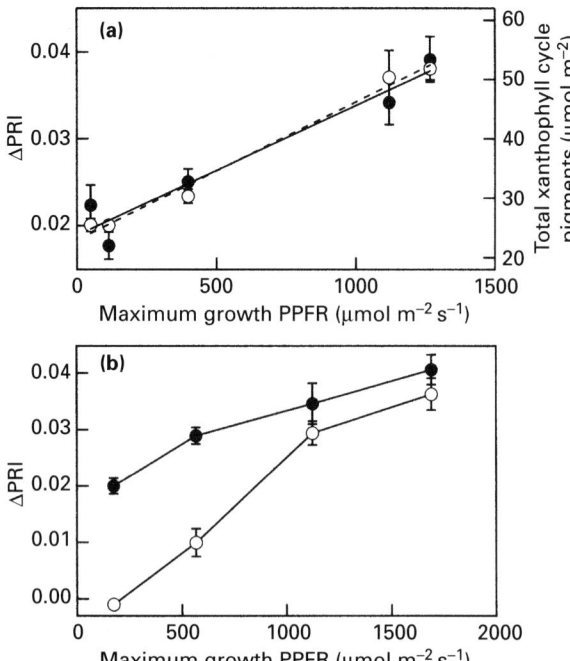

Fig. 6. (a) Photochemical reflectance index (ΔPRI; closed symbols) and content of total xanthophyll cycle pigments (violaxanthin + antheraxanthin + zeaxanthin; open symbols) as a function of maximum PPFR during growth of *Helianthus annuus* (sunflower) leaves. ΔPRI was sampled by exposing dark-adapted leaves to 2000 μmol photons m^{-2} s^{-1} using the reflectometer halogen source. (b) ΔPRI versus maximum growth PPFR for a second set of sunflower plants exposed to four growth light levels (deep shade in glasshouse to full sun outdoors). ΔPRI was sampled either by exposing dark-adapted leaves to 2000 μmol photons m^{-2} s^{-1} using the reflectometer halogen source (saturating light, solid symbols), or by comparing reflectance at dawn and solar noon under normal growth light conditions (ambient light, open symbols). In (a) and (b), each point represents a mean (±1 SE) for three leaves on three separate plants.

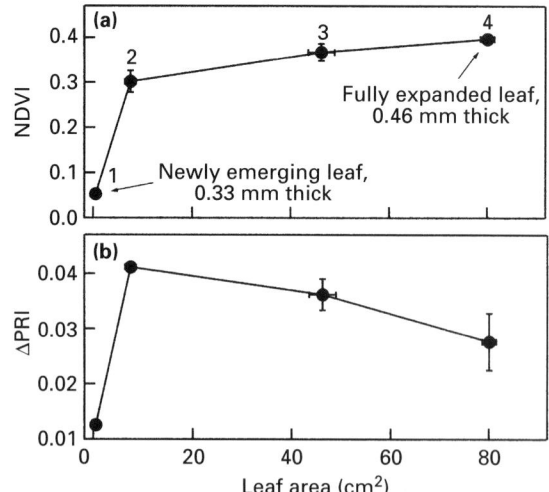

Fig. 7. (a) Relative chlorophyll content (NDVI) and (b) xanthophyll content (ΔPRI) as a function of leaf area for a developmental sequence of four *Helianthus annuus* leaves. Leaf thickness and number are indicated in the figure: leaf 1 is a newly emerging leaf, and leaf 4 is a fully expanded leaf. Each point represents a mean (±1 SE) of three leaves from three separate plants, all grown in full sun.

their xanthophyll cycle pigments for photo-protection, leaves grown in full sun appeared to be converting the majority of their xanthophyll cycle pigments to the photoprotective (de-epoxidized) forms during solar noon.

Effect of developmental state

To examine the effect of developmental state on the amount of leaf pigment, reflectance was sampled on a sequence of leaves of *H. annuus* in full sun. On initial emergence (leaf area approx. 0.5 cm^2), amounts of chlorophyll (measured as NDVI) were very low, but increasing rapidly so that by leaf 2 (9% full expansion), chlorophyll was at near-maximum amounts (Fig. 7a). Xanthophyll cycle pigment content (assessed as ΔPRI using the saturating light method) showed a similarly rapid pattern of increase during early leaf expansion (leaves 1–2, 0.6–9% full expansion, Fig. 7b). However, unlike chlorophyll content, which continued to increase with leaf expansion, xanthophyll pigment content apparently declined slightly with further leaf growth as chlorophyll content continued to increase (leaves 2–4).

Pigment contents during leaf development were also examined for the drought-tolerant, sclerophyllous evergreen *Q. agrifolia*. Unlike the fast-growing sunflower leaves which rapidly attained maximal chlorophyll content, the amounts of chlorophyll in oak approached maximal values gradually over a 24-leaf sequence (Fig. 8a). Xanthophyll cycle pigment amounts, assessed as ΔPRI, rose in a similar gradual manner, with a slight decline evident between leaves 4 and 6 (9–27% full expansion; see Fig. 8b, arrows). This delayed development of xanthophyll pigments corresponded to increased

methods showed a near-linear variation with maximum growth light intensity (Fig. 6b). Using the saturating light method, ΔPRI values of leaves grown in full sun were approximately twice those of leaves in full shade (solid symbols, Fig. 6b); this indicated that the xanthophyll pigment pool roughly doubled in size in sun leaves, which was consistent with our earlier finding (Fig. 6a).

The ambient light sampling approach revealed that only a portion of the total xanthophyll pigments were actually in the de-epoxidized form under typical midday light conditions. In this case, the amount of xanthophyll cycle pigment conversion (assessed by the difference in PRI between dawn and midday) ranged from essentially zero in deep shade and increased steadily with midday light levels (open symbols, Fig. 6b). At the highest growth light levels, the ΔPRI values determined using this ambient light method approached the maximum ΔPRI values determined using the saturating light method (compare solid and open symbols, Fig. 6b). Thus, unlike the shade leaves which did not appear to use any of

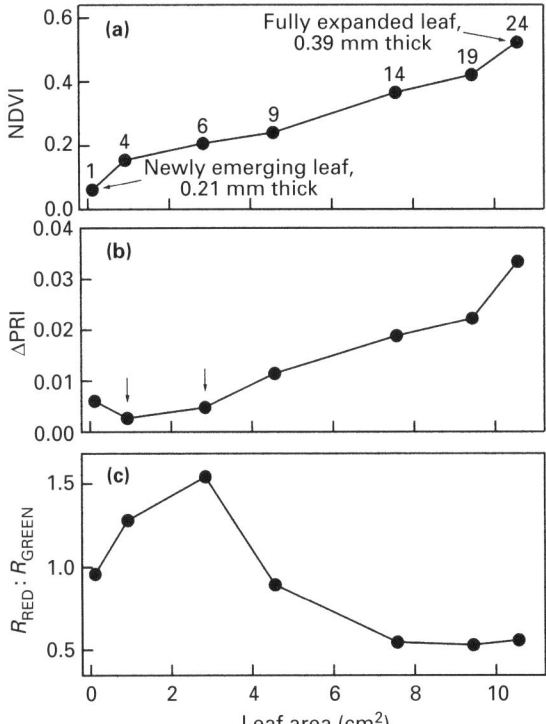

Fig. 8. (a) Relative chlorophyll content (NDVI), (b) xanthophyll content (ΔPRI) and (c) anthocyanin content ($R_{\mathrm{RED}}:R_{\mathrm{GREEN}}$) as a function of leaf area, for a developmental sequence of 24 *Quercus agrifolia* (oak) leaves from a single plant grown in full sun. Leaf thickness and leaf number are indicated on the figure; leaf 1 is a newly emerging leaf, and leaf 24 is a fully expanded leaf.

amounts of anthocyanin pigments (assayed by $R_{\mathrm{RED}}:R_{\mathrm{GREEN}}$) during early leaf development (Fig. 8c). Amounts of xanthophyll and anthocyanin pigments during leaf development (leaves 1–24) were negatively correlated ($r = -0.83$) and this correlation was particularly strong during the early phases of leaf expansion (leaves 1–19, $r = -0.90$).

DISCUSSION

These comparisons of reflectance indices to extracted pigments demonstrate that it is possible to obtain accurate pigment estimates in intact leaves with a reflectometer. This sampling method is rapid, portable, non-destructive, applicable to small leaf regions (<1 mm diameter), and capable of evaluating simultaneously several pigment classes, offering several potential advantages over wet chemical methods. The ability to sample the same leaf over time, non-invasively, offers new prospects for assessing changing pigment amounts during ontogenesis or senescence. The ability to sample small leaves is particularly applicable to conditions of early leaf development. The use of reflectance indices to assay changing chlorophyll content during leaf senescence has been reported previously (Gitelson & Merzlyak, 1994a,b; Merzlyak & Gitelson, 1995). This study extends the earlier work by demonstrating the use of

reflectance for assessing a wider range of pigment types during early leaf development and in contrasting light regimes.

Comparison of reflectance indices with extracted pigment levels (Fig. 5) suggests species-specific relationships influenced by leaf structural properties. This indicates a need for empirical calibration when using reflectance indices. However, our comparison of indices with conventional, wet chemical methods has not directly accounted for the potential effect of leaf structural properties (e.g. varying specific leaf weight) which could cause an apparent diversion between methods, particularly when area-based expressions are used as in this study. Healthy Douglas fir needles are typically thicker than sunflower leaves, and this could partly explain the higher area-based chlorophyll content for a given reflectance-based estimate of chlorophyll (Fig. 5a). It is possible that optical methods effectively measure the near-surface pigment levels (and thus the photochemically active levels) more accurately than wet chemical methods, which estimate an average, area-based leaf pigment content not weighted by cell layer. In this sense, reflectance-based methods may be better than destructive methods for representing the amounts of optically effective pigment. Further studies incorporating detailed examination of leaf structure and involving both area-based and volume- or mass-based content could help clarify these possibilities.

The simultaneous estimation of several leaf pigment classes using two-band reflectance indices is admittedly a simplistic approach to a complex problem. Part of the strength of these indices resides in their simplicity, they effectively reduce a multidimensional spectrum to a single set of numerical values. However, the index approach also has several potential drawbacks. One problem is that narrowband indices require careful attention to instrument calibration; slight errors in spectral calibration can lead to significant errors in the resulting index calculations. This problem was avoided with careful attention to wavelength calibration during sampling (see the Materials and Methods section). Another drawback is that any two-band index is necessarily confounded by other overlying factors that also influence apparent leaf reflectance properties at either of the two wavelengths. For example, changes in leaf 'brightness' due to changing surface properties (e.g. cuticle thickness or density of leaf hairs), or associated with sampling artefacts (e.g. drift in lamp output or improper sampling of the reference standard), will necessarily influence apparent reflectance properties independently of pigment content. These effects probably account for some of the scatter when relating reflectance indices to extracted pigment content (Fig. 5). However, by combining two bands, the indices presented here are internally normalized, reducing the impact of these artefacts.

Additionally, division of a difference by a sum (Eqns 1, 3) has the effect of further reducing possible artefacts (Gamon *et al.*, 1992). This normalized difference formulation is often used in remote sensing studies that have additional confounding effects due to changing illumination, atmospheric conditions, or the presence of many scene components (Gamon *et al.*, 1995; Gamon & Qiu, 1999).

Many potential alternatives exist to the simple index approach presented here: these include continuum removal (Clark & Roush, 1984); spectral feature-fitting methods (Gao & Goetz, 1994, 1995); hierarchical foreground / background analysis (Pinzon *et al.*, 1998); derivative spectrometry (Demetriades-Shah, 1990; Wessman, 1990); and spectral mixture analysis (Adams *et al.*, 1986; Wessman *et al.*, 1990; Ustin *et al.*, 1993). Stepwise multiple regression has been widely used for analysing reflectance features associated with absorption by biochemical constituents (e.g. Card *et al.*, 1988; Wessman *et al.*, 1988; Johnson *et al.*, 1994; Martin & Aber, 1997), and the advantages and disadvantages of this method have been discussed (Curran, 1989; Curran & Kupiec, 1995; Grossman *et al.*, 1996). Another alternative would be to use all the information present by modelling a reflectance spectrum, derived from a series of absorbing compounds with known absorption coefficient spectra, and modified by scattering coefficients related to leaf structure. Jacquemoud & Baret (1990) have explored this modelling approach with the assumptions that chlorophyll and carotenoid levels vary in concert and are the only leaf pigments. However, our study indicates that during leaf development the amounts of different leaf pigments do not always co-vary (Figs 7, 8), and similar conclusions can be drawn from studies of leaf senescence (Merzlyak & Gitelson, 1995). Furthermore, most existing leaf models do not consider the occasional presence of anthocyanins, which can significantly influence the leaf reflectance spectrum (Fig. 2b). Further development of these models to include the separate effects of different pigment groups on leaf reflectance might provide an improvement over the two-band indices presented here. Clearly, more work is needed to fully evaluate the relative merits of alternative means of interpreting leaf spectral features.

A primary focus of this study was the exploration of simple reflectance indices as indicators of pigment content. However, this sampling method can also reveal functional leaf properties not readily detectable with conventional destructive sampling. A good example is provided by the multiple applications of ΔPRI. If determined with saturating light, ΔPRI can be used to estimate total xanthophyll pigment pool size, which can also provide an index of light history during growth. Alternatively, ΔPRI can be measured using ambient light to stimulate the xanthophyll cycle, providing an estimate of de-

epoxidation under typical light conditions experienced by the leaf. The midday epoxidation state can be a potent indicator of other photosynthetic properties and correlates with actual midday photosynthetic rates in leaves of contrasting nutrient status and functional type (Gamon *et al.*, 1997).

The increasing xanthophyll pigment content with increasing growth light is consistent with previous reports of higher amounts of xanthophyll pigment in sun than in shade leaves (Thayer & Björkman, 1990; Demmig-Adams & Adams, 1992). It is remarkable that leaves grown in deep shade had xanthophyll amounts approximately half those in full sun (Fig. 6). These high amounts of xanthophyll cycle pigments appear to be an unnecessary investment; reflectance measured under actual midday light conditions (open symbols, Fig. 6b) reveal that no detectable xanthophyll cycle photoconversion occurs at this low light level, demonstrating that there is little excess light in need of dissipation under these conditions. These high constitutive levels of xanthophyll pigments in shade-grown leaves may represent a protective, 'bet-hedging' strategy for avoiding the otherwise damaging effects of potential sunflecks, which are common in many shaded environments (Chazdon, 1988; Pearcy, 1988; Pearcy & Pfitsch, 1994). It is also possible that sun-adapted species such as sunflower are genetically predisposed to have a high content of photoprotective pigments. Further studies with obligate shade species could help clarify the reasons for high xanthophyll levels in shade-grown leaves.

ΔPRI appeared to be a particularly robust measure of xanthophyll cycle pigment activity, as judged by its ability to account for almost all the variability in xanthophyll cycle pigment content (high r^2 value, Fig. 5c). One reason for the particularly good results with this index is that it was applied as an active probe of pigment conversion. Used in this way, it follows the actual production of de-epoxidized pigments, i.e. zeaxanthin and to a lesser extent antheraxanthin (Yamamoto, 1979). This deliberate application of the internal light source to force the conversion of xanthophyll pigments is analogous to widely used spectrophotometric methods of assaying biochemical activity by measuring changing absorbance. This non-destructive sampling is made possible by the unique design of the reflectometer, which incorporates a built-in light source, and offers a new way to study photoregulatory processes involving the xanthophyll cycle in the intact leaf.

The use of reflectance indices to monitor dynamic pigment changes during leaf development provided another powerful application of reflectance. One notable finding of this study was the contrasting patterns of pigment development between leaf types. *H. annuus* (a fast-growing herbaceous annual) attained high chlorophyll and xanthophyll pigment levels very soon after leaf emergence. The slight decline in the amounts of xanthophyll pigments with full leaf expansion was associated with an increase in the amount of chlorophyll, presumably reflecting a reduced need for photoprotection upon the development of full photosynthetic competence in this rapidly growing species. By contrast, *Q. agrifolia* exhibited gradual development of chlorophyll and carotenoid pigment content, paralleling the slow leaf expansion in this species; this delayed pigment development appears to be typical of evergreen leaves in general (Sestak, 1985). The contrasting trajectories of pigment development for sunflower and oak may well be linked to a larger suite of functional characters that define contrasting evergreen and annual 'strategies' (Grime, 1979). For example, relative to shorter-lived leaves, evergreen leaves have lower photosynthetic rates and higher investments in structural and biochemical protection (Reich *et al.*, 1991, 1992, 1995; Gamon *et al.*, 1997). More work with additional species is needed to confirm the generality of the links between pigments and these other functional characteristics, and the leaf reflectometer is ideally suited for such a survey.

Notable in the oak leaves were the high amounts of anthocyanin pigments during early leaf expansion. The exact role of anthocyanins is unclear: both photoprotective (Gould *et al.*, 1995) and defensive (Coley & Aide, 1989; Coley & Barone, 1996) functions have been attributed to these pigments. In this study, the complementary patterns of xanthophyll and anthocyanin pigmentation during early leaf development suggest that anthocyanins provide a critical, photoprotective role before xanthophyll pigments reach final levels. Other flavonoids are generally viewed as having a 'sunscreen' function under conditions of high UV irradiance (Robberecht *et al.*, 1983; Day *et al.*, 1992). It is possible that anthocyanins, which are often located in epidermal vacuoles, serve a similar photoprotective function by absorbing visible light when leaves have not fully developed photosynthetic competence.

The results of this study indicate that rapid assessment of several pigments in the intact leaf is now possible using spectral reflectance. Because the method is non-destructive, it allows repeated sampling of changing optical properties during leaf development and senescence. Because it readily allows sampling on most leaves, the leaf reflectometer offers a powerful tool for comparing the optical properties of contrasting leaf types. For example, differing levels of photoprotection in sun and shade leaves can be readily detected. Additionally, comparisons of annual and evergreen leaves reveal contrasting patterns of pigment development that may be characteristic of broader functional groupings. Similar applications of reflectance can be attempted at larger scales of aggregation, including canopy, stand and landscape levels (Gamon & Qiu, 1999). A particularly exciting application of reflect-

ance may be in the definition and detection of contrasting functional types at these multiple scales of aggregation. In this way, fundamental leaf properties revealed by leaf-level reflectance measurements can be examined for their contribution to larger scale ecological processes.

ACKNOWLEDGEMENTS

Our thanks to R. Johnson and D. Horvath, whose proficient technical skills allowed the development of the prototype reflectometer, to Dennis Kimura for providing Fig. 1, and also to M. Doyle and J. Distelbrink of PP Systems, USA, for their technical assistance and continued interest in this project. D. Sims and M. Doyle provided helpful suggestions on the manuscript. Funding for this work was provided by NASA and NSF. This is publication no. 15 of the Center for Spatial Analysis and Remote Sensing at California State University, Los Angeles.

REFERENCES

Adams JB, Smith MO, Johnson PE. 1986. Spectral mixture modeling: a new analysis of rock and soil types at the Viking Lander site. *Journal of Geophysical Research* **91**: 8098–8112.

Adams WW III, Winter K, Schreiber U, Schramer P. 1990. Photosynthesis and chlorophyll fluorescence characteristics in relationship to changes in pigment and elemental compositions of leaves of *Platanus occidentalis* L. during autumnal leaf senescence. *Plant Physiology* **92**: 1184–1990.

Arnon DI. 1949. Copper enzymes in isolated chloroplasts. Polyphenoloxidase in *Beta vulgaris*. *Plant Physiology* **24**: 1–15.

Bartlett DS, Whiting GJ, Hartman JM. 1990. Use of vegetation indices to estimate intercepted solar radiation and net carbon dioxide exchange of a grass canopy. *Remote Sensing of Environment* **30**: 115–128.

Britton G, Young AJ. 1993. Methods for isolation and analysis of carotenoids. In: Young A, Britton G, eds. *Carotenoids in photosynthesis*. London, UK: Chapman & Hall, 409–457.

Butler WL. 1964. Absorption spectroscopy *in vivo*: theory and application. *Annual Review of Plant Physiology* **15**: 151–170.

Card DH, Peterson DL, Matson PA, Aber JD. 1988. Prediction of leaf chemistry by the use of visible and near-infrared reflectance spectroscopy. *Remote Sensing of Environment* **26**: 123–147.

Carter GA. 1994. Ratios of leaf reflectances in narrow wavebands as indicators of plant stress. *International Journal of Remote Sensing* **15**: 697–703.

Chazdon RL. 1988. Sunflecks and their importance to forest understorey plants. *Advances in Ecological Research* **18**: 1–52.

Clark RN, Roush TL. 1984. Reflectance spectroscopy: quantitative analysis techniques for remote sensing applications. *Journal of Geophysical Research* **89**: 6329–6340.

Coley PD, Aide TM. 1989. Red coloration of tropical young leaves: a possible anti-fungal defence? *Journal of Tropical Ecology* **5**: 283–300.

Coley PD, Barone JA. 1996. Herbivory and plant defenses in tropical forests. *Annual Review of Ecology and Systematics* **27**: 305–335.

Curran PJ. 1989. Remote sensing of foliar chemistry. *Remote Sensing of Environment* **30**: 271–278.

Curran PJ, Kupiec JA. 1995. Imaging spectrometry: a new tool for ecology. In: Danson FM, Plummer SE, eds. *Advances in environmental remote sensing*. Chichester, UK: Wiley, 71–88.

Day TA, Vogelmann TC, DeLucia EH. 1992. Are some plant life forms more effective than others in screening out ultraviolet-B radiation? *Oecologia* **92**: 513–519.

Demetriades-Shah TH, Steven MD, Clark JA. 1990. High resolution derivative spectra in remote sensing. *Remote Sensing of Environment* **33**: 55–64.

Demmig-Adams B, Adams WW III. 1992. Carotenoid composition in sun and shade leaves of plants with different life forms. *Plant, Cell and Environment* **15**: 411–419.

Filella I, Serrano L, Serra J, Peñuelas J. 1995. Evaluating wheat nitrogen status with canopy reflectance indices and discriminant analysis. *Crop Science* 35: 1400–1405.

Frank HA, Cua A, Chynwat V, Young A, Gosztola D. 1994. Photophysics of the carotenoids associated with the xanthophyll cycle in photosynthesis. *Photosynthesis Research* 41: 389–395.

Fukshansky L. 1981. Optical properties of plants. In: Smith H, ed. *Plants and the daylight spectrum.* London, UK: Academic Press, 21–40.

Gamon JA, Field CB, Bilger W, Björkman O, Fredeen AL, Peñuelas J. 1990. Remote sensing of the xanthophyll cycle and chlorophyll fluorescence in sunflower leaves and canopies. *Oecologia* 85: 1–7.

Gamon JA, Field CB, Goulden M, Griffin K, Hartley A, Joel G, Peñuelas J, Valentini, R. 1995. Relationships between NDVI, canopy structure, and photosynthetic activity in three Californian vegetation types. *Ecological Applications* 5: 28–41.

Gamon JA, Filella I, Peñuelas I. 1993. The dynamic 531-nanometer Δ reflectance signal: a survey of twenty angiosperm species. *Current Topics in Plant Physiology* 8: 172–177.

Gamon JA, Peñuelas J, Field CB. 1992. A narrow-waveband spectral index that tracks diurnal changes in photosynthetic efficiency. *Remote Sensing of Environment* 41: 35–44.

Gamon JA, Serrano L, Surfus JS. 1997. The photochemical reflectance index: an optical indicator of photosynthetic radiation use efficiency across species, functional types, and nutrient levels. *Oecologia* 112: 492–501.

Gamon JA, Qiu H-L. 1999. Ecological applications of remote sensing at multiple scales. In: Pugnaire FI, Valladares F, eds. *Handbook of functional plant ecology.* New York, USA: Marcel Dekker, 805–846.

Gao BC. 1996. NDWI – a normalized difference water index for remote sensing of vegetation liquid water from space. *Remote Sensing of Environment* 58: 257–266.

Gao BC, Goetz AFH. 1994. Extraction of dry leaf spectral features from reflectance spectra of green vegetation. *Remote Sensing of Environment* 47: 369–374.

Gao BC, Goetz AFH. 1995. Retrieval of equivalent water thickness and information related to biochemical components of vegetation canopies from AVIRIS data. *Remote Sensing of Environment* 52: 155–162.

Gausman HW. 1985. *Plant leaf optical properties.* Lubbock, TX, USA: Texas Tech University Press.

Gausman HW, Allen WA, Escobar DE, Rodriquez RR, Cardenas R. 1971. Age effects of cotton leaves on light reflectance, transmittance and absorption and on water content and thickness. *Agronomy Journal* 63: 465–469.

Gitelson A, Merzlyak MN. 1994a. Spectral reflectance changes associated with autumn senescence of *Aesculus hippocastanum* L. and *Acer platanoides* L. leaves. Spectral features and relation to chlorophyll estimation. *Journal of Plant Physiology* 143: 286–292.

Gitelson AA, Merzlyak MN. 1994b. Quantitative estimation of chlorophyll *a* using reflectance spectra. Experiments with autumn chestnut and maple leaves. *Journal of Photochemistry and Photobiology B* 22: 247–252.

Gould KS, Kuhn DN, Lee DW, Oberbauer SF. 1995. Why leaves are sometimes red. *Nature* 378: 241–242.

Govaerts YM, Jacquemoud S, Verstraete MM, Ustin SL. 1996. Three-dimensional radiation transfer modeling in a dicotyledon leaf. *Applied Optics* 35: 6585–6598.

Grime JP. 1979. *Plant strategies and vegetation processes.* Chichester, UK: Wiley.

Grossman YL, Ustin SL, Jacquemoud S, Sanderson EW, Schmuck G, Verdebout J. 1996. Critique of stepwise multiple linear regression for the extraction of leaf biochemistry information from leaf reflectance data. *Remote Sensing of Environment* 56: 182–193.

Harborne JB. 1984. *Phytochemical methods: a guide to modern techniques of plant analysis.* London, UK: Chapman & Hall.

Hatfield JL, Asrar G, Kanemasu ET. 1984. Intercepted photosynthetically active radiation estimated by spectral reflectance. *Remote Sensing of Environment* 14: 65–75.

Heath RL. 1994. Possible mechanisms for the inhibition of photosynthesis by ozone. *Photosynthesis Research* 39: 439–451.

Jacquemoud S, Baret F. 1990. Prospect: a model of leaf optical properties spectra. *Remote Sensing of Environment* 34: 75–91.

Jacquemoud S, Ustin SL, Verdebout J, Schmuck G, Andreoli G, Hosgood B. 1996. Estimating leaf biochemistry using the PROSPECT leaf optical properties model. *Remote Sensing of Environment* 56: 194–202.

Johnson LF, Hlavka CA, Peterson DL. 1994. Multivariate analysis of AVIRIS data for canopy biochemical estimation along the Oregon Transect. *Remote Sensing of Environment* 47: 216–230.

Khamis S, Lamaze T, Lemoine Y, Foyer C. 1990. Adaptation of the photosynthetic apparatus in maize leaves as a result of nitrogen limitation. *Plant Physiology* 94: 1436–1443.

Kumar M, Monteith JL. 1981. Remote sensing of crop growth. In: Smith H, ed. *Plants and the daylight spectrum.* London, UK: Academic Press, 133–144.

Lichtenthaler HK. 1987. Chlorophylls and carotenoids: pigments of photosynthetic biomembranes. *Methods in Enzymology* 148: 349–382.

Mancinelli AL, Yang C-PH, Lindquist P, Anderson OR, Rabino I. 1975. Photocontrol of anthocyanin synthesis. *Plant Physiology* 55: 251–257.

Martin ME, Aber JD. 1997. High resolution remote sensing of forest canopy lignin, nitrogen, and ecosystem processes. *Ecological Applications* 72: 431–443.

Merzlyak MN, Gitelson A. 1995. Why and what for the leaves are yellow in autumn? On the interpretation of optical spectra of senescing leaves (*Acer platanoides* L.). *Journal of Plant Physiology* 145: 315–320.

Murray JR, Hackett WP. 1991. Dihydroflavonol reductase activity in relation to differential anthocyanin accumulation in juvenile and mature phase *Hedera helix* L. *Plant Physiology* 97: 343–351.

Pearcy RW. 1988. Photosynthetic utilisation of lightflecks by understory plants. *Australian Journal of Plant Physiology* 15: 223–238.

Pearcy RW, Pfitsch WA. 1994. The consequence of sunflecks for photosynthesis and growth of forest understory plants. In: Schulze E-D, Caldwell MM, eds. *Ecophysiology of photosynthesis.* Berlin, Germany: Springer-Verlag, 343–359.

Peñuelas J, Baret F, Filella I. 1995a. Semi-empirical indices to assess carotenoids/chlorophyll *a* ratio from leaf spectral reflectance. *Photosynthetica* 31: 221–230.

Peñuelas J, Filella I. 1998. Visible and near-infrared reflectance techniques for diagnosing plant physiological status. *Trends in Plant Science* 3: 151–156.

Peñuelas J, Filella I, Biel C, Serrano L, Save R. 1993. The reflectance at the 950–970 nm region as an indicator of plant water status. *International Journal of Remote Sensing* 14: 1887–1905.

Peñuelas J, Filella I, Gamon JA. 1995b. Assessment of photosynthetic radiation-use efficiency with spectral reflectance. *New Phytologist* 141: 291–296.

Peñuelas J, Gamon JA, Fredeen AL, Merino J, Field CB. 1994. Reflectance indices associated with physiological changes in nitrogen- and water-limited sunflower leaves. *Remote Sensing of Environment* 48: 135–146.

Pinol J, Filella I, Ogaya R, Peñuelas J. 1998. Ground-based spectroradiometric estimation of live fine fuel moisture of Mediterranean plants. *Agricultural and Forest Meteorology* 90: 173–186.

Pinzon JE, Ustin SL, Castenada CM, Smith MO. 1998. Investigation of leaf biochemistry by hierarchical foreground/background analysis. *IEEE Transactions on Geoscience and Remote Sensing* 36: 1–15.

Reich PB, Kloeppel BD, Ellsworth DS, Walters MB. 1995. Different photosynthesis–nitrogen relations in deciduous hardwood and evergreen coniferous species. *Oecologia* 104: 24–30.

Reich PB, Uhl C, Walters MB, Ellsworth DS. 1991. Leaf lifespan as a determinant of leaf structure and function among 23 Amazonian tree species. *Oecologia* 86: 16–24.

Reich PB, Walters MB, Ellsworth DS. 1992. Leaf life-span in relation to leaf, plant, and stand characteristics among diverse ecosystems. *Ecological Monographs* 62: 365–392.

Robberecht R, Caldwell MM, Billings WD. 1983. Leaf

ultraviolet optical properties along a latitudinal gradient in the arctic–alpine life zone. *Ecology* **61**: 612–619.

Sestak Z. 1985. Chlorophylls and carotenoids during leaf ontogeny. In: Sestak Z, ed. *Photosynthesis during leaf development*. Dordrecht, The Netherlands: Dr W Junk, 76–106.

Thayer SS, Björkman O. 1990. Leaf xanthophyll content and composition in sun and shade determined by HPLC. *Photosynthesis Research* **23**: 331–343.

Ustin SL, Smith MO, Adams JB. 1993. Remote sensing of ecological processes: a strategy for developing and testing ecological models using spectral mixture analysis. In: Ehleringer JR, Field CB, eds. *Scaling of physiological processes: leaf to globe*. San Diego, CA, USA: Academic Press, 339–357.

Wellburn AR. 1994. *Air pollution and climate change: the biological impact*, *2nd edn.* Harlow, Essex, UK: Longman Scientific and Technical.

Wessman CA. 1990. Evaluation of canopy biochemistry. In: Hobbs RJ, Mooney HA, eds. *Remote sensing of biosphere functioning*. Berlin, Germany: Springer-Verlag, 135–156.

Wessman CA, Aber JD, Peterson DL, Melillo JM. 1988. Remote Sensing of canopy chemistry and nitrogen cycling in temperate forest ecosystems. *Nature* **335**: 154–156.

Yamamoto HY. 1979. Biochemistry of the xanthophyll cycle in higher plants. *Pure and Applied Chemistry* **51**: 639–648.

Yoder BJ, Pettigrew-Crosby RE. 1995. Predicting nitrogen and chlorophyll content and concentration from reflectance spectra (400–2500 nm) at leaf and canopy scales. *Remote Sensing of Environment* **53**: 199–211.

Relationships between photosynthesis, nitrogen and leaf structure in 14 grass species and their dependence on the basis of expression

E. GARNIER*, J.-L. SALAGER, G. LAURENT AND L. SONIÉ

Centre d'Ecologie Fonctionnelle et Evolutive (CNRS-UPR 9056), 1919, Route de Mende, 34293 Montpellier Cedex 5, France

Received 6 November 1998 ; accepted 8 April 1999

SUMMARY

The relationships between leaf structure, nitrogen concentration and CO_2 assimilation rate (A) were studied for 14 grass species grown in the laboratory under non-limiting nutrient conditions. Structural features included leaf thickness and density, and the proportion of leaf volume occupied by different types of tissue (mesophyll, epidermis, vessels and sclerenchyma). Relationships were assessed for data expressed per unit leaf area and fresh mass. The latter was found to be closely related to leaf volume, which allowed us to use A per unit leaf fresh mass (A_{fm}) as a surrogate of A per unit leaf volume. Assimilation rate per unit leaf area (A_a) was positively correlated with leaf thickness and with the amount of mesophyll per unit leaf area; the relationship with leaf nitrogen content per unit area was only marginally significant. A_{fm} was negatively correlated with leaf thickness and positively with fresh mass-based leaf organic nitrogen concentration. A multiple regression involving these two variables explained 81% of the variance in A_{fm}. The value of A_{fm} was also significantly related to the proportion of mesophyll in the leaf volume, but surprisingly the correlation was negative. This was because thin leaves with high A_{fm} and nitrogen concentration had proportionally more mechanically supportive tissues than thick ones; as a consequence, they also had a lower proportion of mesophyll. These data suggest that, in addition to leaf nitrogen, leaf thickness has a strong impact on CO_2 assimilation rate for the grass species studied.

Key words: leaf anatomy, leaf nitrogen, leaf thickness, grasses, light penetration, mesophyll, assimilation rate, mechanical support.

INTRODUCTION

Differences in photosynthetic rate among species may be attributed to differences in biochemical, morphological and/or anatomical features of their leaves (e.g. Nobel, 1983; Sharkey, 1985). In particular, a positive correlation is usually observed between the net assimilation rate of CO_2 and the nitrogen (N) content of the leaf (Field & Mooney, 1986; Evans, 1989), which may be explained by the fact that up to 75% of leaf organic N is in the chloroplasts, most of it in the photosynthetic machinery (Evans & Seemann, 1989). The influence of leaf structure on photosynthesis has also been the focus of much research, but only a few generalities have emerged. This may be the consequence of the

number of, and interactions between, the parameters involved (e.g. Austin *et al.*, 1982).

To our knowledge, few studies have tried to relate the overall structure of a leaf, its N content and its assimilation rate for a relatively large number of species (but see Niinemets, 1999). The aim of the present study is to do so for 14 grass species which have been shown to differ substantially in their leaf N concentration (Garnier & Vancaeyzeele, 1994) and structure (Garnier & Laurent, 1994). Differences in leaf structure include thickness, density, and the proportion of the leaf volume occupied by various types of tissue (mesophyll, epidermis, sclerenchyma and vascular bundles). The present analysis considers the interplay between these structural and biochemical features, and how these affect CO_2 assimilation rate.

A first step was to choose a basis of expression for the rate of CO_2 assimilation (A). In physiological

*Author for correspondence (fax +33 4 67 41 21 38; e-mail garnier@non.cefe.cnrs-mop.fr).

research, A is traditionally expressed per unit leaf area (A_a), which allows the flux of carbon to be related to that of incoming photons. However, the size of the assimilatory apparatus of the leaf depends not only on its area, but also on other structural features such as its thickness and/or the proportion of its volume occupied by the different tissues (particularly mesophyll). Differences in A_a may thus be simply due to differences in the amount of photosynthetic tissue per unit area. For example, A_a is generally found to be higher in sun than in shade leaves, but this is associated with thicker leaves in the former (Charles-Edwards & Ludwig, 1975; Nobel, 1977; McMillen & McClendon, 1983; Sims & Pearcy, 1992), and thus with more assimilatory tissue per unit leaf area. To separate this structural effect from, say, biochemical effects, Charles-Edwards & Ludwig (1975) proposed that A be expressed on a leaf volume basis. When this is done, differences in A between sun and shade leaves can disappear (Charles-Edwards & Ludwig, 1975; Sims & Pearcy, 1992) or be reversed (McMillen & McClendon, 1983). Other bases of expression may be chosen to account for this size effect (leaf dry mass, e.g. Enríquez *et al.*, 1996; Reich *et al.*, 1997; N content, e.g. Field & Mooney, 1986; Poorter & Evans, 1998), but considering that 'the rate at which living systems operate is a function of substrate and/or product concentration' (i.e. amount of components per unit volume), Charles-Edwards *et al.* (1972) argued that plant processing rates in general should be expressed as volume. Taking an engineering perspective on leaf form and function, Roderick *et al.* (1999) arrived at a similar conclusion. Leaf area (usual in plant physiology) and leaf volume thus appear to be relevant for expression of assimilation rate.

However, leaf volume is not easy to measure. It is usually calculated from other dimensions of the leaf, as the product of leaf surface area × leaf thickness (often measured at one single location along the leaf blade). Leaf saturated fresh mass ('fresh mass' hereafter) is thus sometimes used as an estimate of leaf volume. This is done, for example, when leaf thickness is calculated as the ratio between fresh mass (~volume) and surface area (McMillen & McClendon, 1983; Atkin *et al.*, 1996; see Sims *et al.*, 1998 for a validation for soybean). Using fresh mass rather than volume has at least two advantages: it is much easier to measure, so easy that it can be routinely incorporated as a standard measure in large-scale comparative experiments (Wilson *et al.*, 1999); and it does not imply any hypothesis on the homogeneity of leaf thickness over the whole blade. Whether using one instead of the other is a valid approximation will be examined for the 14 grass species studied here.

Having chosen the basis of expression, the specific questions addressed in this study were: which leaf structural parameters (thickness, density, proportion of tissues, etc) affect net CO_2 assimilation rate?; is leaf N related to leaf structure, and how does it affect A?; and how are the relationships between leaf structure, leaf N and A modified when different bases of expression are used (leaf area vs leaf fresh mass basis)? These questions were examined for 14 grass species grown in the laboratory under standardized conditions.

MATERIALS AND METHODS

Species, germination and growth conditions

This study was conducted on 14 C_3 grass species (Poaceae) belonging to the same subfamily, the Pooideae (Watson & Dallwitz, 1992). Seven species were annuals: *Brachypodium distachyon* L., *Bromus hordeaceus* L., *Bromus madritensis* L., *Hordeum murinum* L., *Lolium rigidum* Gaud., *Poa annua* L. and *Avena barbata* Pott. Seven species were perennials: *Brachypodium phoenicoides* L., *Bromus erectus* Huds., *Bromus ramosus* Huds., *Hordeum secalinum* Schreb., *Lolium perenne* L., *Poa pratensis* L. and *Dactylis glomerata* L. (subsp. *hispanica* Roth.). Differences between life forms have been discussed elsewhere (Garnier, 1992; Garnier & Laurent, 1994; Garnier & Vancaeyzeele, 1994), and will not be dealt with in this paper.

Plants were grown from seeds in a water-culture system in a growth room at ambient atmospheric CO_2 concentration (not controlled). Light was provided by four metal halide lamps (Osram HQI-T 400 W), and the mean ($\pm$SE) photon flux of photosynthetically active radiation (PAR) at seedling height was 545 ± 15 μmol m^{-2} s^{-1} ($n = 108$). The plants received a photoperiod of 16:8 h (light:dark) at an air temperature of 22:16°C (light:dark), and rh $> 55\%$ (i.e. a vapour pressure deficit (VPD) < 1.2 kPa during the day).

The water-culture system consisted of a 100 dm^3 tank filled with the complete nutrient solution described by Koch *et al.* (1987), where nitrate concentration was set at 500 mmol m^{-3}; to avoid any substantial drop in the concentration of ions during the experiments this solution was completely renewed every week. Germination, transplantation and growing procedures were similar to those described by Garnier (1992). Due to limited space, the plants were cultivated in four successive batches. To check for the repeatability of the growing conditions, six individuals of *Bromus hordeaceus* were grown in each batch at a fixed place in the culture system. Their mean relative growth rate was calculated from periodic measurements of total fresh mass. The four values obtained ranged from 0.247 to 0.289 g g^{-1} d^{-1}, and were not significantly different from one another ($P > 0.05$ from an analysis of covariance). These values did not differ significantly from those obtained previously in a leaf anatomy

Table 1. *Leaf CO_2 assimilation rate expressed per unit leaf area (A_a), leaf fresh mass (A_{fm}, leaf volume (A_{vol}), leaf dry mass (A_{dm}) and leaf nitrogen ($A/[N]_{lf}$), for 14 grass species grown under standard conditions in the laboratory*

Species	ID	A_a (µmol m⁻² s⁻¹)	A_{fm} (nmol g⁻¹ s⁻¹)	A_{vol} (nmol cm⁻³ s⁻¹)	A_{dm} (nmol g⁻¹ s⁻¹)	$A/[N]_{lf}$ (µmol mol⁻¹ s⁻¹)
Brachypodium distachyon	1	14.5 (0.7)	156 (9)	121 (6)	508 (24)	130 (2)
Brachypodium phoenicoides	2	15.8 (1.2)	86 (7)	88 (7)	319 (25)	121 (17)
Bromus hordeaceus	3	17.4 (0.6)	91 (3)	89 (3)	509 (17)	134 (9)
Bromus erectus	4	15.2 (1.1)	74 (6)	82 (6)	321 (23)	106 (13)
Bromus madritensis	5	19.6 (0.6)	87 (2)	109 (3)	537 (16)	181 (6)
Bromus ramosus	6	15.6 (0.5)	122 (4)	90 (3)	403 (14)	123 (12)
Hordeum murinum	7	14.4 (0.4)	82 (3)	70 (2)	489 (13)	141 (7)
Hordeum secalinum	8	16.0 (0.5)	93 (4)	95 (3)	492 (16)	136 (8)
Lolium rigidum	9	15.6 (0.8)	79 (3)	73 (4)	524 (28)	153 (14)
Lolium perenne	10	17.2 (0.5)	105 (3)	80 (2)	546 (16)	148 (9)
Poa annua	11	14.2 (1.4)	118 (9)	126 (12)	613 (59)	155 (3)
Poa pratensis	12	15.1 (0.8)	131 (9)	122 (6)	539 (28)	152 (3)
Avena barbata	13	19.4 (0.5)	89 (3)	86 (2)	561 (15)	134 (9)
Dactylis glomerata	14	15.8 (0.6)	104 (7)	105 (4)	516 (20)	146 (9)
CV (%)		11	23	19	18	13

ID, identification of species on figures. Data presented are mean (SE in parenthesis) of 10 (four for $A/[N]_{lf}$) replicate values per species. CV is the coefficient of variation across species, calculated using the mean value of each species (i.e. not taking into account intra-specific variation).

experiment (range 0.230–0.269 g g⁻¹ d⁻¹; Garnier & Laurent, 1994).

Gas-exchange measurements

Twenty days after transplanting the plants into the culture system, CO_2 exchange was measured on the youngest mature leaf (i.e. whose ligule had appeared) of the main tiller, on 10 replicate plants per species. These measurements were conducted with a portable leaf chamber analyser (Analytical Development Company, London, UK, model LCA-2, used with a Parkinson leaf chamber for grass leaves). Environmental conditions in the leaf chamber were maintained close to those of the growth environment. The average PAR was 604 ($\pm$12) µmol m⁻² s⁻¹, and air temperature was 22.1 ($\pm$0.4)°C. Air flow through the chamber was between 4.25 and 4.45 cm³ s⁻¹, and the VPD of the air leaving the chamber was between 0.8 and 1.1 kPa.

Harvests and chemical analyses

Immediately after the gas-exchange measurements, the portion of the leaf which was inserted in the chamber was severed from the blade; its saturated fresh mass was measured, and its area was determined with an area meter (Delta-T Devices, Cambridge, UK, model MK2). Dry mass was measured after oven-drying for 48 h at 70°C.

For each species, two or three leaves (severed portion plus remaining part) were then pooled to obtain four replicate batches of leaf samples. These batches were ground individually, and their total N concentration was measured with an elemental

analyser (Carlo Erba Instruments, Milan, Italy, model EA 1108). Nitrate was extracted from subsamples of the ground material placed in polypropylene vials with 10 cm³ 0.1 M HCl, and kept at 4°C for 48 h. The nitrate content of each sample was assayed colorimetrically according to Henriksen & Semer-Olsen (1970). Organic N concentration was then calculated as the difference between total N and nitrate concentrations.

Treatment of data

The raw data for net assimilation rate were calculated for leaves on which gas exchange was measured for: area (A_a); saturated fresh mass (A_{fm}); dry mass (A_{dm}); leaf organic N ($A/[N]_{lf}$), the so-called photosynthetic nitrogen-use efficiency (Field & Mooney, 1986). Data were also calculated on a leaf volume basis (A_{vol}) using the average volume values obtained previously (Garnier & Laurent, 1994), which were calculated as the product of leaf thickness and surface area. Data on global leaf structural parameters in the two sets of experiments were in good agreement, as shown by the relationships between the specific leaf area (SLA) and the leaf percentage dry matter (PDM) measured on the two occasions:

$$SLA_{ps} = 0.98\ SLA_{ana} + 0.66$$

($r = 0.82$, $P < 0.001$, $n = 14$), and

$$PDM_{ps} = 0.94\ PDM_{ana} + 0.03$$

($r = 0.78$, $P < 0.001$, $n = 14$),

($_{ps}$ and $_{ana}$ refer to the current photosynthesis and anatomy (Garnier & Laurent, 1994) experiments, respectively). The anatomical data (leaf thickness, density, proportion, and amount of different tissues in the leaf) were all taken from Garnier & Laurent (1994).

Correlations between the five bases of expression of A are presented (Table 2), but for reasons outlined in the introduction, relationships between anatomical features and assimilation rate will be examined only for A_a and A_{fm} (Figs 2, 3) and A_{vol} (Table 3). Correlations between parameters were assessed as Pearson's correlation coefficients among species' means of characters.

RESULTS

Photosynthetic rates, leaf fresh mass and volume

Leaf net CO_2 assimilation rate varied approximately twofold when expressed as leaf fresh mass, volume or dry mass (Table 1); when expressed on a leaf area basis, all values were within a much narrower range (between 14.2 and 19.6 µmol m^{-2} s^{-1}, Table 1); variability in $A/[N]_{lf}$ among species was intermediate (compare the coefficients of variation in Table 1). Table 2 shows the correlations among the different bases of expression. The only significant ones were between (i) A per unit fresh mass and A per unit volume, and (ii) A per unit dry mass and $A/[N]_{lf}$; that between A per unit dry mass and A per unit volume was marginally so ($P = 0.07$).

There was a close association between the fresh mass of a leaf and its volume (Fig. 1a). With the exception of four species, this translated into a good agreement between assimilation rate expressed as volume and as fresh mass (Fig. 1b). Although this relationship was not very close ($r = 0.73$, $P < 0.01$), the correlations between photosynthetic rate and anatomical features were qualitatively similar for assimilation rate data expressed on these two bases (Table 3). The significance level differed slightly depending on the basis of expression: in the case of the proportion of leaf volume occupied by mesophyll, the conclusion as to whether the relationship was significant with a P value fixed at 0.05 differed between the two bases.

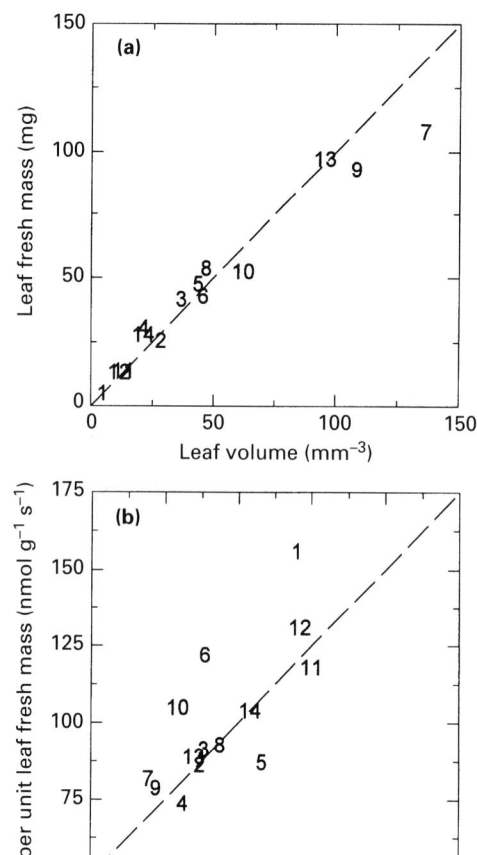

Fig. 1. Relationships between (a) the volume of a leaf and its fresh mass ($r = 0.98$, $P < 0.001$); (b) volume-based photosynthetic rate and fresh mass-based photosynthetic rate ($r = 0.73$, $P < 0.01$), for 14 grass species grown under standard conditions in the laboratory. Numbers identify species (see Table 1 for species codes). The dashed line represents the 1:1 relationship.

Table 3. *Pearson's correlation coefficients between various leaf structural parameters (taken from Garnier & Laurent, 1994) and CO_2 assimilation rate expressed per unit leaf area (A_a), per unit leaf fresh mass (A_{fm}) and per unit leaf volume (A_{vol})*

Independent variable	A_a	A_{fm}	A_{vol}
Leaf thickness	0.54*	−0.73**	−0.90***
Leaf density	−0.31ns	0.30ns	0.45†
Proportion mesophyll	0.41†	−0.46*	−0.43†
Proportion NP mesophyll	0.51*	−0.56*	−0.53*
Proportion epidermis	−0.29ns	0.31ns	0.22ns
Proportion vessels	−0.21ns	0.29ns	0.30ns
Proportion sclerenchyma	−0.29ns	0.20ns	0.31ns

NP mesophyll is for non-parietal mesophyll (i.e. mesophyll minus proportion of mesophyll occupied by cell walls). Data are for 14 grass species grown under standard conditions in the laboratory. Significance level: ***, $P < 0.001$; **, $P < 0.01$; *, $P < 0.05$; †, $P < 0.10$; ns, not significant ($P > 0.10$).

Table 2. *Pearson's correlation coefficients between net CO_2 assimilation rate expressed on different bases*

	A_a	A_{fm}	A_{vol}	A_{dm}	$A/[N]_{lf}$
A_a	1				
A_{fm}	−0.36ns	1			
A_{vol}	−0.16ns	0.73**	1		
A_{dm}	0.21ns	0.30ns	0.40†	1	
$A/[N]_{lf}$	0.33ns	0.05ns	0.35ns	0.74**	1

A_a is per unit leaf area; A_{fm} is per unit leaf fresh mass; A_{vol} is per unit leaf volume; A_{dm} is per unit leaf dry mass; $A/[N]_{lf}$ is per unit leaf nitrogen. Significance levels: **, $P < 0.01$; †, $P < 0.10$; ns, not significant ($P > 0.10$).

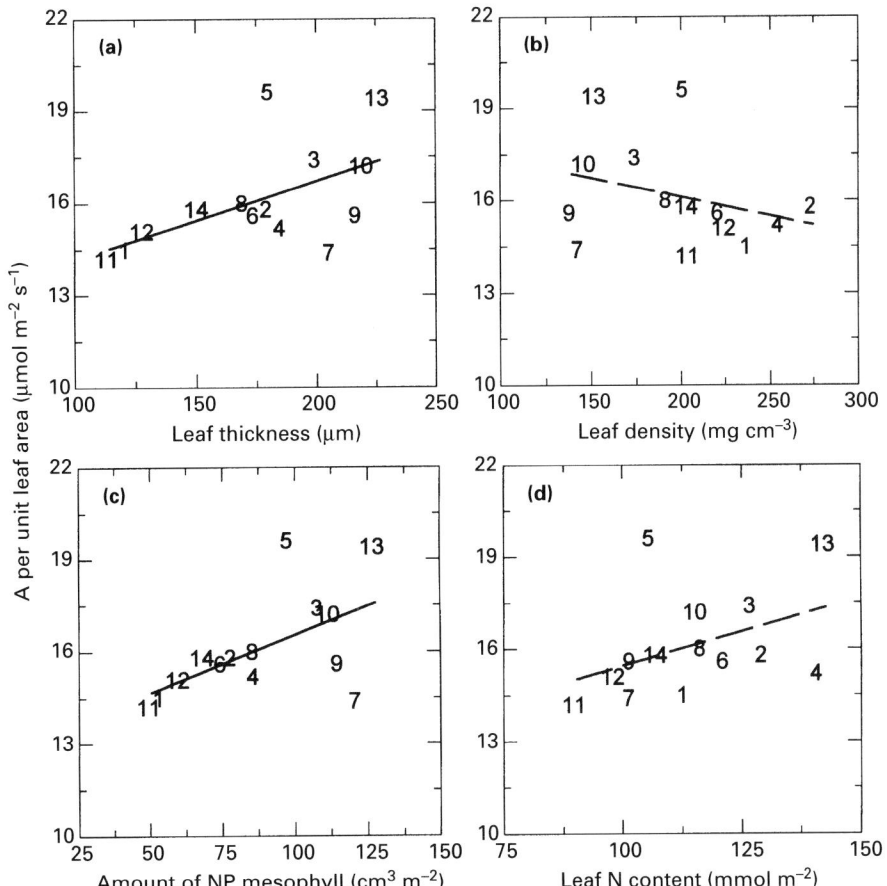

Fig. 2. Relationships between area-based photosynthetic rate (*A*) and (a) leaf thickness (*r* = 0.54, *P* <0.05); (b) leaf density (*r* = −0.31, not significant); (c) amount of non-parietal mesophyll per unit leaf area (*r* = 0.56, *P* <0.05); (d) leaf organic nitrogen content per unit leaf area (*r* = 0.41, *P* = 0.07), for 14 grass species grown under standard conditions in the laboratory. Numbers identify species (see Table 1 for species codes). A solid line indicates a significant relationship; broken line a ns relationship.

Hereafter, relationships between assimilation rate and leaf anatomy are examined only for *A* expressed per unit leaf area and as fresh mass.

Photosynthetic rate and leaf structure

Area basis. A_a was positively correlated to leaf thickness (Table 3, Fig. 2a), and to the proportion of the leaf volume occupied by non-parietal mesophyll (i.e. mesophyll minus the proportion of mesophyll occupied by cell walls (Garnier & Laurent, 1994), Table 3). A_a was also positively correlated (*P* <0.05) to the amount of non-parietal mesophyll per unit leaf area (Fig. 2c). Relationships between A_a and density, as well as those between A_a and the proportions of other tissues, were not significant (Table 3, Fig. 2b). Finally, A_a was only marginally related to leaf organic N content per unit leaf area (*P* = 0.07, Fig. 2d).

Fresh mass basis. Leaf thickness was the structural feature which correlated best with assimilation rate expressed as fresh mass (Table 3), the thicker the leaf, the lower the A_{fm} (Fig. 3a). The relationship between A_{fm} and leaf density was not significant (Fig. 3b). When the histological structure of the leaf

was considered, the only significant (negative) correlations were those between A_{fm} and the proportion of leaf volume occupied by the mesophyll and non-parietal mesophyll (Table 3, Fig. 3c). A_{fm} was also strongly and positively related to fresh mass-based leaf N concentration (Fig. 3d). A multiple regression combining leaf thickness (LT) and fresh mass-based N concentration ($[N]_{lfm}$) as independent variables explained 81% (*P* <0.001) of the variance of A_{fm} in our data set:

$$A_{fm}(\text{nmol g}_{fm}^{-1}\text{ s}^{-1}) = -0.195\text{ LT}(\mu\text{m})$$
$$+87.4\,[N]_{lfm}(\text{mmol g}_{fm}^{-1})+71.4 \quad \text{Eqn 1}$$

This analysis indicates the important role of leaf thickness in the determination of assimilation rate expressed both as area and fresh mass. The way in which the different leaf characteristics measured covary with thickness are considered below.

Correlates of leaf thickness

Thick leaves were found to have a significantly higher proportion of non-parietal mesophyll but a lower proportion of 'dense' tissues (i.e. sum of vessels plus sclerenchyma) than thin leaves (Fig.

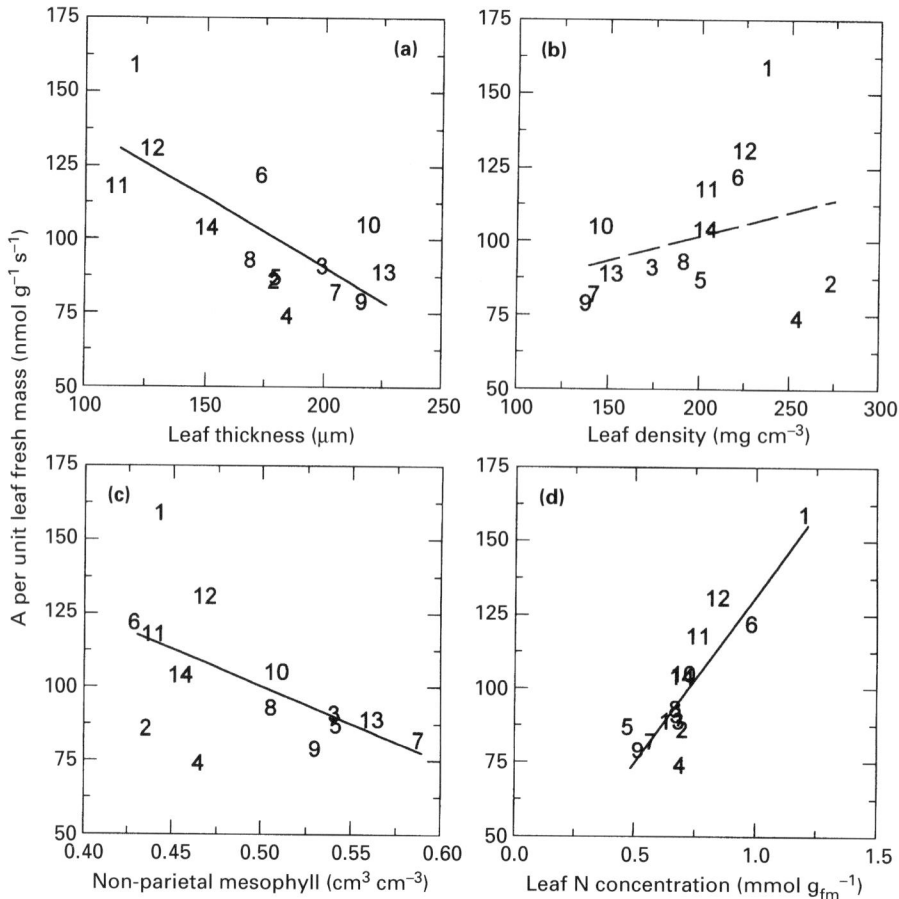

Fig. 3. Relationships between fresh mass-based photosynthetic rate (A) and (a) leaf thickness ($r = -0.73$, P <0.01); (b) leaf density ($r = 0.30$, not significant); (c) the proportion of non-parietal mesophyll in the leaf volume ($r = -0.56$, P <0.05); (d) leaf organic nitrogen concentration per unit leaf fresh mass ($r = 0.88$, P <0.001), for 14 grass species grown under standard conditions in the laboratory. Numbers identify species (see Table 1 for species codes). A solid line indicates a significant relationship; a broken line indicates a ns relationship.

4a,b). The proportions of these two types of tissue were negatively related to each other ($r = -0.52$, P < 0.05; compare Fig. 4a and b). As expected, the amount of tissue per unit leaf area was higher in thicker leaves (Fig. 4c,d). The relationship with the amount of non-parietal mesophyll was very close ($r = 0.96$, P <0.001; Fig. 4c), as the proportion of this tissue was also high in thick leaves (Fig. 4a).

Thick leaves were also found to have more N per unit area because they had a larger amount of tissue per unit area, but the relationship was not particularly close (Fig. 5a). By contrast, leaf thickness was negatively and more strongly related to fresh mass-based leaf N concentration (Fig. 5b). Finally, there was also a positive relationship between the thickness of a leaf and its area ($r = 0.71$, P <0.01; not shown).

DISCUSSION

Expression of CO_2 assimilation rate

Comparisons between bases of expression. A lower range of variation of A given on a leaf area basis

compared with other bases of expression (Table 1) seems to be a general finding in interspecific comparisons. This is the case when A_a is compared with A_{dm} (Field & Mooney, 1986; Atkin *et al.*, 1996; Poorter & Evans, 1998; Niinemets, 1999) or to $A/[N]_{lf}$ (Field & Mooney, 1986; Poorter & Evans, 1998). The greater ranges of variation were found for A expressed per unit fresh mass and volume (Table 1), but since A has seldom been expressed on either of these bases in previous studies, it is not known how general this finding is. Assimilation rates expressed on these different bases appear to be unrelated in most cases (Table 2) indicating that the choice of a particular basis to express data is of clear importance for the interpretation of results.

A detailed analysis of differences between bases of expression and of causes leading to the (lack of) observed correlations is beyond the scope of this paper, except for that between A_{fm} and A_{vol}, which is discussed in the following section.

Leaf fresh mass and volume. The close association between leaf fresh mass and volume observed at the

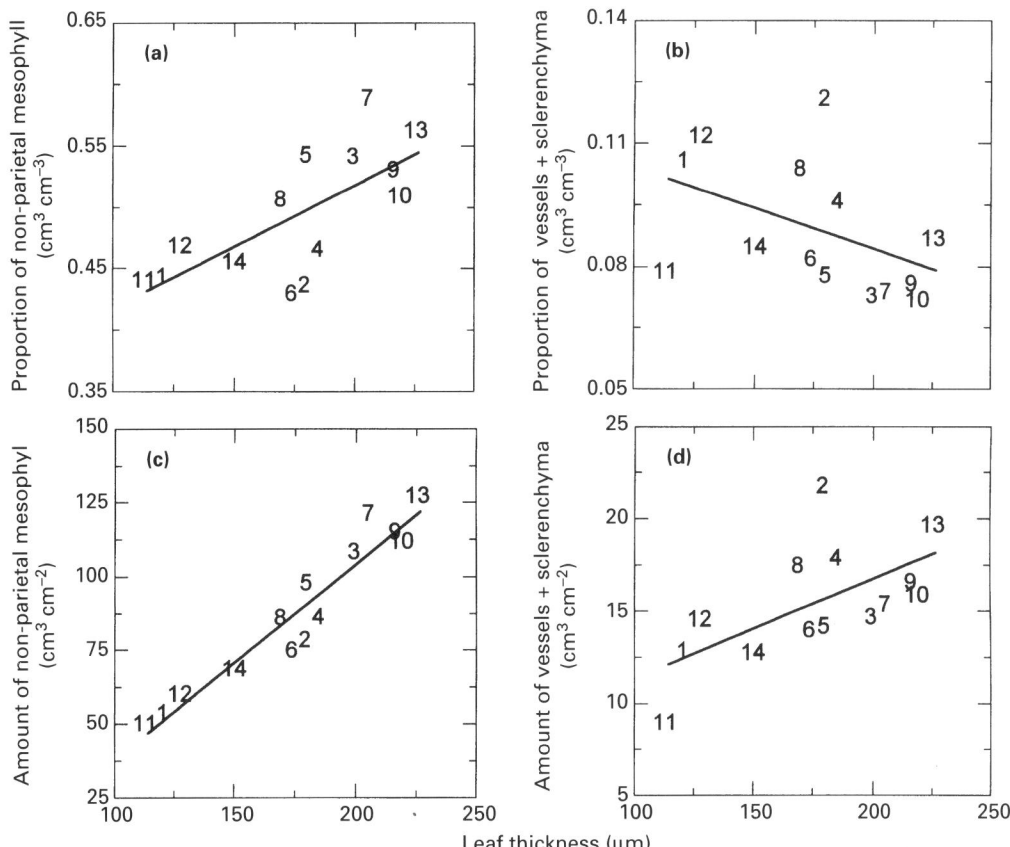

Fig. 4. Relationships between leaf thickness and (a) the proportion of non-parietal mesophyll in the leaf volume ($r = 0.70$, $P < 0.01$); (b) the proportion of 'dense' tissues (sclerenchyma plus vessels) in the leaf volume ($r = -0.45$, $P < 0.05$); (c) the amount of non-parietal mesophyll per unit leaf area ($r = 0.96$, $P < 0.001$); (d) the amount of 'dense' tissues (sclerenchyma plus vessels) per unit leaf area ($r = 0.62$, $P < 0.01$), for 14 grass species grown under standard conditions in the laboratory. Numbers identify species (see Table 1 for species codes). A solid line indicates a significant relationship.

interspecific level (Fig. 1a) has also been found when the variation in leaf structure of a single species was produced by varying different environmental conditions, such as light, N availability and atmospheric CO_2 (Sims *et al.*, 1998). Assuming that the leaf is composed of three phases – air, liquid and solid – (cf. Roderick *et al.*, 1999), the fresh mass of a leaf is equal to $(M_a + M_l + M_s)$, and its volume to $(V_a + V_l + V_s)$, where $_a$, $_l$ and $_s$ refer to the air spaces, liquid and solid phases, respectively. Given that the density of the air is negligible compared with that of the liquid and solid phases (i.e. $M_a \approx 0$), any increase in the volume of the leaf brought about by an increase only in the volume of air spaces will not affect its fresh mass. Therefore if two species have leaves with a similar assimilation rate per unit fresh mass but different proportions of air spaces, their assimilation rates per unit volume will be different.

The proportion of the leaf volume occupied by air spaces has been found to vary between 5 and 40% in 14 grass species, with most of the values (11 out of 14) between 10 and 20% (van Arendonk & Poorter, 1994; data are not available for the species studied here). Although assimilation rates expressed on fresh mass and volume bases were relatively well correlated

(Fig. 1b), such differences in the volume of air spaces between species are likely to explain a substantial part of the scatter observed in this correlation. Despite this scatter, the relationships between assimilation rate and the structural features of leaves were qualitatively similar using the two bases of expression (Table 3). Taken together, this validates the use of leaf fresh mass as a surrogate for leaf volume, at least in grasses. Whether this holds for a broader range of species remains to be tested.

Leaf structure, nitrogen and assimilation rate

Area basis. Contrary to what is usually found for sun and shade leaves, the positive relationship between A_a and leaf thickness observed here (Table 3, Fig. 2a) does not appear to be general in interspecific comparisons. It has been found in some instances (Nobel, 1977; McMillen & McClendon, 1983; Niinemets, 1999) but not in others (Körner & Diemer, 1987; Kebede *et al.*, 1994). Here, thicker leaves have a larger amount of non-parietal mesophyll tissue per unit area (Fig. 4c). This is for two reasons: the absolute amount of material is larger in these leaves (cf. Fig. 4c and d); and independent of

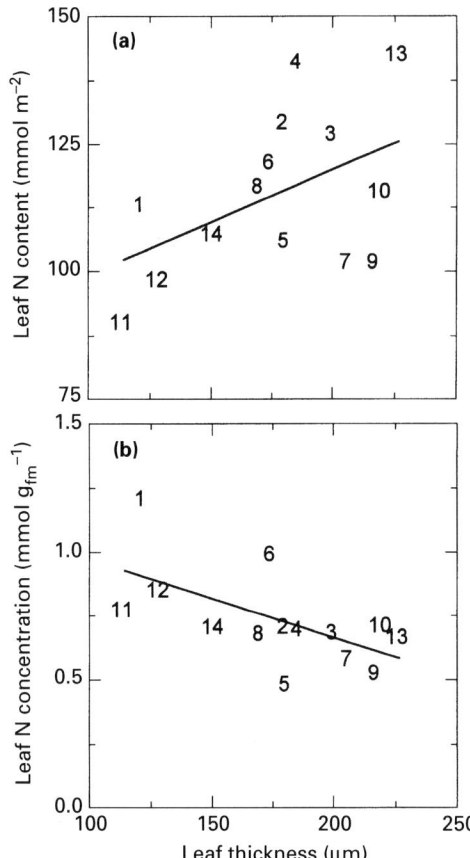

Fig. 5. Relationships between leaf thickness and (a) leaf organic nitrogen content per unit leaf area ($r = 0.48$, $P < 0.05$); (b) leaf organic nitrogen concentration per unit leaf volume ($r = -0.61$, $P < 0.01$), for 14 grass species grown under standard conditions in the laboratory. Numbers identify species (see Table 1 for species codes). A solid line indicates a significant relationship.

this, the proportion of mesophyll tissue is higher in thick leaves (Fig. 4a). A combination of these two factors leads to the very close relationship between leaf thickness and amount of mesophyll per unit leaf area. Thus the relationship between A_a and leaf thickness may be merely a consequence of the larger amount of mesophyll tissue per unit leaf area in thick leaves, also expressed in the relationship between the latter and A_a (Fig. 2c).

However, Nobel (1977) suggested that, rather than the amount of mesophyll, the parameter that relates best to A_a is the surface area of mesophyll cells exposed to internal air space, expressed per unit leaf area (S_m): the larger this surface, the greater the internal leaf area across which CO_2 can diffuse in the liquid phase, thereby increasing the transfer conductance to the sites of carboxylation. Since S_m increases with leaf thickness (Nobel, 1977; Patton & Jones, 1989; Kebede *et al.*, 1994), this could explain why A_a increases with leaf thickness for the species studied here.

A larger amount of mesophyll tissue per unit leaf area could also be expected to accompany a higher N content per area, but this was not the case (no

significant relationship between the two variables: $r = 0.34$, $P = 0.25$). More surprising was the lack of a significant relationship between A_a and leaf N content per unit area (Fig. 2d). Although a causal link between the two is usually postulated, when both variables are expressed on a unit leaf area basis this relationship is sometimes very loose (Field & Mooney, 1986; Evans, 1989; Peterson *et al.*, 1999) or even non-existent (Körner & Diemer, 1987). Here the relationship between A and leaf N appeared to be much tighter when expressed on a volume basis (Fig. 3d).

In summary, leaf thickness and amount of mesophyll per unit leaf area were the factors best explaining the variation in A_a for the species studied here. However the proportion of the variance explained (31% at best) was relatively low.

Fresh mass basis. A negative correlation between leaf thickness and assimilation rate comparable to that observed here (Fig. 3a) has been found in a number of studies where assimilation rate was expressed as leaf dry mass (A_{dm}). This was the case for 10 species of tropical trees (Kitajima, 1992, measurements on cotyledons); Enríquez *et al.* (1996) found the same on a broad range of photosynthetic organisms (from cyanobacteria to long-lived terrestrial plants), but did not in Niinemets (1999) a survey of woody plants. Data presented by Reich *et al.* (1997), showing a positive association between specific leaf area (SLA: the ratio of leaf area to leaf dry mass) and assimilation rate for a large number of terrestrial plants, may be interpreted in a similar way, bearing in mind that SLA actually depends on two components: SLA $= 1/($leaf thickness $\times$ leaf density$)$ (Witkowski & Lamont, 1991). It is not known whether such a negative association is generally found when assimilation rate is expressed on a leaf fresh mass basis (as for data presented in Fig. 3a), as A_{fm} is not available for many studies. However, there is no reason why the hypotheses put forward to explain this correlation when assimilation rate is expressed as dry mass do not apply when it is expressed as fresh mass.

Three sets of non-exclusive hypotheses could help explain the negative association between leaf thickness and assimilation rate per unit leaf fresh mass.

- Leaf organic-N concentration per unit fresh mass ($[N]_{lfm}$) was lower in thick than in thin leaves (Fig. 5b). Since it was also positively and strongly related to A per unit fresh mass (Fig. 3d), this implies that A_{fm} and leaf thickness were negatively related. This points to a lower concentration of the enzyme Rubisco in thick leaves (Field & Mooney, 1986; Evans, 1989), which may limit their capacity for CO_2 fixation. Interestingly, the lower $[N]_{lfm}$ of thick leaves is seen with a higher proportion of non-parietal meso-

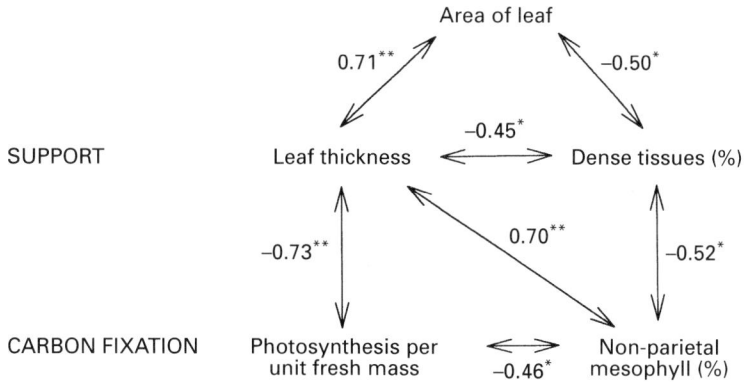

Fig. 6. Diagram representing the overall scheme for assimilation and support functions of the leaf. Percentage dense tissues refers to the sum of the percentages of vascular and sclerenchyma tissues in the leaf. Pearson's correlation coefficient is given for each relationship, with level of significance: **, $P < 0.01$; *, $P < 0.05$. See text for further explanations.

phyll (compare Figs 4a and 5b), indicating that the volume of mesophyll is not a good indicator of the volume of chloroplasts in the leaf.

- Leaf absorptance depends strongly on chlorophyll content per unit leaf area (Agustí *et al.*, 1994; Syvertsen *et al.*, 1995; Evans, 1998), but this does not seem to vary systematically with leaf thickness (Pammenter *et al.*, 1986; Kebede *et al.*, 1994). However, when expressed per unit mass, the average chlorophyll concentration is lower in thick than in thin leaves (Pammenter *et al.*, 1986; Agustí *et al.*, 1994; Kebede *et al.*, 1994), and/or in leaves with a low specific leaf area (Poorter & Evans, 1998). Since light absorption per unit mass was found to be positively related to chlorophyll *a* concentration (Agustí *et al.*, 1994), on average, a unit mass or volume is likely to absorb less light in thick than in thin leaves. How this affects assimilation rate depends on the light gradients within the leaf (Vogelmann, 1993) and on the coupling between light absorption and CO_2 fixation profiles within the leaf (see e.g. Evans, 1999).
- The path length of CO_2 within the leaf is supposed to be longer in thick leaves, inducing a higher resistance to CO_2 diffusion in the gaseous phase across the leaf (Parkhurst, 1994; Syvertsen *et al.*, 1995; Evans & von Caemmerer, 1996).

The ways in which leaf thickness is likely to influence assimilation rate are thus relatively complex, and it is not possible here to determine the relative importance of each of the features discussed.

Combining leaf thickness and N concentration allowed us to explain 81% of the variance in assimilation rate per unit fresh mass among the 14 grass species studied (Eqn 1). This confirms the results of Enríquez *et al.* (1996) and Reich *et al.* (1997), who have obtained comparable relationships (with all variables log-transformed) for a much broader range of species. It thus appears that, whatever the range of species compared, leaf as-

similation rate per unit leaf (dry or fresh) mass can be predicted relatively accurately by a combination of structural (leaf thickness and/or specific leaf area) and biochemical (chlorophyll and/or organic N concentration) characteristics.

The percentage of variance explained contrasts markedly with that obtained when A is expressed on a leaf area basis. This supports the expression of assimilation as leaf fresh mass rather than area, when it is to be related to leaf structure and chemical composition. In particular, the strong relationship between A and leaf N when expressed on a fresh mass basis agrees with Charles-Edwards *et al.*'s (1972) suggestion that a better explanation of processing rates may be achieved when these are expressed on a volume basis (see also Roderick *et al.*, 1999). Peterson *et al.* (1999) recently suggested that discrepancies between bases of expression could be resolved when the coefficient used to convert data from one basis to the other (1/SLA in their case) was statistically controlled. The data presented here are certainly amenable to the same type of analysis, but the coefficient involved would be leaf thickness. Indeed, $A_a = A_{fm} \times$ (fresh mass/area), and if we assume that fresh mass $\approx$ volume, then the fresh mass: area ratio is an estimate of leaf thickness.

Leaf assimilation rate and mechanical support

A counter-intuitive result of the present study is the negative association observed between A_{fm} and the proportion of leaf volume occupied by the mesophyll (with or without cell walls: Table 2, Fig. 3a). As assimilation rate takes place in the mesophyll, a positive association between the two was expected. Such a negative relationship may be interpreted as a side effect of leaf thickness on A_{fm}: our data show that leaves having proportionally more mesophyll are also thicker (Fig. 4a), and for reasons already discussed, their assimilation rate is lower.

This positive relationship between leaf thickness and the proportion of mesophyll may be best understood if the overall size and structure of the leaf are taken into account, together with the different functions it performs. An overall scheme is presented in Fig. 6 where carbon fixation and mechanical support are considered. The latter is assumed to be insured by leaf thickness and dense tissues (vessels and sclerenchyma). Our data show that larger leaves tend to support themselves by greater thickness, rather than by a higher proportion of dense tissues (Figs 4b, 6). By contrast, smaller and thinner leaves tend to support themselves by a higher proportion of vessels plus sclerenchyma (Figs 4b, 6). As a consequence, these leaves also have a lower proportion of mesophyll (compare Fig. 4a and b); for the reasons already discussed, this does not prevent them from having a high rate of CO_2 assimilation. This chain of correlations leads to a negative association between the area of a leaf and A per unit fresh mass ($r = -0.55$, $P < 0.05$, not shown, but see Fig. 6). This is also frequently found when A is expressed per unit leaf area (e.g. Evans & Dunstone, 1970; Austin *et al.*, 1982; Chapin *et al.*, 1989).

CONCLUSIONS

The relationships between leaf structure, N and assimilation rate differed when different bases of expression were used. When A was expressed per unit leaf area, the only significant relationships found were between leaf thickness and the amount of non-parietal mesophyll per unit leaf area; the percentage of variance explained by the parameters measured was relatively low (31% at best). When A was expressed on a fresh mass basis (found to be a proper estimate of volume) the two features that best explained A_{fm} were leaf thickness (negative relationship) and organic N concentration (positive relationship); 81% of the variance in A was explained by these two parameters The proportion of mesophyll in the leaf volume was the only histological parameter considered which was significantly correlated with A_m, but the correlation was negative. It is suggested that such a counter-intuitive result may be best understood if the overall size and structure of the leaf are taken into account, together with two of the main functions it performs, carbon fixation and support. This perspective stresses the central role of leaf thickness for the functioning of the leaf, in addition to that of leaf N.

ACKNOWLEDGEMENTS

Many thanks to Jacques Fabreguettes and François Jardon for the improvement and maintenance of the growing system, and to Alain Gojon and Bruno Touraine (INRA Montpellier, Biochimie et Physiologie Moléculaire des Plantes) who gave us advice on the nitrate measurements and access to their auto-analyser. Comments by John Evans, Hendrik Poorter and Jim Syvertsen contributed greatly to the improvement of the manuscript.

REFERENCES

Agustí S, Enríquez S, Frost-Christensen H, Sand-Jensen K, Duarte CM. 1994. Light harvesting in the plant kingdom. *Functional Ecology* **8**: 273–279.

Atkin OK, Botman B, Lambers H. 1996. The causes of inherently slow growth in alpine plants: an analysis based on the underlying carbon economies of alpine and lowland *Poa* species. *Functional Ecology* **10**: 698–707.

Austin RB, Morgan CL, Ford MA, Bhagwat SG. 1982. Flag leaf photosynthesis of *Triticum aestivum* and related diploid and tetraploid species. *Annals of Botany* **49**: 177–189.

Chapin FS III, Groves RH, Evans LT. 1989. Physiological determinants of growth rate in response to phosphorus supply in wild and cultivated *Hordeum* species. *Oecologia (Berlin)* **79**: 96–105.

Charles-Edwards DA, Charles-Edwards J, Sant FI. 1972. Models for mesophyll cell arrangement in leaves of ryegrass (*Lolium perenne* L.). *Planta* **104**: 297–305.

Charles-Edwards DA, Ludwig LJ. 1975. The basis of expression of leaf photosynthetic activities. In: Marcelle R, ed. *Environmental and biological control of photosynthesis*. The Hague, The Netherlands: Dr W. Junk, 37–44.

Enríquez S, Duarte CM, Sand-Jensen K, Nielsen SL. 1996. Broad-scale comparison of photosynthetic rates across phototrophic organisms. *Oecologia (Berlin)* **108**: 197–206.

Evans JR. 1989. Photosynthesis and nitrogen relationships in leaves of C₃ plants. *Oecologia (Berlin)* **78**: 9–19.

Evans JR. 1998. Photosynthetic characteristics of fast- and slow-growing species. In: Lambers H, Poorter H, Van Vuuren MMI, eds. *Inherent variation in plant growth. Physiological mechanisms and ecological consequences.* Leiden, The Netherlands: Backhuys Publishers, 101–119.

Evans JR. 1999. Leaf anatomy enables more equal access to light and CO_2 between chloroplasts. *New Phytologist* **143**: 93–104.

Evans JR, von Caemmerer S. 1996. Carbon dioxide diffusion inside leaves. *Plant Physiology* **110**: 339–346.

Evans JR, Seemann JR. 1989. The allocation of protein nitrogen in the photosynthetic apparatus: costs, consequences, and control. In: Briggs WR, ed. *Photosynthesis.* New York, USA: Alan R Liss, 183–205.

Evans LT, Dunstone RL. 1970. Some physiological aspects of evolution in wheat. *Australian Journal of Biological Sciences* **23**: 725–741.

Field CB, Mooney HA. 1986. The photosynthesis–nitrogen relationship in wild plants. In: Givnish TJ, ed. *On the economy of plant form and function.* Cambridge, UK: Cambridge University Press, 25–55.

Garnier E. 1992. Growth analysis of congeneric annual and perennial grass species. *Journal of Ecology* **80**: 665–675.

Garnier E, Laurent G. 1994. Leaf anatomy, specific mass and water content in congeneric annual and perennial grass species. *New Phytologist* **128**: 725–736.

Garnier E, Vancaeyzeele S. 1994. Carbon and nitrogen content of congeneric annual and perennial grass species: relationships with growth. *Plant, Cell and Environment* **17**: 399–407.

Henriksen A, Selmer-Olsen AR. 1970. Automatic methods for determining nitrate and nitrite in water and soil extracts. *Analyst* **95**: 514–518.

Kebede H, Martin B, Nienhuis J, King G. 1994. Leaf anatomy of two *Lycopersicon* species with contrasting gas exchange properties. *Crop Science* **34**: 108–113.

Kitajima K. 1992. Relationship between photosynthesis and thickness of cotyledons for tropical tree species. *Functional Ecology* **6**: 582–589.

Koch GW, Winner WE, Nardone A, Mooney HA. 1987. A system for controlling the root and shoot environment for plant growth studies. *Environmental and Experimental Botany* **27**: 365–377.

Körner Ch, Diemer M. 1987. *In situ* photosynthetic responses to light, temperature and carbon dioxide in herbaceous plants from low and high altitude. *Functional Ecology* **1**: 179–194.

McMillen GG, McClendon JH. 1983. Dependence of photosynthetic rates on leaf density thickness in deciduous woody plants grown in sun and shade. *Plant Physiology* **72**: 674–678.

Niinemets Ü. 1999. Components of leaf dry mass per area – thickness and density – alter leaf photosynthetic capacity in reverse directions in woody plants. *New Phytologist.* (In press.)

Nobel PS. 1977. Internal leaf area and cellular CO_2 resistance: photosynthetic implications of variations with growth conditions and plant species. *Physiologia Plantarum* **40**: 137–144.

Nobel PS. 1983. *Biophysical plant physiology and ecology.* San Francisco, CA, USA: WH Freeman.

Pammenter NW, Drennan PM, Smith VR. 1986. Physiological and anatomical aspects of photosynthesis of two *Agrostis* species at a sub-antarctic island. *New Phytologist* **102**: 143–160.

Parkhurst DF. 1994. Diffusion of CO_2 and other gases into leaves. *New Phytologist* **126**: 449–479.

Patton L, Jones MB. 1989. Some relationships between leaf anatomy and photosynthetic characteristics of willows. *New Phytologist* **111**: 657–661.

Peterson AG, CMEAL participants. 1999. Reconciling the apparent difference between mass– and area–based expressions of the photosynthesis–nitrogen relationship. *Oecologia (Berlin)* **118**: 144–150.

Poorter H, Evans JR. 1998. Photosynthetic nitrogen-use efficiency of species that differ inherently in specific leaf area. *Oecologia (Berlin)* **116**: 26–37.

Reich PB, Walters MB, Ellsworth DS. 1997. From tropics to tundra: global convergence in plant functioning. *Proceedings of the National Academy of Sciences, USA* **94**: 13730–13734.

Roderick ML, Berry SL, Noble IR. 1999. The relationship between leaf composition and morphology at elevated CO_2 concentrations. *New Phytologist* **143**: 63–72.

Sharkey TD. 1985. Photosynthesis in intact leaves of C_3 plants: physics, physiology and rate limitations. *Botanical Review* **51**: 53–105.

Sims DA, Pearcy RW. 1992. Response of leaf anatomy and photosynthetic capacity in *Alocasia macrorrhiza* (Araceae) to a transfer from low to high light. *American Journal of Botany* **79**: 449–455.

Sims DA, Seemann JR, Luo Y. 1998. Elevated CO_2 concentration has independent effects on expansion rates and thickness of soybean leaves across light and nitrogen gradients. *Journal of Experimental Botany* **49**: 583–591.

Syvertsen JP, Lloyd J, McConchie C, Kriedemann PE, Farquhar GD. 1995. On the relationship between leaf anatomy and CO_2 diffusion through the mesophyll of hypostomatous leaves. *Plant, Cell and Environment* **18**: 149–157.

van Arendonk JJCM, Poorter H. 1994. The chemical composition and anatomical structure of leaves of grass species differing in relative growth rate. *Plant, Cell and Environment* **17**: 963–970.

Vogelmann TC. 1993. Plant tissue optics. *Annual Review of Plant Physiology and Plant Molecular Biology* **44**: 231–251.

Watson L, Dallwitz MJ. 1992. *The grass genera of the world.* Wallingford, UK: CAB International.

Wilson PJ, Thompson K, Hodgson JG. 1999. Specific leaf area and leaf dry matter content as alternative predictors of plant strategies. *New Phytologist* **143**: 155–162.

Witkowski ETF, Lamont BB. 1991. Leaf specific mass confounds leaf density and thickness. *Oecologia (Berlin)* **88**: 486–493.

New Phytol. (1999), 143, 101-112

Leaf structure and specific leaf mass: the alpine desert plants of the Eastern Pamirs, Tadjikistan

VLADIMIR I. PYANKOV[1], ALEXANDRA V. KONDRATCHUK[1] AND BILL SHIPLEY[2]*

[1] *Department of Plant Physiology, Urals State University, Ekaterinburg 620083, Russia*
[2] *Department of Biology, University of Sherbrooke, Sherbrooke, Quebec J1K 2R1, Canada*

Received 5 October 1998; accepted 11 February 1999

SUMMARY

This study examines interrelationships between eight leaf attributes (specific leaf mass, area, dry mass, lamina thickness, mesophyll cell number per cm², mesophyll cell volume, chloroplast volume, and number of chloroplasts per mesophyll cell) in field-grown plants of 94 species from the Eastern Pamir Mountains, at elevations between 3800 and 4750 m. Unlike most other mountain areas, the Eastern Pamirs, Karakorum system, Tadjikistan provide localities where low temperatures and radiation combine with moisture stress at high altitudes. For all the attributes measured, significant differences were found between plants with different mesophyll types. Leaves with dorsiventral palisade structure (dorsal palisade, ventral spongy mesophyll cells) had thicker leaves with larger but fewer mesophyll cells, containing more and larger chloroplasts. These differences in mesophyll type are reflected in differences in the total surface of mesophyll cells per unit leaf area (A_{mes}/A) or volume (A_{mes}/V). Plants with isopalisade leaf structure (palisade cells under both dorsal and ventral surfaces) are more commonly xerophytes and their increased values of A_{mes}/A and A_{mes}/V decrease CO_2 mesophyll resistance, which is an important adaptation to drought. Path analysis shows the critical importance of mesophyll cell volume in leading to the covariance between the different leaf attributes and hence to specific leaf mass (SLM), even though mesophyll cell volume is not itself strongly correlated with SLM. This is because mesophyll cell volume increases SLM through its effects on leaf thickness and chloroplast number per cell, but decreases SLM through its negative effect on mesophyll cell density.

Key words: alpine vegetation, mesophyll structure, path analysis, structural equation modelling, specific leaf area, SLA, specific leaf mass, SLM.

INTRODUCTION

Much effort has been expended in characterizing the physiological functions, for example gas exchange or the economy of nitrogen (N) and carbon (C), of leaves from species with different habitat preferences. These physiological processes occur in structures (organelles, cells, tissues) which have specific spatial relationships to one another within the leaf. Because of this, leaf anatomy can be expected to affect the physiological processes of the leaf and, ultimately, the ecological and evolutionary success of plants growing in different environments. For instance, the ratio of the surface area of the mesophyll cells to the surface area of the leaf (A_{mes} : A) is correlated with the rate of CO_2 assimilation (Nobel et al., 1975; Longstreth & Nobel, 1979; Nobel & Walker, 1985; Patton & Jones, 1989). The ratio of the surface area of the chloroplasts to the surface area of the leaf (A_{chl} : A) is related to the internal resistance for CO_2 diffusion (Laisk et al., 1970; Araus et al., 1986; Evans et al., 1994; Evans & von Caemmerer, 1996). This suggests that the size, number and layering of mesophyll cells within the leaf, as well as the size and number of chloroplasts within each mesophyll cell, could exert strong control over basic physiological functions such as photosynthesis and transpiration. Presumably, different combinations of size and number of these cells and organelles will lead to different physiological outcomes in different environments.

Author for correspondence (fax 001 819 821 8049; e-mail bshipley@courrier.usherb.ca).

One morphological attribute of leaves that presumably includes such size–number trade-offs is specific leaf mass (SLM). The SLM is known to be correlated with relative growth rate under some environmental conditions (Poorter & Remkes, 1990); with leaf gas exchange (Mooney *et al.*, 1978; Field & Mooney, 1983; Ellsworth & Reich, 1992); with seedling regeneration (Shipley *et al.*, 1989; Maranon & Grubb, 1993); and with leaf palatability (Lucas & Pereira, 1990; Choong *et al.*, 1992). Thus, SLM is implicated in ecological processes (growth and survival) that are key to evolutionary fitness.

The SLM may be expressed as the product of leaf thickness and leaf tissue density (Wiltkowski & Lamont, 1991; Garnier & Laurent, 1994). Leaf thickness and leaf tissue density are often themselves correlated (Shipley, 1995) and can both be affected by the spatial arrangement of cells within the leaf. For example, at a given laminar thickness, and for cells packed tightly, many small mesophyll cells in several layers would lead to a greater leaf density than fewer large mesophyll cells arranged in fewer layers. These speculations lead to the two main questions explored in this paper. How do different combinations of cell and organelle size compared with number vary between the leaves of different plant species? Is it possible to model quantitatively how size and number interact to determine SLM?

In order to answer these questions, we studied 94 species of alpine plants found on a high mountain plateau in the Eastern Pamirs. This area is distinguished from most other mountain systems by an extremely dry continental climate that has led to the formation of a habitat type called mountain deserts (Ikonnikov, 1963). The coefficient of aridity of the Eastern Pamirs is close to the aridity of hot deserts of Central Asia (Pyankov & Mokronosov, 1991). The area is characterized by low temperatures, high insolation and UV radiation, reduced partial pressure of CO_2, and a short growing season.

Previous studies of the leaf morphology and mesophyll structure of alpine species, conducted in New Zealand (Körner *et al.*, 1986), the Alps (Körner & Diemer, 1994; Körner *et al.*, 1989), the Caucasus (Goryshyna & Hetsuriany, 1980), and Tien-Shan (Miroslavov & Kravkina, 1990a,b, 1991), have documented the most common features of adaptation of plants to high elevations. These are small thick leaves, increased SLM, an increased number of layers of palisade cells, and larger sizes of mesophyll cells. All of these studies were carried out at elevations below 3500 m and in comparatively humid ecosystems, with annual precipitation of >700 mm. It is likely that the unique environmental conditions and high altitudes of the Eastern Pamirs have affected the structure of the photosynthetic apparatus of plants occurring in this region. Preliminary studies of plants from the Eastern Pamirs (Pyankov & Kondratchuk, 1995) have documented

different types of mesophyll (dorsiventral, iso-palisade, homogeneous and succulent), and quantitative parameters of leaf structure (Pyankov & Kondratchuk, 1998).

MATERIALS AND METHODS

The study was conducted in the summers of 1989 and 1990, in the vicinity of the Pamirs Biology Station of the Academy of Sciences of Tajikistan. The station is located at an altitude of 3860 m in the village of Chechekty, 25 km from the town of Murgab. The Eastern Pamirs (39° 05′–37° 20′ N, 75° 10′–75° 40′ E) belongs to the Karakorum Mountain Range formed in the quaternary period. According to floristic assignment, the Eastern Pamirs belong to the high-mountain Tibetan province of the Central Asian sub-region of the Afro-Asian desert region (Lavrenko, 1962). Summers in the Eastern Pamirs are short, dry and cold. The frost-free period at a height of 3860 m continues for 25–50 d; freezing (from −1 to −5°C) and snow may occur during this period (Ladygina & Litvinova, 1966). The average yearly temperature is *c.* −2°C, the average summer temperature is 7.6°C, and average July temperatures range between 8.2 and 11.2°C. The average low temperature in July at a height of 3860 m does not exceed 5°C, whereas the height of 4100–4200 m marks the border of nightly frosts. The annual precipitation varies between 70 and 120 mm, and in the summer months (June–Aug.) average monthly precipitation ranges from 12–20 mm. The annual number of days with air humidity <30% varies between 200 and 250. The height of snow in winter does not exceed 10 cm, and the margin of snow during summer is at 4800–5100 m. The average July temperature at 5000 m is *c.* 0°C.

Leaves were collected at elevations between 3800 and 4750 m in late June and early July, mainly at the stage of flowering. We investigated 94 species from 32 families (Table 1) occurring in three elevation belts: sub-alpine, alpine and nival (4700–5000 m). To smooth the variability in the material, five to 10 leaves were taken from five to 10 plants belonging to ecotypes typical of the species studied. The anatomical characteristics were determined according to Mokronosov (1978) and Pyankov *et al.* (1998).

The leaf thickness of laminar leaves, or the diameter of cylindrical leaves (e.g. *Gypsophila capituliflora*) or assimilatory shoots (e.g. *Ephedra regeliana*), and the size of mesophyll cells, were measured directly in the field using leaf cross-sections stored in Tris–HCl–sorbitol buffer pH 7.4, a Biolam D-13 light microscope (LOMO, Russia), and an AM-9-2 eyepiece micrometer (GSZ, Russia). Chloroplast dimensions were measured using photographs of the same cross-sections using an MFN-11 camera attachment (LOMO) and Mikrat-300 films (AO Tasma-Kholding, Russia). The photographs were subsequently projected onto a screen and the plastid dimensions measured. To determine the number of chloroplasts in the cell and the number of cells per unit leaf area, samples were fixed with 3.5% glutaraldehyde in phosphate buffer pH 7.0. The number of cells per unit leaf area was determined from samples macerated in 20% KOH. The number of chloroplasts per cell was determined in a mixture of 5% CrO_3 with 1 M HCl after heating on a water bath at 50–60°C for 15–20 min. The cell volume in the palisade mesophyll was calculated as the volume of a cylinder with a coefficient depending on the cell length:width ratio (Tselniker, 1978). The cell volume was determined according to the Chezare relation for a rotation ellipsoid. The following equations were used:

$$S_{cell} = [\pi D(2L + D)/2]$$
$$V_{cell} = \pi(d)^2 LK$$

(S_{cell} and V_{cell} are the surface area and volume of the cell; L and D are the length and width of the cell, respectively; d is $(D/2)$; and K is the cell shape coefficient. This coefficient was found empirically (Tselniker, 1978) and has a good correlation ($r = 0.95$) with the cell length:width ratio. Specifically, $K = 0.38 + 0.117L/D$.)

Chloroplast surface and volume were calculated assuming a rotation ellipsoid geometrical model.

$$S_{chl} = 4\pi(l\ d^2)^{2/3}$$

$$V_{chl} = 4/3\ \pi\ l\ d^2$$

(S_{chl} and V_{chl} are the surface area and volume of the chloroplast; and l and d are half the length and width of the chloroplast, respectively).

Determinations of leaf area, leaf thickness and mass per unit area were obtained from 10 replicates of five to 10 samples of each species. The number of chloroplasts in cells, as well as cell and chloroplast dimensions, were determined by macerating the tissues obtained from five to 10 fixed leaf fragments of 30 replicates per species. The number of cells in macerated tissues was used to calculate the number of cells per unit leaf area, and was determined in 20 replicates of cell suspension with the use of 90 or 225 square cells of the Goryaev haemocytometer, depending on cell concentration in suspension. Chloroplast number per unit of leaf area was calculated by multiplying chloroplast number per cell by the number of cells (palisade and spongy) per unit leaf area. Total indices of the surfaces of palisade cells and chloroplasts per unit leaf area, A_{mes}/A and A_{chl}/A were calculated by multiplying the surfaces of elements (cell, chloroplast) by their number in unit leaf area. This procedure kept the standard errors of the mean below 5%.

Species were classified according to field abundance based on plant occurrence in the Eastern Pamirs (Ikonnikov, 1963). There were 27 rare, 27 occasional and 40 common species. The mesophyll type was classified for each species as follows. Dorsiventral mesophyll (dv, 46 species) is formed from two cell types, palisade and spongy. The palisade cell arrangement under the upper epidermis was in two or three layers, having a cylindrical shape with hemispheres on each end. Spongy cells (two to four layers) were on the ventral surface and had an irregular shape. Isopalisade mesophyll (ip, 24 species) had two to four layers of palisade cells under both dorsal and ventral surfaces, but some species (e.g. *Krascheninnikovia ceratoides*) had one or two layers of spongy mesophyll between them. Homogeneous mesophyll of grasses (hg, 12 species) consisted only of spongy mesophyll. The remaining 12 species had other types of mesophyll structure such as the homogenous dicot type (spongy cells only) as well as succulents and *Alliums* with a large cell volume; the latter included too few species per type to be statistically analysed. These classifications are indicated in Table 1.

Statistical analysis

In order to better approximate multivariate normality, data were transformed to their natural logarithms. The multivariate patterns of correlation in the leaf attributes were summarized using principal components analysis based on the correlation matrix in order to remove the gross size-related effects. The importance of each leaf attribute in defining each of the principal components was determined from its loading on that axis. The results were summarized using a biplot (Gabriel, 1971). This type of diagram plots the position of each species on the first two axes, and then plots the loading of each attribute on the same graph as an arrow. The length of the arrow, when projected vertically (axis 1) or horizontally (axis 2), is proportional to its correlation with that axis. Thus an arrow parallel to axis 1 would load onto axis 1 but not onto axis 2; an arrow parallel to axis 2 would load onto axis 2 but not onto axis 1; an arrow with a 45° angle to axis 1 would load equally strongly on both axes.

A more detailed model of how the variables interact as a system, and an inferential test of this multivariate hypothesis, can be obtained using structural equation modelling in the form of path models. Path models were fitted using the *EQS* package (Bentler, 1995) based on maximum likelihood techniques; goodness-of-fit was evaluated using the maximum likelihood chi-square statistic (Bollen, 1989) or the Satorra–Bentler correction of this statistic, which is more robust to non-normality of the data (Bentler, 1995). This test consists of comparing the observed covariance matrix with the covariance matrix predicted by the model. Data that contradict the predicted patterns of covariance, and

Table 1. *Ninety-four species occurring in an extremely cold mountain desert in the Eastern Pamir Mountains, taxonomy follows Czerepanov (1995)*

Species	Family	Mesophyll type	Ecological group	Rarity
Allium platyspathum	Alliaceae	al	m	r
Allium sp.	Alliaceae	al	m	r
Lomatocarpa albomarginata	Apiaceae	dv	m	r
Ajania tibetica	Asteraceae	ip	x	c
Artemisia leucotricha	Asteraceae	ip	x	c
Artemisia pamirica	Asteraceae	ip	xm	c
Artemisia rhodantha	Asteraceae	ip	x	c
Erigeron heterochaeta	Asteraceae	dv	m	c
Erigeron poncinsii	Asteraceae	ip	x	c
Leontopodium ochroleucum	Asteraceae	dv	m	c
Ligularia alpigena	Asteraceae	dv	m	r
Pyrethrum djilgense	Asteraceae	ip	xm	r
Saussurea salsa	Asteraceae	dv	xm	r
Senecio krascheninnikovii	Asteraceae	ip	x	r
Serratula procumbens	Asteraceae	ip	x	r
Taraxacum dissectum	Asteraceae	dv	m	r
Taraxacum leucantum	Asteraceae	ip	m	c
Christolea crassifolia	Brassicaceae	s	xm	c
Draba korshinskyi	Brassicaceae	h	m	c
Draba pamirica	Brassicaceae	dv	m	r
Smelowskia pectinata	Brassicaceae	ip	xm	c
Sophiopsis annua	Brassicaceae	ip	xm	r
Macrotomia euchroma	Boraginaceae	ip	xm	c
Lindelofia pterocarpa	Boraginaceae	ip	xm	c
Lonicera semenovii	Caprifoliaceae	dv	m	r
Gypsophila capituliflora	Caryophyllaceae	dv	xm	o
Silene graminifolia	Caryophyllaceae	dv	m	o
Stellaria winkleri	Caryophyllaceae	dv	m	r
Krascheninnikovia ceratoides	Chenopodiaceae	ip	x	c
Chenopodium foliosum	Chenopodiaceae	dv	xm	o
Suaeda olufsenii	Chenopodiaceae	s	xm	o
Rhodiola gelida	Crassulaceae	s	xm	o
Rhodiola heterodonta	Crassulaceae	s	xm	c
Rhodiola pamiroalaica	Crassulaceae	s	xm	c
Carex melanantha	Cyperaceae	hg	m	c
Carex orbicularis	Cyperaceae	hg	m	c
Carex pseudofoetida	Cyperaceae	hg	m	c
Carex stenocarpa	Cyperaceae	hg	m	r
Carex dimorphotheca	Cyperaceae	hg	xm	c
Kobresia capilliformis	Cyperaceae	hg	m	c
Ephedra regeliana	Ephedraceae	ip	x	o
Astragalus tibetanus	Fabaceae	ip	xm	c
Hedysarum minjanense	Fabaceae	ip	xm	c
Oxytropis chiliophylla	Fabaceae	ip	xm	c
Oxytropis incanescens	Fabaceae	ip	xm	r
Oxytropis globiflora	Fabaceae	ip	m	o
Corydalis stricta	Fumariaceae	dv	xm	o
Gentiana karelinii	Gentianaceae	h	m	c
Gentiana leucomelaena	Gentianaceae	h	m	o
Swertia marginata	Gentianaceae	h	m	c
Geranium himalayense	Geraniaceae	dv	xm	r
Dracocephalun heterophyllum	Lamiaceae	dv	xm	c
Dracocephalum paulsenii	Lamiaceae	dv	xm	c
Lloydia serotina	Liliaceae	ip	m	c
Acantholimon diapensioides	Limoniaceae	ip	x	c
Parnassia laxmannii	Parnassiaceae	dv	m	o
Plantago arachnoidea	Plantaginaceae	ip	x	o
Achnatherum splendens	Poaceae	hg	m	r
Calamagrostis anthoxanthoides	Poaceae	hg	m	c
Elymus nutans	Poaceae	hg	m	o
Hordeum turkestanicum	Poaceae	hg	xm	c
Leymus pubescens	Poaceae	hg	xm	c
Stipa orientalis	Poaceae	hg	xm	c
Oxyria digyna	Polygonaceae	dv	xm	r

Table 1 (*cont.*)

Species	Family	Mesophyll type	Ecological group	Rarity
Knorringia pamiricum	Polygonaceae	ip	xm	o
Bistorta viviparum	Polygonaceae	dv	m	c
Rheum spiciforme	Polygonaceae	dv	xm	o
Androsace akbaitalensis	Primulaceae	dv	xm	c
Glaux maritima	Primulaceae	dv	m	o
Papaver involucratum	Papaveraceae	dv	m	r
Primula algida	Primulaceae	dv	m	o
Primula macrophylla	Primulaceae	dv	m	c
Primula pamirica	Primulaceae	dv	m	r
Clematis tangutica	Ranunculaceae	dv	m	o
Halerpestes sarmentosa	Ranunculaceae	dv	m	o
Oxygraphis glacialis	Ranunculaceae	dv	m	o
Ranunculus krasnovii	Ranunculaceae	dv	m	o
Ranunculus pseudohirculus	Ranunculaceae	dv	m	c
Ranunculus rufosepalus	Ranunculaceae	dv	m	o
Comarum salesovianum	Rosaceae	dv	m	o
Pentaphylloides dryadanthoides	Rosaceae	dv	xm	o
Potentilla anserina	Rosaceae	dv	m	o
Potentilla malacotricha	Rosaceae	dv	xm	r
Potentilla moorcroftii	Rosaceae	dv	xm	o
Potentilla multifida	Rosaceae	dv	m	c
Potentilla pamirica	Rosaceae	dv	xm	o
Sibbaldia tetrandra	Rosaceae	dv	m	r
Saxifraga hirculus	Saxifragaceae	dv	m	r
Pedicularis ludwigii	Scrophulariaceae	dv	m	c
Scrophularia pamirica	Scrophulariaceae	dv	xm	o
Myricaria squamosa	Tamaricaceae	dv	xm	r
Valeriana fedtschenkoi	Valerianaceae	dv	xm	r
Viola tianschanica	Violaceae	dv	m	r
Zygophyllum rosowii	Zygophyllaceae	s	xm	r

Field occurence: r, rare; o, occasional; c, common. Ecological groups. x, xerophyte, m, mesophyte; xm, intermediate. Mesophyll type: al, *Allium*; dv, dorsiventral; ip, isopalisade; s, succulent; h, homogeneous; hg, homogeneous type of grass.

therefore the hypothesized causal structure of the data, will produce a significant chi-square value, indicating that the model must be rejected. A well-fitting model will produce a non-significant chi-square value. A more detailed explanation of this method in a biological context is given by Shipley & Meziane (1998).

RESULTS

Table 2 lists the estimated values of each leaf attribute for each of the 94 species, and Fig. 1 shows the bivariate relationships between these leaf attributes. A principal components analysis of seven leaf attributes captured 75% of the variance in only two principal components (Fig. 2). The first component (axis 1) contrasted the number of mesophyll cells per square centimetre with a set of highly correlated attributes relating to cell size: mesophyll and chloroplast volume, the number of chloroplasts per mesophyll cell, and leaf thickness. The second component (axis 2) reflects differences in leaf surface area and dry mass, thus capturing differences in SLM that are linearly uncorrelated with the other

variables. Because of the loadings, species with higher SLM occur at the bottom of the graph and species with low SLM occur at the top.

There were no significant differences between the mesophytes, xerophytes and intermediates (meso/xerophytes), nor between groups defined on field abundance, in any single leaf attribute, based on non-parametric ANOVA (Kruskal–Wallis test, $P > 0.05$.) Significant differences were found when comparing species with different types of mesophyll structure (Kruskal–Wallis test, $P < 0.05$) for every variable shown in Fig. 3. Species with the dorsiventral arrangement tended to have thicker leaves with fewer but larger mesophyll cells. Because chloroplast volume did not differ between the groups, the larger mesophyll cells of species with the dorsiventral arrangement resulted in a larger number of chloroplasts per mesophyll cell (Fig. 3). The multivariate pattern (Fig. 2) largely confirms these observations, but shows that there was still substantial overlap between groups. The isopalisade and (especially) the homogeneous grasses tend to have leaves formed by a larger number of smaller mesophyll cells. The dorsiventral species, and the 12

Table 2. *Average values of leaf attributes measured for 94 species from the Eastern Pamir Mountains*

SLM mg dm^{-2}	A_L dm^2	M_L Mg	L_{thick} μm	D_{mes} cm^{-2}	$V_{mes.cell}$ 10^3 μm^3	V_{chl} μm^3	N_{chl}/M_{cell}	A_{mes}/A_L m^2 m^{-2}
1164	14.87	17317	280	320.49	60.62	29.10	113	16.57
357	3.14	1121	259	25.10	81.50	40.32	63	0.90
550	1.14	627	626	524.29	38.06	75.17	31	13.93
1101	0.24	260	267	1092.92	10.04	33.63	44	24.07
1000	1.00	1000	340	2292.47	8.60	38.47	23	28.77
704	0.69	485	376	482.88	18.49	54.54	48	16.05
880	0.62	546	452	1281.71	14.59	62.86	22	21.12
595	1.38	820	424	156.96	22.99	45.14	66	6.35
690	1.07	740	203	1145.80	5.05	27.25	30	15.24
799	0.55	440	200	1194.55	6.09	39.87	38	25.72
639	49.20	31430	814	256.27	154.64	45.28	56	8.74
1048	1.54	1614	354	622.65	10.71	31.27	56	16.72
753	7.60	5726	438	552.59	31.98	36.11	55	16.20
599	0.78	468	174	2243.93	2.96	32.26	38	41.57
822	7.14	5866	356	271.19	13.77	45.13	32	3.30
601	4.96	2980	378	295.41	32.79	35.15	52	7.91
665	2.18	1449	490	555.19	30.33	64.92	36	15.40
1155	1.82	2102	751	574.87	55.33	47.38	124	45.14
293	0.34	100	345	170.07	17.97	41.30	58	5.72
654	0.14	90	254	265.53	14.57	52.65	59	10.71
409	1.27	519	237	1273.39	3.33	28.14	22	12.48
794	1.05	833	324	1871.81	6.79	42.27	27	29.37
1250	3.41	4262	456	721.03	30.99	38.86	54	21.79
614	19.39	11910	589	639.27	37.22	44.08	39	15.17
779	0.46	360	251	882.17	6.00	23.94	31	10.91
1177	0.19	220	881	414.98	45.86	31.79	69	13.89
579	1.73	90	355	432.15	29.63	66.94	44	15.32
440	0.20	90	338	306.14	21.88	35.43	31	4.95
934	0.93	869	349	706.45	13.97	49.33	51	23.32
663	1.16	770	614	318.39	43.80	54.32	47	10.40
1826	0.21	380	1085	204.74	459.78	33.91	63	6.55
963	0.37	360	1136	96.01	298.19	64.11	185	13.75
1032	0.59	609	1389	47.38	60.18	26.16	214	4.32
440	0.41	180	855	132.91	165.44	31.85	80	5.18
602	7.77	4678	274	3351.98	3.38	26.38	13	19.25
821	3.84	3150	273	5423.59		32.71	28	73.78
521	1.73	901	325	1979.91	1.85	28.91	17	15.09
440	4.90	2156	284	1331.70	3.95	26.73	21	12.19
518	1.54	797	333	2420.26	2.26	29.02	18	19.78
408	5.76	2351	152	1558.01	1.89	21.94	16	9.35
761	0.94	714	1130	389.81	6.83	36.60	39	8.18
415	3.45	1431	243	1033.01	4.31	21.95	34	13.22
492	6.95	3423	238	1189.69	2.92	25.43	35	17.42
364	8.06	2933	298	904.28	8.91	34.55	34	15.69
1017	0.76	772	314	908.20	18.89	50.27	39	23.09
501	4.30	2152	276	1043.45	6.73	43.07	35	21.83
920	9.52	8760	299	750.74	8.79	90.22	40	14.84
424	0.18	76	363	251.66	16.31	65.85	25	4.91
660	0.07	46	455	228.14	31.18	32.01	81	9.03
548	4.09	2241	471	584.43	35.74	30.44	47	13.05
477	7.50	3580	267	724.54	36.19	46.99	32	14.52
685	2.82	1932	365	629.82	13.90	27.98	51	14.44
643	0.29	186	213	1690.55	3.40	33.04	22	18.70
1656	1.38	2280	666	725.36	18.02	65.94	50	28.61
1669	0.02	26	530	1645.41	11.22	39.51	43	39.82
526	3.41	1795	396	364.30	29.96	34.81	35	6.56
762	1.18	899	436	622.45	17.29	40.30	43	15.08
1193	8.11	9671	240	1957.27	3.39	33.70	27	26.67
503	2.23	1121	402	321.19	6.30	25.40	32	8.47
502	5.86	2940	229	221.63		50.24	37	5.33
461	2.04	939	163	65.95	9.28	27.67	31	0.89
959	13.50	12940	420	268.74		30.23	65	8.17
351	1.34	4670	130	728.46	0.50	29.15	12	4.11
445	1.67	744	294	338.34	87.78	56.48	143	34.36
1163	1.27	1477	662	1049.22	13.08	38.18	63	36.38

Table 2 (*cont.*)

SLM mg dm^{-2}	A_L dm^2	M_L Mg	L_{thick} μm	D_{mes} cm^{-2}	$V_{mes.cell}$ 10^3 μm^3	V_{chl} μm^3	N_{chl}/M_{cell}	A_{mes}/A_L m^2 m^{-2}
813	2.75	2240	443	611.59	11.43	39.59	31	10.59
1164	70.89	82538	687	682.55	19.79	64.82	42	22.14
391	0.20	78	377	172.16	39.78	49.37	45	5.06
763	0.16	120	499	180.22	32.46	72.65	95	14.48
721	1.93	1390	326	332.36	13.06	48.63	66	14.19
1135	0.59	670	631	551.83	19.84	37.03	60	17.65
1062	13.38	14202	387	382.10	21.74	39.89	69	14.89
395	2.73	1078	377	319.08	44.85	44.36	54	10.35
822	5.18	4258	345	406.97	17.99	37.40	108	23.71
587	0.09	50	584	121.51	14.77	37.84	67	4.44
685	2.32	1589	325	229.81	29.83	30.80	97	10.55
461	1.36	626	379	352.29	39.29	39.21	81	15.87
616	1.82	1122	407	620.06	33.41	43.13	45	16.70
547	1.68	919	413	309.42	36.18	40.21	92	16.24
349	1.02	356	231	480.98	3.07	26.77	35	7.25
880	0.64	563	135	4132.94	1.30	33.86	11	22.22
781	4.95	3867	261	1858.73	6.16	31.68	24	21.89
812	1.97	1600	163	1338.91	3.07	30.08	19	11.61
1274	1.60	2039	237	1482.79	2.96	24.40	49	29.51
1017	4.09	4159	282	1345.08	9.62	33.22	41	27.41
1288	0.98	1267	212	4063.42	1.89	29.34	22	41.14
890	0.16	140	177	3024.16	2.89	39.74	23	38.39
862	0.45	390	443	184.58	58.81	43.56	84	9.31
701	2.42	1700	304	356.11	52.19	34.80	73	13.38
738	3.58	2640	466	913.89	35.49	41.66	49	25.82
1100	0.18	198	291	503.65	21.55	36.87	54	14.68
514	0.43	222	411	179.98	81.20	58.00	50	6.48
625	0.70	437	332	767.91	11.18	35.50	40	16.01
1608	0.66	1061	803	831.60	72.91	52.47	109	61.45

Variables: specific leaf mass (SLM); leaf surface area (A_L), mass (M_L) and thickness (L_{thick}); mesophyll density (D_{mes}), volume of average mesophyll cell ($V_{mes.cell}$) and average chloroplast (V_{chl}); number of chloroplasts per mesophyll cell (N_{chl}/M_{cell}); mesophyll area per leaf area (A_{mes}/A_L).

species having other types of mesophyll arrangement, tended to have fewer, larger mesophyll cells, thicker leaves and more, larger chloroplasts per mesophyll cell.

DISCUSSION

Size versus number

In this study we found that mesophyll cell size and number per cm^2 were strongly but negatively correlated (Fig. 1). A possible biological explanation for this size compared with number trade-off comes from the observation that leaf thickness was positively correlated with mesophyll cell volume, but negatively correlated with mesophyll cell density. Since lamina thickness will be determined largely by the average size of the mesophyll cells and the number of layers of such cells within the leaf, this negative correlation probably arises from the functional requirement for leaf thickness to be constrained by environmental conditions, especially by irradiance level. The thickness of the leaf lamina will determine the efficiency with which photons can be trapped and used by the leaf. Leaves with larger

mesophyll cells, and therefore more and larger chloroplasts per cell, will tend to absorb more incoming photons, leaving less residual energy available to penetrate into other mesophyll cells deeper in the leaf lamina. This restricts the number of layers of mesophyll cells that can maintain a positive C gain. If lamina thickness is constrained within set limits, then the leaf must trade off mesophyll size and density. The effect of increasing the number (and therefore decreasing the size) of the mesophyll cells would be to increase the total surface area of the mesophyll. This would affect photosynthetic rate, as the ratio of the surface area of the mesophyll cells to the surface area of the leaf (A_{mes} : A) is correlated with the rate of CO_2 assimilation (Nobel *et al.*, 1975; Longstreth & Nobel, 1979; Nobel & Walker, 1985; Patton & Jones, 1989).

Such a trade-off with respect to the size and number of chloroplasts was not observed within mesophyll cells; larger mesophyll cells had both more and larger chloroplasts. No explanation suggests itself for this, but the total volume of chloroplasts within a mesophyll cell is a very small proportion of the total mesophyll cell volume. The question of the relative advantages and disadvantages

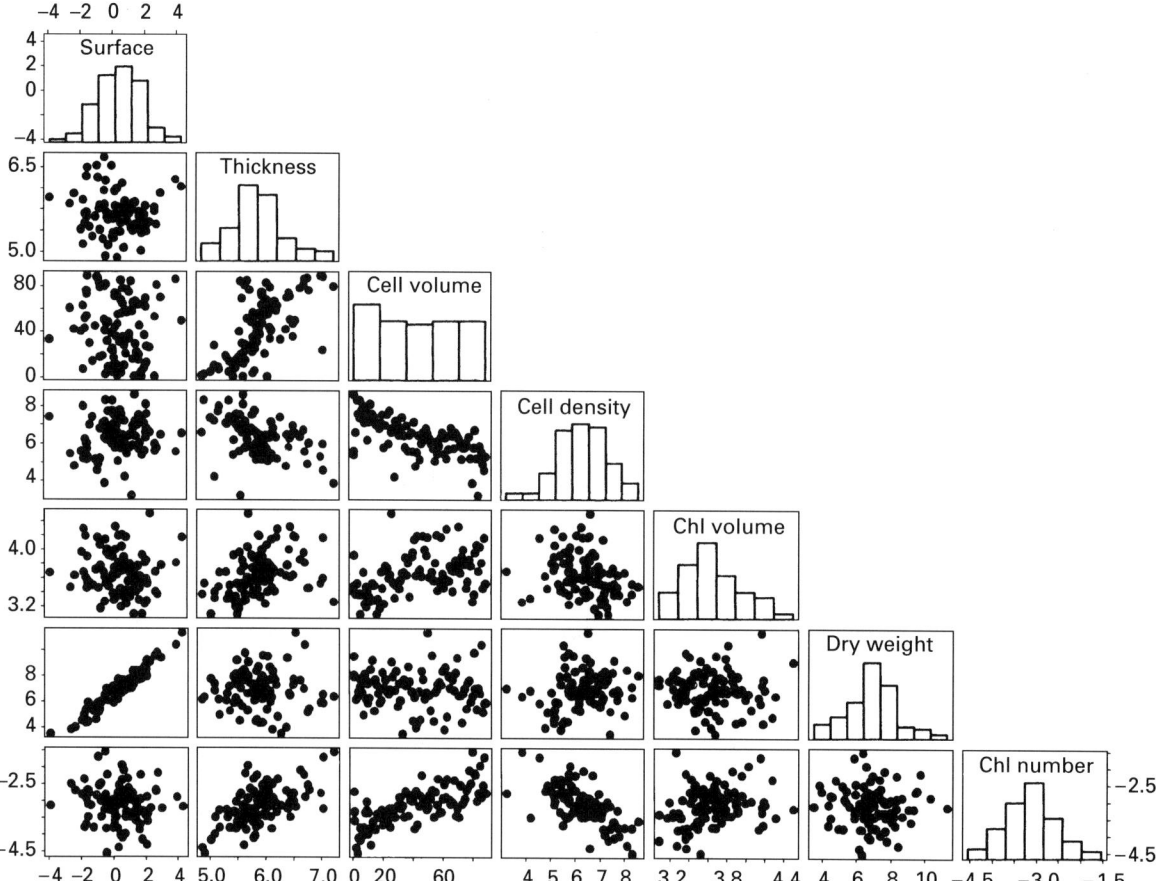

Fig. 1. Scatterplot matrix of seven leaf attributes ($\log_e$ transformed) based on means of 94 species of field-collected plants. Diagonal panels show the distribution of each variable; off-diagonal panels show scatterplots: *x* axis, column variable; *y* axis, row variable.

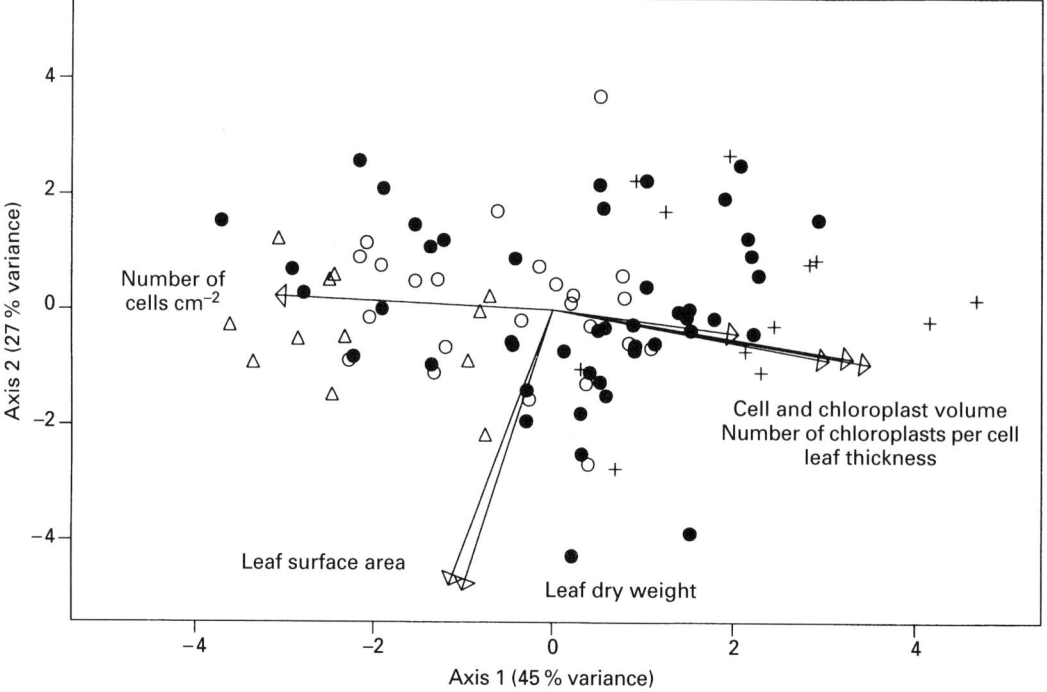

Fig. 2. Biplot showing the species scores (symbols) of 94 species of field-collected plants on the first two axes of a principal components analysis of seven leaf attributes ($\log_e$ transformed). Loadings of the leaf attributes (multiplied by 7 for visual clarity) on the first two axes are shown by arrows. The three different mesophyll types are: dv, dorsiventral (closed circles); ip, isopalisade (open circles); hg, homogeneous grasses (open triangles); other types (crosses).

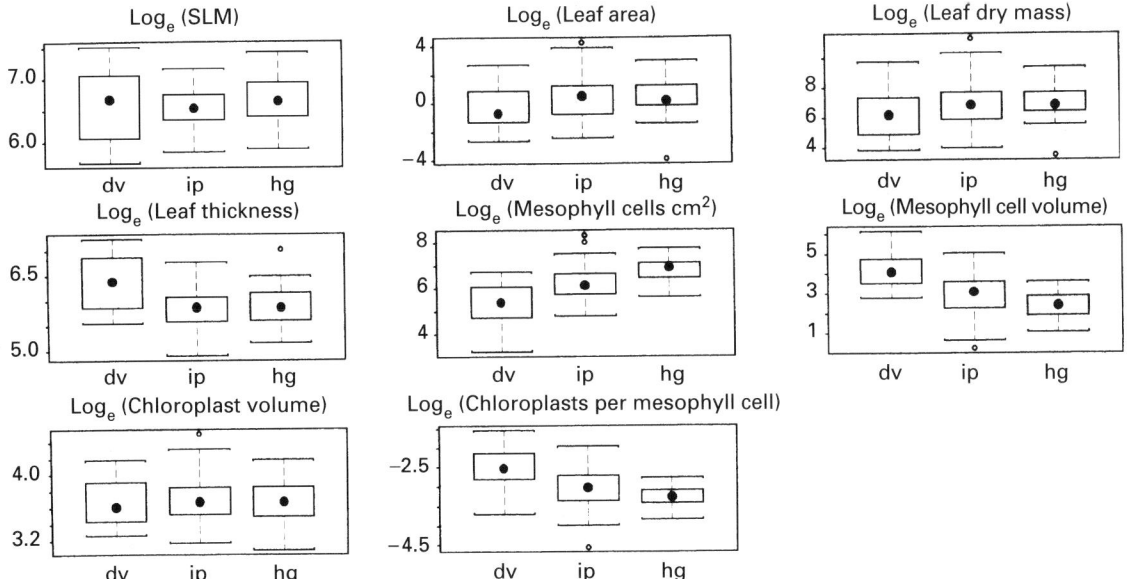

Fig. 3. Box plots of seven leaf attributes ($\log_e$ transformed) grouped according to mesophyll structure (dv, dorsiventral; ip, isopalisade; hg, homogeneous grasses). Solid dots show the mean, the box gives ± 1 quantile, bars show 95% quantiles, and outliers are shown by open circles. Significant differences in the means exist between groups for every variable.

of different combinations of mesophyll cell and chloroplast size versus number in different environmental contexts is important, but relatively unstudied. For example, given the necessary scaling of surface area to volume of such chloroplasts, larger chloroplast size would probably affect the diffusion of CO_2 into chloroplasts.

Size versus number in relation to specific leaf mass

The SLM is the product of leaf thickness and leaf tissue density. Leaf tissue density (leaf dry mass per volume of leaf tissue, not including intercellular spaces) is often measured as the proportional water content of the leaf; that is, the proportion of leaf fresh weight that is due to leaf water (Garnier & Laurent, 1994; Shipley, 1995). This is because most of the volume of most leaf cells (except for xylem or those forming structural tissues) is occupied by cytoplasm. We can therefore expect that leaf tissue density will be strongly affected by the size and number of mesophyll cells. Most of the mass of a mesophyll cell is concentrated in its cell wall, and is therefore proportional to its surface area and cell wall thickness. Most of the water in the cell is in the cytoplasm, and the mass of water is proportional to the volume of the mesophyll cell. It follows that larger mesophyll cells would generally have a larger proportion of water (cytoplasm) to dry mass (cell wall). This means that larger mesophyll cells would contribute to a larger proportional water content and a lower leaf tissue density. Increasing the number of such cells would not change this proportionality.

Although conceptually different, leaf thickness and leaf tissue density are not biologically independent. For instance Meziane & Shipley (1999)

grew 22 herbaceous species typical of open, sunny, lowland habitats, under controlled conditions in hydroponic sand culture in four different environments: high (1100) and low (200) irradiance (μmol m^{-2} s^{-1} PAR) crossed with a high (1:1) and low (1:6 dilution of the hydroponic solution) nutrient concentration. They found that mesophyll and lamina thickness increased with increasing irradiance and with increasing nutrient supply. Conversely, leaf water content decreased with increasing irradiance, but increased with increasing nutrient supply. Lamina thickness and leaf water content were negatively correlated across species at low irradiance but the two were largely uncorrelated at high irradiance. This is consistent with the explanation given above. At the low photosynthetic photon fluence density (PPFD) (200 μmol m^{-2} s^{-1} PAR) self-shading within the leaf would have constrained cell layering, generating the negative correlation; presumably the high PPFD used in that experiment largely overcame self-shading within the leaves.

Modelling specific leaf mass

The path model of SLA proposed by Shipley (1995) essentially expressed SLA as leaf thickness and leaf tissue density, measured by proportional water content. The model is shown in Fig. 4a in qualitative form (excluding the thickness of the protruding midvein for which we have no measures in this paper); note that the original model dealt with specific leaf area, not mass, so the signs of the path coefficients are reversed in Fig. 4a. This model was shown to provide a good fit to data obtained from field-collected herbaceous plants typical of open sunny habitats in south-eastern Canada. As previously

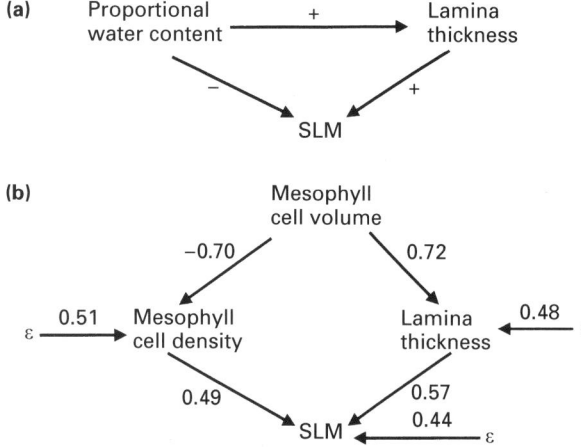

Fig. 4. (a) Path model proposed by Shipley (1995). (b) First path model proposed in this study. Model parameters are based on standardized variables. Error variables (ε) have a variance of 1. This model produced a good fit to the empirical data.

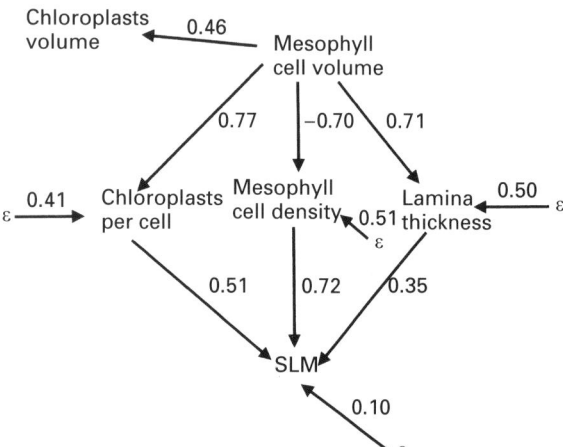

Fig. 5. Second path model proposed in this study. Model parameters are based on standardized variables. Error variables (ε) have a variance of 1. This model produced a good fit to the empirical data.

argued, leaf proportional water content will be partly determined by the average volume of the mesophyll cells. An effect from mesophyll volume to mesophyll density was included because there is a trade-off between average mesophyll cell volume and the number of these cells per cm² cross-section. Therefore the path model shown in Fig. 4b was proposed.

This model produced a good fit to the data (χ^2 = 3.50, 2 df, P = 0.17) and, using the Satorra–Bentler chi-square statistic (S–$B\chi^2$) which corrects for non-normality (Bentler, 1995), indicates an even better fit (S–$B\chi^2$ = 2.02, 2 df, P = 0.36). This model points to the central role of mesophyll cell volume in relating the measured variables to SLM. Increasing mesophyll cell volume increases leaf thickness but decreases the number of mesophyll cells in cross-section. This trade-off between size and number results in leaf thickness being negatively correlated with mesophyll cell density (r = −0.44). The SLM

increases both with increasing leaf thickness and with increasing mesophyll cell density. This means that increasing mesophyll cell volume increases SLM by increasing leaf thickness, but decreases SLM by decreasing mesophyll cell density. The result is that the overall correlation between mesophyll cell volume and SLM is almost zero (r = 0.07).

Although cell dry mass would be mostly determined by cell wall mass, the mass of cellular organelles such as chloroplasts would also contribute. From the empirical data, it is known that both chloroplast number and size are positively correlated with mesophyll cell volume. Fig. 5 therefore presents an enlarged version of the path model in Fig 4b. This model also gives a good fit to the data (S–$B\chi^2$ = 9.98, 8 df, P = 0.27) and indicates that an increasing number of chloroplasts per mesophyll cell also increases SLM. Again, average mesophyll cell volume is a key variable in the model. Although one might think that chloroplast volume and the number of chloroplasts per mesophyll cell should affect SLM, no empirical evidence was detected for this. Including a path from chloroplast volume to SLM in the model shown in Fig. 5 does not affect the overall fit of the model, but this path is not significantly different from zero (t = 0.16, P = 0.87). All of these path models should be interpreted simply as hypotheses that are consistent with the available data. Independent data from species occurring in different habitats are required to provide strong tests of the models.

The ecological relevance of specific leaf mass and its components

Shipley's (1995) field data, obtained from herbaceous species from open sunny habitats typical of south-eastern Canada, had an average SLM of 362 mg dm⁻², with most being between 252 and 645 mg dm⁻². Typical values of herbaceous species from the forest understorey in south-eastern Canada (unpublished) range from 169 to 305 mg dm⁻² with an average of 225 mg dm⁻², and for 160 understorey dicot species from the Middle Urals (boreal forest) the modal class for this parameter was 200–350 mg dm⁻² (Barinov, 1988; Pyankov *et al.*, 1998; V. I. Pyankov *et al.*, unpublished). Körner & Diemer (1994) showed the average SLM in their mountain plants increased from 436 to 606 mg dm⁻² as altitude increased from 600 to 2600 m in the Austrian Alps. The modal class of SLM for arctic plants (*c.* 100 species) was 400–600 mg dm⁻² (Barinov, 1988). By contrast, the SLM values recorded for the species in this paper, growing in extremely cold and dry habitats in the Eastern Pamir Mountains, average 706 with the first and third quartiles being 518 and 952 mg dm⁻². The fact that no significant differences in SLM or other leaf attributes were detected in these Eastern Pamir species when combined into ecological groups (mesophytes, xerophytes or intermediates) or degree of rarity, is probably partly because all of the species are adapted to the extreme conditions of this area. It may also be due to the fact that there was substantial heterogeneity among species within these groups, because every ecological or rarity group included plants belonging to different life forms, altitudinal ranges and taxonomic affiliations (Pyankov & Kondratchuk, 1998). The one classification that did produce significant differences was that based on the organization of the mesophyll. Species with dorsiventral palisade had larger average mesophyll cells. Consistent with the path diagrams, this translated into thicker leaves with fewer mesophyll cells per square centimetre and more chloroplasts per mesophyll cell, but only small differences in SLM. Because of the scaling of surface area to volume and the partial trade-off between mesophyll cell size and number, dorsiventral species tend to have a lower $A_{mes} : A$ ratio. This would increase CO_2 mesophyll resistance, resulting in poorer drought tolerance. What, then, might be the adaptive advantage of the dorsiventral organization with its large mesophyll cells? This cannot be answered, but increased irradiance increases the development of the palisade and can even change leaves from dorsiventral into isopalisade, as demonstrated on leaves in different parts of the crown (Tselniker, 1978) or plants developing in spring or summer as irradiance increases (Goryshyna, 1989). Perhaps these species are trading off drought tolerance and the ability to photosynthesize efficiently at high PPFD. If this is the case then the different paths from mesophyll cell volume to SLM may represent these different selective pressures. Further research is needed to test these hypotheses.

ACKNOWLEDGEMENTS

Sergey S. Ikonnikov (Komarov Botanical Institute, St Petersburg) helped in the identification of species. This study was partially financed by grants from the Russian Foundation of Basic Research and from the High Education of Russia 'Universities of Russia' (V.I.P and A.V.K.), and the Natural Sciences and Engineering Research Council of Canada (B.S.).

REFERENCES

Araus J, Alegre L, Tapia L, Calafel R, Serret M. 1986. Relationships between photosynthetic capacity and leaf structure in several shade plants. *American Journal of Botany* **73**: 1760–1770.

Barinov MG. 1988. *Mesostructure of photosynthetic apparatus of plants from different climatic zones.* Syvktyvkar: Komi Sientific Center of Urlas Branch Academy of Sciences of the USSR.

Bentler PM. 1995. *EQS Structural Equations Program Manual.* Encino, CA, USA: Multivariate Software, Inc.

Bollen KA. 1989. *Structural equations with latent variables.* New York, USA: Wiley.

Choong MF, Lucas PW, Ong JSY, Pereira B, Tan HTW,

Turner IM. 1992. Leaf fracture toughness and sclerophylly: their correlations and ecological implications. *New Phytologist* **121**: 597–610.

Czerepanov SK. 1995. *Vascular plants of Russia and adjacent states (the former USSR).* Cambridge, UK: Cambridge University Press.

Ellsworth DS, Reich PB. 1992. Leaf mass per area, nitrogen content and photosynthetic carbon gain in *Acer saccharum* seedlings in contrasting forest light environments. *Functional Ecology* **6**: 423–435.

Evans J, von Caemmerer S. 1996. Carbon dioxide diffusion inside leaves. *Plant Physiology* **110**: 339–346.

Evans J, von Caemmerer S, Setchell BA, Hudson GS. 1994. The relationships between CO_2 transfer conductance and leaf anatomy in transgenic tobacco leaves. *Australian Journal of Plant Physiology* **21**: 475–495.

Field C, Mooney HA. 1986. The photosynthesis–nitrogen relationship in wild plants. In: Givnish J, ed. *On the economy of plant form and function.* Cambridge, UK: Cambridge University Press, 25–55.

Gabriel KR. 1971. The biplot graphic display of matrices with application to principal component analysis. *Biometrika* **58**: 453–467.

Garnier E, Laurent G. 1994. Leaf anatomy, specific leaf mass and water content in congeneric annual and perennial grass species. *New Phytologist* **128**: 725–736.

Goryshyna TK. 1989. *Photosynthetic apparatus and adaptation of plant to environment.* Leningrad, Russia: Leningrad University Publishers.

Goryshyna TK, Hetsuriany LD. 1980. Peculiarities of photosynthesis apparatus in leaves of some plants of subalpine meadow, Central Caucasus. *Soviet Journal of Ecology* **11**: 21–26 (in Russian).

Ikonnikov SS. 1963. *Definitorium of the Pamirs plants.* Donish: Dushanbe.

Körner C, Diemer M. 1994. Evidence that plants from high altitudes retain their greater photosynthetic efficiency under elevated CO_2. *Functional Ecology* **8**: 58–68.

Körner C, Bannister P, Mark AF. 1986. Altitudinal variation in stomatal conductance, nitrogen content and leaf anatomy in different plant life forms in New Zealand. *Oecologia* **69**: 577–588.

Körner C, Neumayer M, Pelaez Mendez-Riedl S, Smeet-Scheel A. 1989. Functional morphology of mountain plants. *Flora* **182**: 353–383.

Ladygina GM, Litvinova NP. 1966. Studies of vegetable communities in the Pamirs high mountains. *Botanicheskii Zhurnal* **51**: 792–800 (in Russian).

Laisk A, Oja V, Rahi M. 1970. Diffusion resistance of leaves in connection with their anatomy. *Fiziologiya Rastenii* **47**: 40–48.

Lavrenko EM. 1962. *Basic features of botanical geography of Euro-Asian and North Africa deserts.* Moscow/Leningrad, Russia: Academy of Sciences of the USSR (in Russian).

Longstreth DJ, Nobel PS. 1979. Salinity effects on leaf anatomy. Consequences for photosynthesis. *Plant Physiology* **63**: 700–703.

Lucas PW, Pereira B. 1990. Estimation of the fracture toughness of leaves. *Functional Ecology* **4**: 819–822.

Maranon T, Grubb PJ. 1993. Physiological basis and ecological significance of the seed size–relative growth rate relationship in Mediterranean annuals. *Functional Ecology* **7**: 591–599.

Meziane D, Shipley B. 1999. Interacting determinants of specific leaf area in 22 herbaceous species: effects of irradiance and nutrient availability. *Plant, Cell and Environment* **22** (In press.)

Miroslavov EA, Kravkina IM. 1990a. Comparative anatomy of leaves of mountain plants at different elevations. *Botanicheskii Zhurnal* **75**: 368–375 (in Russian).

Miroslavov EA, Kravkina IM. 1990b. The structural adaptation of chloroplasts and mitochondria to high mountain and extreme northern conditions. *Soviet Journal of Ecology* **21**: 36–42 (in Russian).

Miroslavov EA, Kravkina IM. 1991. Comparative analysis of chloroplasts and mitochondria in leaf chlorenchyma from mountain plants grown at different altitudes. *Annals of Botany* **68**: 195–200.

Mokronosov AT. 1978. Mesostructure and functional activity of photosynthetic apparatus. In: Mokronosov AT, ed. *Mesostructure and functional activity of photosynthetic apparatus.* Sverdlovsk, Russia: Urals State University, 5–31 (in Russian).

Mooney HA, Ferrer PJ, Slatyer RP. 1978. Photosynthetic capacity and carbon allocation patterns in diverse growth forms of *Eucalyptus. Oecologia* **50**: 109–112.

Nobel PS, Walker DB. 1985. Structure of photosynthetic leaf tissue. In: Barber J, Baker NR, eds. *Photosynthetic mechanisms and the environment.* Amsterdam, The Netherlands: Elsevier Science, 501–536.

Nobel PS, Zaragoza LJ, Smith WK. 1975. Relation between mesophyll surface area, photosynthetic rate and illumination level during development of *Plectrantus parviflorus* Henkel. *Plant Physiology* **55**: 1067–1070.

Patton L, Jones MB. 1989. Some relationships between leaf anatomy and photosynthetic characteristics of willow. *New Phytologist* **111**: 657–661.

Poorter H, Remkes C. 1990. Leaf area ratio and net assimilation rate of 24 wild species differing in relative growth rate. *Oecologia* **83**: 553–559.

Pyankov VI, Ivanova LA, Lambers H. 1998. Quantitative anatomy of photosynthetic tissues of plant species of different functional types in arboreal vegetation. In: Lambers H, Poorter H, Van Vuuren MMI, eds. *Inherent variation in plant growth. physiological mechanisms and ecological consequences.* Leiden, The Netherlands: Backhuys Publishers, 71–87.

Pyankov VI, Kondrachuk AV. 1995. Specific features of structural organization of photosynthetic apparatus of plants of the Eastern Pamirs. *Proceedings of the Academy of Sciences of Russia* **344**: 712–716 (in Russian).

Pyankov VI, Kondrachuk AV. 1998. Structure of the photosynthetic apparatus in woody plants from different ecological and altitudinal groups in Eastern Pamir. *Russian Journal of Plant Physiology* **45**: 567–578.

Pyankov VI, Mokronosov AT. 1991. Physiology-biochemical basis of ecological differentiation of desert plants and problems of phytoamelioration in arid ecosystems. *Problem of reclamation of desert* **3–4**: 161–170 (in Russian)

Shipley B. 1995. Structured interspecific determinants of specific leaf area in 34 species of herbaceous angiosperms. *Functional Ecology* **9**: 312–319.

Shipley B, Meziane D. 1998. The statistical modelling of plant growth and its components using structural equations. In: Lambers H, Poorter H, Van Vuuren MMI, eds. *Inherent variation in plant growth. physiological mechanisms and ecological consequences.* Leiden, The Netherlands: Backhuys Publishers, 393–408.

Shipley B, Keddy PA, Moore DRJ, Lemky K. 1989. Regeneration and establishment strategies of emergent macrophytes. *Journal of Ecology* **77**: 1093–1110.

Tselniker YL. 1978. The physiological basis of tolerance of tree plants to shadow. Moscow, Russia: Nauka (in Russian).

Wiltkowski ETF, Lamont BB. 1991. Leaf specific mass confounds leaf density and thickness. *Oecologia* **88**: 486–493.

New Phytol (1999), **143**, 143–154

Research review
Low-light carbon balance and shade tolerance in the seedlings of woody plants: do winter deciduous and broad-leaved evergreen species differ?

MICHAEL B. WALTERS[1]* AND PETER B. REICH[2]

[1] *Department of Forestry, Michigan State University, East Lansing, MI 48824, USA*
[2] *Department of Forest Resources, University of Minnesota St Paul, MN 55108, USA*

Received 9 December 1998; accepted 8 April 1999

SUMMARY

A popular conceptual model asserts that shade tolerance is characterized by morphological and physiological traits that enhance the net rate of carbon capture in low light. We tested this model by quantitatively reviewing growth, leaf lifespan, CO_2 exchange and morphological data from 76 studies on woody seedlings grown under conditions of low light. Data were placed into three tolerance categories (intolerant, intermediate, tolerant), two light categories (less than 4% and 4–12%) and two leaf phenology categories (broad-leaved evergreen and winter deciduous). For both evergreen and deciduous groups, intolerant species had traits conferring greater growth potential than tolerant species in both light categories. These traits included greater leaf mass ratio, leaf area ratio, specific leaf area and mass-based photosynthetic rates above light compensation. However, in 0–4% light, growth rates were similar for intolerant and tolerant species, because low light together with higher respiration rates for intolerant species limited the expression of their growth potential differences. Deciduous and evergreen intolerant species were similar in many respects. However, both intermediate and tolerant deciduous species had markedly lower leaf mass ratios and higher root mass ratios than intermediate and tolerant evergreen species. In addition, deciduous species and intolerant evergreens must cope with as much as sixfold higher leaf turnover rates than tolerant evergreen species. Thus, rather than maximizing growth rates in low light, tolerant evergreen species minimize biomass loss through long leaf lifespans and low respiration rates. Tolerant deciduous species also minimize biomass losses by minimizing whole-plant respiration rates but they accomplish low biomass turnover though low leaf mass ratio and not low leaf turnover rates. Furthermore, unlike most tropical evergreens, tolerant deciduous species can gain large fractions of their total growing season carbon during short periods when the overstory is leafless and then allocate this carbon to storage (as reflected by high root mass ratios) rather than new leaves. In conclusion, we found no support for the low-light-enhanced carbon capture model of shade tolerance as viewed strictly from the perspective of physiological growth capacity. This can be explained by the disadvantages to net growth and survival of maintaining a high growth potential at low light, because high growth potential results in greater rates of whole-plant respiration, tissue turnover, herbivory and mechanical damage and in decreased storage. Thus, shade tolerance can be characterized by traits that maximize survival and net growth, where net growth includes losses to all agents.

Key words: Low light, carbon balance, shade tolerance, woody plants, deciduous, evergreen.

INTRODUCTION

Competition in a given environment favours plants whose form and physiology maximize their net rate of carbon gain there.

T. J. Givnish (1988)

*Author for correspondence (tel +1 517 355 1762; fax +1 517 432 1143; e-mail mwalters@pilot.msu.edu).

Givnish's (1988) statement encapsulates a fundamental working hypothesis underlying many studies of shade tolerance. Often it is interpreted as 'compared with shade-intolerant species, tolerant species have greater low-light growth rates and the physiological and morphological traits required to enhance low-light growth'. Among these traits are lower respiration rates, higher photosynthetic rates in low light, lower leaf light compensation points (the light level at which net leaf photosynthesis is zero) and higher apparent quantum yields (the slope of the graph of net photosynthesis against light, at low light intensity), higher leaf area per unit leaf dry mass (SLA), and higher whole-plant leaf fractions (leaf area ratio, leaf mass ratio).

Until recently, the fundamental assumption that tolerant species possess enhanced carbon-capture abilities under conditions of low light intensity had persisted despite experimental evidence that survival, not growth, is associated with observationally based shade tolerance classifications (Baker, 1945; Shirley, 1945; Kitajima, 1994; Pacala *et al.*, 1996; Walters & Reich, 1999). There might be several reasons for the persistence of this assumption, including: (1) a focus on components of growth (e.g. photosynthetic rates), without placing them in a whole-plant context (well elucidated by Givnish (1988)), (2) too small a collective body of information from which to draw general conclusions, and (3) mixed results, with some studies reporting greater growth for intolerant than tolerant species in low light (Baker, 1945; Shirley, 1945; Kitajima, 1994; Walters & Reich, 1999) and others reporting the opposite (Pompa & Bongers, 1988; Denslow *et al.*, 1990; Walters & Reich, 1996).

In the past decade, the development of whole-plant approaches that integrate growth analysis with physiological and biochemical approaches (Poorter & Remkes, 1990; Poorter *et al.*, 1990) have helped inspire numerous multiple-species whole-plant investigations of shade tolerance (Kitajima, 1994; Ducrey, 1994; Osunkoya *et al.*, 1994; Walters & Reich, 1996, 1999; Veenendaal *et al.*, 1996; Poorter, 1998; Reich *et al.*, 1998a,b). The large amount of information resulting from these studies has spawned interest in quantitatively reviewing these data to test general notions of shade tolerance. In one such synthesis of tropical tree seedlings, Veneklaas & Poorter (1998) show that intolerant species have growth advantages and traits associated with high growth potential (e.g. greater leaf area ratios) in both gap and understory light environments.

In this study we shall further test the enhanced low-light carbon-capture hypothesis of shade tolerance by quantitatively reviewing data on growth and on growth-related morphology, CO_2 exchange and leaf turnover characteristics, for woody plants classified as tolerant, intermediate and intolerant. We shall compare two species groups of contrasting leaf phenologies, broad-leaved evergreen species, mostly from the moist and wet tropics, and winter deciduous species from temperate and boreal regions. These groups were chosen for comparison because data were abundant. Furthermore, we speculated that climate-induced differences in leaf lifespan and linked morphological and CO_2 exchange traits (Reich *et al.*, 1992, 1997, 1999) could promote different strategies for coping with shade.

Specifically, we asked the following questions: 'Compared with intolerant species, do shade-tolerant tree seedlings have traits that enhance low-light, whole-plant carbon gain?' and 'Do carbon balance strategies for tolerant and intolerant species differ for tropical evergreen and winter deciduous species?'

METHODS

Literature data set

The Appendix includes all of the raw data used in this review and the references they were obtained from (www.journals.cup.org). We compiled our own unpublished data and all published data that we could find for woody broad-leaved evergreen (BROAD-LEAVED EVERGREEN) and winter deciduous (WINTER DECIDUOUS) species from closed forest ecosystems grown in less than 12% of full light. BROAD-LEAVED EVERGREEN species were all dicots from the moist or wet tropics except for two Japanese temperate species (Cao & Ohkubo, 1998). Except for two conifer *Larix* species, WINTER DECIDUOUS species were all temperate and boreal dicots. Compiled characteristics included relative growth rate (RGR), growth-related morphology (RMR, SMR, LMR, SLA, LAR), leaf-level CO_2 exchange characteristics (A_{max}, QY, LCP, R_d) and leaf lifespan (see Table 1 for definitions and units of all abbreviations and symbols). Unlike other parameters, lifespan data were collected for all light environments, because data at low light (i.e. less than 12%) were scarce. Data were taken directly from tables, calculated from composite parameters (e.g. SLA from LMR and LAR) and, when it could be accomplished with high precision, estimated from graphs. We did not estimate LCP, QY and RD from net photosynthesis-PPFD response curves because it could not be accomplished with high precision. When data were available, plant mass values were recorded and used to adjust RGR and whole-plant morphology for differences in mass, because these traits commonly exhibit ontogenetic drift (Coleman *et al.*, 1993; Walters *et al.*, 1993b; Veneklaas & Poorter, 1998). Plant mass for RGR was estimated at the midpoint in the time interval used to calculate RGR from the equation RGR = [ln(final mass) − ln(initial mass)]/ time, where time was changed from the total time for the RGR determination to the midpoint time (i.e. total time/2). If the authors reported either final mass or initial mass, then midpoint mass could be calculated with this equation. Plant mass for morphology characteristics (RMR, SMR, LMR, LAR) was for the harvest used to determine these characteristics.

Data were placed into groupings of 0–4% and 4–12% of open-sky light. These levels were chosen because they approximate closed forest understory

Table 1. *Parameters of leaf CO_2 exchange, leaf and whole-plant morphology and light, with their acronyms and units of measure*

Parameter	Symbol or abbreviation	Units or definition
Light-saturated CO_2 assimilation (area basis)	$A_{max,area}$	μmol CO_2 m^{-2} s^{-1}
Light-saturated CO_2 assimilation (mass basis)	$A_{max,mass}$	nmol CO_2 g^{-1} s^{-1}
Light compensation point	LCP	PPFD at which leaf net CO_2 exchange is zero
Apparent quantum yield	QY	mmol CO_2 mol^{-1} PPFD
Dark respiration (area basis)	$R_{d,area}$	μmol CO_2 m^{-2} s^{-1}
Dark respiration (mass basis)	$R_{d,mass}$	nmol CO_2 g^{-1} s^{-1}
Specific leaf area	SLA	m^2 leaf kg^{-1} leaf dry mass
Leaf mass ratio	LMR	g leaf g^{-1} plant
Leaf area ratio	LAR	m^2 leaf kg^{-1} plant
Stem mass ratio	SMR	g stem g^{-1} plant
Root mass ratio	RMR	g root g^{-1} plant
Photosynthetic photon flux density	PPFD	μmol quanta m^{-2} s^{-1}

and tree-fall gap light environments respectively (Canham *et al.*, 1990). The percentage of open-sky light was chosen because it was more commonly reported than measured total daily PPFD. In situations where only total daily PPFD values were reported, we assumed 100% of open-sky light corresponded to a PPFD of 40 and 34 mol m^{-2} d^{-1} for temperate (M. B. Walters and P. B. Reich, unpublished) and tropical (Raich, 1989; Moad, 1992; Rich *et al.*, 1993) studies respectively. We also recorded whether studies were conducted outdoors (natural forest understory and open shade house) or indoors (glass house and growth chamber).

Species were placed into three shade-tolerance categories: tolerant, mid-tolerant and intolerant. These categories were based primarily on observations by the original authors and published observational tolerance classifications (Baker, 1949; Burns & Honkala, 1990). In a few cases they were based on low-light survival data (Moad, 1992). We excluded species if placement into one of the three tolerance groups was ambiguous. Here we define shade tolerance as the ability to survive in deep shade. The relative abilities of tree species to maintain seedling and sapling populations in forest understories is probably the basis of observational shade tolerance classifications, and traits in addition to juvenile survival (e.g. a large seed) might partly underlie these patterns. However, recent studies have found a strong association between observational shade-tolerance ranks and seedling/sapling survival in deep shade (Kobe *et al.*, 1995; Kobe, 1999; Walters & Reich, 1996, 1999), suggesting that our largely subjective tolerance categories should approximate low-light survival categories.

Analysis

For mean characteristic values of each combination of light, leaf phenology and tolerance categories we considered species as our experimental units. Thus all characteristics were reduced to species means

before the calculation of category means. We did this so that group means were not biased towards species that were better represented in the data set than others (e.g. *Acer saccharum*). After data had been reduced to species means, ANOVA (JMP statistical software, SAS Institute, Cary, NC, USA) was used to test the effects of leaf phenology, tolerance and light categories and their interactions on characteristics of CO_2 exchange and leaf morphology. For those characteristics that varied with mass (i.e. RGR, RMR, SMR, LMR and LAR) we also considered individual data experimental units for better estimations of the ontogenetic trends in these characteristics. These models included plant mass and leaf phenology, tolerance and light categories and their interactions. We pooled the four-way interaction term because it was always insignificant at $P > 0.25$ (Bancroft, 1964). Mass was logarithm-transformed before inclusion in these models, to decrease heteroscedasticity and/or increase evenness in data distribution over the range of masses. Plant mass was a significant predictor of RGR ($P < 0.05$), but interactions of plant mass with phenology, tolerance and light categories were generally weak ($P > 0.15$); we therefore normalized RGR means to a common plant mass. For RMR, SMR, LMR and LAR, mass and many of its interactions with phenology and tolerance classes were significant. Furthermore, with mass and its interactions in the model, light and its interactions were not significant. We therefore pooled the two light categories and analysed models with leaf phenology, tolerance, mass and their interactions as factors. We then presented leaf-phenology-tolerance groups as continuous functions of total plant mass and report significant tolerance-mass interaction terms (i.e. separate slopes tests). We also tested for outdoor-indoor experiment effects by including it as a nominal variable in a full factorial with species and tolerance groupings.

Means $\pm$ SE are presented for each light group. Because data distributions were logarithmic, leaf lifespan means and SE are back-transformed logarithmic values. Using JMP statistical software (SAS Institute, Cary, NC, USA), tests of significance for differences in means among tolerant, intermediate and intolerant groups within light and species groups were made with Tukey-Kramer honestly significant difference (HSD) ($\alpha = 0.05$).

Carbon balance simulations

We developed growing-season, whole-plant growth simulations in dynamic forest understory light environments for seedlings of two winter deciduous species, intolerant *Betula papyrifera* and tolerant *A. saccharum*. These simulations were based on CO_2 exchange and morphology characteristics for greenhouse grown seedlings (M. B. Walters and P. B. Reich, unpublished), and measurements of the phenology of the understory photosynthetic activity of *A. saccharum* in relation to overstory leaf phenology (Gill *et al.*, 1998).

In a greenhouse at the University of Minnesota, newly germinated *B. papyrifera* and *A. saccharum* seedlings were transplanted in mid-February into pots filled with washed silica sand and placed in shade treatments that transmitted $2.77 \pm 0.15\%$ (mean $\pm$ SE) of the open-sky light outside the greenhouse (daily mean, 37.8 mol m^{-2} d^{-1} PPFD). Seedlings were supplied daily with modified Hoagland's solution containing 3.4 mM nitrogen. To determine ontogenetic changes in LMR for young seedlings (see Table 4), six seedlings per treatment were harvested at 3, 9 and 19 wk of age. Seedlings were separated into leaves, stems and roots, oven-dried at 70°C, mass and carbon content (Table 4; NA 1500 Analyser; Carlo Erba, Milan, Italy) were determined, and LMR was calculated. For other four-month-old seedlings in these treatments, we measured leaf net CO_2 exchange to PPFD (0.3–600 µmol quanta m^{-2} s^{-1}) over several days on three or more plants per species with a LI-6200 photosynthesis system (LI-COR Inc., Lincoln, NB, USA). Before sunrise on the day after completion of photosynthesis measurements, seedlings were moved to a dark room, harvested and root systems were washed free of sand; the seedlings were then separated into leaves, roots and stems immediately before respiration measurements. Measurements were made on whole, detached leaf, stem and root systems at 21.0 ± 0.5°C (mean $\pm$ SD) and 365 µl l^{-1} CO_2 with two LCA-3 portable gas exchange systems (Analytical Development Corporation, Hoddesdon, Herts., UK). After measurements of CO_2 exchange, leaf areas were measured, tissues were oven-dried at 70°C and leaf, root and stem mass were measured. Plots of leaf-level net CO_2 exchange against PAR for each species were fitted with the model of Hanson *et al.* (1987) with the use of the nonlinear platform of JMP statistical software (SAS Institute, Cary, NC, USA) (Table 4). This function was modified to

whole-plant net CO_2 exchange against PPFD responses by incorporating respiration rates, photoperiod length and allocation to leaves, roots and stems as $(A_{mass} \times LMR) - R_W$, where A_{mass} is mass-based photosynthesis as a function of PPFD and R_W ($R_W = LMR \times$ leaf-$R_{d,mass} \times$ (h of darkness/24) + SMR $\times$ stem-R_{mass} + RMR $\times$ root-R_{mass}) is whole-plant respiration rate. By using this function we estimated whole-plant CO_2 exchange in forest understories with and without canopy leaves from 10-min averages of instantaneous PPFD recorded diurnally over 10 d in September and October in two forest understories, dominated by *A. saccharum*, one with approx. 2% and one with approx. 4% canopy openness (LAI 2000; LI-COR Inc.). Values were transformed from CO_2 exchange to RGR by using our measured carbon content values. Daily values were summed for a 169-d growing season, with or without a 2-wk period during which the overstory was leafless while understory seedlings were photosynthetically active (Gill *et al.*, 1998). Seedling leaf turnover was assumed to occur only at the end of the growing season, and was estimated from late-season LMR and estimates of pre-leaf-senescence mass translocation from leaves (Reich *et al.*, 1991) as LMR $\times$ [1 − (proportion re-translocated)] $\times$ plant mass.

RESULTS

Leaf lifespan, leaf CO_2 exchange and SLA

Mean leaf lifespan was similar between tolerance categories for WINTER DECIDUOUS species. In contrast, intermediate and tolerant BROAD-LEAVED EVERGREEN

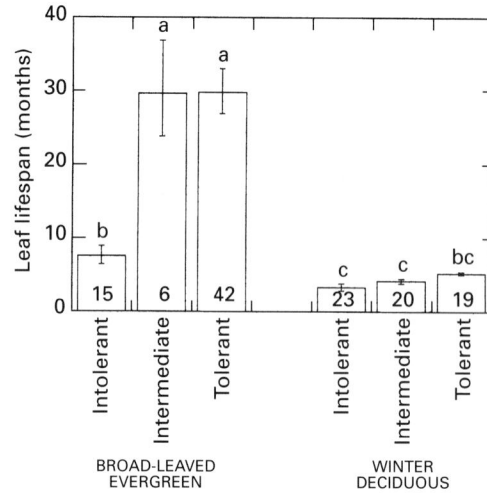

Fig. 1. Leaf lifespan for intolerant, intermediate, and tolerant species. Means $\pm$ SE are back-transformed logarithmic means. Transformed means that do not share a common letter are significantly different ($P < 0.05$, Tukey-Kramer HSD). The number of species (n) in each category is indicated at the base of each column. Data sources are Coley (1988), Koike (1988), Williams *et al.* (1989), Moad (1992), Tinoco-Ojanguren & Pearcy (1992), Reich *et al.* (1999) and P. B. Reich (unpublished).

Table 2. *Results of ANOVAs and ANCOVAs*

(a) ANOVA with light (Lt, 0–4% and 4–12%), leaf phenology (Lf, BROAD-LEAVED EVERGREEN, WINTER DECIDUOUS) and tolerance (T, intolerant, intermediate, tolerant) categories as factors. Each datum in each light category is a species mean. Category means and n for each characteristic are given in Table 3.

Characteristic	Lt	Lf	T	Lt × Lf	Lt × T	Lf × T	Lt × Lf × T
LCP	ns	**	ns	ns	ns	ns	ns
QY	ns	***	ns	ns	ns	ns	ns
$A_{max,area}$	***	ns	***	ns	ns	ns	ns
$A_{max,mass}$	ns	ns	***	ns	ns	ns	ns
$R_{d,area}$	ns	**	ns	ns	ns	*	ns
$R_{d,mass}$	ns	ns	***	ns	ns	ns	ns
SLA	***	***	***	**	*	ns	ns
LMR	*	***	***	ns	ns	***	ns
LAR	***	**	***	ns	ns	ns	ns
RGR	***	***	***	***	ns	ns	ns

(b) ANCOVA with the logarithm of plant mass (M), leaf phenology (Lf) and tolerance classification (T) as factors. Each datum in each light category is a single observation; thus a species may be represented more than once. The least squares linear regression lines, and n for S × T treatments as a function of mass are in Fig. 2.

Characteristic	M	Lf	T	M × Lf	M × T	Lf × T	M × Lf × T
LAR	***	ns	***	ns	ns	ns	ns
LMR	***	***	**	ns	**	***	*
SMR	***	ns	***	***	ns	ns	ns
RMR	ns	***	**	***	ns	*	ns

*, $P < 0.05$; **, $P < 0.01$; ***, $P < 0.001$. Underlining indicates the most significant factor for each characteristic.

Table 3. CO_2 *exchange and morphology characteristics in 0–4% light (a) and 4–12% light (b)*

Characteristic	BROAD-LEAVED EVERGREEN			WINTER DECIDUOUS		
	Intolerant	Intermediate	Tolerant	Intolerant	Intermediate	Tolerant
(a) 0–4% light						
LCP	5.4 ± 1.0^a (7)	9.3 ± 1.3^a (10)	6.6 ± 1.6^a (15)	10.7 ± 2.4^a (5)	11.7 ± 6.0^a (4)	11.5 ± 4.0^a (4)
QY_{area}	42 ± 0^a (3)	47 ± 7^a (5)	41 ± 4^a (11)	27 ± 2^a (5)	33 ± 7^a (4)	31 ± 8^a (4)
$A_{max,area}$	3.8 ± 0.4^a (15)	3.7 ± 0.5^a (11)	3.2 ± 0.2^a (31)	4.1 ± 0.5^a (8)	4.2 ± 0.5^a (6)	3.1 ± 0.4^a (12)
$A_{max,mass}$	178 ± 19^a (13)	181 ± 20^a (7)	78 ± 8^b (25)	153 ± 39^{ab} (5)	118 ± 29^{ab} (5)	118 ± 17^{ab} (10)
$R_{d,area}$	0.22 ± 0.03^a (9)	0.40 ± 0.08^a (11)	0.35 ± 0.05^a (22)	0.51 ± 0.20^a (6)	0.33 ± 0.05^a (5)	0.27 ± 0.06^a (7)
$R_{d,mass}$	14.8 ± 2.8^{ab} (7)	12.9 ± 1.7^{abc} (7)	6.7 ± 1.0^c (18)	20.2 ± 5.5^a (3)	6.5 ± 2.7^{bc} (3)	8.5 ± 1.5^{bc} (5)
LMR	0.48 ± 0.02^a (18)	0.50 ± 0.04^a (10)	0.49 ± 0.01^a (35)	0.56 ± 0.03^a (7)	0.33 ± 0.07^b (5)	0.35 ± 0.03^b (9)
SLA	52.5 ± 3.2^a (30)	35.2 ± 3.2^{cd} (16)	26.1 ± 1.7^d (49)	52.3 ± 5.5^{ab} (9)	34.7 ± 4.7^{bcd} (10)	37.5 ± 1.7^c (23)
LAR	23.9 ± 2.1^{ab} (18)	13.5 ± 1.1^c (9)	14.1 ± 2.7^c (29)	33.5 ± 32^a (7)	14.4 ± 6.8^{bc} (5)	13.7 ± 2.0^c (9)
RGR	10.3 ± 1.6^a (19)	6.5 ± 2.2^a (14)	5.3 ± 1.5^a (25)	14.7 ± 4.5^a (7)	12.5 ± 10.0^a (3)	8.7 ± 2.9^a (8)
(b) 4–12% light						
LCP	8.7 ± 1.0^a (8)	7.5 ± 0.8^a (10)	5.9 ± 0.7^a (15)	10.4 ± 2.1^a (4)	11.4 ± 6^a (3)	8.2 ± 2.4^a (3)
QY_{area}	44 ± 3^a (10)	46 ± 2^a (6)	39 ± 4^a (11)	37 ± 2^a (4)	36 ± 6^a (3)	42 ± 5^a (4)
$A_{max,area}$	5.2 ± 0.6^{ab} (10)	5.6 ± 0.4^a (10)	3.8 ± 0.3^b (18)	5.3 ± 0.4^{ab} (5)	5.7 ± 1.0^{ab} (3)	4.5 ± 0.6^{ab} (4)
$A_{max,mass}$	163 ± 40^{ab} (7)	174 ± 9^{ab} (4)	87 ± 19^b (10)	238 ± 34^a (4)	186 ± 31^{ab} (2)	129 ± 15^{ab} (3)
$R_{d,area}$	0.28 ± 0.06^{bc} (10)	0.31 ± 0.06^{bc} (9)	0.18 ± 0.03^c (17)	0.63 ± 0.16^a (5)	0.50 ± 0.07^{ab} (4)	0.31 ± 0.10^{abc} (5)
$R_{d,mass}$	10.5 ± 3.2^b (7)	14.3 ± 1.6^{ab} (4)	5.0 ± 1.7^b (10)	25.3 ± 6.7^a (4)	10.0 ± 4.0^a (3)	9.8 ± 1.9^a (4)
LMR	0.46 ± 0.05^{ab} (7)	0.42 ± 0.03^{ab} (12)	0.44 ± 0.03^{ab} (23)	0.50 ± 0.03^a (11)	0.27 ± 0.07^b (7)	0.34 ± 0.04^{ab} (11)
SLA	30.9 ± 32^{bc} (21)	24.4 ± 23^{cd} (15)	22.0 ± 15^d (39)	47.2 ± 51^a (11)	33.6 ± 26^{abcd} (9)	38.8 ± 21^{ab} (14)
LAR	15.8 ± 3.7^{ab} (8)	8.3 ± 1.0^b (17)	10.7 ± 1.1^b (24)	22.9 ± 3.2^a (10)	10.8 ± 4.2^b (6)	13.2 ± 1.1^b (11)
RGR	14.8 ± 3.7^{bc} (9)	10.2 ± 1.7^{bc} (11)	8.7 ± 0.9^c (26)	45.6 ± 7.8^a (8)	33.6 ± 23.0^{ab} (3)	22.5 ± 2.5^{bc} (8)

Values are means ± SE, with the number of species used in parentheses. Means without a common letter within a row are significantly different ($P < 0.05$, Tukey–Kramer HSD).

species had fourfold and sixfold greater leaf lifespans than intolerant BROAD-LEAVED EVERGREEN and tolerant WINTER DECIDUOUS species respectively (Fig. 1). $A_{max,area}$ was greater in higher (4–12%) than in lower (0–4%) light, but it was only marginally lower ($P > 0.05$) for tolerant than intermediate and intolerant species in both light environments. In contrast, $A_{max,mass}$ did not differ between light

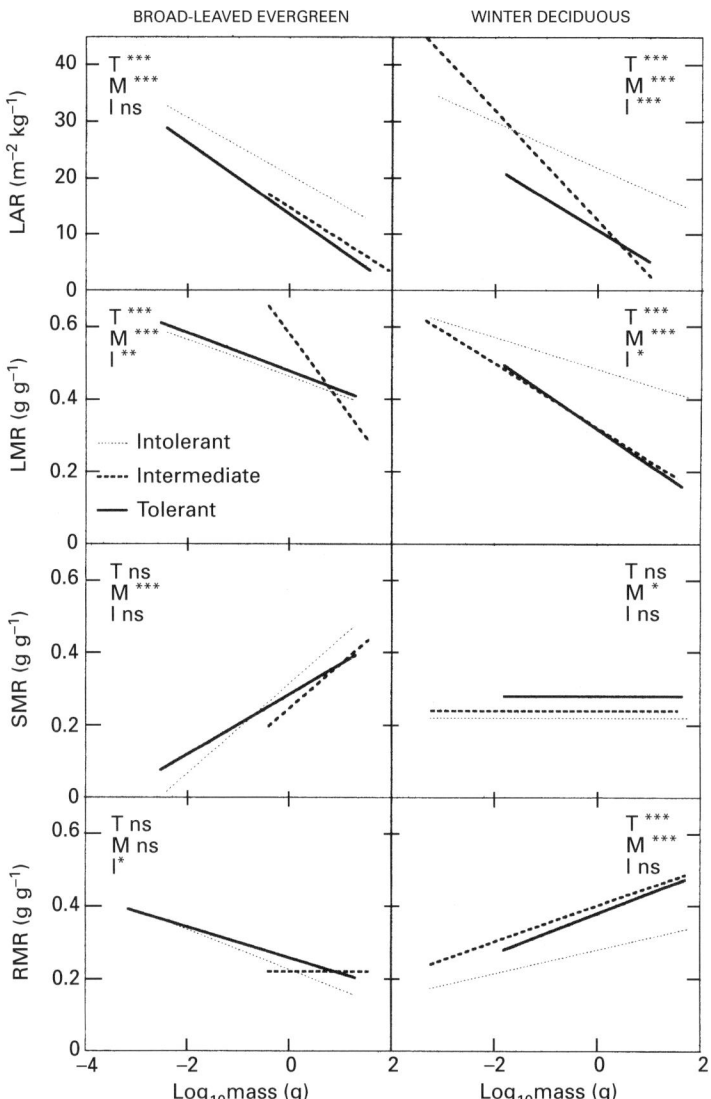

Fig. 2. Relationships of leaf area ratio (LAR), leaf mass ratio (LMR), stem mass ratio (SMR) and root mass ratio (RMR) to total plant mass. Data are for seedlings in less than 12% light. The lines are least-squares regressions and are shown over the range of the data from which they were developed. Data for each regression ranged from 21 to 81 and averaged 37. Insignificant regressions ($P > 0.05$) are shown as flat lines. For each leaf phenology group significant effects of $\log_{10}$mass (M), tolerance group (T) and the M × T interaction (I, i.e. a separate slopes test) are indicated as follows: ns, $P \geqslant 0.05$; *, $P < 0.05$; **, $P < 0.01$; ***, $P < 0.001$.

environments, but tolerant species had lower $A_{max,mass}$ values than intolerant and intermediate species, especially for BROAD-LEAVED EVERGREEN species (Tables 2, 3). $R_{d,area}$ did not differ significantly among tolerance categories, whereas $R_{d,mass}$ was approx. twofold greater for intolerant than tolerant species ($P < 0.05$ in 0–4% light only). Intermediate BROAD-LEAVED EVERGREEN species had $A_{max,mass}$ and $R_{d,mass}$ values that were similar to those of intolerant BROAD-LEAVED EVERGREEN species, whereas intermediate WINTER DECIDUOUS species had $A_{max,mass}$ and $R_{d,mass}$ values that were more similar to those of tolerant WINTER DECIDUOUS species. Differences in mass-based rates of both A_{max} and R_d between tolerance classes were generally larger than for area-based rates and these differences were greater in 0–4% than in 4–12% light. This occurred because SLA was greater for intolerant than for

tolerant species in both light environments; these differences were greatest in 0–4% light (Tables 2, 3).

LCP and QY did not differ between tolerance groups, but they were greater for WINTER DECIDUOUS than TROPICAL EVERGREEN species (Tables 2, 3). Thus there was no evidence that A_{area} differed between tolerance classifications at low PPFD. However, because the slope of the plot of A_{mass} against PPFD at low PPFD equals QY × SLA, and because SLA was greater for intolerant than for tolerant species, especially in 0–4% light, intolerant species might have greater A_{mass} values than tolerant species at any PPFD above LCP.

Whole-plant morphology and RGR

For LMR there was a strong interaction between leaf phenology and tolerance categories (Table 2).

LMR was similar for all tolerance categories for BROAD-LEAVED EVERGREEN species in both light environments (Table 3). In contrast, and especially in 0–4% light, tolerant and intermediate WINTER DECIDUOUS species had much lower LMR than both intolerant WINTER DECIDUOUS species and all tolerance categories of BROAD-LEAVED EVERGREEN species. LAR was greater for intolerant than for intermediate and tolerant species in both leaf phenology groups. For intolerant BROAD-LEAVED EVERGREEN species, greater LAR was due to greater SLA (LAR = LMR × SLA). For intolerant WINTER DECIDUOUS species, greater LAR was due to both greater LMR and greater SLA.

RGR was similar between leaf phenology and tolerance groups in 0–4% light (Table 3). RGR increased with light; these increases were greatest for intolerant WINTER DECIDUOUS species such that their RGR in 4–12% light was significantly greater than for any other tolerance-leaf-phenology group. Initial mass was a significant negative covariate predictor of RGR in both 0–4% light ($P < 0.01$) and 4–12% light ($P < 0.05$); however, normalizing for plant mass changed means only slightly (data not shown).

Almost all data for LMR, SLA and LAR for BROAD-LEAVED EVERGREEN species were from outdoor (forest understories, open shadehouse) experiments, whereas for WINTER DECIDUOUS species one-third of the SLA and LAR data and one-half of the LMR and RGR data were from indoor (greenhouse, growth chamber) experiments. For WINTER DECIDUOUS species, values for SLA, LAR, LMR and RGR were lower outdoors than indoors ($P < 0.05$ for all characteristics; ANOVA). However, trends between tolerance categories for WINTER DECIDUOUS species and between WINTER DECIDUOUS and BROAD-LEAVED EVERGREEN species were similar for outdoor and outdoor plus indoor data. Therefore outdoor data are included in the complete data set but are not presented separately.

Changes in plant morphology with mass

LMR and LAR declined similarly with mass in the BROAD-LEAVED EVERGREEN and WINTER DECIDUOUS species groups (Table 2b, Fig. 2). However, changes in SMR and RWR with mass differed markedly between leaf phenology groups. For BROAD-LEAVED EVERGREEN species, as mass increased and LMR decreased, SMR increased and RMR decreased to relatively low values. For WINTER DECIDUOUS species, as mass increased and LMR decreased, RMR increased to relatively high values, especially for intermediate and tolerant species, whereas SMR remained constant. Furthermore, LMR-mass relationships clearly show that LMR is smaller and RMR greater for intermediate and tolerant WINTER DECIDUOUS species than for any other tolerance-leaf-

phenology category and that differences between these groups increase with mass.

DISCUSSION

For both BROAD-LEAVED EVERGREEN and WINTER DECIDUOUS groups, intolerant species had greater growth potential than tolerant species, as indicated by their greater LAR, SLA and mass-based photosynthetic rates above LCP. However, despite greater growth potential for intolerants, realized growth rates were similar for tolerant and intolerant groups in 0–4% light. Intolerant species did not realize significant growth advantages over tolerant species until higher light levels (4–12%) were reached, and then only for WINTER DECIDUOUS species. Thus, there was no evidence of low-light growth advantages for tolerant species as has long been hypothesized (Baker, 1945; Björkman, 1981; Givnish, 1988), but neither was there strong evidence for the opposite. We also found large differences between tolerant species of WINTER DECIDUOUS and BROAD-LEAVED EVERGREEN groups in SLA and in biomass allocation. Most strikingly, intermediate and tolerant WINTER DECIDUOUS species had lower LMR and higher RMR than any other leaf-phenology-tolerance category.

Of the several possible reasons for differences in the patterns that we observed among tolerance and leaf phenology categories, we shall discuss three. These are: (1) the greater growth capacities exhibited by intolerant than tolerant species in low light can occur only if respiration rates are high and characteristics conferring protection from damage are low; (2) tolerant BROAD-LEAVED EVERGREEN species minimize biomass turnover through low leaf turnover, whereas tolerant WINTER DECIDUOUS species minimize turnover through low LMR; and (3) for tolerant WINTER DECIDUOUS species, predictable high light periods followed by low light then winter dormancy favours allocation to storage over continuous production of new leaves.

Low-light growth potential, respiration rates and protection

Greater low-light growth potential for intolerant than for tolerant species combined with a strong positive association between low-light survival and shade tolerance classifications (Baker, 1945; Shirley, 1945; Kitajima, 1994; Kobe *et al.*, 1995; Kobe, 1999; Walters & Reich, 1996, 1999) suggests a trade-off between growth potential and low light survival (Walters *et al.*, 1993a; Kitajima, 1994; Kobe *et al.*, 1995; Pacala *et al.*, 1996). This trade-off might occur owing to the combination of (1) selective premium's being placed on energy conservation traits if RGR is severely constrained by low light, because a given amount of energy or biomass loss consumes a greater proportion of total growth or storage when growth

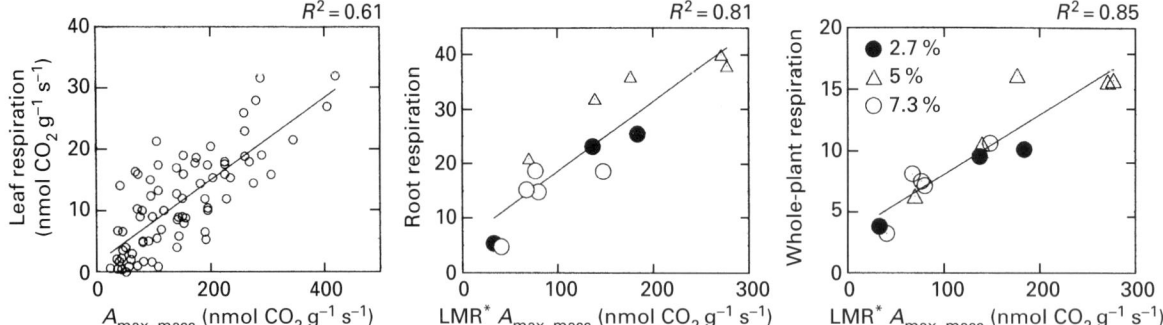

Fig. 3. Relationships between respiration and photosynthesis on both leaf and whole-plant bases. Lines and r^2 values are from least-squares regressions ($P < 0.0001$ in all cases). Data for leaf respiration against $A_{max, mass}$ are from our data compilation; all data are therefore from less than 12% light. Relationships between root and whole-plant respiration against whole-plant, light saturated photosynthetic rate (i.e. LMR $\times A_{max, mass}$) are for seedlings from 5% (Reich *et al.*, 1998a), 2.7% and 7.3% light (M. B. Walters & P. B. Reich, unpublished), where each result is a species mean.

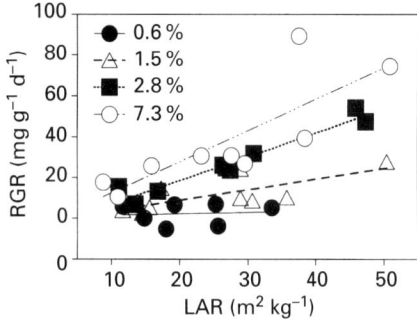

Fig. 4. Plot of RGR against LAR at four low-light levels for tree seedlings of nine WINTER DECIDUOUS and evergreen conifer species. Each result is a species mean. Lines are from least-squares regressions (graph from Walters & Reich, 1999).

rates are low, and (2) direct selection against high, and unused, growth capacity in low light, if excess capacity directly promotes greater energy or biomass loss rates via higher rates of respiration, tissue turnover, mechanical damaging and herbivory.

For example, the high growth potentials of intolerant species might require high whole-plant respiration rates, a characteristic that might be disadvantageous in very low light. High growth potentials require high $A_{max, mass} \times$ LMR (or $A_{max, area} \times$ LAR) (Walters *et al.*, 1993a,b). Not only does leaf $R_{d, mass}$ scale with $A_{max, mass}$ (Fig. 3), but because root metabolism is required to satisfy the soil resource requirements of whole-canopy photosynthesis, root respiration (and stem respiration; not shown) also scales with $A_{max, mass} \times$ LMR (Fig. 3). Thus, in comparison with tolerant species, intolerant species with their higher growth potential have both higher leaf and whole-plant R_d values (Fig. 3). Therefore, unrealized growth potential due to severe light limitations should be trimmed by selection for low respiration rates. There is ample evidence of unrealized physiological growth potential in low light, especially for intolerant species. For example, in high and moderately low light, LAR is the mor-

phological trait most strongly related to RGR (Lambers & Poorter, 1992; Walters *et al.*, 1993a; Walters and Reich, 1996; 1996; Hunt & Cornelissen, 1997) and we reported higher LAR for our intolerant species groups. However, we found little variation in low-light RGR despite large variation in LAR (Fig. 4). This occurs, in part, because very low, and similar, photosynthetic rates (A_{area}) in low light decrease the magnitude of LAR's impact on whole-plant photosynthetic productivity (i.e. LAR $\times A_{area}$ = whole-plant photosynthetic productivity, and A_{area} is small in low light). In addition, because whole-plant R_d scales with LAR, the higher R_d of plants with high LAR might further diminish the slope of RGR-LAR relationships. Thus the advantages of maintaining a high LAR, in terms of RGR, diminish with decreasing light.

High growth potential in low light might also increase the risk of damage and death. In low light, as in high light (Coley, 1988; Reich *et al.*, 1991, 1992), leaf traits that enhance growth potential directly or indirectly (e.g. higher nitrogen (Reich *et al.*, 1992) and water (Roderick *et al.*, 1999) contents, lower fibre, tannins and toughness (Coley, 1988)) might also increase the probability of leaf loss to herbivores and/or mechanical damage (Coley, 1988; Edenius *et al.*, 1993; Braker & Chazdon, 1993). Intolerant species generally had higher SLA and $A_{max, mass}$, both of which scale positively with nitrogen (Reich *et al.*, 1997) and water (Roderick *et al.*, 1999) contents and negatively with leaf toughness (Cornelissen *et al.*, 1999). However, unlike high light, in which intolerant species can outgrow tolerant species despite incurring greater rates of herbivory (Coley, 1988), greater herbivory rates on intolerant species in shade, where growth is strongly constrained, might lead to lower growth and survival for intolerant than for tolerant species. Thus selection for protection from herbivores (and mechanical damage) should trim excess growth capacity in shade, because these traits, such as higher structural carbon and lower nitrogen, should be reflected in lower SLA and

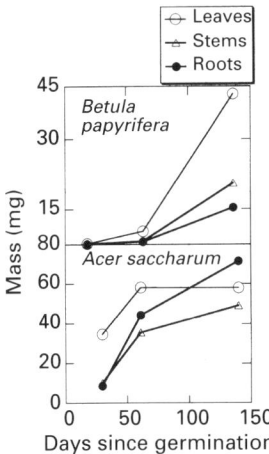

Fig. 5. Changes in the masses of leaves, stems and roots from germination to 150 d of age in 2.8% light for the WINTER DECIDUOUS species, intolerant *B. papyrifera* and tolerant *A. saccharum*. Each result is a mean for six or more seedlings.

$A_{max,mass}$. These traits are clearly exhibited by tolerant BROAD-LEAVED EVERGREEN species.

Biomass turnover

For WINTER DECIDUOUS species, the higher growth of intolerant species than tolerant species is due partly to their higher LMR (Table 2). Because leaves are completely shed at the end of the growing season, a large fraction of biomass gained over the season is lost to leaf turnover. As light levels and growth rates decrease, the quantitative importance of seasonal leaf loss to net annual growth increases. Thus a low LMR for tolerant WINTER DECIDUOUS species might be adaptive in that it minimizes biomass losses to seasonal leaf turnover. This is in sharp contrast to tolerant BROAD-LEAVED EVERGREEN species that have high LMR and minimize biomass turnover rates partly through long leaf lifespans.

Carbon reserves and high-light photosynthesis opportunities

Carbon acquired by photosynthesis can be allocated to structure for the acquisition of carbon (leaves), mineral nutrients and water (roots), support and mechanical integrity (leaves, stems and roots), or it can be allocated to storage (Chapin *et al.*, 1990; Kobe, 1997). General patterns in allocation between storage and structure among intolerant and tolerant species in low light are not known. However, for temperate species, several supportive lines of evidence suggest that, compared with intolerant species, tolerant species favour storage over growth potential in low light. This evidence includes the following: tolerant species have greater RMR (this study); roots often have equal to higher carbon storage concentrations than other organs (DeLucia *et al.*, 1998); tolerant species often (Kobe, 1997), but not

always (DeLucia *et al.*, 1998; Niinemets, 1998) have greater carbon storage concentrations than intolerant species; and tolerant species typically have determinate shoot growth patterns, whereas intolerant species are indeterminate. Determinate growth results in the occurrence of a single leaf flush near the beginning of the growing season (Koike, 1988), with allocation for the rest of the growing season necessarily being to stems and roots (e.g. *A. saccharum*, Fig. 5). Collectively these patterns suggest that juveniles of tolerant species favour increasing whole-plant carbohydrate concentrations over the production of new leaves (Kobe, 1997). After the early cessation of leaf growth, they can accomplish this through greater allocation to storage-compound rich root systems, thus resulting in their high RMR.

In temperate and boreal forests, this strategy might be especially important for shade-tolerant seedlings and saplings growing beneath canopies of early successional trees, because the latter are often broad-leaved deciduous species that flush leaves relatively late in the spring (e.g. *Populus*, *Betula*, *Quercus* spp.). This provides a temporary (1–3 wk) opportunity for high growth rate and storage at high light for shade-tolerant deciduous species that flush leaves earlier in the growing season (e.g. *A. saccharum*) (Gill *et al.*, 1998). For understory seedlings, allocation to storage during this brief period might be critical because it is followed by an extended period of low light, then winter dormancy. In addition, allocation to new leaf structure after the brief high-light period may not be favoured because those leaves will probably have little opportunity for carbon gain. Thus, early leaf flush followed by allocation to storage, and not new leaf structure, might be an effective seedling strategy for tolerant species adapted to succeed earlier successional WINTER DECIDUOUS species. The high RMR of tolerant WINTER DECIDUOUS species might reflect this strategy. In contrast, the high LMR and near-continuous (growing season) leaf production of intolerant WINTER DECIDUOUS species suggest a predisposition to maximize carbon gain in low light even if this compromises survival, since leaf production may occur at the expense of storage. Without predictable opportunities for growth in high light, tolerant BROAD-LEAVED EVERGREEN species might be less prone to allocate large amounts of resources periodically to storage. Instead, they might maintain lower, but less variable, storage reserves, and incur low leaf turnover rates as they grow slowly in the understory. This strategy might be reflected in their low RMR and relatively high LMR and SMR.

Whole-plant carbon balance simulations

For WINTER DECIDUOUS seedlings we used a simple whole-plant carbon balance model to estimate the combined impacts of whole-plant photosynthesis

Table 4. *Seedling parameters used for growth simulations for intolerant* Betula papyrifera *and tolerant* Acer saccharum

Total daily PPFD (mol quanta m^{-2} d^{-1}), leaf on/leaf off
0.37/4.48
0.75/4.02

Plant characteristic	Acer	Betula
Carbon content (mmol g^{-1})	33.3	33.3
LMR (g g^{-1})		
Day 0–40	0.65	0.50
Day 40–100	0.44	0.68
Day 100–164	0.29	0.63
Whole-plant respiration rate (nmol CO^2 g^{-1} s^{-1})		
Day 0–40	4.8	9.2
Day 40–100	4.1	10.7
Day 100–164	3.5	10.3
Autumn leaf retranslocation (%)	15	15

Species	Relationship of A_{mass} to PPFD
A. saccharum	$A_{mass} = 111\{1 - 0.3821\exp[1 - (0.01581\text{PPFD})]\}$
B. papyrifera	$A_{mass} = 308\{1 - 0.03821\exp[1 - (0.0581\text{PPFD})]\}$

Sources are M. B. Walters and P. B. Reich (unpublished), except for light (M. Tobin and P. B. Reich, unpublished) and leaf translocation (Reich *et al.*, 1991).

Table 5. *Whole-season (1 May to 15 October) simulated growth for intolerant* Betula papyrifera *and tolerant* Acer saccharum *in two growth light environments, and with and without a 2-wk period of high light (Gill* et al., *1998)*

| | Simulated growth (mg g^{-1}) | | | |
| | 0.37 and 4.48 mol quanta m^{-2} d^{-1} | | 0.75 and 4.02 mol quanta m^{-2} d^{-1} | |
Model	B. papyrifera	A. saccharum	B. papyrifera	A. saccharum
With 2-wk high-light period	−943	−74	936	456
Without 2-wk high-light period	−1625	−396	382	201
With high-light period, minus leaf turnover	−1473	−321	406	209
Without high-light period, minus leaf turnover	−2155	−643	−148	−46

Leaf turnover was 247 and 530 mg g^{-1} plant for *A. saccharum* and *B. papyrifera* respectively. Plant mass at the end of the growing season was decreased by these proportions in models that included loss to leaf turnover. See model parameters in Table 4 and other details in the Methods section.

and respiration rates, leaf turnover and leaf phenology on the growth of tolerant *A. saccharum* and intolerant *B. papyrifera* in dynamic understory light environments. Parametrization and the assumptions of the model are contained in Table 4. Our simulations showed that both species had net negative growth rates in a forest understory with very low light over the entire growing season (0.37 mol quanta m^{-2} d^{-1}), but growth was less negative for *Acer* because of its lower whole-plant respiration rate (Table 5). At slightly higher, but still very low, light levels (0.75 mol quanta m^{-2} d^{-1}), both species had positive growth, with *Betula* having higher rates than *Acer*. Thus, *Acer* had lower biomass loss rates owing to lower respiration than *Betula* in very low light, but growth rankings reversed at slightly higher light levels. If leaf

turnover at the season's end is subtracted from growth, *Acer* has a net growth rate advantage over *Betula* in both low-light environments because by virtue of its lower LMR *Acer* loses a smaller fraction of its total biomass to leaf turnover than *Betula* (Table 5). Finally, a 2-wk high-light period can prevent excessive growing-season carbon deficits in very low (0.37 mol quanta m^{-2} d^{-1}) light and can account for more than two-thirds of seasonal carbon gain in low (0.75 mol quanta m^{-2} d^{-1}) light (Table 5). If *A. saccharum* takes greater advantage of this high-light period by making a photosynthetically active leaf canopy of high LMR earlier in the growing season than *B. papyrifera*, this could lead to an even greater growth advantage for *A. saccharum* over *B. papyrifera*. At the leaf level, the importance of this spring 'window of light' to carbon balance has been shown in the field for both understory shrubs and trees, including *A. saccharum* (Harrington *et al.*, 1989; Gill *et al.*, 1998).

Although differences in seedling characteristics between leaf phenology, light and tolerance categories were compelling, we had few data for seedlings more than 1 yr old, and almost none for seedlings greater than 3 yr old (Walters & Reich, 1997; Niinemets, 1998; DeLucia *et al.*, 1998). Thus, we do not know whether biomass allocation and growth strategies change for older seedlings and saplings. As Grubb (1998) points out, shade-tolerant temperate trees might be characterized by a shift in their adaptive syndrome from conservative growth traits as juveniles to traits promoting rapid growth as adults. Furthermore, biomass allocation patterns provide a limited picture of adaptive differences in form between tolerance groups, because potentially important traits such as height (Grime & Jeffery, 1965), volume occupancy (Küppers, 1985) and efficiency of leaf display (Givnish, 1988) are dependent on both biomass allocation and the morphology and architecture of stems and leaves.

Conclusions

Our data clearly demonstrate that young seedlings of shade-tolerant species do not have low-light carbon capture advantages over shade-intolerant species, perhaps because they maximize biomass and energy conservation at the expense of enhanced low-light carbon capture. In a quantitative review of tropical tree seedling data (mostly BROAD-LEAVED EVERGREEN species), Veneklaas & Poorter (1998) obtained results similar to ours. Our results build on those of Veneklaas & Poorter (1998) in two major ways. First, we found that the morphologies of WINTER DECIDUOUS species differ markedly from those of BROAD-LEAVED EVERGREEN species. For both intermediate and tolerant groups, WINTER DECIDUOUS species had lower LMR and higher RMR, SLA and leaf turnover rates than BROAD-LEAVED EVERGREEN species. Thus, in part, tolerant BROAD-LEAVED EVERGREEN species minimize low-light biomass losses through long lifespans of their leaves, whereas tolerant WINTER DECIDUOUS species minimize biomass loss though low LMR. Second, with their low-light carbon balance simulations (2.16 mol quanta m^{-2} d^{-1}) and literature survey for low (0–5%) light, Veneklaas & Poorter (1998) found that intolerant species realize greater growth rates than tolerant species, owing partly to their greater LAR. We also found that intolerant species had a greater LAR in low (0–4%) light. However, in our carbon balance simulations in very low light (0.37 and 0.75 mol quanta m^{-2} d^{-1}) we found that, despite their lower LAR, tolerant WINTER DECIDUOUS species can realize carbon balance advantages over intolerant species through lower whole-plant respiration and lower biomass turnover.

Furthermore, tolerant WINTER DECIDUOUS species might have greater access than intolerant species to seasonal high-light 'windows' when the overstory is leafless. This is a carbon gain opportunity that is not periodically available to tolerant BROAD-LEAVED EVERGREEN species. The differences that we found in whole-plant carbon balance strategies for WINTER DECIDUOUS and BROAD-LEAVED EVERGREEN woody plant seedlings were striking. How these patterns differ for plants from other biomes (e.g. drought deciduous, temperate and boreal evergreens) and for different plant forms (e.g. herbs) remains largely unknown.

ACKNOWLEDGEMENTS

We thank Eric Garnier for inviting us to the symposium; Hendrik Poorter and two anonymous referees for improving the manuscript markedly; and Kim Walters, who worked very hard on the primary data table.

REFERENCES

Baker FS. 1945. Effects of shade upon coniferous seedlings grown in nutrient solutions. *Journal of Forestry* **43**: 428–435.

Baker FS. 1949. A revised shade tolerance table. *Journal of Forestry* **47**: 179–181.

Bancroft TA. 1964. Analyses and inferences for incompletely specified models involving the use of preliminary tests of significance. *Biometrics* **20**: 427–442

Björkman O. 1981. Responses to different quantum flux densities. In: Large OL, Nobel PS, Osmond CB, Ziegler H, eds. *Encyclopedia of plant physiology*, vol. 12A (*Plant Physiological Ecology I*). Berlin, Germany: Springer Verlag, 57–107.

Braker E, Chazdon R.L 1993. Ecological, behavioral and nutritional factors influencing use of palms as host plants by a neo-tropical forest grasshopper. *Journal of Ecology* **9**: 183–197.

Burns RM, Honkala BH. 1990. *Silvics of North America*, vols 1 and 2. (Agricultural Handbook no. 654.) Washington, D.C., USA: United States Department of Agriculture.

Canham CD, Denslow JS, Platt WJ, Runkle JR, Spies TA, White PS. 1990. Light regimes beneath closed canopies and tree-fall gaps in temperate and tropical forests. *Canadian Journal of Forestry Research.* **20**: 620–631.

Cao K, Ohkubo T. 1998. Allometry, root/shoot ratio and root architecture in understory saplings of deciduous dicotyledonous trees in Central Japan. *Ecological Research* **13**: 217–227.

Chapin FS III, Schulze ED, Mooney, HA. 1990. The ecology and economics of storage in plants. *Annual Review of Ecological Systems* **21**: 423–447.

Coleman JS, McConnaughay KDM, Bazzaz FA. 1993. Elevated CO$_2$ and plant nitrogen-use - is reduced tissue nitrogen concentration size-dependent. *Oecologia* **93**: 195–200.

Coley PD. 1988. Effects of plant growth rate and leaf lifetime on the amount and type of anti-herbivore defense. *Oecologia* **74**: 531–536.

Cornelissen JHC, Pérez-Harguindeguy N, Díaz S, Grime JP, Marzano B, Cabido M, Vendramini F, Cerabolini B. 1999. Leaf structure and defence control litter decomposition rate across species and life forms in regional floras on two continents. *New Phytologist* **143**: 000–000.

DeLucia EH, Sipe TW, Herrick J, Maherali H. 1998. Sapling biomass allocation and growth in the understory of a deciduous hardwood forest. *American Journal of Botany* **85**: 955–963.

Denslow JS, Schultze JC, Vitousek PM, Strain BR. 1990. Growth responses of tropical shrubs to tree-fall gap environments. *Ecology* **71**: 165–179.

Ducrey M. 1994. Influence of shade on photosynthetic gas exchange of 7 tropical rain-forest species from Guadeloupe (French West Indies). *Annales des Sciences Forestières* **51**: 77–94.

Edenius L, Danell K, Bergstrom R. 1993. Impact of herbivory and competition on compensatory growth in woody-plants following winter browsing by moose on scots pine. *Oikos* **66**: 286–292

Gill DS, Amthor JS, Bormann FH. 1998. Leaf phenology, photosynthesis, and the persistence of saplings and shrubs in a mature northern hardwood forest. *Tree Physiology* **18**: 281–289.

Givnish TJ. 1988. Adaptation to sun and shade: a whole plant perspective. *Australian Journal of Plant Physiology* **15**: 63–92.

Grime JP, Jeffery DW. 1965. Seedling establishment in vertical gradients of sunlight. *Journal of Ecology* **53**: 621–642.

Grubb PJ. 1998. A reassessment of the strategies of plants which cope with shortages of resources. *Perspectives in Plant Ecology, Evolution and Systematics* **1**: 3–31.

Hanson, PJ, McRoberts, RE, Isebrands JG, Dixon RK. 1987. An optimal sampling strategy for determining CO_2 exchange rate as a function of photosynthetic photon flux density. *Photosynthetica* **21**: 98–101.

Harrington RA, Brown BJ, Reich PB. 1989. Ecophysiology of exotic and native shrubs in southern Wisconsin. 1. Relationship of leaf characteristics, resource availability, and phenology to seasonal patterns of carbon gain. Oecologia **80**: 356–367

Hunt R, Cornelissen JHC. 1997. Components of relative growth rate and their interrelations in 59 temperate plant species. *New Phytologist* **135**: 395–417.

Kitajima K. 1994. Relative importance of photosynthetic traits and allocation patterns as correlates of seedling shade tolerance of 13 tropical trees. *Oecologia* **98**: 419–428.

Kobe RK. 1997. Carbohydrate allocation to storage as a basis of interspecific variation in sapling survivorship and growth. *Oikos* **80**: 226–233.

Kobe RK. 1999. Light gradient partitioning among tropical tree species through differential seedling mortality and growth. *Ecology* **80**: 187–201

Kobe RK, Pacala SW, Silander JA, Canham CD. 1995. Juvenile tree survivorship as a component of shade tolerance. *Ecological Applications* **3**: 517–532

Koike T. 1988. Leaf structure and photosynthetic performance as related to the forest succession of deciduous, broad-leaved trees. *Plant Species Biology* **3**: 77–87.

Küppers M. 1985. Carbon relations and competition between woody species in a central European hedgerow. *Oecologia* **66**: 343–352.

Lambers H, Poorter H. 1992. Inherent variation in growth rate between higher plants: a search for physiological causes and ecological consequences. *Advances in Ecological Research* **23**: 187–261.

Moad AS. 1992. *Dipterocarp juvenile growth and understory light availability in Malaysian tropical forest.* PhD thesis, Harvard University, MA, USA.

Niinemets Ü. 1998. Growth of young trees of *Acer platanoides* and *Quercus robur* along a gap-understory continuum: interrelationships between allometry, biomass partitioning, nitrogen and shade tolerance. *International Journal of the Plant Sciences* **159**: 318–330.

Osunkoya OO, Ash JE, Hopkins MS, Graham AW. 1994. Influence of seed size and seedling ecological attributes on shade tolerance of rain-forest tree species in northern Queensland. *Journal of Ecology* **82**: 149–163.

Pacala SW, Canham CD, Saponara J, Silander JA Jr., Kobe RK, Ribbens E. 1996. Forest models defined by field measurements. II. Estimation, error analysis and dynamics. *Ecological Monographs* **66**: 1–43.

Pompa J, Bongers F. 1988. The effect of canopy gaps on growth and morphology of seedlings of rainforest species. *Oecologia* **75**: 625–632.

Poorter H, Remkes C. 1990. Leaf-area ratio and net assimilation ratio of 24 wild species differing in relative growth rate. *Oecologia* **83**: 553–554.

Poorter H, Remkes C, Lambers H. 1990. Carbon and nitrogen economy of 24 wild species differing in relative growth rate. *Plant Physiology* **94**: 621–627.

Poorter H, Van der Werf A. 1998. Is inherent variation in RGR determined by LAR at low irradiance and by NAR at high irradiance? A review of herbaceous species. In: Lambers H, Poorter H, Van Vuuren MMI, eds. *Inherent variation in plant growth: physiological mechanisms and ecological consequences.* Leiden, The Netherlands: Backhuys Publishers.

Poorter L. 1998. *Seedling growth of Bolivian rain forest tree species in relation to light and water availability.* PhD thesis, Utrecht University, The Netherlands.

Raich JW. 1989. Seasonal and spatial variation in the light environment in a tropical dipterocarp forest and gaps. *Biotropica* **21**: 299–302.

Reich PB, Walters MB, Ellsworth DS. 1991. Leaf age and season influence the relationships between leaf nitrogen, leaf mass per area and photosynthesis in maple and oak trees. *Plant, Cell and Environment* **14**: 251–259.

Reich PB, Walters MB, Ellsworth DS. 1992. Leaf life-span in relation to leaf, plant, and stand characteristics among diverse ecosystems. *Ecological Monographs* **62**: 365–392.

Reich PB, Walters MB, Ellsworth DS. 1997. From tropics to tundra. Global convergence in plant functioning. *Proceedings of the National Academy of Sciences, USA* **94**: 13730–13734.

Reich PB, Walters MB, Tjoelker MG, Vanderklein DW, Buschena C. 1998a. Photosynthesis and respiration rates depend on leaf and root morphology and nitrogen concentration in nine boreal tree species differing in relative growth rate. *Functional Ecology* **12**: 327–338.

Reich PB, Tjoelker MG, Walters MB, Vanderklein D, Buschena C. 1998b. Close association of RGR, leaf and root morphology, seed mass and shade tolerance in seedlings of nine boreal tree species grown in high and low light. *Functional Ecology* **12**: 395–405.

Reich PB, Ellsworth DS, Walters MB, Vose JM, Gresham C, Volin JC, Bowman WD. 1999. Generality of photosynthetic and associated leaf trait relationships: a test across six biomes. *Ecology* (in press).

Rich PM, Clark DB, Clark DA, Oberbauer SF. 1993. Longterm study of solar radiation regimes in a tropical wet forest using quantum sensors and hemispherical photography. *Agricultural and Forest Meteorology* **65**: 107–127

Roderick ML, Berry SL, Noble, IR. 1999. The relationship between leaf composition and morphology at elevated CO_2 concentrations. *New Phytologist* **143**: 000–000.

Shirley HL. 1945. Reproduction of upland conifers in the lake states as affected by root competition and light. *The American Midland Naturalist* **33**: 537–611.

Tinoco-Ojanguren C, Pearcy RW. 1992. Dynamic stomatal behavior and its role in carbon gain during light flecks of a gap phase and an understory *Piper* species acclimated to high and low light. *Oecologia* **92**: 222–228.

Veenendaal EM, Swaine MD, Lecha RT, Walsh MF, Abebrse IK, Owusu-Afriyie K. 1996. Responses of West African forest tree seedlings to irradiance and soil fertility. *Functional Ecology* **10**: 501–511.

Veneklaas EJ, Poorter L. 1998. Growth and carbon partitioning of tropical tree seedlings in contrasting light environments. In: Lambers H, Poorter H, Van Vuuren MMI, eds. *Inherent variation in plant growth: physiological mechanisms and ecological consequences.* Leiden, The Netherlands: Backhuys Publishers: 337–362.

Walters MB, Kruger EL, Reich PB. 1993a. Relative growth rate in relation to physiological and morphological traits for northern hardwood seedlings: species, light environment and ontogenetic considerations. *Oecologia* **96**: 219–231.

Walters MB, Kruger EL, Reich PB. 1993b. Growth, biomass distribution and CO_2 exchange of northern hardwood seedlings in high and low light: relationships with successional status and shade tolerance. *Oecologia* **94**: 7–16.

Walters MB, Reich PB. 1996. Are shade tolerance, survival, and growth linked? Low light and nitrogen effects on hardwood seedlings. *Ecology* **77**: 841–853.

Walters MB, Reich PB. 1997. Growth of *Acer saccharum* seedlings in deeply shaded understories of northern Wisconsin: effects of nitrogen and water availability. *Canadian Journal of Forestry Research* **27**: 237–247.

Walters MB, Reich PB. 1999. Seed size, nitrogen supply and growth rate affect tree seedling survival in deep shade. *Ecology* (in press).

Williams K, Field CB, Mooney HA. 1989. Relationships among leaf construction cost, leaf longevity, and light environments in rainforest plants of the genus *Piper*. *American Naturalist* **133**: 198–211.

New Phytol. (1999), **143**, 155–162

Specific leaf area and leaf dry matter content as alternative predictors of plant strategies

PETER J. WILSON, KEN THOMPSON* AND JOHN G. HODGSON

Unit of Comparative Plant Ecology, Department of Animal and Plant Sciences, The University, Sheffield S10 2TN, UK

Received 5 October 1998; accepted 10 February 1999

SUMMARY

A key element of most recently proposed plant strategy schemes is an axis of resource capture, usage and availability. In the search for a simple, robust plant trait (or traits) that will allow plants to be located on this axis, specific leaf area is one of the leading contenders. Using a large new unpublished database, we examine the variability of specific leaf area and other leaf traits, the relationships between them, and their ability to predict position on the resource use axis. Specific leaf area is found to suffer from a number of drawbacks; it is both very variable between replicates and much influenced by leaf thickness. Leaf dry-matter content (sometimes referred to as tissue density) is much less variable, largely independent of leaf thickness and a better predictor of location on an axis of resource capture, usage and availability. However, it is not clear how useful dry matter content will be outside northwest Europe, and in particular in dry climates with many succulents.

Key words; leaf area, leaves, tissue density, leaf thickness, functional types, CSR, resource use, ecosystem function.

INTRODUCTION

One aim of plant ecological theory is to understand the constraints and opportunities that have shaped the evolution of plants. Another is to predict the behaviour of species, communities and, ultimately, whole ecosystems in response to competitors and changes in climate and land use (MacGillivray et al., 1995; Cornelissen & Thompson, 1997; Díaz et al., 1998; Fraser & Grime, 1998; Grime, 1998; Cornelissen et al., 1999). At the same time, plant ecologists have begun to realize that these objectives are unlikely to be achieved without some form of simplification. All these demands have driven a search for a limited number of key traits that will enable long lists of Latin names to be reduced to short lists of plant traits, while retaining the largest possible fraction of the information in the original list (Díaz & Cabido, 1997; Grime et al., 1997; Westoby, 1998; Hodgson et al., 1999; Walters & Reich, 1999). Another imperative is that candidate traits should be easy to measure; anything difficult or time-consuming will never be measured for enough

species. However, expensive equipment that yields quick results, especially in the field, might repay the investment (see, for example, Gamon & Surfus, 1999).

Opinion has begun to coalesce around three major axes of variation, although it must be admitted that not everyone agrees about the identity of the axes (Westoby, 1998), or even believes that this degree of reduction is possible (Grubb, 1998). One of these axes will concern the capacity for competitive dominance and will consist largely, if not entirely, of plant size (Gaudet & Keddy, 1988; Hodgson et al., 1999). Another axis might concern response to disturbance. Here opinion is more divided, although some combination of life history and number, size and dispersal capacity of seeds seems most likely to prevail (Greene & Johnson, 1993; Willson, 1993; van Dorp et al., 1996; Askew et al., 1997).

The third axis relates to the capacity to exploit resource-rich and resource-poor environments. This axis reflects the inevitable trade-off between those traits that are advantageous in resource-rich environments, including rapid rates of resource acquisition and loss and high rates of tissue turnover, and the more conservative strategy prevailing in chronically resource-poor environments. Very many traits are

*Author for correspondence (tel 0114 2224314; fax 0114 2220015; e-mail ken.thompson@sheffield.ac.uk).

entrained in this axis (Grime *et al.*, 1997), which is referred to here as a 'resource use' axis, although it must be stressed that it embraces far more than a simple axis of resource availability or concentration.

In the search for a single trait that captures the core of this axis, it is clear that specific leaf area (SLA, leaf area per unit dry mass) is the leading contender (Poorter & Remkes, 1990; Lambers & Poorter, 1992; Reich *et al.*, 1992; Cornelissen *et al.*, 1996; but see Van der Werf *et al.*, 1998). The case for SLA as the key trait has recently been made eloquently by Westoby (1998). The prevailing view is that SLA reflects the expected return on previously captured resources, and that high-SLA leaves are productive (Poorter & Van der Werf, 1988; Van der Werf *et al.*, 1988) but are necessarily also short-lived and vulnerable to herbivores (Coley *et al.*, 1985; Grime *et al.*, 1996). They therefore work best in resource-rich environments. In contrast, low-SLA leaves work better in resource-poor environments where retention of captured resources is a higher priority.

However, SLA is not without problems, of two kinds. A practical difficulty is how to deal with plants with vertical leaves or, worse still, no leaves. Sometimes these problems are combined, e.g. the erect, leafless stems of many *Juncus* species. Should one measure the area of one side or both sides of, for example, an *Iris* leaf? What is leaf area in *Ulex* or most cacti? Westoby (1998) acknowledges these problems and admits that there is no easy solution. The second problem arises from the fact that SLA can vary as a result of change in leaf thickness, composition, or both. Inevitably, therefore, there is a relatively close relationship between SLA and leaf thickness, but the latter can vary for reasons not directly connected with the axis that SLA is supposed to measure. An example in the British flora is *Oxalis acetosella*, which is very slow-growing (Grime & Hunt, 1975) and arguably the most shade-tolerant higher plant in the European flora (Ellenberg *et al.*, 1992). In common with many other species of deep shade, however, it also has thin leaves and therefore a surprisingly high SLA, indeed one of the highest values in the British flora (62 mm^2 mg^{-1}; unpublished data). Clearly in *Oxalis*, and presumably in shade plants in general, SLA is likely to be a poor indicator of position on the resource use axis.

So, while accepting that no single trait will do this job perfectly, we can suggest that an alternative to SLA would have two desirable properties; (1) it would not require any measurement of area, and (2) it would be relatively independent of leaf thickness. In the rest of this paper, and bearing in mind these two objectives, we shall examine the reliability and usefulness of a number of leaf characters as indicators of the resource use axis in a large cross section of the British flora. We focus particularly on the relative merits of SLA and leaf dry matter content. Although we cannot avoid joining the debate on the control of plant growth rate, the emphasis of the paper is on the detection of robust, simple predictors of plant strategy, *sensu* Grime (1979) and Westoby (1998).

METHODS

Leaf dry matter content, SLA, leaf thickness, maximum leaf area and leaf width were measured for 769 species, representing about half the native British flora plus a small number of naturalized aliens. Leaf area, width and thickness were not measured for some woody species. As far as possible, samples were collected from the Unit of Comparative Plant Ecology 'Central England' survey area, centred on Sheffield and comprising an area of 3000 km^2, covering lowlands in the east and uplands in the west, with three contrasted geological strata in each half of the area. Leaves were also collected from a minority of species not found in the Sheffield area. Such species, including coastal and montane plants, were collected from a variety of locations from Scotland to Devon.

All material was collected from robust, well-grown plants growing in unshaded habitats. Senescent or damaged leaves were avoided. Up to three collections of each species were made, where possible from sites geographically widely spread within the survey area, to include as much intra-specific variability due to geology and climate as possible. The locations of the collections were recorded. A minimum of eight fully expanded leaves were collected from several plants where possible, but if only one plant was available or the plant was very rare a smaller sample was collected. In some members of the Orchidaceae, only two or three leaves could be collected. Leaves were returned to the laboratory in sealed plastic bags. Nomenclature follows Stace (1991).

Leaf dry matter content

As soon as possible (usually on the day of collection), leaves (or, for smaller plants, branches and whole plants) were placed between damp papers, resealed in the plastic bags and maintained at 5°C in the dark overnight to produce a standard degree of turgor. Preliminary work had shown that turgor did not change further after one night. The sample of leaf lamina was blotted dry with tissue paper to remove any surface water, then immediately weighed to produce a value for saturated weight. The sample was dried in a paper envelope at 80°C for 3 d and reweighed to produce a value for dry weight. Values for dry matter content were calculated as dry weight, expressed as a percentage of saturated weight. Note that Ryser (1996) and Westoby *et al.* (1998) refer to dry matter content as tissue density. Tissue density is dry weight per unit volume, but owing to the very close correlation between saturated weight and volume, dry matter content and tissue density are closely correlated (Garnier & Laurent, 1994).

Measurements were based, where possible, on three leaves. Any petiole was removed and, for compound leaves, the individual leaflets were removed to include only laminar material. For leaves with massive midrib support structures such as *Petasites hybridus*, a sample of lamina was excised from a leaf.

Specific leaf area and maximum leaf area

Measurements were made on the same samples as were used for water content. After measurement of fresh weight, the area of the sample was measured with a Delta-T Devices (Cambridge, UK) area measurement system. The area of the largest leaf (specifically included in the sample) was also measured to produce a value for maximum leaf area. In particularly large-leaved species, samples of lamina were excised for measurement of SLA and the maximum leaf area was measured separately on an entire leaf or via extrapolation in species such as *Pteridium aquilinum*. The dry weight and measured leaf area of the samples were used to calculate specific leaf area as mm^2 mg^{-1} d. wt.

Leaf thickness

Leaf thickness was measured with a Verdict analogue thickness gauge mounted on a steel sheet and contacting leaves on a glass slide. One measurement was made on each of five leaves to an accuracy of 0.01 mm. Preliminary work had shown that replicate measurements on the same leaf varied very little (see also Shipley, 1995). Where adequate material was not available fewer leaves were measured. Thickness was measured on the leaf lamina between veins. The mean leaf thickness for each sample of five leaves was calculated.

Leaf width

Leaf width was measured with a ruler at a point representing the maximum width of leaf lamina supported by a single major vein, e.g. in compound leaves this was the width of a single leaflet; on palmate leaves this was the width of the broad base of one of the digits.

Analysis of data

For the data set containing all individual sample values, homoscedasticity was checked for each variable and where necessary the data were transformed to logarithms to base 10. For each variable, hierarchical analysis of variance was performed with the NESTED command in MINITAB (version 11 for Windows). The taxonomic hierarchy of Stace (1991) was used, in which conifers, ferns and angiosperms are regarded as classes and monocots and dicots as subclasses. Results are expressed as percentages of the total variance at each level of the hierarchy, with the final level being replicates nested within each species. Subsets of the data (seed plants, angiosperms or dicots only; see Table 1) were also analysed. When analysing the full data set, major differences between higher taxonomic ranks often obscured differences at lower levels.

To examine further the high between-replicate variance in SLA, the hierarchical analysis of variance was repeated for SLA, dry matter content and leaf thickness × dry matter content (Table 2). Because $1/SLA$ = leaf thickness × tissue density, and tissue density $\approx$ dry matter content, leaf thickness × dry matter content is algebraically equivalent to $1/SLA$, although unlike SLA it requires no measurement of leaf area. This analysis, unlike that in Table 1, was confined to those angiosperms for which all three variables were known. The effect of this restriction was to remove from the analysis a small number of species, chiefly trees and shrubs, for which leaf thickness was unknown.

The distributions of mean species data were checked for normality and transformed where necessary. Dry matter content and SLA were square-root transformed. Other variables were transformed to logarithms to base 10. We then examined relationships between individual variables by calculating Pearson product moment correlation coefficients between all the leaf traits. For three variables (SLA, dry matter content and leaf thickness) we also conducted separate phylogenetically independent analyses, by using a modified version of the CAIC (Comparative Analysis by Independent Contrasts (Purvis & Rambaut, 1995)) package. CAIC calculates the difference (or contrast) in the traits of interest between extant pairs of species; this contrast represents the amount of evolutionary divergence since they speciated from their common ancestor. In addition, CAIC calculates contrasts at internal nodes of the phylogeny. Because we do not know what the ancestral species at these nodes were like, values at nodes are averages of the species (or nodes) that evolved from them. In principle it is possible to weight these averages by the branch lengths, but in practice we assumed that all branch lengths were equal. A dichotomous phylogeny with n species yields $(n-1)$ contrasts, but in practice phylogenies contain polytomies and therefore yield fewer than $(n-1)$ contrasts. The important point is that each contrast is independent of all the others. We used a molecular phylogeny, which is therefore quite independent of the morphological traits under investigation. Further details of the program and of the phylogeny employed can be found in Hodkinson *et al.* (1998). We then analysed the standardized linear contrasts by linear regression forced through the origin (Purvis & Rambaut, 1995).

Finally, we attempted to answer the question; how well do SLA and dry matter content predict plant strategies defined by other criteria? Here we were fortunate in being able to draw on the results of the UCPE Integrated Screening Programme (ISP). Standardized procedures were used to measure 67 traits in 43 common species of the British flora; the most consistent patterns in the resulting data matrix were uncovered by means of multivariate analyses (Grime *et al.*, 1997). Principal components analyses (PCA) revealed a strong 'axis 1', closely related to a number of traits, including leaf palatability, relative yield under low nutrient regimes, leaf tensile strength, concentrations of major nutrients in leaves, decomposition rates, leaf longevity and specific leaf area. Location of any of the 43 species could thus be defined by its score on this leading PCA axis.

The further analysis of these data, including protocols for locating species on all three axes of the CSR triangle, is described elsewhere (Hodgson *et al.*, 1999), and only the briefest summary is presented here. Seven easily determined 'soft' traits were selected as potential predictors of location on the competition (C), resource use (S) and disturbance (R) axes of the strategic triangle of Grime (1979), although here we are concerned only with the S axis. These traits were canopy height, dry matter content, flowering period, flowering start, lateral spread, leaf dry weight and specific leaf area. Some of these traits have been described previously; others are self-explanatory. Flowering period is the normal duration of flowering period in months, whereas canopy height, flowering start and lateral spread are categorical. The categories were defined after extensive trial-and-error tests and are described in Hodgson *et al.* (1999). These seven traits were selected after a preliminary examination of a much longer list of potential candidates. Leaf thickness was among these candidate traits but was discarded because it was found, either alone or in combination with SLA or dry matter content, to make no significant contribution to any regression (for any axis).

Clearly, not all soft traits were relevant to the S axis, but nevertheless all were offered to a multiple regression with objectively defined position (of the 43 ISP species) on the S axis as the dependent variable. Predictor variables were transformed where necessary and all were also offered in the form of squares, but no interaction terms were constructed. Graminoids (grasses, sedges and rushes) on the one hand, and the remaining (principally dicotyledonous) species on the other, were analysed separately.

RESULTS

Variability of leaf traits

Table 1 contains values for the percentage of the total variance within each morphological trait that was accounted for at each level of the taxonomic hierarchy and by replicates within each species. There were often large amounts of variance at the level of class, highlighting obvious differences between conifers, ferns and angiosperms. When the angiosperms are analysed separately, SLA and dry matter content reveal rather different patterns. Much of the variation in dry matter content occurs between families, whereas the largest source of variation in SLA is between replicates. Indeed, SLA shows much higher levels of variation between replicates than the other four leaf traits.

Table 2 shows the results of the hierarchical analysis of variance of SLA, dry matter content and the product of dry matter content and leaf thickness (equivalent to 1/SLA). Because many woody species (which tend to be in distinct families and for which leaf thickness was not measured) are not included in this analysis, between-replicate variance of all traits is higher than in Table 1. Between-replicate variance in dry matter content × leaf thickness is much lower than in SLA (variance components of 1/SLA and SLA are of course identical).

Correlation of leaf traits

Table 3 shows Pearson product moment correlation coefficients between all the leaf traits. The strongest correlation, not surprisingly, is between maximum leaf area and leaf width. Leaf dry matter content and SLA are both correlated with leaf thickness, but the strongest correlation is with SLA. Note, however, that both dry matter content and SLA are negatively correlated with leaf thickness. Species with thin leaves apparently also tend to have high-density leaf tissues, although the relationship is weak.

The analysis of phylogenetically independent contrasts shows that, within clades, increasing SLA is associated with both decreased dry matter content and leaf thickness, although the relationship with the former is stronger (Fig. 1). Dry matter content and leaf thickness are not related. The apparent cross-species relationship between these two traits must therefore be due to one or a few differences between higher taxa. However, an examination of the contrasts revealed that the negative cross-species correlation between dry matter content and leaf thickness could not be attributed to a single phylogenetic node.

Usefulness of leaf traits in defining strategies

The best predictor regression equations of position on the S axis were, for graminoids, $54.6 - 1.666H_c^2 + 1.069C_{DM} - 2.732SLA + 1.722S_l^2$, where H_c is canopy height, C_{DM} is dry matter content and S_l is lateral spread, and for dicots $-39.52 - 7.581H_c + 2.633C_{DM} - 0.3510W_L^2$, where W_L is leaf dry weight.

Note that these regressions place a plant on the first PCA axis identified by Grime *et al.* (1997).

Table 1. *Results of the hierarchical ANOVA for all the measured leaf traits*

| Character... | Dry matter content | | SLA | | Leaf thickness | | Maximum leaf area | | Leaf width | |
Hierarchical level	All plants	Angiosperms	All plants	Angiosperms	All plants	Seed plants	All plants	Angiosperms	All plants	Dico-tyledons
Division	0		0		18		0		0	
Class	56		74		0	0	58		12	
Sub-class	0	0	0	0	0	0	0	0	16	
Order	8	18	0	0	0	0	7	20	8	4
Family	19	44	6	25	25	31	5	7	4	8
Genus	6	13	5	19	23	28	14	34	28	46
Species	5	11	6	25	21	26	13	31	25	33
Replicate	6	14	8	32	13	15	3	8	6	10

Figures are the percentage of the total variance in that variable accounted for at each level of the taxonomic hierarchy. Headings define the part of the total data set used for the analysis. SLA, specific leaf area.

Table 2. *Results of the hierarchical ANOVA for specific leaf area (SLA), dry matter content and leaf thickness × dry matter content*

Character... Hierarchical level	SLA	Dry matter content	Leaf thickness × dry matter content
Sub-class	6	18	12
Order	0	1	0
Family	10	34	21
Genus	20	15	17
Species	26	15	27
Replicate	38	18	24

Figures are the percentage of the total variance in that variable accounted for by each level of the taxonomic hierarchy. The analysis is confined to those angiosperms for which all three variables are known. $n = 1198$.

Table 3. *Values of Pearson's correlation coefficient (r) between the measured leaf traits*

Variable	Dry matter content	SLA	Leaf thickness	Leaf width
SLA				
r	−0.549			
n	767			
P	***			
Leaf thickness				
r	−0.289	−0.478		
n	731	730		
P	***	***		
Leaf width				
r	−0.230	0.099	0.021	
n	697	696	696	
P	***	**	ns	
Max. leaf area				
r	−0.028	−0.011	0.067	0.713
n	761	760	726	693
P	ns	ns	ns	***

n is the number of species used in the correlation; P is the significance of correlation (ns, $P > 0.05$; **, $P < 0.01$; ***, $P < 0.001$). SLA, specific leaf area.

Further manipulation of these raw scores is necessary to produce a standardized S-axis score (Hodgson *et al.*, 1999). Other than size, the significant predictor variables are dry matter content alone for dicots and dry matter content combined with SLA for graminoids.

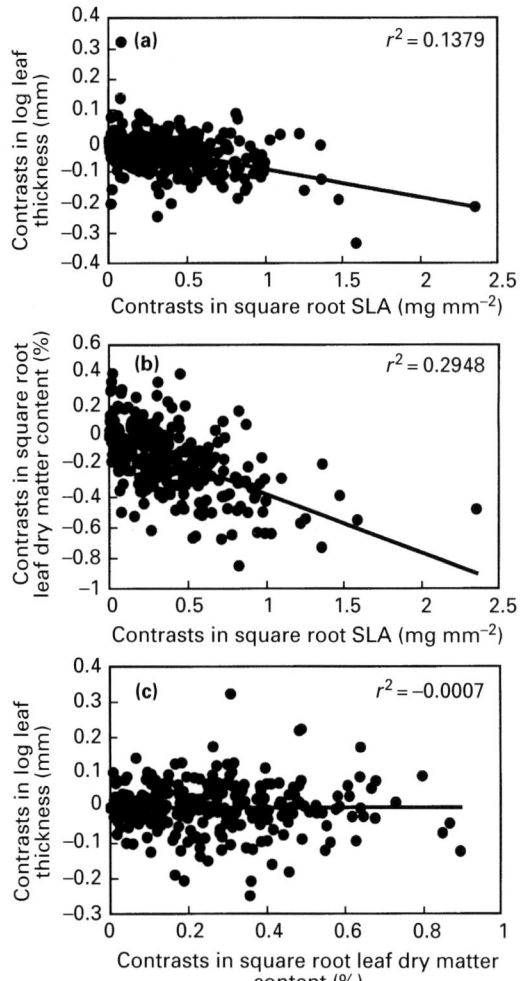

Fig. 1. Phylogenetically independent contrasts in (a) square root of specific leaf area (SLA) and log of leaf thickness, (b) square root of SLA and square root of leaf dry matter content, (c) square root of leaf dry matter content and log of leaf thickness. $n = 264$ contrasts. Note that regressions are forced through the origin and therefore r^2 values must be interpreted with caution.

DISCUSSION

Two striking findings, of direct relevance to the objective of discovering robust predictors of position on a resource use axis, emerge from this analysis. The first is that despite efforts to avoid the effects of sub-optimal habitats in general and shade in particular, it seems to be much harder to make a reliable measurement of SLA than of other leaf traits; the second is that dry matter content is a better predictor than SLA of objectively defined position on a resource use axis.

However, we consider first a different question; what is the relationship between SLA and its two components, leaf thickness and dry matter content? In our flora, it seems that SLA is related either to leaf thickness or to dry matter content, but not normally to both (Fig. 1); that is, there is no tendency for evolutionary changes in leaf thickness to be associated, one way or the other, with changes in dry

matter content (Fig. 1c). Westoby *et al.* (1998), in addressing this question in the Australian flora, came to a similar conclusion. In their study, decreases in SLA were due mainly to changes in leaf thickness or dry matter content in about half the contrasts examined, and due to changes in both traits in the other half. Westoby *et al.* (1998) looked at pairs of species deliberately chosen to represent contrasts along either rainfall or soil-fertility gradients. Our contrasts were not chosen to represent any particular gradient, but it is fair to assume that contrasts in SLA in the British flora owe most to soil nutrient gradients (Grime *et al.*, 1997; Thompson *et al.*, 1997), or at least that, in the uniformly damp climate of central England, they do not represent a rainfall gradient. The question of how plant traits vary along gradients of different forms of 'stress' (*sensu* Grime, 1979) is a relatively new and promising area of enquiry. We suspect that variation in leaf traits along gradients of rainfall might be quite different in floras (unlike Australia's) with a substantial proportion of succulents.

Returning to the primary objective of this paper, the very large between-replicate variation in SLA could be a 'real' effect, or a consequence of experiment error, or both. A real effect could also operate on a variety of levels. There might simply be a lot of genetic variability in SLA, i.e. plants from different populations of the same species, even when grown under identical conditions, might exhibit large differences in SLA. Variation in SLA might also be caused by environmental variability, both temporal and spatial. Even if plants are always sampled from unshaded habitats, it is hard to control for the effects of self-shading. This could potentially have very large effects, because sun and shade leaves on the same plant might differ in SLA by a factor of two (Popma & Bongers, 1988; Cornelissen, 1992; Dong, 1993). Shipley & Meziane (1998) have shown that both mineral nutrition and irradiance can have complex and interactive effects on SLA. However, shade cannot be the whole story, because shade also has large effects on leaf thickness, but leaf thickness is much less variable than SLA (Table 1). However, small differences in leaf thickness might have large consequences for SLA. SLA is more strongly dependent than dry matter content on leaf thickness (Table 3, Fig. 1). Sampling plants from dry and moist habitats might also affect SLA, although this would also be expected to have a large effect on dry matter content. Leaf age might have large effects, and old leaves might have lower or higher SLA than young leaves (Garnier & Freijsen, 1994 and references therein).

Errors seem to be most likely to arise from measurement of leaf area. On every occasion that leaf areas were measured, the machine was calibrated with known areas of appropriate size, but no effort was made to validate measured areas by independent means. Many leaves are not truly flat, and it is easy to see how different amounts of pressure applied to crinkly or wavy leaves might influence measured leaf area. Small leaves in particular might also suffer from 'edge effects'. In our study all measurements were made by a very small number of observers and used a uniform protocol, but errors could also arise through differences of interpretation between different observers. For example, if leaf surfaces are curved, as they often are in small or thick leaves, much depends on whether the projected area is measured or whether some allowance is made for curvature. Between-replicate variance in leaf thickness × dry matter content, which is algebraically equivalent to 1/SLA, is much less than in SLA itself (Table 2). This result clearly implicates errors in the measurement of leaf area as a major cause of high between-replicate variance in SLA.

Westoby (1998) was clearly aware of these problems and advised that the design of the SLA protocol needed to take explicit account of possible sources of variation. Unfortunately we currently have no idea of how important the different sources of variation are, nor how this varies between floras. For example, important sources of error in floras with many evergreens and succulents might be much less important in mesic, deciduous floras.

Regardless of difficulties of measurement, dry matter content is a better predictor of the resource use axis. In part this might be due to the 'Oxalis effect' referred to earlier. Not surprisingly, SLA is more influenced than dry matter content by leaf thickness. As pointed out by several authors (Witkowski & Lamont, 1991; Garnier *et al.*, 1997; Westoby, 1998; Westoby *et al.*, 1998), SLA can vary through differences in thickness or in tissue composition. Dry matter content focuses explicitly on one of these sources of variation, whereas SLA attempts to integrate their effects. The superiority of dry matter content suggests that variation in leaf composition is a more important correlate of resource capture and usage than is leaf thickness. Other work supports this view. Ryser (1996) showed that leaf life span is closely related to dry matter content, and moreover that interspecific rankings of dry matter content seem to be unaffected by nutrient supply or competition (Ryser & Notz, 1996). Garnier & Laurent (1994) found that leaf water content was tightly linked to the volume of protoplasm in the leaf, whereas Niemann *et al.* (1992) found that growth rate was positively correlated with investment in cytoplasmic compounds (especially protein) and negatively correlated with cell wall compounds. However, we must not forget that we are here concerned with predicting not growth rate but location on 'axis 1' of Grime *et al.* (1997), which is itself not closely correlated with growth rate. It is not surprising that some of the traits with the highest loading on axis 1 (leaf toughness, longevity, de-

composition, palatability and nutrient content) are strongly dependent on leaf composition, and particularly on the balance between structure and cytoplasm. Many of these important axis 1 traits are also of the foremost importance to ecosystem function, whereas relative growth rate is perhaps best regarded as a 'secondary' trait that varies largely as an indirect consequence of these more basic traits.

In conclusion, dry matter content seems to be the best single variable for locating plant species on a resource use axis; furthermore, it is relatively easy to measure. However, this conclusion must be qualified by two important caveats. First, it has proved difficult in practice to apply the protocol described here to plants from arid climates. Specifically, we have yet to devise an adequate method for ensuring full hydration of leaves from such plants (S. Díaz, pers. comm.). Second, the rationale underlying the use of dry matter content is that leaf water content is linked to leaf protein content. This rationale might not apply to succulents. Our limited experience with succulents in the British flora (*Sedum* spp., *Umbilicus rupestris*) suggests that they have low to moderate SLA but (not surprisingly) very low dry matter content. If succulents have anatomically distinct water-storage tissues, it might prove possible to make some simple allowance for this. Alternatively, as with British graminoids, it might be that the strategy is best predicted by some combination of dry matter content and SLA.

ACKNOWLEDGEMENTS

P. J. W. was funded by a NERC CASE studentship. We also acknowledge the assistance of Stuart Band and the long-term support of NERC. This paper was much improved by the comments of Hans Cornelissen, Sandra Díaz and two anonymous referees.

REFERENCES

Askew AP, Corker D, Hodkinson DJ, Thompson K. 1997. A new apparatus to measure the rate of fall of seeds. *Functional Ecology* **11**: 121–125.

Coley PD, Bryant JP, Chapin III FS. 1985. Resource availability and plant anti-herbivore defense. *Science* **230**: 895–899.

Cornelissen JHC. 1992. Seasonal and year to year variation in performance of *Gordonia acuminata* seedlings in different light environments. *Canadian Journal of Botany* **70**: 2405–2414.

Cornelissen JHC, Castro-Diez P, Hunt R. 1996. Seedling growth, allocation and leaf attributes in a wide range of woody plant species and types. *Journal of Ecology* **84**: 755–765.

Cornelissen JHC, Perez-Harguindeguy N, Diaz S, Grime JP, Marzano B, Cabido M, Vendramini F, Cerabolini B. 1999. Leaf structure and defence control litter decomposition rate across species and life forms in regional floras on two continents. *New Phytologist* **143**: 191–200.

Cornelissen JHC, Thompson K. 1997. Functional leaf attributes predict litter decomposition rate in herbaceous plants. *New Phytologist* **135**: 109–114.

Díaz S, Cabido M. 1997. Plant functional types and ecosystem function in relation to global change. *Journal of Vegetation Science* **8**: 463–474.

Díaz S, Cabido M, Casanoves F. 1998. Plant functional traits and environmental filters at the regional scale. *Journal of Vegetation Science* **9**: 113–122.

Dong M. 1993. Morphological plasticity of the clonal herb *Lamiastrum galeobdolon* (L.) Ehrend. & Polatschek in response to partial shading. *New Phytologist* **124**: 291–300.

Ellenberg H, Weber HE, Düll R, Wirth V, Werner W, Paulißen D. 1992. Zeigwerte von Pflanzen in Mitteleuropa. *Scripta Geobotanica* **18**. Göttingen: Erich Goltze.

Fraser LH, Grime JP. 1998. Primary productivity and trophic dynamics investigated in a North Derbyshire dale. *Oikos* **80**: 499–508.

Gamon J, Surfus JS. 1999. Assessing leaf pigment content and activity with a reflectometer. *New Phytologist* **143**: 105–117.

Garnier E, Cordonnier P, Guillerm J-L, Sonié L. 1997. Specific leaf area and leaf nitrogen concentration in annual and perennial grass species growing in Mediterranean old-fields. *Oecologia* **111**: 490–498.

Garnier E, Freijsen AHJ. 1994. On ecological inference from laboratory experiments conducted under optimum conditions. In: Roy J, Garnier E, eds. *A whole plant perspective on carbon–nitrogen interactions.* The Hague, The Netherlands: SPB Academic Publishing, 267–292.

Garnier E, Laurent G. 1994. Leaf anatomy, specific mass and water content in congeneric annual and perennial grass species. *New Phytologist* **128**: 725–736.

Gaudet CL, Keddy PA. 1988. A comparative approach to predicting competitive ability from plant traits. *Nature* **334**: 242–243.

Greene DF, Johnson EA. 1993. Seed mass and dispersal capacity in wind-dispersed diaspores. *Oikos* **67**: 69–74.

Grime JP. 1979. *Plant strategies and vegetation processes.* Chichester, UK: John Wiley.

Grime JP. 1998. Plant functional types and ecosystem processes. *Mededelingen van de KNAW* 1998: 104–108.

Grime JP, Cornelissen JHC, Thompson K, Hodgson JG. 1996. Evidence of a causal connection between anti-herbivore defence and the decomposition rate of leaves. *Oikos* **77**: 489–494.

Grime JP, Hunt R. 1975. Relative growth rate: its range and adaptive significance in a local flora. *Journal of Ecology* **63**: 393–422.

Grime JP, Thompson K, Hunt R, Hodgson JG, Cornelissen JHC, Rorison IH, Hendry GAF, Ashenden TW, Askew AP, Band SR, Booth RE, Bossard CC, Campbell BD, Cooper JEL, Davison AW, Gupta PL, Hall W, Hand DW, Hannah MA, Hillier SH, Hodkinson DJ, Jalili A, Liu Z, Mackey JL, Matthews N, Mowforth MA, Neal AM, Reader RJ, Reiling K, Ross-Fraser AM, Spencer RE, Sutton F, Tasker DE, Thorpe PC, Whitehouse J. 1997. Integrated screening validates primary axes of specialisation in plants. *Oikos* **79**: 259–281.

Grubb PJ. 1998. A reassessment of the strategies of plants which cope with shortages of resources. *Perspectives in Plant Ecology, Evolution and Systematics* **1**: 3–31.

Hodgson JG, Wilson PJ, Hunt R, Grime JP, Thompson K. 1999. Allocating C-S-R plant functional types; a soft approach to a hard problem. *Oikos* **85**: 282–296.

Hodkinson DJ, Askew AP, Thompson K, Hodgson JG, Bakker JP, Bekker RM. 1998. Ecological correlates of seed size in the British flora. *Functional Ecology* **12**: 762–766.

Lambers H, Poorter H. 1992. Inherent variation in growth rate between higher plants; a search for physiological causes and ecological consequences. *Advances in Ecological Research* **23**: 188–242.

MacGillivray CW, Grime JP and the ISP team. 1995. Testing predictions of resistance and resilience of vegetation subjected to extreme events. *Functional Ecology* **9**: 640–649.

Niemann GJ, Pureveen JBM, Eijkel GB, Poorter H, Boon JJ. 1992. Differences in relative growth rate in 11 grasses correlate with differences in chemical composition as determined by pyrolysis mass spectrometry. *Oecologia* **89**: 567–573.

Poorter H, Remkes C. 1990. Leaf area ratio and net assimilation rate of 24 wild species differing in relative growth rate. *Oecologia* **83**: 553–559.

Poorter H, Van der Werf A. 1998. Is inherent variation in RGR determined by LAR at low irradiance and by NAR at high irradiance? A review of herbaceous species. In: Lambers H, Poorter H, Van Vuuren MMI, eds. *Inherent variation in plant growth, physiological mechanisms and ecological consequences.* Leiden, The Netherlands: Backhuys, 309–336.

Popma J, Bongers F. 1988. The effect of canopy gaps on growth and morphology of seedlings of rain forest species. *Oecologia* **75**: 625–632.

Purvis A, Rambaut A. 1995. Comparative analysis by independent contrasts (CAIC): an Apple Macintosh application for analysing comparative data. *Computer Applications in Biosciences* **11**: 247–251.

Reich PB, Walters MB, Ellsworth DS. 1992. Leaf life-span in relation to leaf, plant, and stand characteristics among diverse ecosystems. *Ecological Monographs* **62**: 365–392.

Ryser P. 1996. The importance of tissue density for growth and life span of leaves and roots: a comparison of five ecologically contrasted grasses. *Functional Ecology* **10**: 717–723.

Ryser P, Notz R. 1996. Competitive ability of three ecologically contrasting grass species at low nutrient supply in relation to their maximal relative growth rate and tissue density. *Bulletin of the Geobotanical Institute ETH Zurich* **62**: 3–12.

Shipley B. 1995. Structured interspecific determinants of specific leaf area in 34 species of herbaceous angiosperms. *Functional Ecology* **9**: 312–319.

Shipley B, Meziane D. 1998. The statistical modelling of plant growth and its components using structural equations. In: Lambers H, Poorter H, Van Vuuren MMI, eds. *Inherent variation in plant growth, physiological mechanisms and ecological consequences.* Leiden, The Netherlands: Backhuys, 393–408.

Stace C. 1991. *New flora of the British Isles.* Cambridge, UK: Cambridge University Press.

Thompson K, Parkinson JA, Band SR, Spencer RE. 1997. A comparative study of leaf nutrient concentrations in a regional herbaceous flora. *New Phytologist* **136**: 679–689.

Van der Werf A, Geerts RHEM, Jacobs FHH, Korevaar H, Oomes MJM, De Visser W. 1998. The importance of relative growth rate and associated traits for competition between species during vegetation succession. In: Lambers H, Poorter H, Van Vuuren MMI, eds. *Inherent variation in plant growth, physiological mechanisms and ecological consequences.* Leiden, The Netherlands: Backhuys, 489–502.

van Dorp D, van den Hoek WPM, Daleboudt C. 1996. Seed dispersal of six perennial grassland species measured in a wind tunnel at varying wind speed and height. *Canadian Journal of Botany* **74**: 1956–1963.

Walters M, Reich P. 1999. Low-light carbon balance and shade tolerance in the seedlings of woody plants: do winter deciduous and broad-leaved evergreen species differ? *New Phytologist* **143**: 143–154.

Westoby M. 1998. A leaf–height–seed (LHS) plant ecology strategy scheme. *Plant and Soil* **199**: 213–227.

Westoby M, Cunningham S, Fonesca CR, Overton JM, Wright IJ. 1998. Phylogeny and variation in light capture area deployed per unit investment in leaves: designs for selecting study species with a view to generalising. In: Lambers H, Poorter H, Van Vuuren MMI, eds. *Inherent variation in plant growth, physiological mechanisms and ecological consequences.* Leiden, The Netherlands: Backhuys, 539–566.

Willson MF. 1993. Dispersal mode, seed shadows, and colonization patterns. *Vegetatio* **107/108**: 261–280.

Witkowski ETF, Lamont BB. 1991. Leaf specific mass confounds leaf density and thickness. *Oecologia* **88**: 486–493.

New Phytol. (1999), 143, 163–176

A comparison of specific leaf area, chemical composition and leaf construction costs of field plants from 15 habitats differing in productivity

HENDRIK POORTER* AND ROB DE JONG

Plant Ecophysiology, Utrecht University, PO Box 800.84, 3508 TB Utrecht, The Netherlands

Received 25 October 1998; accepted 12 April 1999

SUMMARY

Laboratory experiments have shown a large difference in specific leaf area (SLA, leaf area:leaf mass) between species from nutrient-poor and nutrient-rich habitats, but no systematic difference in the construction costs (the amount of glucose required to construct 1 g biomass). We examined how far these patterns are congruent with those from field-grown plants. An analysis was made of the vegetation in a range of grasslands and heathlands differing in productivity. The SLA of the dominant species in 15 different habitats was determined, as well as chemical composition and construction costs of bulk samples of leaves. SLA in the field was generally lower than in the laboratory, but showed consistency in that the ranking across species remained the same. Species from highly productive habitats had higher SLA than those from sites of low productivity, although individual species sometimes deviated substantially from the general trend. Construction costs were similar for plants from different habitats. This was mainly due to the positive correlation between an expensive class of compounds (proteins) and a cheap one (minerals).

Key words: chemical composition, construction costs, productivity, specific leaf area, comparative approach, vegetation, field-grown plants, laboratory-grown plants.

INTRODUCTION

Plant species show wide variation in their potential relative growth rate (RGR) when grown under near-optimal conditions (Grime & Hunt, 1975; Chapin, 1980; Poorter, 1989). In particular, plant species normally found in nutrient-poor, unproductive habitats have low potential RGR, whereas species generally occurring in nutrient-rich, productive habitats tend to have inherently high RGR. Given the strong correlation between a species' potential RGR and its occurrence in specific habitats, it has been suggested that potential RGR has been the target of selection in these cases (Grime & Hunt, 1975; Chapin, 1980). Alternatively, it may be not RGR so much as one of the components underlying RGR that has been the target of selection (Lambers & Dijkstra, 1987; see also Grime, 1979; Coley *et al.*,

1985). Tilman (1988) suggested that the allocation of biomass is different between species from different habitats, with species in productive environments (where relatively strong competition for light occurs) having a high allocation to leaves and stems, and species from unproductive environments (where there is relatively stronger competition for nutrients) allocating inherently more biomass to roots. In an analysis of the inherent variation in growth characteristics of 24 herbaceous species grown at a non-limiting supply of nutrients, there was indeed some correlation between the potential RGR of a species and the leaf mass fraction (the fraction of biomass allocated to leaves; Poorter & Remkes, 1990). However, this correlation was much weaker than that between the potential RGR and the specific leaf area (SLA, leaf area:leaf mass) of the species. This conclusion was corroborated in a survey of 111 published comparative growth experiments on herbaceous species, where again SLA was the predominant factor explaining variation in RGR

*Author for correspondence (fax +31 30 251 8366; e-mail h.poorter@bio.uu.nl).

(Poorter & Van der Werf, 1998). Based partly on these observations, it has been suggested that SLA, or more precisely factors pertaining to a complex of traits related to SLA, could have been the target of selection (Poorter, 1989; Van der Werf *et al.*, 1993; Poorter & Garnier, 1999). Generally, high-SLA species are characterized by high concentrations of nitrogen; high rates of CO_2 and N uptake per unit leaf and root mass, respectively; and a high rate of photosynthesis per unit leaf N (Lambers & Poorter, 1992). These species are geared for a high rate of resource acquisition. Low-SLA species, on the other hand, generally have high values for dry matter content (dry mass:fresh mass); high concentrations of cell walls and secondary compounds; and greater leaf and root longevity (Coley *et al.*, 1985; Choong *et al.*, 1992; Ryser, 1996; Reich, 1998). These species seem to be geared for the conservation of acquired resources (for reviews see Aerts & Chapin, 1999; Poorter & Garnier, 1999).

Most of the above-mentioned analyses on the causes of interspecific variation in growth rate have been carried out in growth rooms or glasshouses, with a relatively high supply of nutrients and water and a relatively low irradiance (Garnier & Freijsen, 1994). Inferences on a possible role of SLA and the suite of related traits as important determinants of plant functioning remain rather speculative if they are based on observations of plants grown under controlled conditions. Therefore, the first question to analyse is how far SLA data from laboratory-grown plants are indicative of SLA values of plants growing in the field. Are the values similar, and does the ranking across species remain the same? The second question relates to the presumed correlation between SLA and habitat productivity. Most of the suppositions regarding a positive correlation between SLA (or potential RGR) and habitat productivity have been based on qualitative impressions (e.g. Grime & Hunt, 1975) or were inferred from semi-quantitative data such as the nitrogen numbers of Ellenberg (Poorter & Remkes, 1990; Fichtner & Schulze, 1992; Van der Werf *et al.*, 1998). It remains to be seen whether there is indeed a positive correlation between SLA and a more quantitative estimate of habitat productivity. Moreover, if any relationship does exist, it is of interest both to assess the form of the relationship (linear, saturating), and to investigate how far individual species may deviate from the general trend. To this end, we determined the SLA of around 70 species from a range of Western European grasslands and from some ericaceous vegetation types. These habitats differ in productivity, due mainly to variation in nutrient availability.

A third concern addressed in this paper regards the construction costs of leaves (the amount of glucose required to construct all of the chemical constituents of 1 g of leaf). Plant species from unproductive habitats tend to accumulate many carbon-based secondary compounds, including lignin and tannins (Coley *et al.*, 1985; Lambers & Poorter, 1992). These compounds have high specific costs of construction, and it has been suggested that leaves of species accumulating these compounds therefore have high construction costs (Miller & Stoner, 1979). However, leaves of these species generally also have low concentrations of proteins, another group of compounds that are costly to produce (Penning de Vries *et al.*, 1974). Thus it may well be that variation in construction costs between species is only small (Chapin, 1989). For species grown with a non-limiting nutrient supply under controlled conditions, no relationship was observed between the potential RGR of a species and the construction costs of the leaves (Poorter & Bergkotte, 1992). Few data are available for leaf construction costs of plant species growing *in situ* in habitats differing in productivity. Therefore the third aim of this paper is to quantify the construction costs of leaves in productive and unproductive habitats. To determine the underlying reasons for possible variation in construction costs, a proximate analysis was carried out of the quantitatively important compounds in the leaf biomass of these vegetation types. For these analyses, all leaves were bulked from within a stand, precluding an assessment of variation between species within a vegetation type.

MATERIALS AND METHODS

Design of the study

Fifteen habitats expected to differ in productivity were selected, all in the central or southern part of the Netherlands. The vegetation in 11 of these habitats is grassland (generally mown once or twice a year). In addition, two ruderal vegetation types in highly disturbed areas and one dry and one wet ericaceous vegetation were chosen (Table 1). The first sampling period was relatively early in the growing season (late April, 1993) and included eight of the habitats. The second sampling period was when the vegetation approaches peak biomass (early July, 1993) and included all 15 habitats. At each time, species were selected that occurred relatively frequently in the habitat or had relatively high biomass. The number of species sampled per habitat varied between one and eight, with an average of five.

For each of the species selected, five individual shoots were randomly chosen, and the youngest fully expanded leaf as well as the oldest green and viable leaf were selected. Hereafter, these leaves will be referred to as 'young' and 'old', respectively. Two categories of leaves were taken to ensure that at least two phases in the life span of the leaves were represented in the analysis. In the case of species

Table 1. *Characterization of the 15 selected sites, topographical location and average estimated above-ground biomass produced per year (EABP ± SD, expressed as percentage of the mean; n = 2)*

Habitat or vegetation type	Location	Coordinates (°E, °N)	Also sampled in first period	EABP (g m^{-2} yr^{-1})
1. Drifting sand dune	Kootwijkerzand (Gld)	5° 47′, 52° 11′	*	90 ± 22%
2. Quaking fen	Westbroekse zodde (Utr)	5° 07′, 52° 10′	*	100 ± 3%
3. Grass heath	Loobosch (Gld)	5° 44′, 52° 10′		130 ± 4%
4. Wet heath	Luttenbergerven (Ovr)	6° 21′, 52° 24′	*	200 ± 29%
5. Dry heath	Wolfhezer heide (Gld)	5° 47′, 52° 00′	*	220 ± 9%
6. Ruderal (trampled)	Uithof (Utr)	5° 10′, 52° 05′	*	280 ± 11%
7. Dry open grassland (south-facing slope)	Berghofweide (Lim)	5° 53′, 50° 50′	*	300 ± 12%
8. Chalk grassland (north-facing slope)	Gerendal (Lim)	5° 52′, 50° 51′		330 ± 10%
9. Poor haymeadow	Luttenbergerven (Ovr)	6° 21′, 52° 24′	*	430 ± 2%
10. Roadside	Rhijnauwen (Utr)	5° 10′, 52° 05′	*	690 ± 9%
11. Ruderal (not trampled)	Westbroek (Utr)	5° 08′, 52° 09′		730 ± 14%
12. Along ditch (never mown)	Lunetten (Utr)	5° 09′, 52° 04′		990 ± 4%
13. Roadside	Amerongse Bovenpolder (Gld)	5° 27′, 52° 00′		1020 ± 22%
14. Fertilized meadow	Uithof (Utr)	5° 10′, 52° 05′		1080 ± n.d.
15. Reed marsh	Westbroekse zodde (Utr)	5° 07′, 52° 09′		1090 ± 33%

with very small leaves, a number of leaves of similar plants were combined. A random sample of five leaves was chosen for those species where old and young leaves could not be distinguished because plants had just developed their leaves.

Twelve of the 24 species grown by Poorter & Remkes (1990) under laboratory conditions (*Cynosurus cristatus, Galinsoga parviflora, Geum urbanum, Hypericum perforatum, Origanum vulgare, Poa annua, Rumex crispus, Scrophularia nodosa, Briza media, Lysimachia vulgaris, Phleum pratense* and *Pimpinella saxifraga*) were not among the species harvested in the 15 habitats. To enable a wider comparison between laboratory and field data, SLA values were also collected for the first eight of these species in other habitats, following the sampling scheme described above. Data were included from the literature on the SLA of *Carex flacca* and *Galium aparine*, obtained under exactly the same conditions in the same laboratory by Van der Werf *et al.* (1993) and Den Dubbelden & Verburg (1996), respectively. All SLA values obtained in the laboratory are averages over all viable leaves of the plants.

SLA and productivity

To obtain an impression of the light climate experienced by the leaves, irradiance (µmol quanta m^{-2} s^{-1} in the 400–700 nm range) was determined immediately above each leaf, as well as the irradiance above the vegetation. Measurements were carried out with a LiCor LI 185A (Lincoln, NE, USA). Light measurements were taken between 10:00 and 15:00 with the light sensor in a horizontal position, and no attempt was made to separate the diffuse from

the direct site factor. Following light measurement, leaves were collected, wetted and stored in a cool box. At the end of the day leaves were placed between wet tissues and stored overnight in a fridge, after which they were further processed. By storing them wet, leaves could reach equilibrium with free water, correcting for possible differences between habitats in short-term water availability (Eliáš, 1985). The next day petioles and leaf sheaths were separated from the leaf blades and discarded. In the case of very narrow or cylindrical leaves (*Corynephorus canescens, Deschampsia flexuosa, Festuca ovina, Juncus subnodulosus*), the leaf width in the middle part of the leaf was determined, as well as the leaf length. Leaf blade area was then calculated as half the total intercepting area, following Chen & Black (1992). For *Calluna vulgaris* and *Erica tetralix*, length and width of the minute leaves were measured under a microscope, and the area calculated by multiplying length × width with a correction factor that takes into account the form of the leaves. For all other species, leaf area was determined with a DIAS image analysis system (Delta-T, Cambridge, UK; small leaves) or a LiCor (larger leaves). After determination of the area, leaves were dried for at least 48 h at 70°C and dry mass was determined.

The above-ground standing crop was determined in duplicate in the second period by harvesting all aerial biomass (except litter and the moss layer) of a 0.5 × 0.5 m area. As most of these sites have negligible aerial living biomass during the winter season, and were not mown before our measurements, the near-maximal standing crop is a somewhat crude estimate of the above-ground productivity in these habitats. This is denoted as estimated above-ground biomass produced (EABP). Total net productivity will be higher, both because the root fraction was not considered and because leaf turnover during the growing season was not quantified. Nevertheless, the approach followed should give a fair estimate of the differences in productivity between sites (cf. Aerts & Berendse, 1989). There were, however, three clear exceptions. Firstly, in the case of the *Lolium perenne* meadow, the vegetation was frequently and heavily grazed. Fliervoet (1987) and Meijer (1984) report productivity values of 700 and 1450 g m^{-2} yr^{-1} for non-grazed grasslands of this type. In this study it was assumed that EABP values on the site would have been the average of these two reported values. Any other number over 700 g m^{-2} yr^{-1} does not significantly affect the conclusions drawn in this paper. Secondly, standing crop in the dry and wet heathland systems is approximately constant throughout the year (Aerts & Berendse, 1989). In these cases, the older fraction of the stems and leaves of *Erica tetralix* and *Caluna vulgaris*, produced, it is most likely, in the years before, were separated from the fraction produced in the current year. To arrive at an EABP value, the ericaceous biomass produced in the current year was added to the total biomass of the other plant species harvested in the clipped area.

From each of these bulk samples of vegetation, two independent samples of leaves were randomly selected to obtain sufficient leaf material for the chemical analyses. Thus, whereas the SLA data were collected for each of the species investigated, the chemical analyses are representative for the leaf fraction of the entire vegetation. Dry mass was determined after drying for at least 48 h at 70°C.

Chemical analyses

Each leaf sample of the collected bulk vegetation was ground to pass through a 0.08 mm sieve, and redried. An exact description of the procedures followed for the chemical analyses and the subsequent calculations is given in Poorter & Villar (1997). Carbon and total nitrogen content were measured with an elemental analyser, and ash content was determined by combustion of plant material in a muffle furnace. Ash contains not only minerals, but also oxides of organic acid and nitrate formed during combustion. Therefore we determined ash alkalinity by titration and, in a separate sample, the nitrate content by a colorimetric assay with salicilic acid. Lipids were determined gravimetrically in the chloroform fraction of a chloroform–methanol–water extract. Soluble phenolics were determined in the methanol–water phase with the Folin–Ciocalteu reagent. Soluble sugars were determined in the same extract; insoluble sugars after boiling with 3% HCl. Sugars were determined with anthrone. The residue left over after the chloroform–methanol extraction and the 3% HCl treatment was considered to consist of crude cell walls. From the C and N content of this residue the concentrations of total structural carbohydrates (TSC) and lignin were derived. The only adjustment made to the scheme given by Poorter & Villar (1997) was for minerals, possibly silica, which appeared to be present in the crude cell wall residue. To correct for this, the ash content in the crude cell wall fraction was determined, and the C and N concentration in this fraction adjusted accordingly. All determinations were carried out in duplicate on each of the independent bulk samples.

Calculations and statistics

Protein concentration was calculated by subtracting nitrate-N from total N, and multiplying by 6.25. Organic acid concentration was estimated by subtracting nitrate content (in meq g^{-1}) from ash alkalinity, and multiplying by an average molecular weight of 62.5. Mineral concentration was calculated by multiplying ash alkalinity (in meq g^{-1}) by 30 g eq^{-1} (mass of carbonate), subtracting this value from total ash, and adding the weight of nitrate. Lignin concentration was estimated from the C and N concentration in the crude cell wall residue, assuming a C concentration in lignin of 640 mg g^{-1}, and a C concentration in the (hemi)cellulose complex of 444 mg g^{-1} (Poorter & Villar, 1997). Construction costs (K) were calculated with the following formula:

$$K = (-1.041 + 5.077 C_{om})(1 - M) + (5.325 N_{org})$$

$$\text{(Eqn 1)}$$

(Poorter, 1994), where K is the construction cost (g glucose g^{-1} d. wt), C_{om} the C content of the organic matter (g g^{-1}), and M and N_{org} the mineral and organic N concentration of the total dry mass (g g^{-1}), respectively. This is a slightly modified approach to that of Vertregt & Penning de Vries (1987), which assumes that NO_3^- is the N source for the plant. The cost of protein is lower if plants utilize ammonium. As the extent to which the study species take up nitrate and ammonium was not known, these construction costs should be considered to be maximum values.

Statistical tests were carried out using SPSS. Area per leaf was ln-transformed before testing, because proportional differences were considered to be more important than absolute differences. The contribution of habitats and species within habitats to the average SLA per individual was analysed using a nested ANOVA. Graphical analyses showed that a number of relationships were non-linear. Therefore, the sum of squares due to habitats was further analysed with orthogonal polynomials, ranking habitats based on the estimated productivity of each site. Where the quadratic term was found to be significant, a saturating function was fitted to the observed data, of the form $y = C_1 + C_{max}[x/(x + C_2)]$,

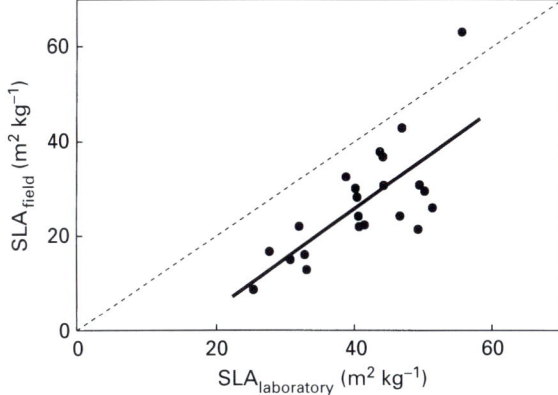

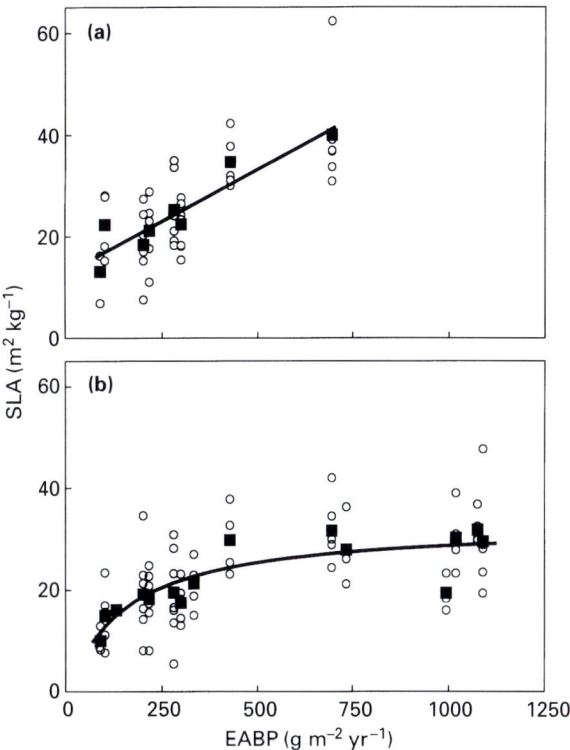

Fig. 1. Relationship between SLA of whole plants grown in the laboratory, and the average SLA of the oldest viable leaf and the youngest fully expanded leaf of the same species grown in the field. Data are for the second sampling period. The broken line is the 1:1 relationship. Total number of species is 22, $r^2 = 0.53$, $P < 0.001$. Average SE is 4.8%, with five independent plants measured in the field. Laboratory data for these species are from Poorter & Remkes, 1990 (20 species); Van der Werf *et al.*, 1993 (*Carex flacca*); Den Dubbelden & Verburg, 1996 (*Galium aparine*).

Fig. 2. The SLA values of the dominant species on a site (open circles) and the average values across species within a habitat (closed squares), plotted against estimated annual above-ground biomass produced (EABP) in a range of habitats. Vegetation types in these habitats consisted of a range of grassland, ruderal and ericaceous stands, measured in (a) late April (eight habitats) and (b) early July (15 habitats). The lines are the linear regressions through the average values per habitat (a: $r^2 = 0.83$, $P < 0.001$; b: $r^2 = 0.92$, $P < 0.001$).

where y is the dependent variable, x the estimated productivity and C_1, C_2 and C_{max} are constants. These constants were estimated with the procedure NLR in SPSS. Where the quadratic term was not significant, data were fitted with a linear regression. Correlations between productivity, SLA, size of leaves and relative irradiance were analysed using path analysis, using standardized coefficients.

RESULTS

SLA and productivity

Generally, recently matured leaves had slightly lower SLA than older leaves (on average 4%); this difference was statistically non-significant. To assess the relationship between the mean SLA of species in

the field and in the laboratory, the SLA values were averaged of the youngest fully grown leaf and the oldest still-viable leaf for each individual. These average values were plotted against average SLA data over all viable leaves of plants of the same species, obtained by Poorter & Remkes (1990), Van der Werf *et al.* (1993) and Den Dubbelden &

Table 2. *ANOVA with average SLA per individual measured during periods 1 (April) and 2 (July) as the dependent variables; habitat, and species nested within habitat, as independent variables*

Factor	Period 1			Period 2		
	SS	df	P	SS	df	P
Habitat	14100	7	***	15200	14	***
linear	11900	1	***	9100	1	***
quadratic	0	1	ns	2200	1	**
rest	2200	5	ns	3900	12	ns
Error (species within habitats)	9700	39		12400	59	
Species within habitats	9700	39	***	12400	59	***
Error (within cells)	1200	183		2500	293	

The sum of squares explained by habitat is further analysed with orthogonal polynomials, ranking habitats on the basis of the estimated above-ground biomass produced at each site. n.s., not significant; **, $P < 0.01$; ***, $P < 0.001$.

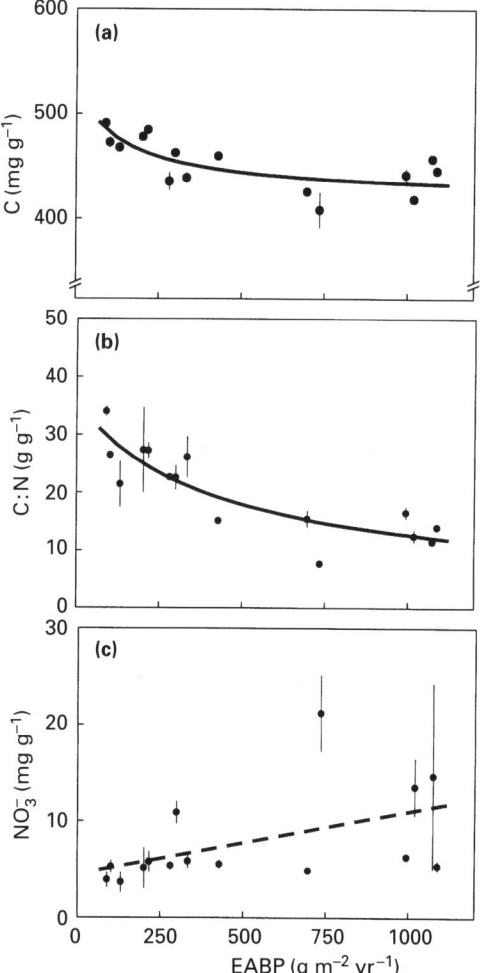

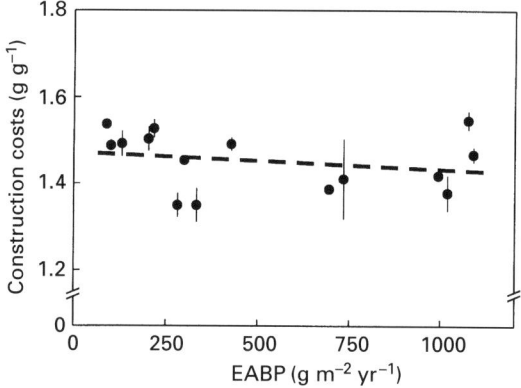

Fig. 4. Construction costs of leaves of different vegetation types plotted against the estimated annual above-ground biomass produced per year in a habitat (EABP). For more information see the legend to Fig. 3.

Fig. 3. (a) Carbon concentration, (b) C:N ratio and (c) nitrate concentration of leaves plotted against estimated annual above-ground biomass produced in a habitat (EABP). Data points are the mean values ($\pm$ SE) from two independent samples of bulked material of all green leaves from all species in that vegetation type. Broken lines indicate a non-significant relationship, continuous straight lines a significant linear regression, and continuous curved lines are saturating curves fitted when the regression analysis showed a significant quadratic component.

Verburg (1996) for plants grown in climate chambers at a daily quantum input of 16 mol m^{-2} d^{-1} (Fig. 1). The relation between the SLAs of field-grown and laboratory-grown plants is positive ($P < 0.001$), but field-grown plants generally have lower SLA values (39% on average).

Subsequently, an analysis was made of a possible relationship between the estimated productivity of a site and the SLA of the dominant species. Overall, EABP values ranged from less than 100 g m^{-2} yr^{-1} for the vegetation of drifting sand dunes, to more than 1000 g m^{-2} yr^{-1} for a stand of *Phragmites australis* (Table 1). Average SLA values per individual were analysed using ANOVA, with habitat as main factor and species nested within habitats. The main effect habitat explained 50–60% of the total variation in SLA (Table 2). SLA was positively

and significantly correlated with the estimated yearly above-ground biomass produced at these sites in both harvesting periods (Fig. 2; see also the first-order polynomial in Table 2). Beyond an EABP value of 500 g m^{-2} yr^{-1}, for which we only have data for the second period, the relationship saturated (Fig. 2b). Variation between species within habitats was considerable, explaining 35–40% of the total variation in SLA (Table 2). This can also be seen from the wide scatter of data points around the mean in Fig. 2.

Construction costs

The C concentration of bulked samples of leaves within a stand correlated negatively with the EABP values of those sites (Fig. 3a, $P < 0.01$), data ranging from 410–490 mg g^{-1}. Total N correlated positively with EABP ($P < 0.01$, data not shown), and, consequently, the C:N ratio was much higher for sites of low productivity (Fig. 3b, $P < 0.001$). Part of the total N (4–12%) was present in the form of nitrate. Although the highest concentrations of nitrate (up to 20 mg g^{-1}) were observed in some of the highly productive sites, no general relationship with EABP was found (Fig. 3c, ns). Leaf construction costs, calculated from the C and organic N values, did show a slightly negative, but non-significant relationship with EABP (Fig. 4, ns). The average value across all habitats was 1.45 g glucose per g leaf dry mass.

Within the group of compounds with low construction costs, highly productive vegetation had markedly higher concentrations of minerals (Fig. 5a, $P < 0.01$) and organic acids (Fig. 5b, $P < 0.05$). Negative correlations were found for the total non-structural carbohydrates (soluble plus insoluble sugars, Fig. 5c, $P < 0.01$) and the amount of total structural carbohydrates (Fig. 5d, $P < 0.05$). Within

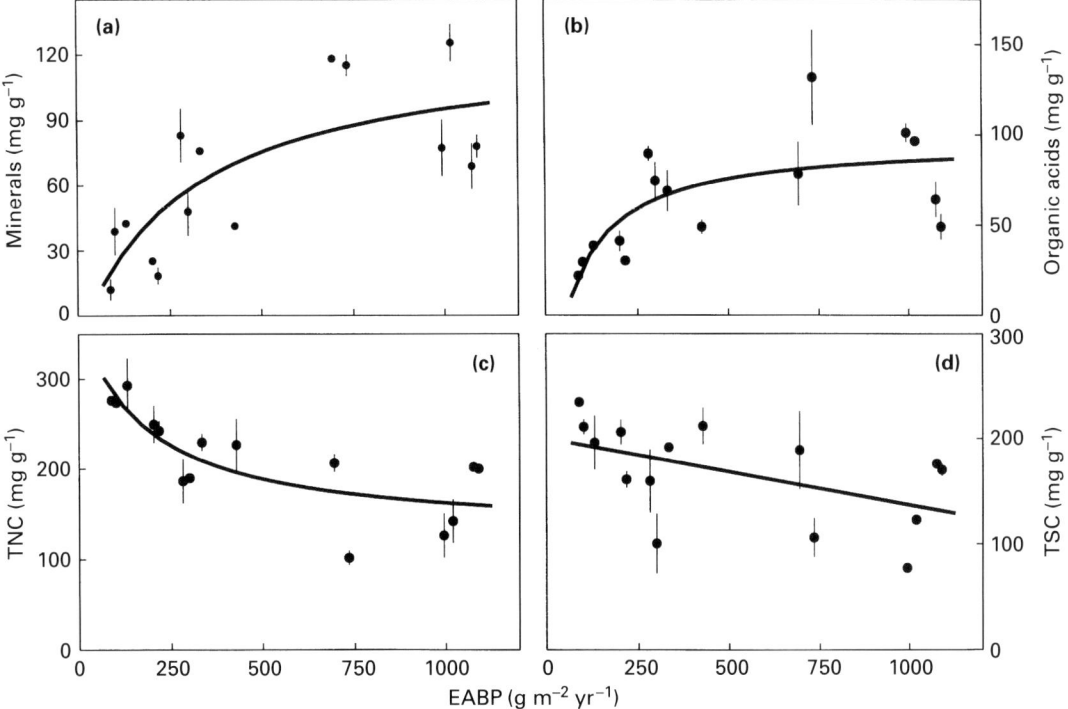

Fig. 5. (a) Mineral concentration, (b) organic acid concentration, (c) concentration of total non-structural carbohydrates (TNC) (starch, fructan, soluble sugars), and (d) concentration of total structural carbohydrates (TSC) (cellulose plus hemicellulose) of leaves plotted against the estimated annual above-ground biomass produced in a habitat (EABP). For more information see the legend to Fig. 3.

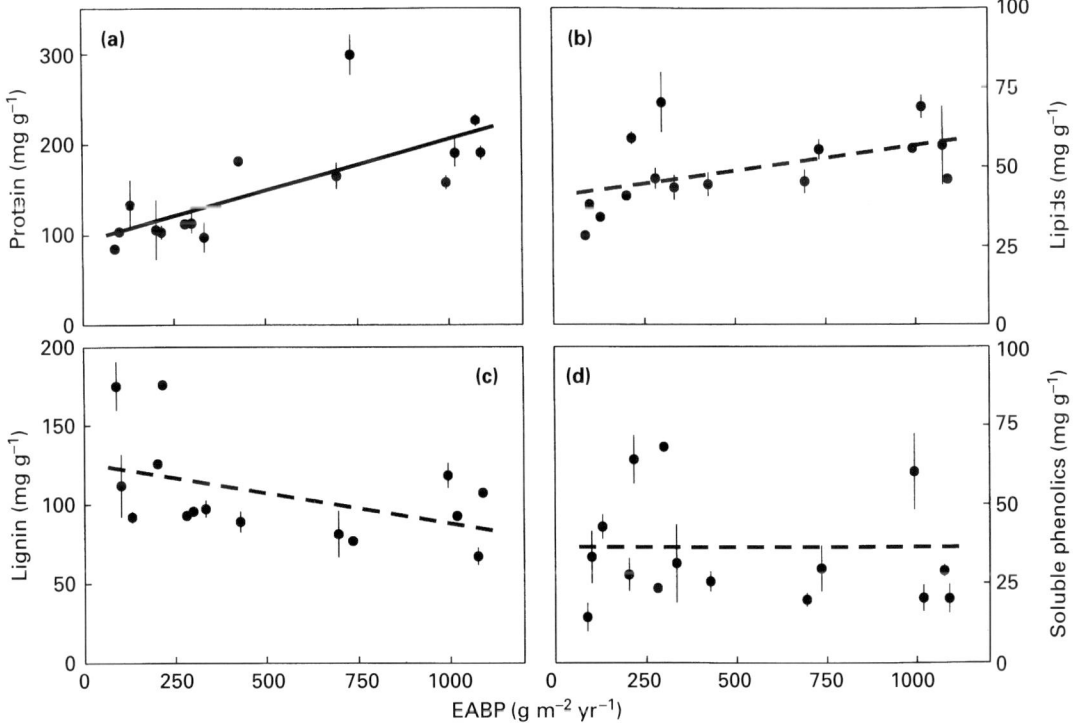

Fig. 6. (a) Protein concentration, (b) lipid concentration, (c) lignin concentration, and (d) concentration of soluble phenolics of leaves plotted against the estimated annual above-ground biomass produced in that habitat (EABP). For more information see the legend to Fig. 3.

the group of expensive compounds, the concentration of protein (Fig. 6a, $P < 0.01$) was significantly higher in the highly productive vegetation types, the concentration of lipids was somewhat higher (Fig. 6b, $0.005 < P < 0.10$) whereas lignin concentrations were somewhat lower (Fig. 6c, $0.05 < P < 0.10$). No clear trend was found for the group of soluble phenolics (Fig. 6d, ns).

DISCUSSION

SLA and productivity

There was a positive relationship between the SLA of field-grown and laboratory-grown plants (Fig. 1), with values for field-grown plants being generally lower. Daily quantum input is an environmental parameter that strongly affects SLA (Chabot *et al.*, 1979). As daily quantum input in growth rooms is generally lower than in the field (Garnier & Freijsen, 1994), the SLA of field plants is indeed expected to be lower. The only exception was for *Galinsoga parviflora* (highest value in Fig. 1), which was collected close to a building where sunlight was blocked for part of the day. Apart from the difference in daily light climate, plants outside will also experience more turbulent air movements than those grown under controlled conditions, another factor that may decrease SLA (Woodward, 1983). More important than the absolute SLA values is the relative ranking between field- and laboratory-grown plants, which remains similar. Clearly, genotype × environment interactions were not large; therefore it is concluded that data from comparative laboratory experiments can be used to infer valid conclusions about SLA ranking in the field. Such conclusions are in line with findings in Mediterranean grasslands (Garnier *et al.*, 1997).

The average SLA of the dominant species in the vegetation was positively correlated with the estimated above-ground productivity of that vegetation (Fig. 2), at least up to an EABP value of around 500 g m^{-2} yr^{-1}, after which the relationship saturated. A positive relationship between SLA and EABP was also found by Fliervoet (1987) in an analysis of the vegetation structure in Western European grasslands. Grubb (1977) reports that with increasing altitude, tropical mountain forests decrease in both SLA and productivity. Contrasting evidence comes from the observation that mature forest stands of evergreen conifers and deciduous hardwood species do not differ in productivity, despite strong differences in SLA (Reich, 1998). Reich (1998) explains this by showing that the evergreen species with the low SLA values allocate a relatively large fraction of their total biomass to foliage. This compensates for the lower biomass gain per g foliage in evergreens, to the extent that productivity per unit ground area is similar to that of the deciduous species with the high SLA values. A similar relationship has not been found in grasslands such as those investigated in this study. On the contrary, Fliervoet (1987) reports that in grasslands with a high EABP, a relatively larger fraction of the above-ground biomass is allocated to stems than in less productive grasslands, and a smaller fraction to leaves.

Is the observed relationship between productivity and SLA caused directly by the commonly suggested

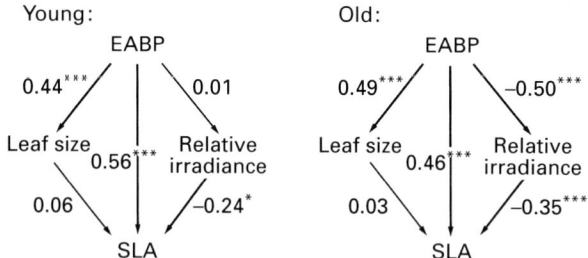

Fig. 7. Path analysis of the SLA data obtained in the second period (early July). Values indicate how much variable *B* changes (expressed in units of SD) when variable *A* is increased by 1 unit of SD. Data were analysed separately for young and old leaves. 'Relative irradiance' is the irradiance at the top of the leaf relative to that above the vegetation. Abbreviations: EABP, estimated annual above-ground biomass produced. Significance levels: *, $P < 0.05$; ***, $P < 0.001$.

compromise between resource capture and conservation? And why is the relationship a saturating one? These are difficult questions to answer. The relationship between average SLA and productivity can be modulated by various factors. Leaves of plant species characteristic of productive environments are generally larger (Grime, 1979; Fliervoet, 1987), and therefore may require relatively more support tissue (Givnish, 1986). As support tissue generally consists of dense, lignified material, it could be envisaged that this has a negative effect on SLA (Grubb, 1998), explaining the saturation above an EABP of 500 g m^{-2} yr^{-1}. On the other hand, there might also be a trend towards higher SLA values in productive environments because of the light climate they experience. If lower leaves developed in a period in which they were more-or-less at the top of the canopy, relatively small differences in SLA between newly matured and old leaves can be expected, caused by seasonal differences in light climate during leaf development. However, specific leaf area can partly adjust to changes in light climate that take place after full development (Pons & Pearcy, 1994). The larger the canopy, the less will be the irradiance experienced by the lower leaves. Therefore, one may expect that in highly productive stands (where a large standing crop develops over time), the difference in SLA between old and new leaves will be greater than in plants growing in open vegetation (cf. Niinemets & Kull, 1994).

To correct for the indirect effects that habitat productivity may have on the SLA of the resident species via leaf size, and for the relative irradiance experienced by each leaf, a path analysis was carried out with these four variables as factors of interest (Fig. 7). As this analysis assumes linear relationships, EABP values were ln-transformed. The analysis confirms the observations of Grime (1979) and Fliervoet (1987) that leaf size is larger for plant species from nutrient-rich sites. No negative effect of leaf size on SLA could be detected, in contrast to the observations of Grubb (1998). Does this imply that larger leaves do not require relatively more support tissue than small leaves, or that the support tissue itself is not very dense? Most of the support tissue in leaves is around the mid-vein, and if this is dense material, it should have a negative but local effect on SLA. Therefore, for a range of dicotyledonous species we compared SLA around the mid-veins with that of the leaf lamina, where veins are expected to be less dominant. In line with the result of Shipley & Meziane (1998), SLA was found to be lower around the mid-vein (on average 18%) than in the rest of the leaf (data not shown). However, as the area around the mid-vein is only a small fraction of the total leaf area, the effect on the SLA of whole leaves is small. This may explain why leaf size alone does not have a dramatic impact on SLA.

To correct for differences in light climate due to position in the canopy, measurements were made of the irradiance above the leaf relative to that above the canopy. For young leaves, generally at the top of the canopy, no significant effect of EABP on relative irradiance was found (Fig. 7). Old leaves, however, were more heavily shaded in productive vegetation types. In both cases there was a negative effect of relative irradiance on the SLA of the leaves. The total effect of productivity on SLA via the light climate is found by multiplying the two effects. For young leaves this effect was negligible; for old leaves the shading effect, expressed in units of standard deviation, increased SLA by 0.17 units of standard deviation. Thus in highly productive vegetation types there may be a small effect of the light environment towards an increase in SLA.

A third factor that may influence SLA is the fact that plants from these sites of low productivity are generally nutrient-limited. Within a given species, nutrient limitation may cause SLA to decrease (e.g. Fichtner & Schulze, 1992; McDonald *et al.*, 1992), although this response is not always observed (e.g. Sims *et al.*, 1998). Nutrient limitation often induces accumulation of total non-structural carbohydrates (TNC) (Poorter & Villar, 1997). Could the lower SLA of plants from environments of low productivity be explained by higher TNC content, analogous to the decrease in SLA for plants grown in elevated CO_2? Considering the leaf fraction of the vegetation as a whole, there was indeed a higher concentration of TNC in the vegetation on low productivity sites (Fig. 5c). We cannot correct for this effect via path analysis as we do not have TNC data for each individual leaf. Assuming that TNC concentration of the plant species for which SLA was measured was the same as for the vegetation as a whole (data in Fig. 5c), we calculated that at most 20% of the difference in SLA between rich and poor sites could be explained by TNC differences. This is an upper limit, as the lower concentration of protein in plants from nutrient-poor sites must also be taken into account. Therefore, it is concluded that this factor cannot be of overriding importance either.

None of the factors discussed above strongly affected the relationship between SLA and productivity. Therefore it is concluded that there is a direct relationship between the SLA of a species and the productivity of the habitats in which these species are generally found. This field survey confirms earlier observations in the growth room, that there are inherent differences in SLA between species from sites differing in productivity. However, two questions remain unanswered. Firstly, we have not been able to explain why the relationship between SLA and productivity saturates above an EABP value of 500 g m^{-2} yr^{-1}. Possibly the differences are strongly related to variation in leaf life span. All of the species in the more productive environments drop most to all of their leaves at the end of the growing season. As there is no strong differentiation in life span between these species (Berendse *et al.*, 1998), differences in investment in defence compounds are possibly also small. A second aspect of the relationship between SLA and productivity is that species' variability in SLA within a given site is large, explaining 35–40% of the total variation. This is in line with conclusions from a field survey of Eliáš (1985) and a growth analysis of Van Andel & Biere (1989) under controlled conditions on a range of co-occurring species. Clearly, more factors than SLA are involved in shaping the performance of plants in various habitats. Does this imply that other factors within the suite of traits related to SLA compensate for the fact that one species has a lower SLA than another? If so, it would be of interest to know the trade-offs involved. Alternatively, there may be factors unrelated to growth rate and SLA that explain the success of species at a given site. The suggestion of Westoby (1998) that plant height and seed mass should be included as simple and biologically independent factors to characterize the strategy of a species offers a promising avenue to further our understanding on this point.

Construction costs

Construction costs of the leaves did not vary systematically with the productivity of the habitat (Fig. 4). Thus we can reject the early hypothesis of Miller & Stoner (1979), who expected evergreens from environments of low productivity to have higher construction costs than more productive deciduous species. The only field data relating construction costs of leaves to soil fertility are those of Merino (1987), who made a qualitative assessment and did not find any effect. Controlled experiments measuring the effect of nutrient availability on leaf construction costs within a given species showed small but significant increases with increased N availability (Shinano *et al.*, 1995; Griffin *et al.*, 1996). When plant species characteristic of low- and high-productivity habitats were grown under controlled conditions at high levels of nutrient supply, no systematic variation in leaf and whole-plant construction costs was found between fast- and slow-growing species (Poorter & Bergkotte, 1992). Chapin (1989) arrived at the same conclusion for various tundra species, all growing in the same habitat. In a review of the literature, Poorter & Villar (1997) concluded that variation in construction costs of leaves is small (within 10%), regardless of the environmental factors assessed.

What causes the construction costs to be constant? To obtain insight into the causes of variation in construction costs, the chemical composition of the biomass has to be known. Variation in leaf chemical composition across habitats is considerable. Leaves sampled in environments of low productivity have high concentrations of C, high C:N ratios, and high concentrations of total non-structural carbohydrates as well as structural carbohydrates (Figs 3a,b, 5c,d). They also have somewhat higher concentrations of lignin (Fig. 6c). This is in line with the carbon–nutrient balance theory (Bryant *et al.*, 1983). Leaves from highly productive sites, on the other hand, have high concentrations of minerals, organic acids, protein and lipids (Figs 5a,b, 6a,b). It is the net balance of all these changes that determines the variation in construction costs. Chapin (1989) concluded that the constancy in construction costs across species was due to a negative correlation between one expensive compound (e.g. lignin) and another (protein). Poorter & Bergkotte (1992), on the other hand, found that the constancy was mainly due to the positive correlation across species between an expensive class of compounds (proteins) and a cheap one (minerals). In both cases the conclusion was drawn from correlations between compounds. However, variation in construction costs depends not only on the direction of changes in compounds, but also on the cost of each of the compounds and the absolute changes across species or habitats.

A quantitative analysis of the changes in chemical composition requires knowledge of the concentrations of all classes of compounds. However, adding up concentrations of the eight classes of constituents, we were not able to account for 100% of the leaf mass, but only for 85±5%. This is in line with previous analyses (87 and 83%, respectively, by Poorter & Bergkotte, 1992 and Poorter *et al.*, 1997), but lower than the 95% reported by Chapin (1989). For the present calculation it was assumed that the missing part could be proportionally distributed over all fractions determined. To determine the reason for the constancy in construction costs, we calculated the differences in chemical composition between vegetation from a typical site of low productivity and a typical site of high productivity.

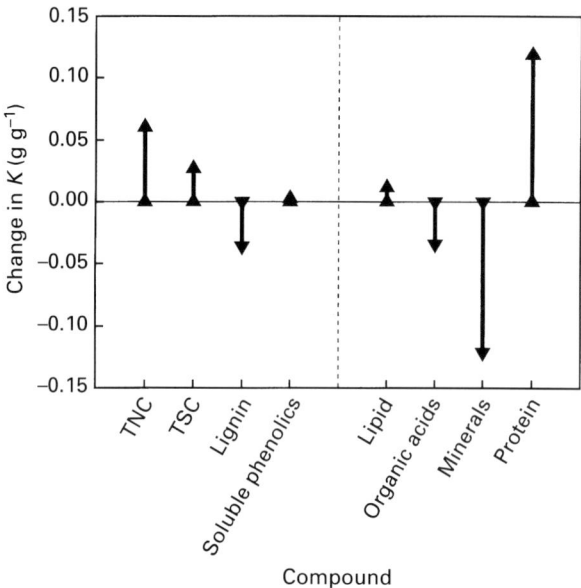

Fig. 8. The effect of differences in the various classes of compounds on the change in construction costs (*K*). Arrows indicate the value by which construction costs increase or decrease if the concentration of each of compound in the leaf biomass of a typical site of low productivity is altered to the concentration of biomass at the typically highly productive site. The four classes of compounds on the left have a higher concentration in sites of low productivity, whereas the four on the right are higher in highly productive sites. Abbreviations: TNC, total non-structural carbohydrates; TSC, total structural carbohydrates.

These values were based on the lowest and highest values of the regression lines in Figs 5 and 6. Let us assume that compound A was present in a higher concentration in the highly productive site. Knowing the difference in concentration (say 50 mg g⁻¹), we then analysed how construction costs of the leaves at the site of low productivity would change if we replaced 50 mg of the total biomass with 50 mg of compound A. That number gives the direction and the absolute change in construction costs due to a difference in that component. A full explanation of the method followed is given in Appendix 2. Results are shown in Fig. 8, and indicate that leaf biomass at highly productive sites tends to have higher construction costs, because of higher concentrations of the expensive compounds protein and lipid, as well as lower concentrations of the cheap compounds TSC and TNC (upward arrows in Fig. 8). On the other hand, these increases are compensated because leaves in the highly productive habitats accumulate less of the expensive lignin, and relatively more of the cheap classes of organic acids and minerals. Quantitatively, the two most important effects are due to the correlated increases in proteins, with high costs, and cheap minerals, with low costs. These two balance each other and are the major reason that construction costs are constant. Changes in other compounds do play a role as well, but are quantitatively less important. Thus it is concluded that these data are in line with the results for fast- and slow-growing species grown at optimum conditions (Poorter & Bergkotte, 1992).

CONCLUSIONS

For the habitats considered in this survey, there is a positive relationship between the SLA of the most common species in a vegetation type and the productivity of the vegetation as a whole. However, within each habitat there is considerable variation in SLA between species. No systematic trend was observed between leaf construction costs and habitat productivity. This was due mainly to the positive correlation in the concentration of an expensive class of compounds (proteins) and a cheap one (minerals).

ACKNOWLEDGEMENTS

Yvonne van Berkel skilfully assisted with the chemical analyses of the samples. Cas Kruitwagen and Maria Schippers provided statistical advice. Hans Lambers made valuable suggestions during the course of the experiment. Cynthia van Rijn, Adrie van der Werf, Peter Ryser, Peter Grubb, Heinjo During, Ian Wright and Eric Garnier made helpful comments on previous versions of this manuscript.

REFERENCES

Aerts R, Berendse F. 1989. Above-ground nutrient turnover and net primary production of an evergreen and a deciduous species in a heath land ecosystem. *Journal of Ecology* **77**: 343–356.

Aerts R, Chapin FS. 1999. The mineral nutrition of wild plants revisited: a re-evaluation of processes and patterns. *Advances in Ecological Research* **29**, in press.

Berendse F, Braakhekke W, Van der Krift T. 1998. Adaptations of plant populations to nutrient-poor environments and their implications for soil nutrient mineralisation. In: Lambers H, Poorter H, Van Vuuren MMI, eds. *Inherent variation in plant growth. Physiological mechanisms and ecological consequences.* Leiden, The Netherlands: Backhuys Publishers, 503–514.

Bryant JP, Chapin FS, Klein DR. 1983. Carbon/nutrient balance of boreal plants in relation to vertebrate herbivory. *Oikos* **40**: 357–368.

Chabot BF, Jurik TW, Chabot JF. 1979. Influence of instantaneous and integrated light-flux density on leaf anatomy and photosynthesis. *American Journal of Botany* **66**: 940–945.

Chapin FS. 1980. The mineral nutrition of wild plants. *Annual Review of Ecology and Systematics* **11**: 233–260.

Chapin FS. 1989. The cost of tundra plant structures: evaluation of concepts and currencies. *American Naturalist* **133**: 1–19.

Chen JM, Black TA. 1992. Defining leaf area index for non-flat leaves. *Plant, Cell and Environment* **15**: 421–429.

Choong MF, Lucas PW, Ong JSY, Pereira B, Tan HTW, Turner IM. 1992. Leaf fracture toughness and sclerophylly: their correlations and ecological implications. *New Phytologist* **121**: 597–610.

Coley PD, Bryant JP, Chapin FS. 1985. Resource availability and plant herbivore defence. *Science* **230**: 895–899.

Den Dubbelden KC, Verburg RW. 1996. Inherent allocation patterns and potential growth rates of herbaceous climbing plants. *Plant and Soil* **184**: 341–347.

Eliáš P. 1985. Leaf indices of woodland herbs as indicators of habitat conditions. *Ekologia (CSSR)* **4**: 289–295.

Fichtner K, Schulze ED. 1992. The effect of nitrogen nutrition on growth and biomass partitioning of annual plants originating from habitats of different nitrogen availability. *Oecologia* **92**: 236–241.

Fliervoet LM. 1987. Characterization of the canopy structure of Dutch grasslands. *Vegetatio* **70**: 105–117.

Garnier E, Cordonnier P, Guillerm JL, Sonié L. 1997. Specific leaf area and leaf nitrogen concentration in annual and perennial grass species growing in Mediterranean old fields. *Oecologia* **111**: 490–498.

Garnier E, Freijsen AHJ. 1994. On ecological inference from laboratory experiments conducted under optimum conditions. In: Roy J, Garnier E, eds. *A whole plant perspective of carbon–nitrogen interactions.* The Hague, Netherlands: SPB Academic Publishing, 267–292.

Givnish TJ. 1986. Biomechanical constraints on crown geometry in forest herbs. In: Givnish TJ. ed. *On the economy of form and function.* Cambridge, UK: Cambridge University Press, 525–583.

Griffin KL, Winner WE, Strain BR. 1996. Construction cost of loblolly and ponderosa pine leaves grown with varying carbon and nitrogen availability. *Plant, Cell and Environment* **19**: 729–738.

Grubb PJ. 1977. Control of growth and distribution on wet tropical mountains: with special reference to minreal nutrition. *Annual Review of Ecology and Systematics* **8**: 83–107.

Grubb PJ. 1998. A reassessment of the strategies of plants which cope with shortages of resources. *Perspectives in Plant Ecology, Evolution and Systematics* **1**: 3–31.

Grime JP. 1979. *Plant strategies and vegetation processes.* Chichester, UK: Wiley.

Grime JP, Hunt R. 1975. Relative growth-rate: its range and adaptive significance in a local flora. *Journal of Ecology* **63**: 393–422.

Lambers H, Dijkstra P. 1987. A physiological analysis of genotypic variation in relative growth rate: can growth confer ecological advantage? In: Van Andel J, Bakker JP, Snaydon RJ, eds. *Disturbance in grasslands.* Dordrecht, Netherlands: Junk Publishers, 237–252.

Lambers H, Poorter H. 1992. Inherent variation in growth rate between higher plants: a search for physiological causes and ecological consequences. *Advances in Ecological Research* **23**: 187–261.

McDonald AJS, Lohammar T, Ingestad T. 1992. Net assimilation rate and shoot area development in birch (*Betula pendula* Roth.) at different steady-state values of nutrition and photon flux density. *Trees* **6**: 1–6.

Meijer WJM. 1984. De stikstofbemesting van zaadteeltgewassen Engels raai, veldbeemd en roodzwenk. Proefstation Lelystad, Report No. 55.

Merino J. 1987. The costs of growing and maintaining leaves of mediterranean plants. In: Tenhunen JD, Catarino FM, Lange OL, Oechel WC, eds. *Plant response to stress.* Berlin, Germany: Springer Verlag, 553–564.

Miller PC, Stoner WA. 1979. Canopy structure and environmental interactions. In: Solbrig OT, Jain S, Johnson GB, Raven PH. eds. *Topics in plant population biology.* New York, USA: Colombia University Press, 428–458.

Niinemets U, Kull K. 1994. Leaf weight per area and leaf size of 85 Estonian woody species in relation to shade tolerance and light availability. *Forest Ecology and Management* **70**: 1–10.

Penning de Vries FWT, Brunsting AHM, van Laar HH. 1974. Products, requirements and efficiency of biosynthetic processes: a quantitative approach. *Journal of Theoretical Biology* **45**: 339–377.

Pons TL, Pearcy RW. 1994. Nitrogen reallocation and photosynthetic acclimation in response to partial shading in soybean plants. *Physiologia Plantarum* **92**: 636–644.

Poorter H. 1989. Interspecific variation in relative growth rate: On ecological causes and physiological consequences. In: Lambers H, Cambridge ML, Konings H, Pons TL, eds. *Causes and consequences of variation in growth rate and productivity of higher plants.* The Hague, Netherlands: SPB Academic Publishing, 45–68.

Poorter H. 1994. Construction costs and payback time of biomass: a whole plant perspective. In Roy J, Garnier E, eds. *A whole plant perspective of carbon–nitrogen interactions.* The Hague, Netherlands: SPB Academic Publishing, 111–127.

Poorter H, Bergkotte M. 1992. Chemical composition of 24 wild species differing in relative growth rate. *Plant Cell and Environment* **15**: 221–229.

Poorter H, Garnier E. 1999. Ecological significance of inherent variation in relative growth rate. In: Pugnaire F, Valladares X, eds. *Handbook of functional plant ecology.* New York, USA: Marcel Dekker, 81–120.

Poorter H, Remkes C. 1990. Leaf area ratio and net assimilation rate of 24 wild species differing in relative growth rate. *Oecologia* **83**: 553–559.

Poorter H, Van der Werf A. 1998. Is inherent variation in RGR determined by LAR at low irradiance and by NAR at high irradiance? A review of herbaceous species. In: Lambers H, Poorter H, Van Vuuren MMI, eds. *Inherent variation in plant growth. physiological mechanisms and ecological consequences.* Leiden, Netherlands: Backhuys Publishers, 309–336.

Poorter H, van Berkel Y, Baxter R, Den Hertog J, Dijkstra P, Gifford RM, Griffin KL, Roumet C, Roy J, Wong SC. 1997. The effect of elevated CO$_2$ on the chemical composition and construction costs of leaves of 27 C$_3$ species. *Plant, Cell and Environment* **20**: 472–482.

Poorter H, Villar R. 1997. The fate of acquired carbon in plants: chemical composition and construction costs. In: Bazzaz FA, Grace J. eds. *Plant resource allocation.* San Diego, USA: Academic Press, 39–72.

Reich PB. 1998. Variation among plant species in leaf turnover rates and associated traits: implications for growth at all life stages. In: Lambers H, Poorter H, Van Vuuren MMI, eds. *Inherent variation in plant growth. physiological mechanisms and ecological consequences.* Leiden, Netherlands: Backhuys Publishers, 467–487.

Ryser P. 1996. The importance of tissue density for growth and life span of leaves and roots: a comparison of five ecologically contrasting grasses. *Functional Ecology* **10**: 717–723.

Shinano T, Osaki M, Tadano T. 1995. Comparison of growth efficiency between rice and soybean at the vegetative growth stage. *Soil Science and Plant Nutrition* **41**: 471–480.

Shipley B, Meziane D. 1998. The statistical modelling of plant growth and its components using structural equations. In: Lambers H, Poorter H, Van Vuuren MMI, eds. *Inherent variation in plant growth. physiological mechanisms and ecological consequences.* Leiden, Netherlands: Backhuys Publishers, 393–408.

Sims DA, Seemann JR, Luo Y. 1998. Elevated CO$_2$ concentration has independent effects on expansion rates and thickness of soybean leaves across light and nitrogen gradients. *Journal of Experimental Botany* **49**: 583–591.

Tilman D. 1988. *Plant strategies and the dynamics of plant communities.* Princeton, USA: Princeton University Press.

Van Andel J, Biere A 1989. Ecological significance of variability in growth rate and plant productivity. In: Lambers H, Cambridge ML, Konings H, Pons T, *Causes and consequences of variation in growth rate and productivity of higher plants.* The Hague, Netherlands: SPB Academic Publishing, 257–267.

Van der Werf A, Geerts RHEM, Jacobs FHH, Korevaar H, Oomes MJM, De Visser W. 1998. The importance of relative growth rate and associated traits for competition between species during vegetation succession. In: Lambers H, Poorter H, Van Vuuren MMI, eds. *Inherent variation in plant growth. physiological mechanisms and ecological consequences.* Leiden, Netherlands: Backhuys Publishers, 489–502.

Van der Werf A, van Nuenen M, Visser A, Lambers H. 1993. Contribution of physiological and morphological traits to a species' competitive ability at high and low nitrogen supply. A hypothesis for inherently fast- and slow-growing monocotyledonous species. *Oecologia* **94**: 434–440.

Vertregt N, Penning de Vries FWT. 1987. A rapid method for determining the efficiency of biosynthesis of plant biomass. *Journal of Theoretical Biology* **128**: 109–119.

Westoby M. 1998. A leaf–height–seed (LHS) plant ecology strategy scheme. *Plant and Soil* **199**: 213–227.

Woodward FI. 1983. The significance of interspecific differences in specific leaf area to the growth of selected herbaceous species from different altitudes. *New Phytologist* **95**: 313–323.

Appendix 1. *List of species sampled in the different habitats, and the average SLA of the youngest fully grown and the oldest still-viable leaf for plants sampled in period 1 (April 1993) and period 2 (July 1993)*

Habitat or vegetation type	Species sampled	SLA (m^2 kg^{-1})	
		Period 1	Period 2
1. Drifting sand dune	*Ammophila arenaria*	6.9	8.2
	Corynephorus canescens	16.2	12.9
	Festuca ovina	16.2	8.7
2. Quaking fen	*Calamagrostis canescens*	28.0	–
	Carex lasiocarpa	18.0	11.1
	Carex nigra	15.2	16.9
	Juncus subnodulosus	–	7.6
	Menyanthes trifoliata	27.8	23.4
	Potentilla palustris	22.4	15.6
3. Grass heath	*Deschampsia flexuosa*	–	16.0
4. Wet heath	*Carex panicea*	20.3	22.9
	Cirsium dissectum	17.0	16.3
	Erica tetralix	7.5	8.0
	Molinia caerulea	24.4	21.3
	Salix repens	15.2	14.2
	Succisa pratensis	15.2	16.3
	Viola palustris	27.4	34.6
5. Dry heath	*Carex pilulifera*	24.6	20.8
	Calluna vulgaris	11.0	8.0
	Deschampsia flexuosa	21.4	17.3
	Galium saxatile	28.8	–
	Genista anglica	17.6	15.5
	Molinia caerulea	23.0	22.7
	Rumex acetosella	–	24.8

Appendix 1. (*cont.*)

Habitat or vegetation type	Species sampled	SLA (m² kg⁻¹)	
		Period 1	Period 2
6. Ruderal and trampled	*Cirsium arvense*	19.2	5.4
	Plantago lanceolata	24.1	13.5
	Plantago major	21.1	16.0
	Potentilla anserina	18.2	30.9
	Trifolium repens	33.6	28.2
	Unidentified grass	35.0	23.2
7. Dry open grassland (south-facing slope)	*Anthoxanthum odoratum*	27.6	–
	Centaurea jacea	25.7	16.5
	Festuca rubra	15.4	13.0
	Leucanthemum vulgare	18.1	14.0
	Plantago lanceolata	24.2	19.4
	Rhinanthus minor	26.5	–
	Trifolium pratense	23.3	23.1
	Unidentified Composite	18.3	18.2
8. Chalk grassland (north-facing slope)	*Brachypodium pinnatum*	–	21.9
	Carex flacca	–	15.0
	Centaurea jacea	–	18.8
	Dactylorrhiza maculata	–	22.0
	Leontodon hispidus	–	27.0
	Ononis repens	–	22.9
	Plantago lanceolata	–	21.4
9. Poor hay meadow	*Agrostis tenuis*	–	25.4
	Alopecurus pratensis	30.0	–
	Anthoxanthum odoratum	32.0	30.0
	Holcus lanatus	31.1	37.8
	Lolium perenne	42.2	–
	Leontodon autumnalis	–	32.7
	Ranunculus repens	34.7	23.1
	Taraxacum officinale	37.8	–
10. Roadside	*Aegopodium podagraria*	39.1	30.0
	Anthriscus sylvestris	30.9	30.0
	Arrhenaterum elatius	36.8	34.5
	Dactylis glomerata	36.9	28.9
	Galium aparine	62.3	42.0
	Urtica dioica	33.7	24.4
11. Ruderal (not trampled)	*Brassica napus*	–	26.1
	Chenopodium album	–	21.2
	Polygonum persicaria	–	36.3
12. Along ditch (never mown)	*Cirsium arvense*	–	16.0
	Epilobium hirsutum	–	19.7
	Phragmites australis	–	18.3
	Urtica dioica	–	23.3
13. Roadside	*Arctium pubescens*	–	31.0
	Elytrigia repens	–	29.7
	Galium aparine	–	39.0
	Heracleum sphondylium	–	23.3
	Urtica dioica	–	27.9
14. Fertilized meadow	*Bromus mollis*	–	29.8
	Dactylis glomerata	–	30.0
	Lolium perenne	–	32.5
	Ramunculus repens	–	29.8
	Taraxacum officinale	–	36.8
15. Reed marsh	*Cirsium vulgare*	–	19.4
	Eupatorium cannabinum	–	28.9
	Galium aparine	–	47.4
	Phragmites australis	–	23.5
	Urtica dioica	–	28.1

Appendix 2. *Calculation of the contribution of various chemical compounds to a difference in the construction costs of plants*

We assume that the chemical composition of a plant can be fully characterized by eight classes of compounds: minerals, organic acids, TNC, TSC, soluble phenolics, protein, lignin and lipids. For each of these classes of compounds (i), the specific cost of construction (S_i) has been estimated, with S_i being the amount of glucose (in g) required to construct 1 g of a given compound, with glucose and minerals as the starting point. For plants, this area was pioneered by Penning de Vries *et al.* (1974). A recent review is given by Poorter & Villar (1997). If the concentration (C) of each class of compounds (i) is known (C_i, given in g g^{-1}), then the construction cost (K) of a plant or plant organ is given by

$$K = \sum_{i=1}^{8} S_i C_i \qquad \text{Eqn A1}$$

Given a difference in chemical composition due to environment or species, the question then arises how we can quantify the effect on K. In the simple case where a plant consists of only two compounds, x and y, and that we consider only two environments, low (L) and high (H), respectively, the construction costs K^L and K^H are given by

$$K^L = C_x^L S_x + C_y^L S_y \qquad \text{Eqn A2}$$

and

$$K^H = C_x^H S_x + C_y^H S_y \qquad \text{Eqn A3}$$

Consequently, the difference in construction costs is

$$K^H - K^L = C_x^H S_x + C_y^H S_y - C_x^L S_x - C_y^L S_y \qquad \text{Eqn A4}$$

which can be rewritten as

$$K^H - K^L = (C_x^H - C_x^L)S_x + (C_y^H - C_y^L)S_y \qquad \text{Eqn A5}$$

This equation correctly calculates the exact difference in construction costs, but the two composing terms do not adequately describe the exact magnitude and direction of the change. To this end, we have to consider the specific construction costs (S) of each class of compound relative to the construction costs of the plant (or organ). How can we introduce that mathematically?

As the concentration of all compounds together is always 1, and there are only two compounds, x and y, it should be the case that

$$(C_x^H - C_x^L) + (C_y^H - C_y^L) = 0 \qquad \text{Eqn A6}$$

and also that

$$(C_x^H - C_x^L)K^L + (C_y^H - C_y^L)K^L = 0 \qquad \text{Eqn A7}$$

Subtracting this from the right-hand side of Equation A5 yields, after some rearranging

$$K^H - K^L = (C_x^H - C_x^L)(S_x - K^L) + (C_y^H - C_y^L)(S_y - K^L) \qquad \text{Eqn A8}$$

Each of these terms gives us the difference in construction costs due to the difference in composition between the L and H plants, using the construction costs of the L plants as a baseline. This approach can be easily extended from two to more compounds.

New Phytol (1999), 143, 177–189

Research review
Leaf life span and nutrient resorption as determinants of plant nutrient conservation in temperate-arctic regions

R. L. ECKSTEIN[1]*, P. S. KARLSSON[1,2] AND M. WEIH[1]†

[1]*Department of Plant Ecology, Uppsala University, Villavägen 14, SE-75236 Uppsala, Sweden*
[2]*Abisko Scientific Research Station, Royal Swedish Academy of Sciences, SE-98107 Abisko, Sweden*

Received 13 November 1998 ; accepted 29 March 1999

SUMMARY

Nutrient conservation plays an important role in plants adapted to infertile environments. Nutrients can be conserved mainly by extending the life span of plant parts and/or by minimizing the nutrient content of those parts that are abscissed. Together these two parameters (life span and resorption) define the mean residence time (MRT) of a nutrient. In this review we summarize available information on nitrogen resorption and life span, and evaluate their relationship to the MRT of nitrogen, both between and within species. Abundant information with respect to nitrogen resorption efficiency and life span is available at the leaf level. By definition, woody evergreen plants have a much longer leaf life span than species of other life-forms. Conversely, differences in resorption efficiency among life-forms or among plants in habitats differing in soil fertility appear to be small. Inter-specific variation in leaf life span is much larger than intra-specific variation (factor of >200 compared with 2, respectively), while resorption efficiency varies by about the same magnitude at both levels (factor of 3.8 compared with 2.7, respectively). The importance of resorption efficiency in determining leaf-level MRT increases exponentially towards and above the maximum resorption efficiency observed in nature. This effect is independent of leaf life span, which may explain the lack of life-form related differences in resorption efficiency. When scaling up from the leaf to the whole-plant level, fundamental differences in turnover rate among different plant organs must be considered. Woody species invest c. 50% of their net productivity into their low-turnover stems, while in herbaceous species the life span of stems is only slightly longer than that of leaves. As a result, nutrient turnover of woody (evergreen and deciduous) plants is generally lower than that of herbaceous species (herbs and graminoids) on a whole-plant basis. At the intra-specific level empirical data show that both biomass life span (i.e. the inverse of biomass loss rate) and resorption efficiency are important sources of variation in MRT. However, we argue that the relative importance of resorption efficiency in explaining variation in MRT is lower at the inter-specific level, whereas the reverse is true for life span. This is because variation in MRT and life span is much larger at the inter-specific level compared with variation in resorption efficiency. Plant traits related to nutrient conservation are discussed with respect to their implications for leaf structure, plant growth, competition, succession and ecosystem nutrient cycling.

Key words: inter- and intra-specific comparisons, leaf structure, life forms, litter, mean residence time, nitrogen, nutrient use efficiency, plant strategies.

INTRODUCTION

At the beginning of this century, scientists first noted that the distribution of leaf life spans was related to habitat nutrient availability (Harper, 1914, as cited in Monk, 1966). Since then, many studies have been conducted to explain the adaptive significance of leaf longevity. Most such studies have focused on the evergreen habit and factors such as leaf lifetime carbon gain and the importance of resorbed nutrients for new growth (Chapin, 1980; Chabot & Hicks, 1982; Kikuzawa, 1991).

*Author for correspondence (tel (46) 18 471 2880; fax (46) 18 553 419; e-mail Lutz.Eckstein@vaxtbio.uu.se).
†Present address: Department of Short Rotation Forestry, SLU, Box 7016, SE-750 07 Uppsala, Sweden.

Table 1. *Abbreviations, terms and units used in this paper in alphabetical order*

Abbreviation	Meaning	Unit
e	Biomass loss rate, i.e. annual biomass loss/mean annual biomass	kg kg^{-1} $year^{-1}$
MRT	Mean residence time	yr
$[N]$	Plant N concentration	mmol N g^{-1}
$[N]_L$	N concentration of litter = resorption proficiency	mmol N g^{-1}
NUE	Nutrient use efficiency	kg $(mol\ N)^{-1}$
N_A	Annual average N pool of the plant	mol N
R_{EFF}	Resorption efficiency	mol N $(mol\ N)^{-1}$
X_A	Mean annual plant biomass	kg

Since the first attempts were made to define plant nutrient use efficiency (NUE) (Chapin, 1980; Vitousek, 1982; Field & Mooney, 1986) there has been rapid progress in relating leaf habit to nutrient use and plant performance. For example, there have been intensive studies focused on the mechanisms behind NUE as defined by Berendse & Aerts (1987). According to their definition, NUE can be broken down into the nutrient productivity (i.e. dry matter productivity per unit nutrient) and the mean residence time (MRT) of the nutrient in question. It has repeatedly been demonstrated that mean residence times of growth-limiting nutrients are typically long in plants growing in infertile habitats (Aerts & de Caluwe, 1995; Eckstein & Karlsson, 1997; Vásquez de Aldana & Berendse, 1997; Berendse, 1998; Garnier & Aronson, 1998). Apparently, high MRT is achieved at the expense of nutrient productivity; MRT and nutrient productivity have commonly been found to be inversely correlated (Aerts, 1990; Eckstein & Karlsson, 1997; Vásquez de Aldana & Berendse, 1997). Therefore, to increase understanding of how plants adapt to nutrient-poor habitats we will further analyse MRT of nitrogen (N) and the traits behind it, largely following a model outlined by Garnier & Aronson (1998). The MRT relates the annual average N pool of the plant (N_A, abbreviations according to Garnier & Aronson, 1998, for units see Table 1) to the annual amount of N lost:

$$MRT = \frac{N_A}{[N]_L.e.X_A} = \frac{[N]}{[N]_L.e} \qquad \text{Eqn 1}$$

($[N]_L$ is the N concentration of litter, e is the relative biomass loss rate and X_A is the mean annual plant biomass).

The relative biomass loss rate, which, in turn, is inversely related to the mean biomass life span (Garnier & Aronson, 1998), is calculated as the biomass lost over 1 yr divided by the mean annual biomass (E. Garnier, pers. comm.; cf. Frissel, 1981).

The average N pool and N losses from the first part of Eqn 1 can be separated into their constituents, i.e. the ratio of plant N concentration $[N]$ to litter N concentration and the relative biomass loss rate (Eqn

1, right part). Under the assumption that mass resorption during senescence can be neglected, Garnier & Aronson (1998) proposed to modify Eqn 1 as follows:

$$MRT = \frac{1}{1 - R_{EFF}}.\frac{1}{e} \qquad \text{Eqn 2}$$

(R_{EFF} is the N resorption efficiency).

However, as will be illustrated later, this assumption is not always valid. Hence, at the whole-plant level MRT is related to the longevity of leaves and other plant parts and to the N resorption efficiency from these parts before they are shed. The patterns with which these traits vary between species and life-forms as well as within species may help us to understand the mechanisms better through which plants have adapted to infertile habitats.

In this paper we will focus on the patterns of variation in MRT, leaf life span and N resorption: are life span and resorption equally important in explaining variation in MRT? We will attempt to answer this question by using inter-specific comparisons, mainly comparing life-forms, as well as by looking at intra-specific variation. Furthermore, we will examine how life span and resorption are related to leaf structure (e.g. leaf area per unit leaf weight), plant performance (growth, defence, competition) and ecosystem processes (decomposition, succession). We have restricted our review to results concerning perennial plants from seasonal environments, that is, from temperate to Arctic areas.

STEADY-STATE ASSUMPTIONS AND THE CALCULATION OF MEAN RESIDENCE TIME

The estimation of mean residence time of nutrients (just as the nutrient productivity) is based on the assumption of steady-state nutrient content in the plant (i.e. nutrient uptake equals nutrient loss; Frissel, 1981; Garnier & Aronson, 1998). This is mainly because only under steady-state conditions is the calculation of an average nutrient pool meaningful. In this case MRT can simply be estimated by calculating the ratio between the average nutrient pool and the annual nutrient losses (Eqn 1; Berendse

Table 2. *Mean residence time of nitrogen* (MRT), *typical values of leaf life span. N resorption efficiencies* (R_{EFF}) *of leaves and stems, and fine-root life span for species of different life-forms*

| | Life-form | | | |
Plant characteristic	Woody evergreen	Woody deciduous	Graminoid	Herb
MRT, subarctic[1] (yr)	3.5 – 11.3 (3)	2.2 – 3.6 (3)	1.2 – 4.3 (3)	1.3 – 3.6 (5)
MRT, temperate (yr)	1.2 – 1.5 (2)[2]	–	0.7 – 4.1 (10)[3]	–
Leaf life span (yr)	1.3 – 17 (34)[4]*	0.1 – 0.7 (65)[5]	0.4 – 0.8 (5)[6]	0.1 – 0.3 (32)[7]
R_{EFF} leaves[8] (%)	46.7 ± 16.4 (108)	54.0 ± 15.9 (115)	41.4 ± 21.4 (33)	58.5 ± 14.2 (31)
R_{EFF} stem[9] (%)	0?	0?	0 – 48 (3)	7 – 61 (5)
Fine-root life span (yr)	1.1 – 2.2 (5)[10]	0.5 – 1.1 (5)[11]	–	0.6 – 1.0 (5)[12]

Except for the R_{EFF} of leaves, where mean ± SD are given, numbers in the table refer to the range (minimum – maximum) of values observed in the respective references, while the number of species considered is given in parentheses
Sources and comments:
– no data.
[1]Eckstein & Karlsson, 1997; *in situ* above-ground *MRT*, 'growth analysis technique'.
[2]Aerts, 1990; *in situ* whole plant *MRT*, 'growth analysis technique'.
[3]Aerts, 1990 'growth analysis technique'; Aerts & De Caluwe, 1994 'growth analysis technique'; Vásquez de Aldana & Berendse, 1997 'Tracer technique'; whole plant *in situ* (Aerts, 1990) or in response to fertiliser treatment (all other references).
[4]Ewers & Schmid, 1981; Aerts, 1989a; Escudero *et al.*, 1992; Karlsson, 1992; Berendse, 1998; *Pinus (6 spp.) leaf life span: 2–45, Ericaceae/Empetraceae (19 spp.) leaf life span: 1.3–4.0. From the literature Ewers & Schmid (1981) report leaf life spans of between 10 and 40 (45) years for *Pinus longaeva*.
[5]Svoboda, 1977; Kikuzawa, 1983, 1984; Karlsson, 1989; Escudero *et al.*, 1992; Reich *et al.*, 1998a.
[6]Aerts & de Caluwe, 1995; Berendse, 1998.
[7]Diemer *et al.*, 1992; Prock & Körner, 1996; Arctic and alpine species.
[8]Aerts, 1996.
[9]Data from Eckstein & Karlsson, 1997.
[10]Nadelhoffer *et al.*, 1985; Aerts *et al.*, 1989; Aerts *et al.*, 1992a.
[11]Nadelhoffer *et al.*, 1985.
[12]Aerts *et al.*, 1992a, Aerts *et al.*, 1992b; Aerts & De Caluwe, 1994.

& Aerts, 1987; Aerts, 1990; Garnier & Aronson, 1998). However, seedlings, especially of fast growing species or under high fertilizer treatments, may violate the steady-state assumption, since their nutrient pool and biomass are steadily increasing. In this case, a valid estimation of *MRT* can be achieved by applying a stable isotope technique, where the fate of a known amount of tracer (e.g. ^{15}N) is monitored. The *MRT* is then calculated as the ratio between the initial amount of tracer and the amount lost during a certain time interval (cf. Vásquez de Aldana & Berendse, 1997). We have indicated the use of these two different approaches in Table 2 and refer to them as 'growth analysis technique' and 'tracer technique', respectively. It has generally been assumed that adult plants under *in situ* conditions (i.e. without addition of fertilizer) satisfy the steady-state assumption. However, in natural stands of woody plants, only *c.* 30–60% of the annual above-ground production is lost as above-ground litter (Nadelhoffer & Raich, 1992; Ericsson, 1994). A litterfall:productivity ratio in the same order of magnitude (*c.* 0.3) was found for mountain birch seedlings by Weih *et al.* (1998). Furthermore, a steady state with respect to nutrient flux or dry matter is much easier to anticipate in models at the population or ecosystem level (Garnier & Aronson, 1998) where mortality and natality of system units (i.e. individual plants) play a role. At the level of the individual plant, periods of nutrient uptake and growth and periods of nutrient and dry matter losses are separated in time and may be controlled by different environmental parameters (Weih & Karlsson, 1997). Hence, the steady-state concept as outlined above appears to be difficult to apply for single plants. The main drawback of calculating *MRT* under non-steady state using growth analysis techniques appears to be that the estimate obtained is representative only for the study period. However, this should not limit the value of comparative studies dealing with the variation of *MRT* among species or the interrelationships among *MRT* and different plant traits.

MEAN RESIDENCE TIME

The longest residence times of N are found among woody evergreen species (Table 2). Here we define evergreens as plants having a leaf life span of > 1 yr (cf. Bell & Bliss, 1977). In the literature, mean residence times for evergreens vary by a factor of nine across studies (Table 2). The *MRT*s of graminoids, for example, are half (Aerts, 1990) to one third (Eckstein & Karlsson, 1997) the magnitude of woody evergreen *MRT*s, irrespective of the plant fractions studied. Variation across studies may be related to species-specific, latitudinal or site differences. Intra-specific variation in *MRT* between

Table 3. *Intra-specific variation in leaf life span, leaf nitrogen resorption efficiency* (R_{EFF}) *and mean residence time of N* (MRT) *for species representing different life-forms*

	Life-form			
Plant characteristic	Woody evergreen	Woody deciduous	Graminoid	Herb
R_{EFF} leaves	1.1 – 2.0 (3)[1]	1.6 – 2.0 (4)[2]	1.0 – 2.7 (4)[3]	1.2 – 1.4 (4)[4]
Leaf life span	1.0 – 2.2 (7)[5]	1.4 (1)[6]	1.0 – 1.5 (4)[7]	1.0 – 1.6 (6)[8]
MRT, subarctic	1.0 – 3.1 (3)[9]	1.2 – 3.3 (4)[10]	1.2 – 1.4 (3)[9]	1.4 – 2.1 (5)[9]
MRT, temperature	–	–	1.0 – 1.5 (9)[11]	–

Variation is expressed as a factor calculated as the ratio between the highest and the lowest value reported in the specific reference. Values in the table refer to the range (minimum–maximum) of variation observed among species. The number of species considered is given in parentheses

Sources and comments:

– no data.

[1]Jonasson, 1983, 1989; Berendse & Jonasson, 1992; Karlsson, 1994; Eckstein & Karlsson, 1997.

[2]Chapin & Shaver, 1989; Berendse & Jonasson, 1992; Nordell & Karlsson, 1995; Eckstein & Karlsson, 1997; Eckstein *et al.*, 1998.

[3]Chapin *et al.*, 1980; Jonasson & Chapin, 1985; data from Eckstein & Karlsson, 1997.

[4]Data from Eckstein & Karlsson, 1997.

[5]Karlsson, 1992; variation between latitudes.

[6]Kudo, 1991; variation along a snow cover gradient.

[7]Aerts & De Caluwe, 1995; variation in response to fertilization.

[8]Diemer *et al.*, 1992; Kudo, 1996; Prock & Körner, 1996; variation between latitudes.

[9]Eckstein & Karlsson, 1997; *in situ* above-ground MRT, variation between sites.

[10]Eckstein & Karlsson, 1997; Weih *et al.*, 1998; above-ground (Eckstein & Karlsson, 1997) and whole plant (Weih *et al.*, 1998) MRT, variation between sites (Eckstein & Karlsson, 1997) and treatments (Weih *et al.*, 1998).

[11]Aerts & De Caluwe, 1994; Vásquez de Aldana & Berendse, 1997; whole plant MRT, variation between fertilizer treatment.

habitats and provenances and in response to experimental manipulations (Table 3) is much smaller than inter-specific variation (factor 3 compared with factor 16, respectively). Fertilizer experiments revealed no significant intra-specific response of MRT to N addition (Aerts & de Caluwe, 1994; Vásquez de Aldana & Berendse, 1997). At the inter-specific level a clear inverse relationship between MRT and N productivity was found (Eckstein & Karlsson, 1997; Vásquez de Aldana & Berendse, 1997). However, when fertilized, species with the highest productivity did not have the lowest MRT (Vásquez de Aldana & Berendse, 1997). By contrast, in an experiment with four congeneric species, the one with the lowest dry matter productivity had the highest N turnover (i.e. the lowest MRT; Aerts & De Caluwe, 1994). Similarly, Weih *et al.* (1998) did not find an inverse relationship between N productivity and MRT in a study comparing different genotypes of mountain birch. Thus results are inconsistent concerning the productivity–MRT relationship, possibly because comparisons were made on the inter-specific level in some cases and on the intra-specific (or congeneric) level in others.

LEAF LIFE SPAN

The typical leaf life span of evergreens ranges between 1.3 and *c*. 17 yr, with conifers having the highest values (Table 2). In woody deciduous species, graminoids and herbs, the length of the growing season and climatic conditions largely determine leaf life span. In regions with a short winter or drought season, the leaf life span of herbs may be close to 1 yr, whereas Arctic and alpine herbs have a leaf life span of *c*. 40 to 90 d (Table 2; Diemer *et al.*, 1992). Leaf life spans of 50–60 d are found in the Mediterranean climate of southern France (M.-L. Navas & E. Garnier, unpublished). There is a positive relationship between leaf life span and length of the annual growth period in perennial herbs, with mean life span and inter-specific variation decreasing from aseasonal to strongly seasonal environments (Diemer, 1998). The leaf life span of graminoids and deciduous trees tends to be higher than that of herbs though there is some overlap (Table 2; Kikuzawa, 1983, 1984; Diemer *et al.*, 1992).

Leaf life span may also be phylogenetically constrained (Rogers & Clifford, 1993). There are several examples of plant genera, including both evergreen and deciduous species, with correspondingly differing leaf life spans. Nonetheless, long leaf life span can be regarded as a phylogenetically primitive character (Axelrod, 1966). Leaves of advanced taxa, among both the major groups of vascular plants and among subclasses and superorders, tend to be shorter lived than those of the more primitive groups (Rogers & Clifford, 1993). This may explain the dominance of members of certain taxonomic groups, namely coniferous and ericaceous species, in infertile environments.

Compared with the > 200-fold inter-specific variation in leaf life span (0.1 compared with ∼40 yr; Ewers & Schmid, 1981; Table 2), the intra-specific variation is much smaller, being about twofold or less (Table 3). These data illustrate that leaf life span is, to a large extent, a genetically determined, species-specific trait. Nevertheless, in a comparison of European evergreens Karlsson (1992) found that leaf longevity was higher at high latitudes (Sweden compared with Austria) and that there were differences between contrasting sites within latitudes. As could be expected, intra-specific variation is largest in evergreens, which as a group also show the largest inter-specific variation. Field experiments indicate that mean leaf survival of evergreens may decrease in response to fertilization and irrigation (Shaver, 1981; Karlsson, 1985; Aerts, 1989a) but maximum leaf longevity is not affected (Aerts, 1989a). In deciduous species the intra-specific variation in leaf life span also depends on climatic conditions or experimentally caused variation in nutrient availability. Thus, the variability observable under natural conditions is mainly between sites with different snow cover (cf. Kudo, 1991, 1996) or between years. For example, over a 30-yr period the leaf longevity of the mountain birch varied by 36 d in an environment where the leaf life span of this species is *c*. 115 d (P. S. Karlsson & B. H. Bylund, unpublished).

NITROGEN RESORPTION

The resorption of leaf nutrients associated with leaf senescence is a highly ordered, energy-consuming process which leads to the withdrawal of nutrients from the leaves to storage sites (Thomas & Stoddart, 1980) or to new growth. The predominance of dinitrogen amino-acids, such as asparagine and glutamine, in autumn leaves (Thomas & Stoddart, 1980) suggests that the breakdown of proteins involves the synthesis of transport units with a minimum C:N ratio (Chapin & Kedrowski, 1983). Nutrient resorption efficiency (R_{EFF}) may be determined by the distribution of N into mobile and insoluble chemical fractions, leaf chemical composition, carbohydrate flux (source–sink relationships) and climatic conditions (Thomas & Stoddart, 1980; Chapin & Kedrowski, 1983; Chapin & Moilanen, 1991; Nordell & Karlsson, 1995; Aerts, 1996).

In general, plants resorb about half of their maximum leaf N pool before abscission (Table 2; Ericson, 1994). Since evergreens are apparently better adapted to infertile environments, one would expect them to resorb nutrients before abscission more efficiently than deciduous species. However, evergreens are slightly less efficient than deciduous species in this respect (Aerts, 1996; Table 2). The negative effect of this low resorption efficiency is outweighed by the positive effect of increased leaf longevity on mean residence time (Escudero *et al.*, 1992; see also discussion following).

There is no support for the hypothesis that on the intra-specific level, plants show a higher resorption efficiency in nutrient-poor habitats than in nutrient-rich ones (Aerts, 1996). Furthermore, there is no life-form-specific difference in the importance of resorbed nutrients for new growth between evergreen and deciduous woody species (Eckstein *et al.*, 1998).

Because of the lack of any consistent pattern of variation in R_{EFF} among plants from contrasting

Table 4. *Resorption proficiency (mmol N g^{-1}) in relation to site fertility. (a) Inter-specific variation among plants of different life-forms. (b) Intra-specific variation in mountain birch trees*

(a) Inter-specific variation (Eckstein & Karlsson, 1997)

Life-form	Site fertility		*Life-form average*
	Low	High	
Woody evergreens	0.29±0.09 (3)	0.33 (1)	*0.30±0.08 (4)*
Woody deciduous	0.53±0.07 (2)	1.13 (1)	*0.73±0.35 (3)*
Graminoids	0.32 (1)	0.64±0.08 (2)	*0.53±0.19 (3)*
Herbs	0.07 (1)	0.37±0.36 (4)	*0.30±0.34 (5)*
Site average	*0.33±0.17 (7)*	*0.52±0.37 (8)*	

(b) Intra-specific variation in the mountain birch (data from Nordell & Karlsson, 1995)

Site	Fertility	N proficiency
Dry lichen heath	Low	0.40±0.12 (31)
Empetrum–Vaccinium vitis-idaea heath	Low	0.41±0.09 (32)
Meadow birch forest	High	0.54±0.14 (31)

Data in the table refer to the mean±SD (*n*). For a description of sites see Eckstein & Karlsson (1997) and Nordell & Karlsson (1995)

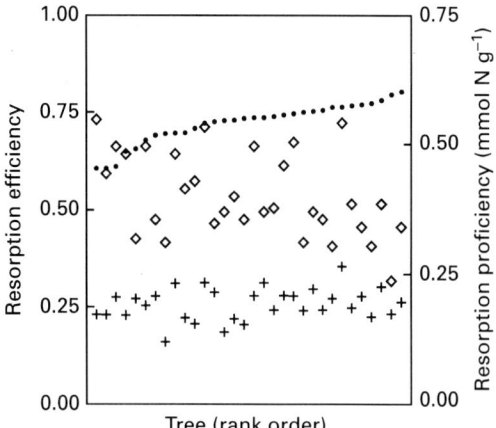

Figure 1. Mean leaf nitrogen (filled circles; mol N [mol N]$^{-1}$) and mass (crosses; g g^{-1}) resorption efficiency of individual adult mountain birch trees during 4 yr (1985–88) and maximum N resorption proficiency (diamonds). The latter refers to the lowest N concentration in senescent leaves measured over a 4 yr period. Trees are rank-ordered according to their 4 yr resorption mean (filled circles). Data from Nordell & Karlsson, 1995.

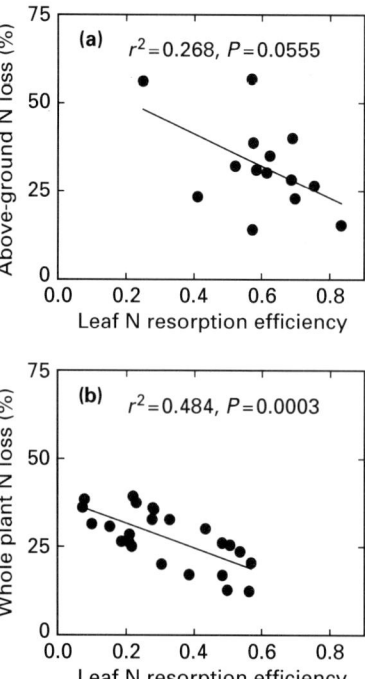

Figure 2. Relationship between plant nitrogen losses by abscised leaves and leaf resorption efficiency (R_{EFF}). (a) Plant N losses expressed as % of plants' total above-ground N before leaf shedding, and calculated for 14 species belonging to four different life-forms (date from Eckstein & Karlsson, 1997). (b) Plant N losses expressed as % of the plants' total above- and below-ground N before leaf shedding, and calculated for individual mountain birch seedlings (data from Weih et al., 1998).

habitats or life-forms, Killingbeck (1996) questioned whether R_{EFF} is the most appropriate measure to quantify resorption and nutrient turnover. As an alternative, Killingbeck (1996) argued for the use of resorption proficiency; the level to which plants can reduce the nutrient contents in senescing leaves before abscission. According to this concept it is the concentration of nutrients in lost tissues rather than the proportion lost that is important for plant nutrient conservation. Thus Eqn 1 provides a biologically more meaningful definition of *MRT* than Eqn 2. It should be noted here that the resorption proficiency concept of Killingbeck (1996) is closely related to the leaf-level definitions of NUE (Vitousek, 1982; Chapin & Shaver, 1989). In terms of nutrient proficiency, woody evergreens show a higher resorption than deciduous species (Killingbeck, 1996; Table 4). This is not surprising since mature evergreen leaves have lower nutrient concentrations than deciduous leaves (Aerts, 1996), but the proportion of nutrients resorbed during senescence does not vary much among life-forms. Though there is an abundance of studies reporting on litter nutrient concentrations, few of them have been designed or analysed with respect to the variation of nutrient proficiency. However, it appears that at the inter-specific level, resorption proficiency varies among life-forms and between sites differing in nutrient availability (Table 4; Craine & Mack, 1998; but see Killingbeck, 1998).

At the intra-specific level, leaf N resorption efficiency varies by a factor of two–three between ramets, years and sites (Table 3; Killingbeck et al., 1990; May & Killingbeck, 1992). In mountain birch differences between individual trees were the most important source of variation in N resorption efficiency, but habitat and climate also affected

resorption (Nordell & Karlsson, 1995). Individuals with a high leaf N content per unit area had a higher N resorption efficiency than individuals with a low leaf N : area ratio (Nordell & Karlsson, 1995). Furthermore, trees with a high mean leaf N resorption tended to have a higher maximum N resorption proficiency (Fig. 1); they reached lower N concentrations in senescent leaves, than individuals with a low mean R_{EFF}. Mountain birch trees from poor sites reduced the N concentration in their senescent leaves to lower levels than trees from a rich site (Table 4). Maximum R_{EFF} observed during the 4 yr of this study did not vary consistently among individuals with low and high mean R_{EFF} (P. S. Karlsson, unpublished). The resorption of N from leaves is an important process for reducing plant N losses, as indicated by the negative relationship between R_{EFF} and the proportion of plant N lost both among and within species (Fig. 2). R_{EFF} may also vary with age; resorption efficiency is consistently lower in mountain birch seedlings than in adult trees (R_{EFF} 0.1–0.6 compared with 0.6–0.8, respectively, Figs 1, 2b).

There are at least two possible explanations for the lack of a consistent pattern in the variation of nutrient resorption efficiency. First, R_{EFF} is not subjected to selection and is unimportant. For the plant's nutrient economy, resorption proficiency is more important than resorption efficiency. Second, the way in which R_{EFF} is usually measured is too crude. Variation between individuals, sites and years as well as the thresholds of complete and incomplete resorption must be considered in order to detect patterns.

RELATIONSHIPS BETWEEN MEAN RESIDENCE TIME, LIFE SPAN AND NITROGEN RESORPTION EFFICIENCY

Theoretical considerations

To evaluate the relative effects of changes in leaf life span and N resorption efficiency (R_{EFF}) on MRT we used Eqn 2 in combination with the typical leaf characteristics (Tables 2, 3). Instead of using the inverse of the relative biomass loss rate (Garnier & Aronson, 1998; Eqn 2), we employed leaf life span in our calculations, because most data in the literature refer to the latter.

First, we kept R_{EFF} constant at a low, intermediate and high value (0.25, 0.50, 0.75, respectively) and varied leaf life span from 0.1 to 10 yr (Fig. 3a). In a second step we kept leaf life span constant at four

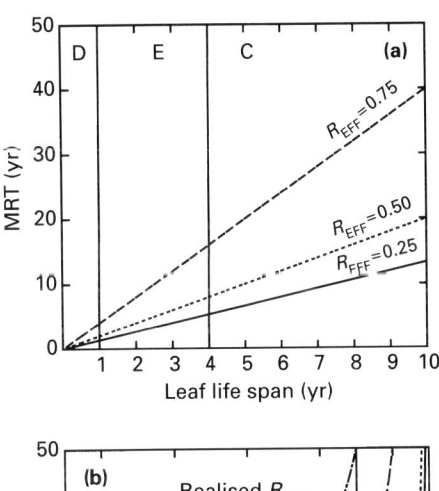

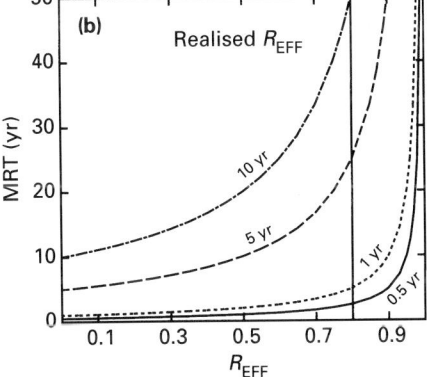

Figure 3. Leaf-level model for the relationship between mean residence time (MRT, yr) and (a) leaf life span for different nitrogen resorption efficiencies (R_{EFF}) and (b) R_{EFF} for different leaf life spans. Vertical lines delimit; (a) the typical leaf life spans (D, deciduous; E, evergreen; C, coniferous species) (cf. Table 2) and (b) the maximum R_{EFF} observed in a wide range of species (*c.* 0.8)(cf. Table 2).

levels representing the range in deciduous and evergreen species (0.5, 1.0, 5.0, 10.0 yr) and varied R_{EFF} from 0 to 0.99 (Fig. 3b) (cf. also Escudero *et al.*, 1992).

The MRT increases linearly with leaf life span, irrespective of R_{EFF} (Fig. 3a). However, the slope of the line increases with increasing R_{EFF}. By contrast, the relative effect of resorption on MRT increases exponentially (or strictly according to the expression $(1-R_{EFF})^{-1}$, cf. Eqn 2) with increasing values of R_{EFF} (Fig. 3a,b). For example, an increase of R_{EFF} from 0.5 to 0.6 raised MRT by 25%, whereas an increase of R_{EFF} from 0.8 to 0.9 doubled MRT (Fig. 3b). As R_{EFF} decreases the same MRT can be achieved through a progressive increase in leaf life span. However, only at resorption efficiencies close to or above the maximum R_{EFF} realized in nature (i.e. *c.* 0.8) did plants with a short leaf life span reach a higher MRT than plants with a high leaf longevity and low R_{EFF}. Thus, the realized R_{EFF} sets a limit for the maximum MRT attainable for a given life span. For example, deciduous species (leaf life span = 1 yr, labelled D in Fig. 3a) may reach a maximum MRT of 5 yr (R_{EFF} = 0.8), whereas ericaceous species (1 yr < leaf life span < 4 yr, E in Fig. 3a) may reach a maximum MRT of 20 yr. Overall, leaf life span appears to have a much greater effect on MRT than resorption efficiency. However, the data in Fig. 3a,b also suggest that improved N resorption is equally beneficial with respect to nutrient conservation for all leaf life spans (i.e. life-forms). This may explain the observed lack of consistent differences between evergreen and deciduous species in the efficiency with which they resorb N from leaves (Aerts, 1996).

Based on the these results and empirical data compiled from the literature (Tables 2, 3), we argue that resorption efficiency is relatively more important in explaining variation in MRT at the intra-specific level than at the inter-specific level. Intra-specific variation in R_{EFF} is of similar magnitude or slightly larger (factor of 1.1–2.7) compared with intra-specific variation in leaf life span (factor of 1.0–2.2; Table 3). For example, consider an intra-specific comparison, where leaf life span may vary by a factor of 1.5 (cf. Table 3). Such a 50% increase in leaf life span leads to a proportionate 50% increase in MRT since the relationship between leaf life span and MRT is linear (cf. Eqn 2, Fig. 3a). Increasing resorption efficiency by a factor of 1.5, from 0.50 to 0.75, would lead to a twofold increase in MRT (Fig. 3a). Thus, the increase in R_{EFF} compensates for a twofold increase in leaf life span, and resorption efficiency appears to have an equal or even stronger effect on MRT compared with leaf life span.

However, among species and life-forms, the variation in leaf life span is much higher (factor of > 200) than the variation in resorption efficiency, which varies by a factor of 1.4 and 3.8 among life-forms and species, respectively (Table 2; Chapin & Kedrowski, 1983; Berendse & Jonasson, 1992). Hence the relative effect of R_{EFF} on MRT is lower when species and life-forms are compared.

Inter-specific patterns

Since the life span of leaves and leaf resorption efficiency are relatively easy to measure the literature is rich with data on these variables. When scaling up to the whole-plant perspective, it is necessary to consider the importance of leaf life span as a determinant of the whole-plant biomass loss rate and biomass life span, as well as the effect of leaf N resorption on whole-plant resorption in different plant life-forms.

There is a fundamental difference between herbaceous (herbs and graminoids) and woody (deciduous and evergreen) species in the allocation to perennial structures: whereas most herbaceous plants die off completely above-ground and store reserves in roots and rhizomes during unfavourable periods (Chapin *et al.*, 1980; Aerts, 1989b), woody species invest a considerable amount of their production in perennial above-ground stems (Shaver & Chapin, 1991). Since these perennial tissues (wood plus bark) have relatively low N concentrations, they consume *c.* 50% of the annual net primary production but only 15% of the annual N uptake (Nadelhoffer *et al.*, 1985). The effect of annual versus perennial stems on plant N turnover rate is illustrated in our comparison of some subarctic plants (Eckstein & Karlsson, 1997). The average above-ground MRT of woody deciduous plants was *c.* 50% higher than that of herbs and graminoids (*c.* 3 compared with 2 yr, cf. Appendix 1 in Eckstein & Karlsson, 1997).

The turnover rate of these different tissues strongly influences the plant biomass loss rate. By definition, there is a large difference in leaf turnover between evergreen and deciduous (woody and herbaceous) species (Table 2). Much less data are available on the life span of roots (Eissenstat & Yanai, 1997). However, the limited data available suggest that root life span is about twofold higher in evergreens than in woody deciduous and graminoid species (Table 2). It should be kept in mind that the present estimates of fine-root turnover and life span may be subject to considerable methodological error (Nadelhoffer *et al.*, 1985; Publicover & Vogt, 1993, Eissenstat & Yanai, 1997), and the available data are strongly biased towards forest ecosystems.

With respect to the turnover rate of woody parts, we expect differences between evergreen and deciduous species to be relatively small. Estimates of relative N losses through above-ground, non-leaf litter (twigs, bark) support this assumption (Nadelhoffer *et al.*, 1985). By contrast, in herbaceous species the life span of supporting stems is probably similar to or only slightly longer than their leaf life span.

Woody and herbaceous species also differ with respect to the resorption efficiency of their organs. Since herbaceous plants die off completely aboveground, nutrients are resorbed from their stems and leaf sheaths (in the case of graminoids) before senescence. Stem resorption efficiency (R_{EFF} stem) lies roughly within the range of leaf resorption efficiencies (Table 2), taking into account that the leaf sheaths of some graminoids may be long lived and thus survive more than one season. There is probably no resorption from woody stems since these tissues serve as storage sites for resorbed nutrients.

To date there is no evidence that nutrients are resorbed from dying roots (Nambiar, 1987; Aerts, 1990; Aerts *et al.*, 1992a).

Intra-specific patterns

To our knowledge, there are only a few species for which information on intra-specific variation in whole-plant nutrient use is available. One such example is the Scandinavian mountain birch (*Betula pubescens* Ehrh. ssp. *czerepanovii* (Orlova) Hämet-Ahti, formerly referred to as *B. p.* Ehrh. ssp. *tortuosa* (Ledeb.) Nyman) (cf. Karlsson & Weih, 1996; Weih, 1998). Therefore, we will use this species to exemplify the intra-specific relationships between mean biomass life span (plant biomass loss rate), N resorption and MRT. Weih *et al.* (1998) presented data on MRT on a whole-plant basis, which was obtained without considering losses due to fine-root turnover. Biomass loss rate was obtained by capturing all senescent leaves from individual seedlings as soon as they were ready to be shed. This loss fraction was then related to the mean annual seedling biomass. At the intra-specific level we assumed that considering below-ground losses would not qualitatively alter the results. However, we estimated the potential effects of root turnover on relative plant biomass loss rate by assuming a root biomass turnover rate of 0.1, 0.5 and 1.0 times the measured leaf biomass turnover rate. These assumptions are justified in mountain birch seedlings in which leaf weight ratio (leaf biomass: plant biomass) is approximately equal to root weight ratio (root biomass: plant biomass), both being *c.* 0.35. Including the assumed root turnover rates in the calculation would increase the plant biomass loss rate (and decrease the MRT) by a factor of 1.1, 1.5 and 1.9, respectively. It should be noted, however, that these birch seedlings were not under steady-state conditions, though they were grown outdoors at natural rates of nutrient availability (Weih, 1998). However, we propose that this may not limit our analysis of the interrelationship between MRT and various plant traits (see the

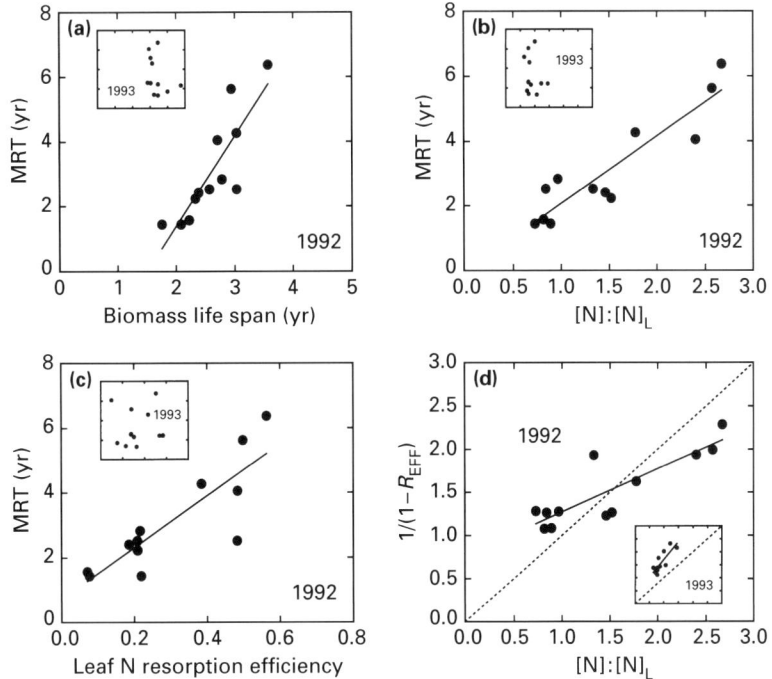

Figure 4. Relationship between whole-plant nitrogen mean residence time (MRT, yr) and (a) mean biomass life span (yr, i.e. the inverse of relative plant biomass loss rate), (b) the ratio between plant N concentration ($[N]$) and litter N concentration ($[N]_L$) and (c) the relative leaf N resorption for mountain birch seedlings in 1992 and 1993 (inset). Linear regression statistics: (a) 1992, $r^2 = 0.715$, $P < 0.001$; 1993, ns; (b) 1992, $r^2 = 0.837$, $P < 0.001$; 1993, ns; (c) 1992, $r^2 = 0.726$, $P < 0.001$; 1993, ns. In (d) the relationship between leaf resorption efficiency and plant N concentration : litter N concentration according to Eqn 1 is shown for 1992 and 1993 (inset). Linear regression statistics: 1993, $r^2 = 0.743$, $P < 0.001$; 1993, $r^2 = 0.664$, $P < 0.01$. Data from Weih *et al.*, 1998.

Steady state assumptions and the calculations of mean residence time section).

Though MRT has mathematically been shown to be related to R_{EFF} (Garnier & Aronson, 1998; Eqn 2), it is conceptually difficult, if not impossible, to estimate the whole-plant R_{EFF}. It appears that R_{EFF} can only be validly defined if individual plant parts (e.g. leaves) are considered. Therefore, here we simply use data on leaf resorption efficiency (Weih *et al.*, 1998). Furthermore, the assumption of insignificant mass losses during senescence underlying the mathematical transformation of Eqn 1 (Garnier & Aronson, 1998) may not always be valid. Indeed, dry matter resorption may be low in small plants (0–10% in woody deciduous species; Eckstein *et al.*, 1998; data from Weih *et al.*, 1998), whereas adult trees may resorb up to 30% of the dry matter from their leaves before abscission (Fig. 1, data from Nordell & Karlsson, 1995).

Whole-plant MRT increased significantly with mean biomass life span ($1/e$) and the ratio of plant N concentration to litter N concentration ($[N] : [N]_L$) (Eqn 1; Fig. 4a,b). There were significant differences in the relationship between MRT and its components between 1992–93 (Fig. 4a–c). These inter-annual differences were probably due to climatic differences between the two study years. In 1992 mountain birch growth was presumably more strongly limited by nutrients than by temperature, whereas this

relation was reversed in the subsequent year (Weih & Karlsson, 1997). This example shows that under *in situ* conditions, the postulated relationships between parameters related to plant nutrient economy (Garnier & Aronson, 1998) may become invalid as soon as the general assumption of nutrient limitation of plant growth is violated. This may happen during certain years, particularly in harsh, high-latitude and/or high-altitude ecosystems where plant growth may be constrained by a number of environmental factors (Chapin & Shaver, 1985).

Garnier & Aronson (1998) proposed that the ratio of plant N : litter N ($[N]:[N]_L$) is directly related to leaf N resorption (compare Eqns 1 and 2). Indeed, a significant correlation between the two traits was found for mountain birch, although the slope of this relationship varied between years, and deviated from the expected slope of 1 (Fig. 4d). This deviation is most probably due to dry matter resorption. Consequently, whole-plant MRT also increased significantly with an increased relative amount of N resorbed (Fig. 4c).

Considered separately, the mean biomass life span ($1/e$) and leaf N resorption (leaf R_{EFF}) explained about equal amounts of the variation in MRT (in 1992). Mean biomass life span only varies by a factor of about two among mountain birch seedlings, whereas leaf R_{EFF} varies by a factor of about seven. The latter may be due to the large variation in leaf

R_{EFF} among individuals (Fig. 1) which, in turn, may lead to a weak relation between resorption figures and plant N turnover. On the other hand, mean biomass life span seems to be directly related to *MRT*. Due to biochemical constraints (Chapin & Kedrowski, 1983; Escudero *et al.*, 1992) N resorption rarely reaches values $> c.$ 0.8. Therefore, the relative amount of biomass lost from the plant is very closely related to the relative amount of N lost.

RELATIONSHIPS BETWEEN LEAF LIFE SPAN, LEAF STRUCTURE AND PLANT TRAITS

Selective forces operative in the adaptation of plants to infertile habitats have important implications not only for leaf life span but also for leaf structure and a number of closely connected plant traits. First, since long-lived leaves may be subjected to extended periods of unfavourable environmental conditions they have to be able to withstand this hostility. Thus long-lived leaves are generally tough and can be characterized by a high fibre:protein ratio (Turner, 1994). Selection for sclerophylly may be reinforced by herbivore pressure (Haukioja *et al.*, 1991; Turner, 1994) and drought (Dunn *et al.*, 1976). Species with long-lived leaves are often characterized by higher concentrations of lignin and other defence substances but lower concentrations of N and P (Coley, 1988). Evidence that the litter decomposition rate depends on the leaf base–cation content and litter lignin : N ratio as well as on leaf N and P concentrations (Aerts, 1997; Cornelissen, 1996; Cornelissen & Thompson, 1997) suggests that there is a link between plant defence and litter decomposability (Grime *et al.*, 1996; Cornelissen *et al.*, 1999). Furthermore, phenolic compounds in pine needle litter have been shown to control litter N release (Northup *et al.*, 1995).

Leaf life span is negatively related to specific leaf area (i.e. leaf area per unit leaf dry matter) (Reich *et al.*, 1992) and positively related to leaf tissue density (Ryser, 1996). In turn, low concentrations of photosynthetic enzymes related to a high tissue density have a direct, negative effect on photosynthetic rate per unit mass (Reich *et al.*, 1992, 1998a,b). Furthermore, species with long-lived leaves were found to have a significantly lower leaf area ratio (i.e. leaf area per unit plant dry matter), and therefore also a lower relative growth rate, compared with species with short-lived leaves (Reich *et al.*, 1992, 1998a,b). Hence, a whole suite of traits related to prolonged leaf life span can be found, particularly in woody evergreen plants (Aerts, 1995; Eckstein & Karlsson, 1997; Reich *et al.*, 1998a,b).

ECOSYSTEM-LEVEL IMPLICATIONS

The plant traits promoting nutrient conservation already outlined, namely low nutrient loss rate, low biomass turnover, high tissue longevity, slow growth,

are also related to plant competition, succession and ecosystem nutrient turnover. As shown by Berendse (1994a), the relative nutrient requirement (i.e. the amount of nutrients lost per unit biomass), is a key trait determining the competitive success of species under different nutrient regimes. In his competition model Berendse (1994a) weighs the ratio of nutrient requirements of two interacting species against the ratio of their abilities to acquire nutrients. At low levels of nutrient availability the species with low nutrient requirements will be competitively superior if this advantage is not compensated for by a greater ability to take up nutrients by the species with higher nutrient requirements. Indeed, competition experiments with *Erica tetralix* and *Molinia caerulea* in Dutch heathland showed that *Erica* was competitively superior to *Molinia* at low levels of nutrient availability (Aerts *et al.*, 1990). Together with nutrient requirement and growth rate, litter decomposability determines the long-term success of species at low and high levels of nutrient supply (Aerts & van der Peijl, 1993; Berendse, 1994b).

There is an increasing body of evidence that dominant plant species enhance soil nutrient processes through feedback effects (Hobbie, 1992; Van Breemen & Finzi, 1998). For example, N mineralization increased from 1 g N m^{-2} yr^{-1} to 13 and 17 g N m^{-2} yr^{-1} during the first 50 yr of succession in two natural chronosequences (Berendse, 1998). Relative biomass loss rate and litter decomposability were those traits explaining the observed increase in mineralization in a modelling approach. Evidence for significantly lower mineralization rate in plots dominated by woody evergreens compared with plots dominated by grasses (Van Vuuren *et al.*, 1992; Berendse, 1998) lends further support to the postulated feedback effect of species on soil processes. These facts emphasize the intimate link between plant traits and ecosystem processes.

In forest canopies, litterfall accounts for 85% of annual above-ground N losses (Chapin & Moilanen, 1991) with falling leaves accounting for 70% of all above-ground litter (O'Neill & DeAngelis, 1981 as cited in Killingbeck, 1996). Hence, the processes regulating annual litter production and litter quality, as already described, are key traits in most, if not all, terrestrial ecosystems (Killingbeck, 1996).

SYNTHESIS

It is apparent that leaf longevity and nutrient turnover through leaf abscission are associated with important aspects of plant and ecosystem performance. However, we have also tried to indicate that in many cases the implications of these can only be understood from a whole-plant perspective. An array of plant characteristics are linked to each other, the most important being mean biomass life span, relative biomass loss rate, productivity/growth rate, nutrient requirements and litter decomposability. The two extreme combinations of these traits are included in the stress-tolerant versus competitive plant strategy (Grime, 1977; Grime *et al.*, 1997), in the stress resistance syndrome (Chapin *et al.*, 1993) and in the *Molinia-* and *Erica*-syndromes (Berendse, 1998).

At the inter-specific level these traits do not vary independently of each other; thus it is not meaningful to attribute plant adaptation to any single one of them (e.g. leaf longevity) in a strict sense. However, it appears evident from the literature that leaf life span (and thus biomass loss rate) is much more variable than resorption efficiency. In evolutionary terms it seems as though there has been strong selection for prolonged life spans and all other physiological traits related to the stress resistance syndrome (Chapin *et al.*, 1993) in response to infertile environments. By contrast, from the lack of patterns in resorption efficiency data, it appears that N resorption is of equal importance for all plants irrespective of life-form or site fertility. The only important exceptions to this general pattern are plants associated with N-fixing symbionts, which resorb little, if any, N before leaf abscission (Côté *et al.*, 1989; Killingbeck, 1993). Nonetheless, resorption proficiency may be a more appropriate measure of the degree to which selection has acted to minimize nutrient loss (Killingbeck, 1996). At present, there is a scarcity of the type of data that would be needed to estimate variation in resorption proficiency among and within species or along gradients of nutrient availability. However, the available data suggest the existence of patterns indicative of differences among life-forms and among plants inhabiting sites differing in nutrient availability (Table 4). This lends support to the hypothesis that there has been strong selection for reduced nutrient loss in infertile environments.

Regarding intra-specific variation in *MRT* and its underlying parameters, our knowledge is more incomplete. At this level empirical data indicate that variation in leaf longevity and variation in resorption efficiency contribute about the same degree to variation in *MRT*. We thus suggest that the relative importance of resorption in explaining variation in *MRT* is higher at the intra-specific level. Resorption proficiency is probably ecologically more important than resorption efficiency. However, there is a strong need for further research particularly regarding intra-specific variation related to the processes of nutrient conservation.

ACKNOWLEDGEMENTS

We are much indebted to E. Garnier for valuable help and comments during the preparation of this paper. Furthermore, we want to thank two anonymous reviewers for their insightful comments that improved the quality of this manuscript. The Abisko Scientific Research Station, the Swedish Royal Academy of Sciences, and the Swedish Natural Science Research Council supported our work reviewed here. D. Tilles corrected the English. All help is gratefully acknowledged.

REFERENCES

Aerts R. 1989a. The effect of increased nutrient availability on leaf turnover and aboveground productivity of two evergreen ericaceous shrubs. *Oecologia* **78**: 115–120.

Aerts R. 1989b. Aboveground biomass and nutrient dynamics of *Calluna vulgaris* and *Molinia caerulea* in a dry heathland. *Oikos* **56**: 31–38.

Aerts R. 1990. Nutrient use efficiency in evergreen and deciduous species from heathlands. *Oecologia* **84**: 391–397.

Aerts R. 1995. The advantages of being evergreen. *Trends in Ecology and Evolution* **10**: 402–407.

Aerts R. 1996. Nutrient resorption from senescing leaves of perennials: are there general patterns? *Journal of Ecology* **84**: 597–608.

Aerts R. 1997. Climate, leaf litter chemistry and leaf litter decomposition in terrestrial ecosystems: a triangular relationship. *Oikos* **79**: 439–449.

Aerts R, Bakker C, de Caluwe H. 1992a. Root turnover as determinant of the cycling of C, N, and P in a dry heathland ecosystem. *Biogeochemistry* **15**: 175–190.

Aerts R, Berendse F, de Caluwe H, Schmitz M. 1990. Competition in heathland along an experimental gradient of nutrient availability. *Oikos* **57**: 310–318.

Aerts R, Berendse F, Klerk NM, Bakker C. 1989. Root production and root turnover in two dominant species of wet heathlands. *Oecologia* **81**: 374–378.

Aerts R, de Caluwe H. 1994. Nitrogen use efficiency of *Carex* species in relation to nitrogen supply. *Ecology* **75**: 2362–2372.

Aerts R, de Caluwe H. 1995. Interspecific and intraspecific differences in shoot and leaf lifespan of four *Carex* species which differ in maximum dry matter production. *Oecologia* **102**: 467–477.

Aerts R, de Caluwe H, Konings H. 1992b. Seasonal allocation of biomass and nitrogen in four *Carex* species from mesotrophic and eutrophic fens as affected by nitrogen supply. *Journal of Ecology* **80**: 653–664.

Aerts R, van der Peijl MJ. 1993. A simple model to explain the dominance of low-productive perennials in nutrient-poor habitats. *Oikos* **66**: 144–147.

Axelrod DI. 1966. Origin of deciduous and evergreen habits in temperate forests. *Evolution* **20**: 1–15.

Bell KL, Bliss LC. 1977. Overwinter phenology of plants in a polar semi-desert. *Arctic* **30**: 118–121.

Berendse F. 1994a. Competition between plant populations at low and high nutrient supplies. *Oikos* **71**: 253–260.

Berendse F. 1994b. Litter decomposability – a neglected component of plant fitness. *Journal of Ecology* **82**: 187–190.

Berendse F. 1998. Effects of dominant plant species on soils during succession in nutrient-poor ecosystems. *Biogeochemistry* **42**: 73–88.

Berendse F, Aerts R. 1987. Nitrogen-use-efficiency: a biological meaningful definition? *Functional Ecology* **1**: 293–296.

Berendse F, Jonasson S. 1992. Nutrient use and nutrient cycling in northern ecosystems. In: Chapin III FS, Jeffries RL, Reynolds JF, Shaver GR, Svoboda J, Chu EW, eds. *Arctic ecosystems in a changing climate.* San Diego, CA, USA: Academic Press, 337–356.

Chabot BF, Hicks DJ. 1982. The ecology of leaf life spans. *Annual Review of Ecology and Systematics* **13**: 229–259.

Chapin III FS. 1980. The mineral nutrition of wild plants. *Annual Review of Ecology and Systematics* **11**: 233–260.

Chapin III FS, Autumn K, Pugnaire F. 1993. Evolution of suites of traits in response to environmental stress. *American Naturalist* **142**: S78–S92.

Chapin III FS, Johnson DA, McKendrick JD. 1980. Seasonal movement of nutrients in plants of differing growth form in an Alaskan tundra ecosystem: implications for herbivory. *Journal of Ecology* **68**: 189–209.

Chapin III FS, Kedrowski RA. 1983. Seasonal changes in nitrogen and phosphorus fractions and autumn retranslocation in evergreen and deciduous taiga trees. *Ecology* **64**: 376–391.

Chapin III FS, Moilanen L. 1991. Nutritional control over nitrogen and phosphorus resorption from Alaskan birch leaves. *Ecology* **72**: 709–715.

Chapin III FS, Shaver GR. 1985. Individualistic growth response of tundra plant species to environmental manipulations in the field. *Ecology* **66**: 564–576.

Chapin III FS, Shaver GR. 1989. Differences in growth and nutrient use among arctic plant growth forms. *Functional Ecology* **3**: 73–80.

Coley PD. 1988. Effects of plant growth rate and leaf lifetime on the amount and type of herbivore defence. *Oecologia* **74**: 531–536.

Cornelissen JHC. 1996. An experimental comparison of leaf decomposition rates in a wide range of temperate plant species and types. *Journal of Ecology* **84**: 573–582.

Cornelissen JHC, Pérez-Harguindeguy N, Díaz S, Grime JP, Marzona B, Cabido M, Vendramini F, Cerabolini B. 1999. Leaf structure and defence control litter decomposition rate across species and life forms in regional floras on two continents. *New Phytologist* **143**: 191–200.

Cornelissen JHC, Thompson K. 1997. Functional leaf attributes predict litter decomposition rate in herbaceous plants. *New Phytologist* **135**: 109–114.

Côté B, Vogel CS, Dawson JO. 1989. Autumnal changes in tissue nitrogen of autumn olive, black alder and eastern cottonwood. *Plant and Soil* **118**: 23–32.

Craine JM, Mack MC. 1998. Nutrients in senesced leaves: comment. *Ecology* **79**: 1818–1820.

Diemer M. 1998. Life span and dynamics of leaves of herbaceous perennials in high-elevation environments: 'news from the elephant's leg'. *Functional Ecology* **12**: 413–425.

Diemer M, Körner C, Prock S. 1992. Leaf life spans in wild perennial herbaceous plants: a survey and attempts at a functional interpretation. *Oecologia* **89**: 10–16.

Dunn EL, Shropshire FM, Song LC, Mooney HA. 1976. The water factor and convergent evolution in Mediterranean-type vegetation. In: Lange OL, Kappen L, Schulze ED, eds. *Water and plant life.* Berlin, Germany: Springer-Verlag, 492–505.

Eckstein RL, Karlsson PS. 1997. Above-ground growth and nutrient use by plants in a subarctic environment: effects of habitat, life-form and species. *Oikos* **79**: 311–324.

Eckstein RL, Karlsson PS, Weih M. 1998. The significance of resorption of leaf resources for shoot growth in evergreen and deciduous woody plants from a subarctic environment. *Oikos* **81**: 569–577.

Eissenstat DM, Yanai RD. 1997. The ecology of root lifespan. *Advances in Ecological Research* **27**: 1–60.

Ericsson T. 1994. Nutrient dynamics and requirements of forest crops. *New Zealand Journal of Forestry Research* **24**: 133–168.

Escudero A, del Arco JM, Sanz IC, Ayala J. 1992. Effects of leaf longevity and retranslocation efficiency on the retention time of nutrients in the leaf biomass of different woody species. *Oecologia* **90**: 80–87.

Ewers FW, Schmid R. 1981. Longevity of needle fascicles of *Pinus longaeva* (Bristlecone Pine) and other North American pines. *Oecologia* **51**: 107–115.

Field C, Mooney HA. 1986. The photosynthesis–nitrogen relationship in wild plants. In: Givnish TJ, ed. *On the economy of plant form and function.* London, UK: Cambridge University Press, 25–55.

Frissel MJ. 1981. The definition of residence times in ecological models. *Ecological Bulletins* **33**: 117–122.

Garnier E, Aronson J. 1998. Nitrogen-use efficiency from leaf to stand level: clarifying the concept. In: Lambers H, Poorter H, van Vuuren MMI, eds. *Inherent variation in plant growth. Physiological mechanisms and ecological consequences.* Leiden, The Netherlands: Backhuys Publishers, 515–538.

Grime JP. 1977. Evidence for the existence of three primary strategies in plants and its relevance to ecological and evolutionary theory. *American Naturalist* **111**: 1169–1194.

Grime JP, Cornelissen JHC, Thompson K, Hodgson JG. 1996. Evidence of a causal connection between anti-herbivory defence and the decomposition rate of leaves. *Oikos* **77**: 489–494.

Grime JP, Thompson K, Hunt R, Hodgson JG, Cornelissen JHC, Rorison IH, Hendry GAF, Ashenden TW, Askew AP, Band SR, Booth RE, Bossard CC, Campbell BD, Cooper JEL, Davison AW, Gupta PL, Hall W, Hand DW, Hannah

MA, Hillier SH, Hodkinson DJ, Jalili A, Liu Z, Mackey JML, Matthews N, Mowforth MA, Neal AM, Reader RJ, Reiling K, Ross-Fraser W, Spencer RE, Sutton F, Tasker DE, Thorpe PC, Whitehouse J. 1997. Integrated screening validates primary axes of specialisation in plants. *Oikos* **79**: 259–281.

Haukioja E, Ruohomaki K, Suomela J, Vuorisala T. 1991. Nutritional quality as a defence against herbivores. *Forest Ecology and Management* **39**: 237–245.

Harper RM. 1914. The 'pocosin' of Pice Co., Ala., and its bearing on certain problems of succession. *Bulletin of the Torrey Botanical Club* **41**: 209–220.

Hobbie SE. 1992. Effects of plant species on nutrient cycling. *Trends in Ecology and Evolution* **7**: 336–339.

Jonasson S. 1983. Nutrient content and dynamics in north Swedish tundra areas. *Holarctic Ecology* **6**: 235–245.

Jonasson S. 1989. Implications of leaf longevity, leaf nutrient re-absorption and translocation for the resource economy of five evergreen plant species. *Oikos* **56**: 121–131.

Jonasson S, Chapin III FS. 1985. Significance of sequential leaf development for nutrient balance of the cotton-sedge, *Eriophorum vaginatum* L. *Oecologia* **67**: 511–518.

Karlsson PS. 1985. Effects of water and mineral nutrient supply on a deciduous and an evergreen dwarf shrub: *Vaccinium uliginosum* L. and *V. vitis-idaea* L. *Holarctic Ecology* **8**: 1–8.

Karlsson PS. 1989. In situ photosynthetic performance of four coexisting dwarf shrubs in relation to light in a subarctic woodland. *Functional Ecology* **3**: 481–487.

Karlsson PS. 1992. Leaf longevity in evergreen shrubs: variation within and among European species. *Oecologia* **91**: 346–349.

Karlsson PS. 1994. The significance of internal nutrient cycling in branches for growth and reproduction of *Rhododendron lapponicum*. *Oikos* **70**: 191–200.

Karlsson PS, Weih M. 1996. Relationships between nitrogen economy and performance in the mountain birch *Betula pubescens* ssp. *tortuosa*. *Ecological Bulletins* **45**: 71–78.

Kikuzawa K. 1983. Leaf survival of woody plants in deciduous broad-leaved forests. 1. Tall trees. *Canadian Journal of Botany* **61**: 2133–2139.

Kikuzawa K. 1984. Leaf survival of woody plants in deciduous broad-leaved forests. 2. Small trees and shrubs. *Canadian Journal of Botany* **62**: 2551–2556.

Kikuzawa K. 1991. A cost-benefit analysis of leaf habit abd leaf longevity of trees and their geographic pattern. *American Naturalist* **138**: 1250–1263.

Killingbeck KT. 1993. Inefficient nitrogen resorption in genets of the actinorhizal nitrogen fixing shrub *Comptonia peregrina*: physiological ineptitude or evolutionary tradeoff? *Oecologia* **94**: 542–549.

Killingbeck KT. 1996. Nutrients in senesced leaves: keys to the search for potential resorption and resorption proficiency. *Ecology* **77**: 1716–1727.

Killingbeck KT. 1998. Nutrients in senesced leaves: reply. *Ecology* **79**: 1820–1821.

Killingbeck KT, May JD, Nyman S. 1990. Foliar senescence in an aspen (*Populus tremuloides*) clone: the response of element resorption to interramet variation and timing of abscission. *Canadian Journal of Forest Research* **20**: 1156–1164.

Kudo G. 1991. Effect of snow-free duration on leaf life-span of four alpine plant species. *Canadian Journal of Botany* **70**: 1684–1688.

Kudo G. 1996. Intraspecific variation of leaf traits in several deciduous species in relation to length of growing season. *Écoscience* **3**: 483–489.

May JD, Killingbeck KT. 1992. Effects of preventing nutrient resorption on plant fitness and foliar nutrient dynamics. *Ecology* **73**: 1868–1878.

Monk CD. 1966. An ecological significance of evergreenness. *Ecology* **47**: 504–505.

Nadelhoffer KJ, Aber JD, Melillo JM. 1985. Fine roots, net primary production, and soil nitrogen availability: a new hypothesis. *Ecology* **66**: 1377–1390.

Nadelhoffer KJ, Raich JW. 1992. Fine root production estimates and belowground carbon allocation in forest ecosystems. *Ecology* **73**: 1139–1147.

Nambiar EKS. 1987. Do nutrients retranslocate from fine roots? *Canadian Journal of Forest Research* **17**: 913–918.

Nordell KO, Karlsson PS. 1995. Resorption of nitrogen and dry matter prior to leaf abscission: variation among individuals, sites and years in the mountain birch. *Functional Ecology* **9**: 326–333.

Northup RR, Yu Z, Dahlgren RA, Vogt KA. 1995. Polyphenol control of nitrogen release from pine litter. *Nature* **377**: 227–229.

O'Neill RV, DeAngelis DL. 1981. Comparative productivity and biomass relations of forest ecosystems. In: Reichle DE, ed. *Dynamic properties of forest ecosystems.* Cambridge, UK: Cambridge University Press, 411–449.

Prock S, Körner C. 1996. A cross-continental comparison of phenology, leaf dynamics and dry matter allocation in arctic and temperate zone herbaceous plants from contrasting altitudes. *Ecological Bulletins* **45**: 93–103.

Publicover DA, Vogt KA. 1993. A comparison of methods estimating fine root production with respect to sources of error. *Canadian Journal of Forest Research* **23**: 1179–1186.

Reich PB, Tjoelker MG, Walters MB, Vanderklein DW, Buschena C. 1998a. Close association of RGR, leaf and root morphology, seed mass and shade tolerance in seedlings of nine boreal tree species grown in high and low light. *Functional Ecology* **12**: 327–338.

Reich PB, Walters MB, Ellsworth DS. 1992. Leaf life-span in relation to leaf, plant, and stand characteristics among diverse ecosystems. *Ecological Monographs* **62**: 365–392.

Reich PB, Walters MB, Tjoelker MG, Vanderklein D, Buschena C. 1998b. Photosynthesis and respiration rates depend on leaf and root morphology and nitrogen concentration in nine boreal tree species differing in relative growth rate. *Functional Ecology* **12**: 395–405.

Rogers RW, Clifford HT. 1993. The taxonomic and evolutionary significance of leaf longevity. *New Phytologist* **123**: 811–821.

Ryser P. 1996. The importance of tissue density for growth and life span of leaves and roots: a comparison of five ecologically contrasting grasses. *Functional Ecology* **10**: 717–723.

Shaver GR. 1981. Mineral nutrition and leaf longevity in an evergreen shrub, *Ledum palustre* ssp. *decumbens. Oecologia* **49**: 362–365.

Shaver GR, Chapin FS III. 1991. Production: biomass relationships and element cycling in contrasting arctic vegetation types. *Ecological Monographs* **61**: 1–31.

Svoboda J. 1977. Ecology and primary production of raised beach communities, Truelove Lowland. In: Bliss LC, ed. *Truelove Lowland, Devon Island, Canada: a high arctic ecosystem.* Edmonton, Canada: The University of Alberta Press, 185–215.

Thomas H, Stoddart JL. 1980. Leaf senescence. *Annual Review of Plant Physiology* **31**: 83–111.

Turner IM. 1994. Sclerophylly: primarily protective? *Functional Ecology* **8**: 669–675.

Van Breemen N, Finzi AC. 1998. Plant–soil interactions: ecological aspects and evolutionary implications. *Biogeochemistry* **42**: 1–19.

Van Vuuren MMI, Aerts R, Berendse F, de Visser W. 1992. Nitrogen mineralization in heathland ecosystems dominated by different plant species. *Biogeochemistry* **16**: 151–166.

Vásquez de Aldana BR, Berendse F. 1997. Nitrogen-use efficiency in six perennial grasses from contrasting habitats. *Functional Ecology* **11**: 619–626.

Vitousek PM. 1982. Nutrient cycling and nutrient use efficiency. *American Naturalist* **119**: 553–572.

Weih M. 1998. *The nitrogen economy of mountain birch as related to environmental conditions and genotype.* PhD thesis, Uppsala University, Sweden.

Weih M, Karlsson PS. 1997. Growth and nitrogen utilization in seedlings of mountain birch (*Betula pubescens* ssp. *tortuosa*) as related to plant nitrogen status and temperature: A two-year study. *Écoscience* **4**: 365–373.

Weih M, Karlsson PS, Skre O. 1998. Intra-specific variation in nitrogen economy among three mountain birch provenances. *Écoscience* **5**: 108–116.

New Phytol. (1999), 143, 191–200

Leaf structure and defence control litter decomposition rate across species and life forms in regional floras on two continents

JOHANNES H. C. CORNELISSEN[1,3]*,
NATALIA PÉREZ-HARGUINDEGUY[2], SANDRA DÍAZ[2],
J. PHILIP GRIME[3], BARBARA MARZANO[4], MARCELO CABIDO[2],
FERNANDA VENDRAMINI[2] AND BRUNO CERABOLINI[5]

[1] Sheffield Centre for Arctic Ecology, Department of Animal and Plant Sciences,
The University, 26 Taptonville Road, Sheffield S10 5BR, UK
[2] Instituto Multidisciplinario de Biología Vegetal, Universidad Nacional de Córdoba,
Casilla de Correo 495, 5000 Córdoba, Argentina
[3] Unit of Comparative Plant Ecology, Department of Animal and Plant Sciences,
The University, Sheffield S10 2TN, UK
[4] V. Asti 16, 14036 Moncalvo AT, Italy
[5] Università degli Studi Milano, III Facoltà di Scienze, Via Ravasi 2, 21100 Varese,
Italy

Received 5 October 1998; accepted 28 January 1999

SUMMARY

There is some evidence that traits of fresh leaves that provide structural or chemical protection ('defence') remain operational in the leaf litter and control interspecific variation in decomposition rate in or on the soil. We tested experimentally whether the negative relationship between foliar defence and litter decomposition rate is fundamental, i.e. whether it is seen consistently across higher plant species and life forms, and whether it is repeated in the floras of geographically and climatically distinct areas separated by an ocean. We employed the published results of two outdoor litter bag experiments, in which we simultaneously compared the relative mass losses ('decomposibility') of leaf litters of a wide range of plant species. One experiment was in Córdoba, Argentina, and included 48 Argentine species typical of the dry, subtropical landscapes along a steep altitudinal gradient. The other was in Sheffield, UK, and hosted 72 British species typical of the temperate–Atlantic landscape there. We linked the two experiments through a similar experiment in Sheffield that hosted litters of subsets of both the Argentine and British species. We also tested fresh leaves of all species from the same areas for tensile strength ('toughness') and relative palatability to generalist herbivorous snails in multi-species 'cafeteria' experiments. Both in Argentina and in Great Britain there were highly significant correlations between leaf palatability ($r = 0.61$; 0.73) or leaf tensile strength ($r = -0.60$; -0.60) and litter mass loss across all species. These relationships could be explained by variation both between and within broad life-form groups. Specific leaf area (area:dry mass) of fresh leaves was consistently correlated only with litter mass loss within British life form groups. We illustrated the possible ecosystem consequences of these relationships by comparing functional traits of British species differing in leaf habit. In comparison with deciduous species, evergreens generally had innately slow growth, which corresponded to their longer-lived leaves of lower specific leaf area, higher tensile strength and lower palatability to generalist invertebrate herbivores. Correspondingly, evergreens produced more resistant leaf litter. Thus, slow-growing evergreens might maintain their position in infertile ecosystems through leaf traits that help them to conserve their nutrients efficiently and to keep nutrient mineralization low, thereby not allowing potentially fast-growing deciduous species to outcompete them.

Key words: decomposition, functional type, herbivory, leaf lifespan, leaf palatability, litter mass loss, SLA, leaf toughness.

*Author for correspondence (tel. +44 114 2220130; fax +44 114 268 2521; e-mail H.Cornelissen@Sheffield.ac.uk).

INTRODUCTION

The turnover of organic matter encompasses some of the key processes of ecosystem functioning. This turnover seems to be under the control of a positive feedback involving plant growth rate and litter decomposition rate (Hobbie, 1992; Reich et al., 1992; Berendse, 1994; Aerts, 1995). In productive ecosystems, plants generally grow faster and their litter is broken down and mineralized more rapidly than in unproductive systems. In the complexity of real ecosystems, however, it has been difficult to disentangle the individual or interactive effects of the soil environment (productivity) from the effects of the component plant taxa or functional types on this feedback (Pastor et al., 1984; Hobbie, 1992). This paper singles out the plant's control over the turnover of organic matter and points towards a critical role for protective leaf traits (against the biotic or abiotic environment) that act against herbivores and allow long leaf lifespans (Coley, 1980; Chabot & Hicks, 1982; Southwood et al., 1986). Such 'defences', which seem to have a strong genetic basis, occupy one side of a fundamental, interspecific trade-off against traits that confer the potential for fast plant growth (Coley et al., 1985; Coley, 1988; Loehle, 1988; Herms & Mattson, 1992; Poorter & Bergkotte, 1992; Chapin et al., 1993; Grime et al., 1997; Cornelissen et al., 1998), the latter itself a major determinant of ecosystem function (Grime, 1979). Another way of approaching the same trade-off (see the Discussion section) is to compare species in terms of emphasis on resource acquisition versus resource conservation (Lambers et al., 1998; Poorter & Garnier, 1999).

The crux of this paper is the contention that leaf traits involved in the interspecific growth–defence trade-off continue to affect the fate of leaves after they have senesced and reached the soil surface as litter. There is some recent interspecific evidence in support of this hypothesis (Cornelissen, 1996; Grime et al., 1996; Cornelissen & Thompson, 1997; Wardle et al., 1998). However, these studies have been limited in terms of the plant types, higher taxa and ecosystems that they represent. Here we present the results of experimental work that tests the hypothesis that growth–defence traits and leaf litter decomposition rate are fundamentally connected in vascular plant species and between broad species groups in terms of life form or higher taxonomy. This was done through standardized multispecies assays on plants from two very distinct regional floras in Argentina and the UK.

METHODS

Study area and plant species

The Argentine species were common representatives of major vegetation types along a steep altitudinal (350–2155 m above sea level) and precipitational (85–996 mm yr^{-1}) gradient near Córdoba in subtropical central–west Argentina (Díaz & Cabido, 1997). The British species (Cornelissen, 1996; Cornelissen & Thompson, 1997) were mostly common features of the semi-natural landscape in the Sheffield area in temperate central England, where the ranges of altitude (70–636 m above sea level) and precipitation (600–1000 mm yr^{-1}) are smaller. The three main groups of species in terms of life form and higher taxonomy (see Table 1) contained 8–29 (mean 19) species each per country in our study. In addition, there were three Argentine bromeliads and four British gymnosperm trees. British herbaceous dicots included three slightly woody procumbent subshrubs (*Empetrum nigrum* L., *Helianthemum nummularium* (L.) Miller and *Thymus polytrichus* A. Kerner ex Borbas) and a perennial scrambler (which partly leans on other plants) that is woody near its base at maturity (*Solanum dulcamara* L.). Results were evaluated only within each continent, partly because of differences in methodology, experimental environment and timing, and partly because one of the main objectives of this paper was to test whether similar patterns could be detected in two distinct floras.

Litter decomposition

Main experiments. Methods for the litter decomposition assay in Sheffield, UK ('UK1') were described in detail by Cornelissen (1996). Its essentials are as follows. Throughout autumn 1993, freshly senesced, undecomposed leaf litter was collected from a typical site for each species, and kept air-dried until processing. For some summer-shedding evergreen species, additional tests revealed similar decomposition rates for litter of the same species collected in summer and in autumn (Cornelissen, 1996). The litters were preweighed after air-drying. For each species, true dry mass was calculated from the water contents of a subsample that was weighed, oven-dried and then reweighed. There were eight replicates (or a minimum of five in smaller leaf collections). After remoistening, all litter bags were incubated simultaneously for 20 winter wk (from 24 Nov. 1993) at 40–50 mm depth in a purpose-built outdoor leaf-mould bed that hosted a variety of litter species and types and a naturally developed decomposer community (details are given in Cornelissen, 1996). After retrieval, drying and reweighing, the percentage (oven-dry) mass loss of each sample in relation to initial mass was used as the decomposition rate.

Results from additional treatments revealed that the ranking of species in decomposibility was little affected by litter bag type (0.3 mm compared with 5 mm mesh), burial duration (8 wk compared with

20 wk) or burial season (winter compared with summer) (Cornelissen, 1996). The results used here are those for a 20 wk burial of 0.3 mm mesh bags in the British winter.

The assay in Córdoba, Argentina ('AR1'), was similar in design, but differed in some details (Pérez-Harguindeguy *et al.*, 1997). Leaf litter of each species was collected from typical sites in the study area during 1997 and stored air-dried until processing. The outdoor decomposition bed was based on local conditions and consisted of a combination of dark earth, fine and coarse peat and a mixture of litters from various collection sites, with earthworms added. Litter bags (0.3 mm mesh) were incubated simultaneously at 100 mm depth in the decomposition bed for 9 wk (from 21 Nov. 1996) during the austral summer. Here, additional treatments also revealed strong correlations between species ranking of percentage litter mass loss at 9 wk compared with 18 wk as well as in 0.3 mm compared with 5 mm mesh bags (Pérez-Harguindeguy *et al.*, 1997).

These decomposition experiments should be seen as outdoor laboratory tests of litter decomposibility; they do not pretend to simulate closely the natural decomposition conditions of the litters involved.

Experiment linking experiments in Great Britain and Argentina. An additional decomposition experiment ('UK2') was performed to link UK1 and AR1, and to assess further the robustness of the species litter decomposibility rankings obtained. Leaf litters of a diverse subset of 16 of the Argentine species were transferred to Sheffield and sealed into 0.3 mm mesh bags, simultaneously with (newly collected) leaf litters from a diverse subset of 14 of the British species. There were eight replicates per species. All samples were processed and incubated as in the UK1 experiment, in leaf-mould of similar quality from the same source in the same decomposition bed, but the incubation period was 15 wk (from 22 Jan. 1998).

Traits of fresh leaves

For the various analyses we always collected fresh, mature, non-senescent sun leaves without significant herbivory symptoms from typical sites for each species. These were stored in clear polythene bags in a refrigerator and used for analysis within a few days.

Leaf palatability. Leaf palatability was quantified by using 'cafeteria' trials in which generalist herbivorous snails were allowed to feed selectively on 100 mm² samples of fresh leaves of a whole range of species that were distributed in random positions on a feeding arena (details of design and replication are given in Grime *et al.*, 1996). In Great Britain we used the snail *Helix aspersa* Mueller. The palatability index for each plant species was calculated as the square root of the mean percentage of area of leaf

samples consumed after 72 h. The palatabilities of the British woody species were assessed in a separate trial, largely similar in design to that used by Grime *et al.* (1996). Mean palatability indices for the woody species (based on the means for six arenas with four replicates per species each) were linked to those in the Grime *et al.* (1996) experiment through linear regression by using marker samples of Whatman grade 540 filter paper and *Ilex aquifolium* used in both trials.

The Argentine assay differed slightly from the British ones, mainly in its snail species (*Otala lactea* (Mueller), which resembles *Helix* in appearance and generalist feeding habits), in replication (10 replicate leaf samples per species with all species distributed at random in a large feeding arena) and duration (48 h).

Leaf tensile strength. Measurements of leaf tensile strength (resistance to tearing) followed Hendry & Grime (1993) in both countries, but it was expressed as force required per unit of width of a leaf sample rather than per cross-sectional area, thus incorporating leaf thickness as a component of tensile strength. Very similar devices were used in Argentina and Great Britain and the two were calibrated against one another. Values were log-transformed for use in analyses.

Specific leaf area. For the British woody species we sampled 15 typical sun leaves per plant in the Sheffield area during the period mid-July to early August 1995 from one to four (average two) plants, each plant from a different site on a different geological substratum. We calculated specific leaf area (SLA) as the quotient of projected area of the fresh leaf and its dry mass. Projected area without the petiole was determined with a Delta-T Area Meter (Burwell, Cambridge, UK) and leaf dry mass after 48 h at 80°C. For six species we collected leaves from the same plant both at the beginning and at the end of the sampling period and, by using linear regression, extrapolated all SLA values to those for the median date, 24 July (J. H. C. Cornelissen & B. Cerabolini, unpublished). For evergreens, we collected 15 leaves from the current year's cohort and 15 from the previous year's cohort; for each plant we took the mean of the two cohort mean SLAs. For the British herbaceous species we used the SLA data previously employed by Cornelissen & Thompson (1997).

The SLA was determined in a similar way in Argentina, except for the following details. A minimum of four mature, fully expanded sun leaves were collected from each of at least six plants per species. Projected leaf area was measured with the combination of a manual image scanner and image analysis software.

Data analysis

Spearman rank correlations (between species) and linear regressions (between life forms) were employed to test relationships between traits of fresh leaves and litter mass loss. Linear regressions were used to transform litter mass loss data from UK1 and AR1 into values matching UK2, in which 16 Argentine and 14 British species were combined.

For comparisons between species groups in terms of leaf habit, we employed a one-tailed Mann–Whitney *U* test for palatability index and one-tailed *t* tests for other traits. All traits were expected to give higher means in deciduous species, except log(tensile strength), which was expected to be lower.

RESULTS

The litter mass losses used for analysis in this paper were published by Cornelissen (1996) and Pérez-Harguindeguy *et al.* (1997), whereas some of the results for the British herbaceous species also featured in Grime *et al.* (1996) and Cornelissen & Thompson (1997).

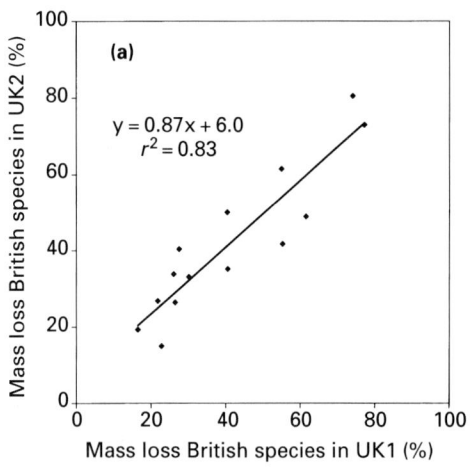

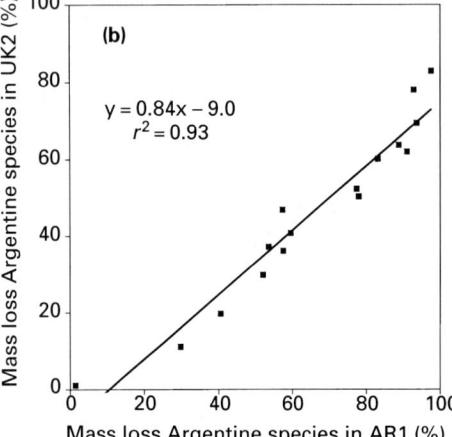

Fig. 1. Percentage leaf litter mass loss of the same species in different experiments: (a) 14 species in UK1 compared with UK2; (b) 16 species in AR1 compared with UK2. Linear regression equations are shown.

The rankings in terms of litter mass loss in UK2 corresponded closely to those in UK1 (Fig. 1a) and AR1 (Fig. 1b). Using the linear regression equations from these relationships (Fig. 1a,b), mass losses of all species in UK1 and AR1 were transformed into UK2 values for further use in this paper (Fig. 2, Tables 1, 2).

Within each of the floras, the palatability of fresh leaves to generalist snails was strongly associated with litter mass loss across all species studied, whereas leaf tensile strength was strongly negatively correlated (Table 1). These relationships had strong foundations at two levels. First, both in Argentina and in Great Britain, the palatability index corresponded to litter mass loss among four broad groups in terms of life form and higher taxonomy (Fig. 2a). Accordingly, leaf tensile strength was negatively associated with litter mass loss among the same life form groups, although the relationship was only marginally significant for the Argentine plants (Fig. 2b). Second, within each of the main life form groups the same relationships were found consistently between species, although correlations were statistically significant only in 7 out of 12 cases (Table 1).

SLA was generally a much poorer correlate of litter mass loss. In the Argentine species no significant correlations appeared, either across all species or among subsets (Table 1, Fig. 2c). There was a weak but significant correlation between all British species (Table 1). This was because graminoid monocots combined high SLA with low litter mass loss (Fig. 2c), whereas correlations between SLA and litter mass loss within each of the life form groups were almost as strong as those for palatability index against litter mass loss (Table 1).

Overall, Argentine leaf litter of a given life form seems to be more decomposible than British litter, whereas fresh leaves seem to be physically tougher and lower in SLA in Argentine than in British plants (Fig. 2). Given the already-described calibration of output for these traits in both countries, these consistent, large differences are likely to be genuine at least to some degree. However, they are beyond the scope of this paper and will not be discussed here.

Within the subset of the British semi-woody and woody plants, deciduous species differed consistently from evergreens in a suite of traits (Table 2). They had the potential for faster growth in a favourable growth-chamber environment, as shown by higher seedling mean relative growth rate (RGR); they had less tough, more palatable leaves of a higher SLA; and their leaf litter was decomposed faster than that of evergreens. These differences were upheld if the gymnosperm trees were excluded from the species set (J. H. C. Cornelissen, unpublished).

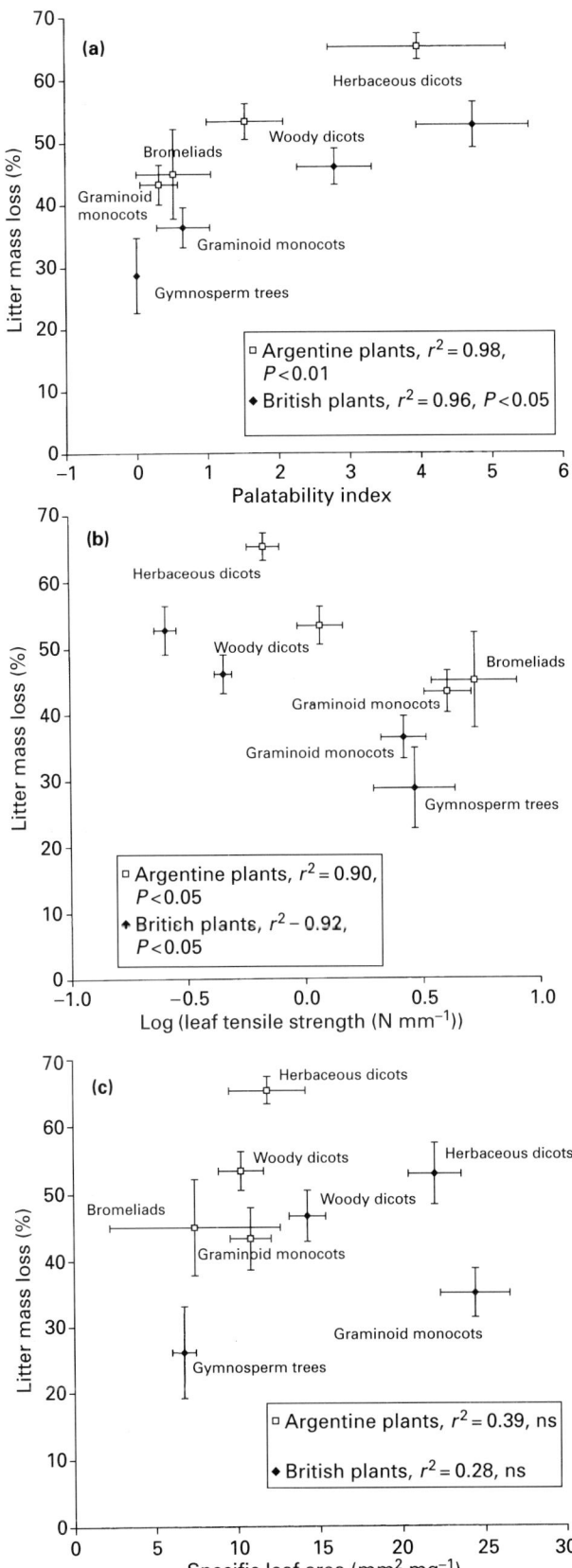

Fig. 2. Percentage leaf litter mass loss of different species assemblages in terms of life form and higher taxonomy as a function of traits of fresh leaves: (a) leaf palatability index; (b) leaf tensile strength; (c) specific leaf area. Linear regressions on the means (SE bars shown) were performed separately for the Argentine and British data sets.

DISCUSSION

Leaf structure and defence as controllers of litter decomposition rate

Methodological considerations. Although biotic and abiotic factors are important determinants of absolute decomposition rates (Swift *et al.*, 1979), it is now well established that litter quality, too, is an important driving force of soil processes (Swift *et al.*, 1979; Cadisch & Giller, 1997). Our UK2 experiment has strongly reinforced previous evidence (Cornelissen, 1996; Pérez-Harguindeguy *et al.*, 1997) that species rankings in terms of litter decomposition rate are robust to methodological and environmental factors. Litters from 16 diverse Argentine species assumed remarkably similar mass loss rankings whether incubated for 8 wk in an Argentine medium exposed to hot and humid weather during the austral summer, or for 15 wk in a British medium during the cool and humid British winter (Fig. 1b). The relatively short incubation periods in our study permitted a simultaneous comparison of the whole range of species. If we had continued the experiments for significantly longer, too large a proportion of the litters would have reached near-maximum mass loss and reduced the resolution at the faster end of the species spectrum. However, the price to pay might have been a relatively poor resolution at the slow end of the species spectrum, where we could only capture the earlier decomposition phase. There is some evidence that rankings of forest species in terms of cumulative mass loss can shift between the early and late phase of decomposition (Berg & Ekbohm, 1991; Berg *et al.*, 1996). Therefore, we should at this stage not extrapolate our results to long-term decomposition in species that produce more persistent litter.

Another word of caution concerns the fact that what we call species traits in this paper are, strictly speaking, the combined expression of both genotypic and environmental factors. Although the leaf traits studied here probably have a strong genetic basis, they will have been modified to some degree by the environment. For instance, low availability of soil nutrient tends to correspond to relatively high concentrations of phenolic compounds (including lignin), some of which would act as defence (Kuiters, 1990; Lambers, 1993; Hartley *et al.*, 1995). Also, a large body of literature has shown that high availability of nutrient to plant roots tends to promote relatively high foliar nutrient concentrations of a given species (see, e.g. Gordon, 1964; Ingestad, 1981) and thus presumably leaf palatability. Nevertheless, because we generally sampled leaves and litters from the species' typical habitats, these are probably more representative of the species than if we had taken them all from plants grown together in an artificial common environment.

Table 1. *Spearman's rank correlation coefficients for relationships between traits of fresh leaves and leaf litter decomposibility within different species assemblages*

Country	Assemblage	Leaf palatability index against percentage leaf litter mass loss			Log(leaf tensile strength) against percentage leaf litter mass loss			Specific leaf area against percentage leaf litter mass loss		
		r	n	P	r	n	P	r	n	P
Argentina	Woody dicotyledons	0.64	20	<0.01	−0.38	20	ns	0.13	20	ns
	Herbaceous dicotyledons	0.50	8	ns	−0.48	8	ns	0.00	8	ns
	Graminoid monocotyledons	0.43	17	ns	−0.49	17	<0.05	0.37	17	ns
	All species	0.61	48	<0.001	−0.60	48	<0.001	0.18	48	ns
UK	Woody dicotyledons	0.59	29	<0.001	−0.51	25	<0.01	0.58	25	<0.01
	Herbaceous dicotyledons	0.68	22	<0.001	−0.33	21	ns	0.44	20	0.06
	Graminoid monocotyledons	0.53	17	<0.05	−0.56	17	<0.05	0.52	17	<0.05
	All species	0.73	72	<0.001	−0.60	67	<0.001	0.37	66	<0.01

Groups with fewer than five species were excluded. *n*, number of species.

Table 2. *Difference between British deciduous and evergreen (semi-)woody species in functional traits*

	Palatability index	Log(tensile strength (N mm⁻¹))	SLA (mm² mg⁻¹)	Mean RGR (d⁻¹)	Leaf litter mass loss (%)
Deciduous species ($n = 23$)(mean ± SE)	3.21 ± 0.63	-0.39 ± 0.04	15.0 ± 1.17	0.111 ± 0.010	48.9 ± 3.9
Evergreen species ($n = 11$)(mean ± SE)	0.12 ± 0.12	-0.021 ± 0.161	9.6 ± 1.19	0.063 ± 0.011	28.0 ± 4.2
P, deciduous against evergreen	<0.001	<0.05	<0.01	<0.01	<0.001

The set of deciduous species included one gymnosperm and the set of evergreens three. Data for seedling mean relative growth rate (RGR), as assessed in a standard assay in a growth chamber environment, are from Cornelissen *et al.* (1996). *n*, number of species.

Our hypothesis tested. Our results have provided strong support for the hypothesis that the link between defences of photosynthetically functional leaves and litter resistance to decomposition is fundamental. Multispecies assays performed on members of broad life-form groups sampled within two contrasting regional floras in Argentina and Great Britain have revealed consistent correspondence between the palatibility of fresh green leaves to generalist herbivorous snails and the rate of leaf litter decomposition. The defences that acted as the most potent inhibitors of decomposition seemed to originate from the physical structure of the living leaf; this could be deduced from the consistent negative trends between leaf tensile strength and rate of litter mass loss in the assays conducted on both continents. Support for this interpretation is available from independent studies (Choong *et al.*, 1992; Wright & Illius, 1995) confirming that high concentrations of lignin and other carbon-rich compounds, particularly when invested in fibres, strengthen living leaves considerably. It is also established that tough litter with a high lignin

concentration or a high lignin : N ratio is resistant to decomposition (see, e.g. Meentemeyer, 1978; Melillo *et al.*, 1982; Taylor *et al.*, 1989; Berg & Ekbohm, 1991; Berg *et al.*, 1993; Gallardo & Merino, 1993; Cornelissen, 1996; Cadisch & Giller, 1997). We demonstrated that, both within and between each of the major groups in terms of life form and higher taxonomy, similar patterns emerged in each continent, variation in both leaf palatability and leaf toughness consistently corresponding (positively and negatively, respectively) to that in litter decomposibility. However, both between and within life forms, decomposibility was more strongly correlated with leaf palatability than with leaf tensile strength. This reinforces existing evidence that, in addition to physical defences, chemical anti-herbivore defences, too, persist in leaf litter to inhibit its decomposition (Satchell & Lowe, 1967; Nicolai, 1988; Kuiters, 1990; Harborne, 1997).

Herbivore species. There is now sufficient evidence to support the view that the relationship between leaf palatability and litter decomposition rate is robust not only to the litter environment but also to the generalist invertebrate herbivore used in palatability trials. Grime *et al.* (1996) showed that it emerged in 43 British, mostly herbaceous species, whether the snail *Helix aspersa* or the cricket *Achaeta domestica* was employed. N. Pérez-Harguindeguy & S. Díaz (unpublished) have found a significant correlation between the palatability of fresh leaves of 53 diverse Argentine species to the grasshopper *Schistocera cancellata* and the litter decomposition rate ($r = 0.36$, $P < 0.01$). The correlation was particularly strong in the 17 graminoid monocots ($r = 0.69$, $P < 0.01$). In addition, Wardle *et al.* (1998) reported consistent positive (although non-significant) trends between leaf palatability to the beetle *Listronotus bonariensis* or the slug *Deroceras reticulatum* and litter decomposition rate in 20 herbaceous dicots in New Zealand. A recurrent pattern emerging from these experiments (Grime *et al.*, 1996; Wardle *et al.*, 1998; N. Pérez-Harguindeguy & S. Díaz, unpublished) seems to be that, in the generalist invertebrate herbivores, Mollusca (e.g. slugs and snails) tend to differentiate 'better' between dicot species in terms of leaf palatability, and insects (e.g. crickets, beetles and grasshoppers) better between monocots. Tentative links have been reported between the digestibility of fresh leaves to large vertebrate herbivores and litter decomposition rate (Scholes & Walker, 1993; Chesson, 1997). Future challenges are to study whether, or to what degree, defences against invertebrate herbivores are similar to those against vertebrate herbivores; how effective these defences are to generalist invertebrate herbivores in comparison with specialist invertebrate herbivores or vertebrate herbivores in real ecosystems; how different types of defence would affect interspecific variation in litter decomposibility; and to what degree species rankings of palatability and decomposibility are modified by stress types and levels in the environments the species grow in (see Herms & Mattson, 1992; Louda & Collinge, 1992).

Specific leaf area. SLA is an important link between plant growth and defence. On one side of this trade-off, SLA tends to be a close correlate and determinant of relative growth rate (see, e.g. Poorter & Remkes, 1990; Reich *et al.*, 1992; Cornelissen *et al.*, 1996; Saverimuttu & Westoby, 1996; Hunt & Cornelissen, 1997). By contrast, SLA tends to be associated negatively with relatively high foliar concentrations of structural (e.g. lignin) or chemical compounds that promote protection against herbivores or the abiotic environment (Choong *et al.*, 1992; Garnier & Laurent, 1994; Van Arendonk & Poorter, 1994; Grime *et al.*, 1997). Indeed, both our Argentine and British data sets revealed strong and significant negative relationships between logarithmic leaf tensile strength and SLA in woody dicots (Spearman's $r = -0.59$; -0.65) and in

herbaceous monocots ($r = -0.80$; -0.65). In contrast, the correlations were low ($r = -0.33$; -0.11) and non-significant in herbaceous dicots. The defence side of the trade-off suggests that SLA should match leaf palatability or leaf toughness in predictive power of litter decomposition rate. Why, then, was this only partly supported by our results? Among life form groups graminoid monocots stood out, particularly in British species, by their combination of tough but high-SLA leaves of low palatability and resistant litter (Fig. 2). These plants seem to be special in that they produce tough leaves with minimal dry matter investments.

Also, SLA was correlated with litter decomposibility within British life form groups but not within Argentine groups. One possible explanation is that some of the Argentine higher-SLA species, for instance *Zizyphus mistol* (Mabberley, 1987), contain specific chemical defences that make their leaves resistant to herbivores and decomposers alike.

Another explanation might be that nutrient resorption from senescing leaves was relatively efficient in some of the high-SLA plants in our Argentine species set, which would have reduced their litter quality. Mass loss during the early decomposition phase tends to be correlated with litter nutrient concentration (see, e.g. Ohlson, 1987; Berg & Ekbohm, 1991; see also Cornelissen & Thompson, 1997). Nutrient resorption is therefore a potentially important variable in the link between traits of living leaves and litter decomposition. Species vary greatly in their nutrient resorption efficiency, but such variation does not seem to correspond consistently to that in plant life form or leaf lifespan (Aerts, 1996; Garnier & Aronson, 1998; Eckstein *et al.*, 1998, 1999). The separate and combined roles of SLA, foliar nutrient concentration and nutrient resorption with respect to interspecific variation in litter decomposibility (Aerts, 1997) are currently under investigation for the regional floras presented here.

A third possible explanation is that small interspecific discrepancies between foliar traits and litter decomposition in this study might be related to the fact that the living leaves were sun leaves whereas the leaf litter presumably consisted of a mixture of sun and shade leaves.

Ecosystem consequences

Ecosystem feedback. Our results have highlighted the importance of plant control over matter cycling in ecosystems, in particular the feedback between plant growth rate, (lack of) defence and litter decomposition rate already mentioned. Interspecific variation in growth–defence traits was evident between plants of deciduous and evergreen leaf habits, the former of which have, on average, shorter leaf lifespans (Kikuzawa, 1991; Reich *et al.*, 1992). Indeed, evergreen species in the British woody flora showed a combination of substantially slower potential seedling growth, lower SLA, lower leaf palatability, greater foliar tensile strength and lower leaf litter decomposibility compared with deciduous species (Table 2). These data quantitatively underpin earlier indications of such differences between evergreen and deciduous species in the temperate zone (Aerts, 1995). Thus, slow-growing evergreens might maintain their position in infertile ecosystems by conserving nutrients and producing poor-quality litter, thus keeping nutrient mineralization low and not allowing potentially fast-growing deciduous species to outcompete them (Pastor *et al.*, 1984; Hobbie, 1992; Reich *et al.*, 1992; Berendse, 1994; Aerts, 1995, 1997; Garnier & Aronson, 1998).

It is not clear to what extent leaf lifespan drives this feedback in the Argentine plants in this study. The same apparent trends between deciduous and evergreen species were much smaller and non-significant in the Argentine woody flora. The palatability index was 1.92 ± 0.77 in 13 deciduous species compared with 0.96 ± 0.64 in six evergreen woody species from Argentina; log(leaf tensile strength) was 0.045 ± 0.139 compared with 0.118 ± 0.137 N m^{-1}; litter mass loss was 52.64 ± 2.91 compared with 47.53 ± 4.35. Likewise, deciduous and evergreen woody species in Mediterranean-type ecosystems of northern latitudes were not consistently different in leaf defence chemistry or litter decomposition rate (Schlesinger & Hasey, 1981; Gallardo & Merino, 1993; Gillon *et al.*, 1994). It seems that, in ecosystems subjected to relatively strong drought stress even in the more favourable season for growth, deciduous plants possess protective leaf traits that resemble those of evergreens.

Changing ecosystems. Plant species feedback on the cycling of matter between plant and soil has consequences for succession. Berendse (1994) found that a higher nutrient use efficiency helped the inherently slow-growing dwarf shrub *Erica tetralix* to outcompete the potentially fast-growing grass *Molinia caerulea* early in the primary succession of infertile Dutch heathlands, when little N had become available through mineralization or external N influx. As succession progressed, *Molinia* gradually outcompeted *Erica*, partly because the former could increase biomass production by monopolizing N made available through the increasing mineralization of its own, more decomposible, leaf litter. By contrast, primary succession on more fertile soils and secondary succession usually see the gradual replacement of faster-growing, poorly defended species by slower-growing, better defended species (see, e.g. Grime, 1979; Coley, 1980). This process can be amplified by preferential herbivory on the more palatable, earlier-successional plants (Pastor *et al.*, 1988; Bryant *et al.* 1991; Brown & Gange, 1992). Thus, secondary succession might be expected to

coincide with slower decomposition rates as far as plant effects are concerned. This was recently supported by ecosystem-level evidence from 50 boreal islands that varied in successional stage owing to increasing lightning frequency with island size (Wardle *et al.*, 1997b). Higher decomposition and N mineralization rates corresponded to larger island size and therefore with earlier succession. This was explained by the decline in leaf and litter quality as later-successional species replaced earlier ones.

This concords with recent work on the effects of the quality of mixed litters on decomposition and N mineralization rates (Wardle *et al.*, 1997a). The functional types to which the litter species belonged, rather than their number, were associated with these properties. Above the ground, too, recent evidence generally points toward a crucial role for functional type composition and functional diversity in ecosystem function (Grime, 1997; Hooper & Vitousek, 1997; Smith *et al.*, 1997; Tilman *et al.*, 1997). Our results might therefore have implications not only for the natural succession of vegetation. Future human-induced changes in land use and climate will probably lead to major changes in the make-up of plant species and functional types in present-day ecosystems (Burrows, 1990; Boyle & Boyle, 1994; Huntley *et al.*, 1997), with profound effects on their functioning (Anderson, 1991; Breymeyer *et al.*, 1996; Chapin *et al.*, 1997; Vitousek *et al.*, 1997). Among such plant-mediated effects we expect to find important, predictable, ecosystem-level changes in plant growth–defence traits as well as in litter decomposition.

ACKNOWLEDGEMENTS

We thank Ingrid Mainland, George Hendry, many staff at IMBIV and the Faculty of Agricultural Studies, Universidad Nacional de Córdoba, and at UCPE, Department of Animal and Plant Sciences, Sheffield University, for their assistance. We acknowledge financial support from the European Union, CONICET (Argentina), the British Council, the Natural Environment Research Council (UK) and Universidad Nacional de Córdoba (Argentina).

REFERENCES

Aerts R. 1995. The advantages of being evergreen. *Tree* **10**: 402–407.
Aerts R. 1996. Nutrient resorption from senescing leaves of perennials: are there general patterns? *Journal of Ecology* **84**: 597–608.
Aerts R. 1997. Nitrogen partitioning between resorption and decomposition pathways: a trade-off between nitrogen use efficiency and litter decomposibility? *Oikos* **80**: 603–606.
Anderson JM. 1991. The effects of climate change on decomposition processes in grassland and coniferous forests. *Ecological Applications* **1**: 326–347.
Berendse F. 1994. Litter decomposability – a neglected component of plant fitness. *Journal of Ecology* **82**: 187–190.
Berg B, Ekbohm G. 1991. Litter mass-loss rates and decomposition patterns in some needle and leaf litter types. Long-term decomposition in a Scots pine forest. VII. *Canadian Journal of Botany* **69**: 1449–1456.

Berg B, Ekbohm G, Johansson MB, McClaugherty C, Rutigliano F, De Santo AV. 1996. Maximum decomposition limits of forest litter types: a synthesis. *Canadian Journal of Botany* 74: 659–672.

Berg B, McClaugherty CA, Johansson MB. 1993. Litter mass-loss rates in late stages of decomposition at some climatically and nutritionally different pine sites. Long-term decomposition in a Scots pine forest. VIII. *Canadian Journal of Botany* 71: 680–692.

Boyle TJB, Boyle CEB. 1994. *Biodiversity, temperate ecosystems, and global change.* (NATO ASI Series 1.) Berlin, Germany: Springer.

Breymeyer AI, Hall DO, Melillo JM, Ågren GI. 1996. Global change: effects on coniferous forests and grasslands. Chichester, UK: John Wiley.

Brown VK, Gange AC. 1992. Secondary plant succession: how is it modified by insect herbivory? *Vegetatio* 101: 3–13.

Bryant JP, Provenza FD, Pastor J, Reichardt PB, Clausen TP, Dutoit JT. 1991. Interactions between woody plants and browsing mammals mediated by secondary metabolites. *Annual Review of Ecology and Systematics* 22: 431–446.

Burrows CJ. 1990. *Processes of vegetation change.* London, UK: Unwin Hyman.

Cadisch G, Giller KE. 1997. *Driven by nature. Plant litter quality and decomposition.* Wallingford, UK: CAB International.

Chabot BF, Hicks DJ. 1982. The ecology of leaf lifespans. *Annual Review of Ecology and Systematics* 13: 229–259.

Chapin III FS, Autumn K, Pugnaire F. 1993. Evolution of suites of traits in response to environmental stress. *American Naturalist* 142 (Suppl.): 78–92.

Chapin III FS, Walker BH, Hobbs RJ, Hooper DU, Lawton JH, Sala OE, Tilman D. 1997. Biotic control over the functioning of ecosystems. *Science* 277: 500–504.

Chesson A. 1997. Plant degradation by ruminants: parallels with litter decomposition in soils. In: Cadisch G, Giller KE, eds. *Driven by nature. Plant litter quality and decomposition.* Wallingford, UK: CAB International, 47–66.

Choong MF, Lucas PW, Ong JSY, Pereira B, Tan HTW, Turner IM. 1992. Leaf fracture toughness and sclerophylly: their correlations and ecological implications. *New Phytologist* 121: 597–610.

Coley PD. 1980. Effects of leaf age and plant life history patterns on herbivory. *Nature* 284: 545–546.

Coley PD. 1988. Effects of plant growth rate and leaf lifetime on the amount and type of anti-herbivore defence. *Oecologia* 74: 531–536.

Coley PD, Bryant JP, Chapin III FS. 1985. Resource availability and plant antiherbivore defense. *Science* 230: 895–899.

Cornelissen JHC. 1996. An experimental comparison of leaf decomposition rates in a wide range of temperate plant species and types. *Journal of Ecology* 84: 573–582.

Cornelissen JHC, Castro-Díez P, Carnelli AL. 1998. Variation in relative growth rate among woody species. In: Lambers H, Poorter H, Van Vuuren MMI, eds. *Inherent variation in plant growth. Physiological mechanisms and ecological consequences.* Leiden, The Netherlands: Backhuys, 363–392.

Cornelissen JHC, Castro-Díez P, Hunt R. 1996. Seedling growth, allocation and leaf attributes in a wide range of woody plant species and types. *Journal of Ecology* 84: 755–765.

Cornelissen JHC, Thompson K. 1997. Functional leaf attributes predict litter decomposition rate in herbaceous plants. *New Phytologist* 135: 109–114.

Díaz S, Cabido M. 1997. Plant functional types and ecosystem function in response to global change: a multiscale approach. *Journal of Vegetation Science* 8: 463–474.

Eckstein RL, Karlsson PS, Weih M. 1998. The significance of resorption of leaf resources for shoot growth in evergreen and deciduous woody plants from a subarctic environment. *Oikos* 81: 567–575.

Eckstein RL, Karlsson PS, Weih M. 1999. Life span and nutrient resorption as determinants of plant nutrient conservation in temperate–arctic regions. *New Phytologist* 143: 177–189.

Gallardo A, Merino J. 1993. Leaf decomposition in 2 mediterranean ecosystems of Southwest Spain – influence of substrate quality. *Ecology* 74: 152–161.

Garnier E, Aronson J. 1998. Nitrogen-use efficiency from leaf to stand level: clarifying the concept. In: Lambers H, Poorter H, Van Vuuren MMI, eds. *Inherent variation in plant growth. Physiological mechanisms and ecological consequences.* Leiden, The Netherlands: Backhuys, 515–538.

Garnier E, Laurent G. 1994. Leaf anatomy, specific mass and water content in congeneric annual and perennial grass species. *New Phytologist* 128: 725–736.

Gillon D, Joffre R, Ibrahima A. 1994. Initial litter properties and decay rate: a microcosm experiment on Mediterranean species. *Canadian Journal of Botany* 72: 946–954.

Gordon AG. 1964. The nutrition and growth of ash, *Fraxinus excelsior*, in natural stands in the English Lake District as related to edaphic site factors. *Journal of Ecology* 52: 169–187.

Grime JP. 1979. *Plant strategies and vegetation processes.* Chichester, UK: John Wiley.

Grime JP. 1997. Biodiversity and ecosystem function: the debate deepens. *Science* 277: 1260–1261.

Grime JP, Cornelissen JHC, Thompson K, Hodgson JG. 1996. Evidence of a causal connection between anti-herbivore defence and the decomposition rate of leaves. *Oikos* 77: 489–494.

Grime JP, Thompson K, Hunt R, Hodgson JG, Cornelissen JHC, Rorison IH, Hendry GAF, Ashenden TW, Askew AP, Band SR, Booth RE, Bossard CC, Campbell BD, Cooper JEL, Davison AW, Gupta PL, Hall W, Hand DW, Hannah MA, Hillier SH, Hodkinson DJ, Jalili A, Liu Z, Mackey JML, Matthews N, Mowforth MA, Neal AM, Reader RJ, Reiling K, Ross-Fraser W, Spencer RE, Sutton F, Tasker DE, Thorpe PC, Whitehouse J. 1997. Integrated screening validates primary axes of specialisation in plants. *Oikos* 79: 259–281.

Harborne JB. 1997. Role of phenolic secondary metabolites in plants and their degradation in nature. In: Cadisch G, Giller KE, eds. *Driven by nature. Plant litter quality and decomposition.* Wallingford, UK: CAB International, 67–74.

Hartley SE, Nelson K, Gorman M. 1995. The effect of fertiliser and shading on plant chemical composition and palatability to Orkney voles, *Microtus arvalis orcadensis*. *Oikos* 72: 79–87.

Hendry GAF, Grime JP. 1993. *Methods in comparative plant ecology.* London, UK: Chapman & Hall.

Herms DA, Mattson WJ. 1992. The dilemma of a plant: to grow or defend? *The Quarterly Review of Biology* 67: 283–335.

Hobbie SE. 1992. Effects of plants on nutrient cycling. *Tree* 7: 336–339.

Hooper DU, Vitousek PM. 1997. The effects of plant composition and diversity on ecosystem processes. *Science* 277: 1302–1305.

Hunt R, Cornelissen JHC. 1997. Components of relative growth rate in 59 British plant species. *New Phytologist* 135: 395–417.

Huntley B, Cramer W, Morgan AV, Prentice HC, Allen JRM. 1997. *Past and future rapid environmental changes: the spatial and evolutionary responses of terrestrial biota.* (NATO ASI Series 1.) Berlin, Germany: Springer.

Ingestad T. 1981. Nutrition and growth of birch and grey alder seedlings in low conductivity solutions and at varied relative rates of nutrient addition. *Physiologia Plantarum* 17: 654–666.

Kikuzawa K. 1991. A cost–benefit analysis of leaf habit and leaf longevity of trees and their geographical pattern. *American Naturalist* 138: 1250–1263.

Kuiters AT. 1990. Role of phenolic substances from decomposing forest litter in plant–soil interactions. *Acta Botanica Neerlandica* 39: 329–348.

Lambers H. 1993. Rising CO_2, secondary plant metabolism, plant–herbivore interactions and litter decomposition – theoretical considerations. *Vegetatio* 104: 263–271.

Lambers H, Chapin III FS, Pons TL. 1998. *Plant physiological ecology.* New York, USA: Springer-Verlag.

Loehle C. 1988. Tree life histories: the role of defenses. *Canadian Journal of Forest Research* 18: 209–222.

Louda SM, Collinge SK. 1992. Plant resistance to insect herbivores: a field hypothesis of the environmental stress hypothesis. *Ecology* 73: 153–169.

Mabberley DW. 1987. *The plant book.* Cambridge, UK: Cambridge University Press.

Meentemeyer V. 1978. Macroclimate and lignin control over litter decomposition rates. *Ecology* 59: 465–472.

Melillo JM, Aber JD, Muratore JF. 1982. Nitrogen and lignin control of hardwood leaf litter decomposition dynamics. *Ecology* 63: 571–584.

Nicolai V. 1988. Phenolic and mineral content of leaves influences decomposition in European forest ecosystems. *Oecologia* **75**: 575–579.

Ohlson M. 1987. Spatial variation in decomposition rate of *Carex rostrata* leaves on a Swedish mire. *Journal of Ecology* **75**: 1191–1197.

Pastor J, Aber JD, McClaugherty CA. 1984. Aboveground production and N and P cycling along a nitrogen mineralization gradient on Blackhawk island, Wisconsin. *Ecology* **65**: 256–268.

Pastor J, Naiman RJ, Dewey B, McInnes P. 1988. Moose, microbes, and the boreal forest. *Bioscience* **38**: 794–800.

Pérez-Harguindeguy N, Díaz S, Cornelissen JHC, Cabido M. 1997. Comparación experimental de la tasa de descomposición foliar de especies vegetales del centro-oeste de Argentina. *Ecología Austral* **7**: 87–94.

Poorter H, Bergkotte M. 1992. Chemical composition of 24 wild species differing in relative growth rate. *Plant, Cell and Environment* **15**: 221–229.

Poorter H, Garnier E. 1999. Ecological significance of inherent variation in relative growth rate and its components. In: Pugnaire FI, Valladares F, eds. *Handbook of functional plant ecology.* New York, USA: Marcel Dekker, 81–120.

Poorter H, Remkes C. 1990. Leaf area ratio and net assimilation rate of 24 wild species differing in relative growth rate. *Oecologia* **83**: 553–559.

Reich PB, Walters MB, Ellsworth DS. 1992. Leaf life-span in relation to leaf, plant and stand characteristics among diverse ecosystems. *Ecological Monographs* **62**: 365–392.

Satchell JE, Lowe DG. 1967. Selection of leaf litter by *Lumbricus terrestris*. In: Satchell JE, Graff O, eds. *Progress in soil biology.* Amsterdam, The Netherlands: North-Holland, 102–119.

Saverimuttu T, Westoby M. 1996. Components of variation in relative growth rate: phylogenetically independent contrasts. *Oecologia* **105**: 281–285.

Schlesinger WH, Hasey MM. 1981. Decomposition of chaparral shrub foliage: losses of organic and inorganic constituents from deciduous and evergreen leaves. *Ecology* **62**: 762–774.

Scholes RJ, Walker BH. 1993. *An African savanna.* Cambridge, UK: Cambridge University Press.

Smith TM, Shugart HH, Woodward FI. 1997. *Plant functional types: their relevance to ecosystem properties and global change.* Cambridge, UK: Cambridge University Press.

Southwood TRE, Brown VK, Reader PM. 1986. Leaf palatability, life expectancy and herbivore damage. *Oecologia* **70**: 544–548.

Swift MJ, Heal OW, Anderson JM. 1979. *Decomposition in terrestrial ecosystems.* Oxford, UK: Blackwell Scientific Publishers.

Taylor BR, Parkinson D, Parsons WJF. 1989. Nitrogen and lignin content as predictors of litter decay rates: a microcosm test. *Ecology* **70**: 97–104.

Tilman D, Knops J, Wedin JD, Reich PB, Ritchie M, Siemann E. 1997. The influence of functional diversity and composition on ecosystem processes. *Science* **277**: 1300–1302.

Van Arendonk JJCM, Poorter H. 1994. The chemical composition and anatomical structure of leaves of grass species differing in relative growth rate. *Plant, Cell and Environment* **17**: 963–970.

Vitousek PM, Mooney HA, Lubchenko J, Melillo JM. 1997. Human domination of earth's ecosystem. *Science* **277**: 494–499.

Wardle DA, Barker GM, Bonner KI, Nicholson KS. 1998. Can comparative approaches based on plant ecophysiological traits predict the nature of biotic interactions and individual plant species effects in ecosystems? *Journal of Ecology* **86**: 405–420.

Wardle DA, Bonner KI, Nicholson KS. 1997a. Biodiversity and plant litter: experimental evidence that does not support the view that enhanced species richness improves ecosystem function. *Oikos* **79**: 247–258.

Wardle DA, Zackrisson O, Hörnberg G, Gallet C. 1997b. The influence of island area on ecosystem properties. *Science* **277**: 1296–1299.

Wright W, Illius AW. 1995. A comparative study of the fracture properties of five grasses. *Functional Ecology* **9**: 269–278.

New Phytol. (1999), **143**, 201-211

Carbon gain in a multispecies canopy: the role of specific leaf area and photosynthetic nitrogen-use efficiency in the tragedy of the commons

FEIKE SCHIEVING AND HENDRIK POORTER*

Department of Plant Ecology and Evolutionary Biology, P.O. Box 800.84, 3508 TB Utrecht, The Netherlands

Received 5 October 1998; accepted 24 March 1999

SUMMARY

Models have been formulated for monospecific stands in which canopy photosynthesis is determined by the vertical distribution of leaf area, nitrogen and light. In such stands, resident plants can maximize canopy photosynthesis by distributing their nitrogen parallel to the light gradient, with high contents per unit leaf area at the top of the vegetation and low contents at the bottom. Using principles from game theory, we expanded these models by introducing a second species into the vegetation, with the same vertical distribution of biomass and nitrogen as the resident plants but with the ability to adjust its specific leaf area (SLA, leaf area:leaf mass). The rule of the game is that invaders replace the resident plants if they have a higher plant carbon gain than those of the resident plants. We showed that such invaders induce major changes in the vegetation. By increasing their SLA, invading plants could increase their light interception as well as their photosynthetic nitrogen-use efficiency (PNUE, the rate of photosynthesis per unit organic nitrogen). By comparison with stands in which canopy photosynthesis is maximized, those invaded by species of high SLA have the following characteristics: (1) the leaf area index is higher; (2) the vertical distribution of nitrogen is skewed less; (3) as a result of the supra-optimal leaf area index and the more uniform distribution of nitrogen, total canopy photosynthesis is lower. Thus, in dense canopies we face a classical tragedy of the commons: plants that have a strategy to maximize canopy carbon gain cannot compete with those that maximize their own carbon gain. However, because of this strategy, individual as well as total canopy carbon gain are eventually lower. We showed that it is an evolutionarily stable strategy to increase SLA up to the point where the PNUE of each leaf is maximized.

Key words: analytical model, canopy, game theory, nitrogen, photosynthetic nitrogen-use efficiency, specific leaf area.

INTRODUCTION

For the past three decades the process of photosynthesis has been studied extensively. This has improved our understanding to such an extent that relatively simple biochemically based models can now adequately describe the processes that occur at leaf level (Farquhar *et al.*, 1980). Similarly, there have been considerable achievements in modelling the carbon gain of the whole canopy. One of the first steps has been to describe the decrease in irradiance with increasing depth of the vegetation by the Lambert–Beer law (Monsi & Saeki, 1953). Field (1983) showed that for maximal carbon gain by a canopy, nitrogen should be distributed so that in each layer of the vegetation the slope of the graph of

*Author for correspondence (fax +31 30 251 8366; e-mail h.poorter@bio.uu.nl).

photosynthesis against nitrogen should be equal. An implication of this condition is that nitrogen distribution within the canopy follows the distribution of light and, consequently, that photosynthetic capacity is lower in lower parts of the vegetation (Field, 1983). Especially in dense stands this has been recurrently found (Werger & Hirose, 1991).

Hirose & Werger (1987) developed a model of the distribution of leaf area, light and nitrogen in the vegetation to evaluate the consequences of different patterns of nitrogen investment for the carbon gain of the canopy. This static model, later refined by Anten *et al.* (1995b), showed that in dense canopies the actual, non-uniform distribution of nitrogen results in a carbon gain 10–30% higher than in a stand with a similar amount of nitrogen, distributed uniformly over the vertical axis (Fig. 1a). However, these models also show that total canopy photo-

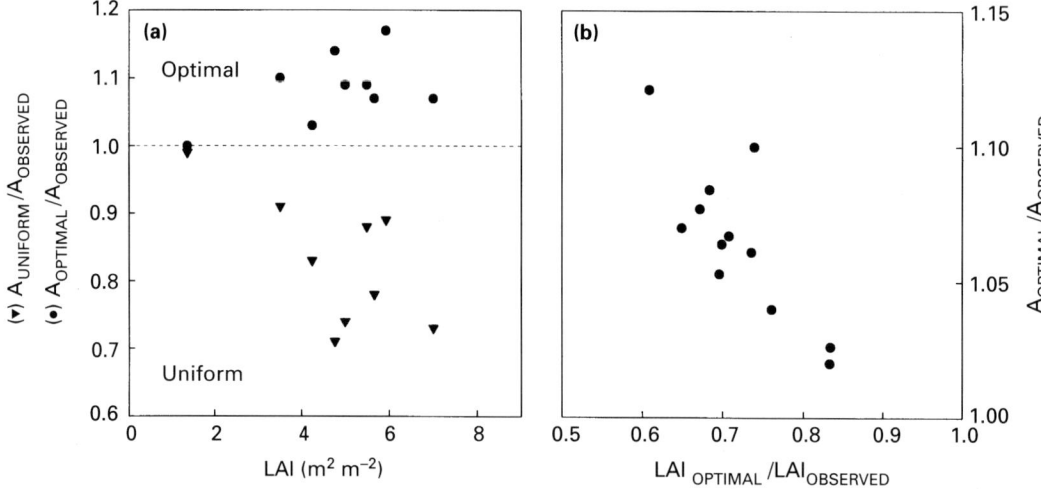

Fig. 1. (a) The rate of photosynthesis of a canopy with a uniform distribution of nitrogen over the vertical axis compared with the actual canopy photosynthesis ($A_{UNIFORM}/A_{OBSERVED}$) (closed triangles) and the values at the optimal nitrogen distribution relative to the observed values ($A_{OPTIMAL}/A_{OBSERVED}$) (closed circles). Data are from herbaceous canopies studied by Hirose & Werger (1987); Pons *et al.* (1989); Schieving *et al.* (1992b); Evans (1993); Anten *et al.* (1995a) and are plotted against the LAI of the stand. (b) Calculated canopy photosynthesis in a stand in which LAI is presumed to be optimized (with respect to canopy photosynthesis) relative to the canopy photosynthesis observed in stands growing naturally ($A_{OPTIMAL}/A_{OBSERVED}$), plotted against the ratio of LAI in an optimized stand compared with the observed LAI. Data are from Schieving *et al.* (1992b) and Anten *et al.* (1995a, 1998).

synthesis could have been even higher than the actual rate for a similar amount of leaf nitrogen per unit ground area. Extra fixation of carbon could be achieved in two ways: (1) by allocating even more nitrogen to the upper layers of the vegetation and less to the lower parts (Fig. 1a), and (2) by decreasing the total leaf area per unit ground area (leaf area index (LAI); Fig. 1b), thus decreasing the respiratory burden imposed by the lower leaves. For each of the two alternatives, the benefits to be gained in herbaceous vegetations are estimated to be 5–10 % (Hirose & Werger, 1987; Pons *et al.*, 1989; Evans, 1993; Anten *et al.*, 1995a). In stands of shrubs or trees the situation might be more complicated, and benefits might be lower (Field, 1983; Leuning *et al.*, 1991).

Why are the nitrogen distribution and LAI actually observed in herbaceous vegetation not optimized for maximal carbon gain? Most papers have studied a monospecific stand, in which all individual plants behave in exactly the same way. However, most natural vegetation consists of a mix of plant species that do not necessarily follow exactly the same strategy (Hikosaka & Hirose, 1997). In this paper we shall analyse the consequences of differing specific leaf area (SLA, leaf area per unit leaf mass) of plants competing for light. For ease of reference we shall define these different plants as being of different species. We shall examine the interaction between the species by using game theory (Maynard Smith, 1982): a model is formulated in which a monostand of a resident species (R), with an SLA that ensures maximal canopy photosynthesis, will be invaded by plants of another species (I). We analyse the

consequences of species I being able to change its SLA at a given vertical distribution of nitrogen and biomass. We shall show that the evolutionarily stable strategy is such that SLA and LAI are higher and nitrogen distribution is less skewed than in a monostand in which the total carbon gain for the stand is maximized. A consequence of these differences is that total canopy carbon gain is lower.

DESCRIPTION OF THE MODEL

The model basically extends the work of Hirose & Werger (1987) and Anten *et al.* (1995a,b). To keep this section brief, we describe the model somewhat loosely; a full and rigorous derivation is given by Schieving (1998). The derivation of some of the equations is given in detail in Appendix 1. Abbreviations and parameters are listed in Table 1.

The assumptions

(1) A canopy is a perfect, homogeneous mixture of two plant species. We shall characterize the total leaf area of the canopy by the variable F_T, the cumulative leaf area of the canopy per m^2 of ground and thus the LAI. As a descriptor of the distribution of total leaf area over the vertical axis, we use F^h, which is the cumulative leaf area above canopy depth h, h being the distance from the top of the canopy.

(2) Leaf angle and absorption coefficient are the same for the two species and constant with vegetation depth. The absorption coefficient is independent of the amount of nitrogen per unit leaf area. For simplicity we consider leaf angle and absorption of both

Table 1. *List of parameters used in the model, and explanation of the abbreviations and units*

Parameter	Definition	Units
a	Leaf absorbance	—
C_{Tj}	Net rate of carbon gain (gross photosynthesis − respiration) per individual of species j, with j = R(esident), I(nvader) or A(verage of all plants)	mol CO_2 s^{-1} per individual
F^h	Cumulative leaf area at canopy depth h	m^2 leaf m^{-2} ground
f^h	Leaf area density at canopy depth h	m^2 leaf m^{-3} space
F_T	Total cumulative leaf area of the canopy	m^2 leaf m^{-2} ground
G_{Tj}	Gross photosynthetic rate per individual of species j	mol CO_2 s^{-1} per individual
g^h	Gross photosynthetic rate per unit leaf area at depth h	mol CO_2 m^{-2} leaf s^{-1}
g^h_{max}	Capacity of the gross photosynthesis-light curve at depth h	mol CO_2 m^{-2} leaf s^{-1}
h	Depth in the canopy, defined as distance from the top	m
i^0	Irradiance in the horizontal plane just above the stand	mol quanta m^{-2} ground s^{-1}
i^h_a	Irradiance absorbed per unit leaf area at canopy depth h	mol quanta m^{-2} leaf s^{-1}
K	Light extinction coefficient	—
m^h_j	Leaf mass density per individual of species j at depth h	g m^{-3}
N_T	Total cumulative leaf nitrogen in the canopy	mol N m^{-2} ground
n^h_m	Leaf nitrogen concentration at depth h	mol N kg^{-1} leaf
n^h_a	Leaf nitrogen content per unit leaf area at depth h	mol N m^{-2} leaf
$n_{a\,min}$	Threshold leaf nitrogen content per unit leaf area for positive g_{max}	mol N m^{-2} leaf
l^h	(Dark) leaf respiration rate per unit leaf mass at depth h	mol CO_2 kg^{-1} leaf s^{-1}
L_{Tj}	(Dark) respiration rate per individual of species j	mol CO_2 s^{-1} per individual
s_g	Slope of the g_{max}-n_a relation	mol CO_2 mol^{-1} N s^{-1}
s_r	Slope of the l–n_a relation	mol CO_2 mol^{-1} N s^{-1}
α	Leaf angle (from horizontal)	degrees
Φ	True quantum yield	mol CO_2 mol^{-1} quanta
Π	(Gross) photosynthetic nitrogen-use efficiency (PNUE)	mol CO_2 mol^{-1} N s^{-1}
θ	Curvature factor of the gross photosynthesis-irradiance curve	—
σ^h	SLA at canopy depth h	m^2 leaf kg^{-1} leaf

species to be equal, but the outcome of the model does not critically depend on this. For a given leaf area distribution and a given irradiance (i) just above the canopy (i^0), we can calculate both the irradiance (i^h) at each depth of the vegetation and, for a given leaf angle and absorbance, the amount of light intercepted by each leaf.

(3) The daily carbon gain of the stand is the difference between the gross assimilation rate (G) and the respiration rate (L) of all the leaves. We consider the area-based assimilation rate for a given leaf of a given species to be related to the amount of nitrogen per leaf area, whereas we assume the mass-based respiration of that leaf to be correlated with the amount of nitrogen per unit leaf mass. Area-based rates and contents can be converted to mass-based values through multiplication by the SLA, σ.

(4) The distribution of leaf mass and organic nitrogen (and consequently the organic nitrogen concentration) can vary throughout the depth of the vegetation, but is the same for the two species. The plants might, however, differ in their SLA distributions. Thus, invading plants play the game by distributing at each depth their leaf mass and nitrogen over a larger or smaller leaf area, thereby changing specific leaf area σ and nitrogen content per unit leaf area.

(5) Water is assumed to be nonlimiting. There are no differences in height between the species. Below-ground processes and interspecific differences in leaf longevity are not taken into account, and neither are acclimation processes at the cellular level due to changes in irradiance. Light intensity above the canopy is constant during the day.

The winner of the game

The winner of the game is the species that forms a monostand which cannot be successfully invaded by any other species that adjusts its SLA. An invasion is considered successful if individuals of invading plants have a greater carbon gain than individuals of the resident plants. The model is not dynamic and does not keep track of changes in the number of individuals of species R and I within or between growing seasons. Rather, the absolute minimum of an equation describing the difference in carbon gain between the I plants and the average plant in the vegetation is sought. In biological terms one could envisage this as the process by which I plants, if successful, completely replace R plants. New invaders then change SLA relative to the new resident plants, and the game is replayed. The end situation is reached if no SLA mutation in any direction and at any height in the vegetation can be found that is more successful within the framework of the model than the mutation analysed penultimately. Plants of this species are considered to have an evolutionarily stable strategy.

The rules

(1) Irradiance within the vegetation decreases with increasing distance from the top, following the Lambert–Beer law. Taking into account the angle of the leaf (α), the intercepted irradiance for leaves at depth h (i_a^h) becomes:

$$i_a^h = i^0 K \exp^{-KF^h}, \quad \text{with} \quad K = a\cos\alpha \qquad \text{Eqn 1}$$

(i^0 is the irradiance in the horizontal plane just above the vegetation, K the light-extinction coefficient and F^h the LAI above canopy depth h for which the irradiance has to be calculated).

(2) Gross leaf photosynthesis is a function of the nitrogen content of the leaf and of the prevailing irradiance. Gross leaf photosynthetic capacity per unit leaf area (g_{max}) is a linear function of the nitrogen content per unit leaf area (n_a), taking into account a threshold value ($n_{a\,min}$):

$$g_{max} = s_g(n_a - n_{a\,min}). \qquad \text{Eqn 2}$$

(3) Whether a leaf achieves the maximum rate of photosynthesis depends on the light climate. A standard non-rectangular hyperbola is used to calculate the actual rate of photosynthesis at each depth h in the vegetation:

$$g^h = \frac{(g_{max}^h + \Phi i_a^h - \sqrt{(g_{max}^h + \Phi i_a^h)^2 - 4\Theta\Phi g_{max}^h i_a^h}}{2\Theta}, \qquad \text{Eqn 3}$$

(θ is the curvature of the non-rectangular hyperbola and Φ the true quantum yield of photosynthesis).

(4) Leaf respiration per unit leaf mass (l) is a linear function of the nitrogen concentration (nitrogen per unit leaf mass):

$$l = s_r n_m. \qquad \text{Eqn 4}$$

No distinction is made between day and night respiration. Rates of photosynthesis and respiration on a leaf mass or area basis can be easily interconverted if the SLA, σ, is known.

(5) Net carbon gain of an individual plant, neglecting respiration of stem and roots, can be written:

$$C_T = G_T - L_T \qquad \text{Eqn 5}$$

with G_T (gross carbon gain) being a function of the leaf area of the plant and the rate of photosynthesis per unit leaf area at each depth in the vegetation. The rate of photosynthesis itself is determined by the nitrogen content (n_a) and the absorbed irradiance (i_a^h) at that depth in the vegetation. Respiration of all leaves of the plant taken together (L_T) is a function of the mass at each height and the nitrogen concentration per unit mass. Thus, total carbon gain per plant is given by the integral:

$$C_T = \int_0^{h_T} dh\, [f^h g(i_a^h, n_a^h) - m^h l(n_m^h)]$$

$$= \int_0^{h_T} dh\, m^h [\sigma^h g(i_a^h, n_a^h) - l(n_m^h)], \qquad \text{Eqn 6}$$

(h_T is the total depth of the vegetation). Because Eqn 6 integrates processes over the depth of the vegetation, the leaf area and mass at depth h are now indicated as leaf area density (f^h; leaf area per unit volume) and leaf mass density (m^h; mass per unit volume), respectively. So far these equations are similar to those given in Anten *et al.* (1995b). Anten *et al.* also give the equations to calculate the optimal distribution of nitrogen and leaf area that leads to a maximal carbon gain per unit ground area.

(6) Having arrived at the 'optimal' distribution of nitrogen and leaf area in a monostand, we then introduce a second species into the model. In the first round of the game, resident plants (R) have the distribution characteristics of the optimal plants. The second species (I) will follow another strategy, by adjusting its SLA. The total carbon gain of an individual of each of the species j, integrated over all layers of the vegetation, is given by:

$$C_{T,j} = \int_0^{h_T} dh\, m^h\, [\sigma_j^h g(i_a^h, n_{a,j}^h) - l(n_m^h)], \qquad \text{Eqn 7}$$

(nitrogen content per unit leaf area is given by $n_{a,j}^h = n_m^h/\sigma_j^h$).

Thus, according to Eqn 7 the difference in performance between plants of species R and I is solely the result of a difference in SLA.

(7) Given the fractions ϕ_R and ϕ_I of plants of species R and I, respectively (with $\phi_R + \phi_I = 1$), the average carbon gain in the stand per plant (C_{TA}) is given by:

$$C_{TA} = \phi_R C_{TR} + \phi_I C_{TI}. \qquad \text{Eqn 8}$$

(8) In its simplest form, a multispecies canopy consists of two species. However, in principle, more species could be involved. Therefore, following the rules of game theory, we evaluate the performance of species I, that is the carbon gain of invader I expressed per individual (C_{TI}), relative to the carbon gain of the average individual in the stand (C_{TA}). In the case that species I has increased SLA in all layers from σ to $(1+\delta).\sigma$, the difference in plant carbon gain with respect to the mean is given by:

$$C_{TI}^\delta - C_{TA}^\delta = C_{TI}^\delta - (\phi_R C_{TR}^\delta + \phi_I C_{TI}^\delta)$$

$$= (1 - \phi_I) C_{TI}^\delta - \phi_R C_{TR}^\delta$$

$$= \phi_R [C_{TI}^\delta - C_{TR}^\delta], \qquad \text{Eqn 9}$$

(C_{Tj}^δ denotes the rate of carbon gain of species j after the small increase in SLA.) Because respiration in our model is related to the nitrogen concentration n_m, which does not differ between residents and invaders, respiration disappears from Eqn 9.

This equation shows what might be expected intuitively: the greatest difference between species I and the mean of the vegetation as a whole (and thus the effect of the δ variation) occurs when the frequency of I plants is low. The difference becomes less pronounced when relatively more I plants are present.

As a matter of convenience we keep δ constant for all layers of the canopy. However, by a calculus-of-variation argument it can be shown that the same results are obtained if δ is any smooth variation function of canopy depth.

(9) For small variations, δ, we can write:

$$C^{\delta}_{TI} - C^{\delta}_{TA} = \frac{d}{d\delta}(C^{\delta}_{TI} - C^{\delta}_{TA}).\delta \qquad \text{Eqn 10}$$

(The derivative $d/d\delta(\ldots)$ is calculated at $\delta = 0$.) The effect of δ on the difference between the carbon gain of the I plants and of the average plant in the vegetation depends on the explicit form of Eqn 10. In Appendix 1 we derive the following equation:

$$\frac{d(C^{\delta}_{TI} - C^{\delta}_{TA})}{d\delta} = \phi_R \int_0^{h_T} dh m^h \sigma^h \left[g(i^h_a, n^h_a) - n^h_a \cdot \frac{\partial g(i^h_a, n^h_a)}{\partial n_a} \right].$$

$$\text{Eqn 11}$$

Thus, for a given change δ, the difference in carbon gain between the I plants and the average plant in the vegetation depends on the fraction of R plants in the stand, the mass at each height, the SLA of species R, and a term in parentheses. As long as the integral is positive, a positive variation δ in the I plants leads to a higher carbon gain per plant for the I-plants than for plants of species R. In the game this could be envisaged as a replacement of R by I plants, I plants becoming the new R plants. This process recurs until any change in δ (positive or negative) causes no further increase in carbon gain. This is the case if the resident plants satisfy at every canopy depth:

$$g - n_a \frac{\partial g}{\partial n_a} = 0 \qquad \text{Eqn 12}$$

In this case, invaders will not be able to replace the resident plants and an evolutionarily stable system is reached.

Although the full model, as described by Schieving (1998), focuses on the effect of enlarging the leaf area at a given leaf biomass, SLA is not used explicitly in the model. State variables are total leaf area and total nitrogen only. To enable the visualization of the changes in SLA, it is necessary to specify the concentration of nitrogen. As a matter of convenience we assumed that the nitrogen concentration of the leaves was 2.5 mmol g^{-1} in all layers of the vegetation. The model is robust in the sense that somewhat different assumptions about the distribution of leaf area, mass and nitrogen do not affect its outcome. For example, if we assume that for each species SLA is equal for all leaves along the axis, and that invading plants can play the game by altering both SLA and leaf nitrogen distribution, we also arrive at Eqn 11.

Table 2. *List of the values of the parameters used in the graphical examples*

Parameter	Value
a	0.8
i^0	1000 µmol quanta m^{-2} s^{-1}
$n_{a\,min}$	40 mmol N m^{-2} leaf
s_g	0.33 mmol CO$_2$ mol^{-1} N s^{-1}
α	0°
Φ	0.063 mol CO$_2$ mol^{-1} quanta
θ	0.8

RESULTS

An optimal monostand

For the convenience of the reader we show the results of the analysis graphically for a given set of parameters listed in Table 2. We started our analysis with a stand consisting only of species R. Following Anten *et al.* (1995b), we calculated, for a given amount of leaf nitrogen per ground area ($N_T = 400$ mmol m^{-2}), the optimal nitrogen distribution for maximal canopy photosynthesis. Thereafter we determined the leaf area index, F_T^*, for which carbon gain of the canopy is maximized. The distribution of nitrogen, expressed as nitrogen content per unit leaf area, is shown by the broken line in Fig. 2a. Usually, such a distribution is shown as a function of cumulative leaf area. However, in the game, the total leaf area alters. Therefore, we choose to plot the relevant variables as a function of the cumulative amount of nitrogen in the vegetation, using equations given by Schieving (1998). The distribution of nitrogen follows the distribution of light in the canopy (Fig. 2b), whereas SLA follows the opposite pattern (Fig. 3a). As expected, the profile of gross photosynthesis follows that of light and nitrogen (Fig. 3b), with low values at the bottom of the canopy.

If we consider the optimal LAI for a range of leaf nitrogen content per ground area, we see that it first increases linearly with N_T, with a slight hint of saturation thereafter (Fig. 4a). Above an N_T of 500 mmol m^{-2}, larger amounts of nitrogen in the canopy do not enhance gross canopy photosynthesis but increase respiration, thereby decreasing net carbon gain (Fig. 4b).

Invasion of the monostand by species I

Let us now assume a stand in which the leaf area game is played. The game starts with the introduction of an infinitesimally small number of plants of species I in the stand of species R. These invading plants have the same distribution of biomass and nitrogen as the R plants after they have achieved the optimal carbon gain. They are well mixed in the vegetation and because they are so few they do not affect the light climate in the stand. In the game

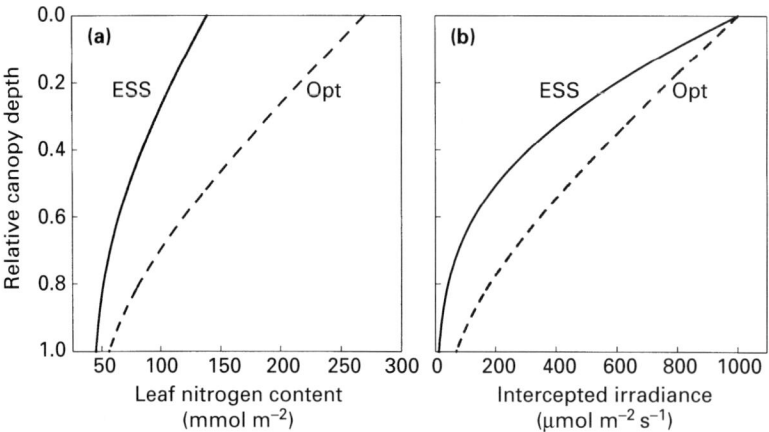

Fig. 2. (a) Leaf nitrogen content per unit area, and (b) intercepted irradiance plotted as a function of relative canopy depth. Optimal stand (Opt, broken line) with a distribution of nitrogen that maximizes canopy photosynthesis; stand in an evolutionarily stable situation (ESS, solid line) at the end of the game. Relative canopy depth is calculated as the cumulative amount of leaf nitrogen in the vegetation above a given point divided by total leaf nitrogen in the vegetation, following Schieving (1998). These graphs show a numerical example for a vegetation with a total leaf nitrogen of 400 mmol m^{-2} ground area. Parameter values used in these simulations are given in Table 2.

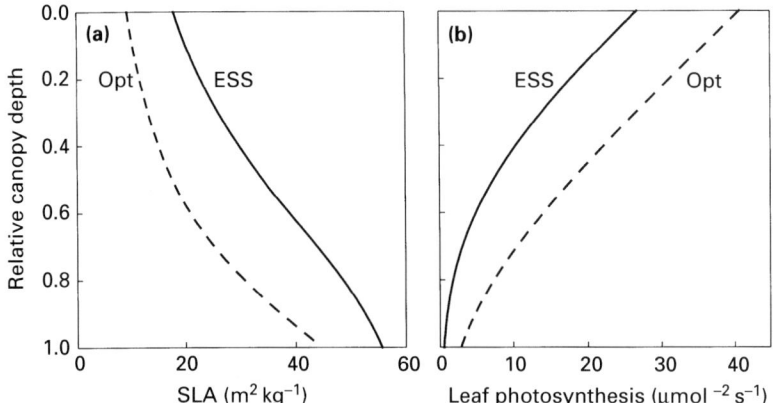

Fig. 3. (a) Specific leaf area (SLA) and (b) rate of gross photosynthesis plotted as a function of relative canopy depth in leaves of an optimal stand (Opt, broken line) and a stand in an evolutionarily stable situation at the end of the game (ESS, solid line). Relative canopy depth is calculated as the cumulative amount of leaf nitrogen in the vegetation above a given point divided by total leaf nitrogen in the vegetation, following Schieving (1998). These graphs show a numerical example for a vegetation with a total leaf nitrogen of 400 mmol m^{-2} ground area. Parameter values used in these simulations are given in Table 2.

theory model these plants have the potential to alter their SLA, and as a first option we assume that they increase SLA at every canopy depth by 10%. As a consequence, nitrogen is diluted over a larger leaf area and the area-based rates of photosynthesis decrease. However, in I plants this is more than compensated for by a greater leaf area per plant. Consequently, in the example of a stand with an N_T of 400 mmol m^{-2}, total carbon gain of individual I plants is 5% higher than that of resident plants. Therefore, the result of the first round is that the invaders perform better in terms of total plant carbon gain and will replace the residents.

The evolutionarily stable end situation

After the first invasion, invading plants have repeatedly outcompeted the R plants and have become the new residents, keeping their own specific SLA distri-

bution. This process continues until invaders cannot find any SLA change at any height in the vegetation to make them perform better than the residents. In this stable end situation, SLA has increased to higher values in all layers of the vegetation (Fig. 3a). By comparison with that of the initial stand, irradiance at the bottom of the canopy and the nitrogen content per unit leaf area are lower, the latter approaching the minimum value of 40 mmol N m^{-2}. Consequently, the rate of leaf photosynthesis has decreased substantially in all layers of the vegetation (Fig. 3b). Contrary to the initial phase, the increase in LAI in the end phase of the game (Fig. 4b) leads, for almost all values of N_T, to a lower rate of canopy photosynthesis than in the optimal stand. Only at very low contents of nitrogen per unit ground area in the vegetation does the leaf area game not have an effect. In such a vegetation the LAI is very low (Fig. 4a) and there is no competition for light. Thus,

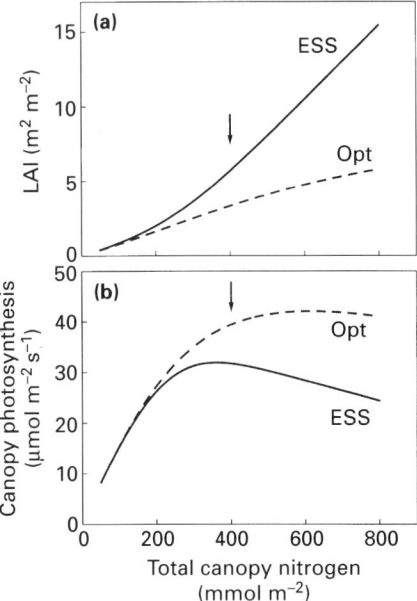

Fig. 4. (a) LAI and (b) canopy photosynthesis for an 'optimal' stand (a monostand in which canopy photosynthesis is maximized (Opt, broken line)) and a stand in an evolutionarily stable situation at the end of the game (ESS, solid line), for a range of leaf nitrogen contents per unit ground area. Arrows indicate the results for an N_T of 400 mmol m^{-2}, used in the simulation presented in Figs 2, 3 and 6.

with the exception of very open stands, the evolutionarily stable end situation has led to an LAI higher, but to canopy photosynthesis lower than in the optimum stand. As density was not changed, this implies that carbon gain of each individual is also lower than in the situation before the game started.

DISCUSSION

Specific leaf area and leaf area index

Specific leaf area is a parameter that is, apart from environmental influences like irradiance, under strong genetic control and differentiates between functional groups of species. Thus, species from infertile habitats, especially evergreens, have an inherently low SLA, whereas species from fertile habitats have high SLAs (Monk, 1966; Poorter & De Jong, 1999). Low SLA in plants growing in nutrient-poor habitats was possibly caused by selection pressure. In such environments it might be a successful strategy to increase the time nutrients are resident in the plant (Berendse & Aerts, 1987; Aerts & Chapin, 1999). This can be achieved in part by increasing the life span of the leaves, which is correlated negatively with SLA (Reich, 1998). Mechanistically, this linkage could be explained by the accumulation of lignin and other secondary compounds in the leaves, which decrease the attractiveness to herbivores, as well as by investment in compounds and structures that facilitate survival

during dry, cold or otherwise unfavourable environmental conditions (Poorter & Garnier, 1999).

Possible evolutionary pressures leading to high SLA have received less attention in the ecological literature. Agricultural models show that before canopy closure there is an advantage in having a high SLA, both in terms of final yield in a monostand (Gutschick, 1988) and in competitive ability in a crop/weed situation (Spitters & Aerts, 1983). However, what happens after canopy closure? Obviously, in fast-developing vegetation competition for light is partly determined by the ability to position leaves high in the vegetation (Barnes *et al.*, 1990). Apart from the insertion point at which a leaf is developing, leaf angle (Hikosaka & Hirose, 1997) and amount of biomass invested in a given leaf, it is possible for the plant to modulate its SLA. Outcomes of the model presented here show that such an evolutionary pressure is possible. All else being equal, a monostand of plants that have distributed leaf nitrogen and area such that total carbon gain is maximized is susceptible to invasion by a species that can increase its SLA. The advantage of the invading species is not due directly to the fact that shade is cast on leaves lower in the vegetation, because this affects leaves of both the resident and invading species equally. Rather, it is the increased light interception resulting from enlargement of their leaves at each layer of the canopy that increases the carbon gain of the invaders (Gutschick & Wiegel, 1988). Moreover, by adjusting SLA they optimize their nitrogen economy, as will be discussed in the next section.

What is the upper limit of SLA? In the model, we constrained SLA indirectly, by setting a minimum nitrogen content per unit leaf area ($n_{a\,min}$). For a given nitrogen content per unit mass, SLA could at most increase until the nitrogen content per unit leaf area equalled $n_{a\,min}$. This seems justified from the perspective that a minimum amount of nitrogen (in the form of DNA, RNA, non-photosynthetic enzymes and other proteins) must be present before a leaf can start to fix carbon. A positive value for $n_{a\,min}$ is supported by a range of observations (Hirose & Werger, 1987; Field, 1988; Pons *et al.*, 1989, 1993) but not all (Evans, 1988). From a biomechanical perspective also, there must be an upper limit of SLA, to prevent a leaf from becoming too frail. At low irradiance SLA can occasionally exceed 80 m^2 kg^{-1} (Corré, 1983). However, these values seem to be extreme, found in experiments in glasshouses or growth rooms. Values for plants in nature are generally lower, at least partly because of more turbulent wind conditions, which decrease SLA (Woodward, 1983).

The effect of increased SLA in a given stand biomass is that LAI increases beyond the optimal value for canopy photosynthesis. This is shown for the numerical example (Fig. 3b), but can also be proved within the analytical framework of the model

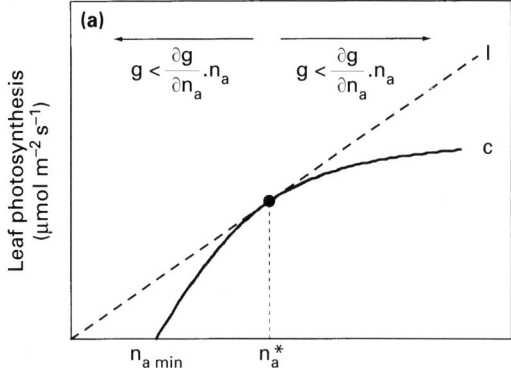

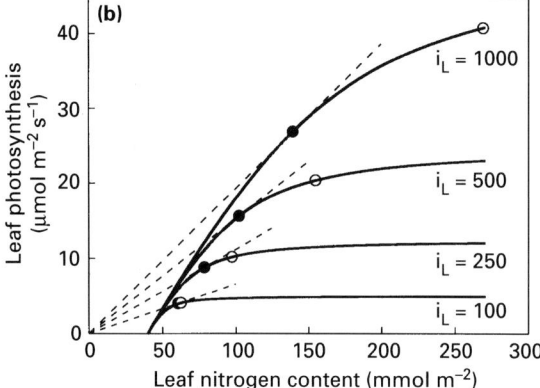

Fig. 5. (a) Continuous curve c; the rate of photosynthesis as a function of leaf nitrogen content per unit leaf area for an intermediate irradiance. Variable $n_{a\,min}$ is the threshold value for photosynthesis; n_a^* is the point at which photsynthetic nitrogen-use efficiency (PNUE) is maximal. At this point the broken line l is tangent to curve c. For more information see discussion section. (b) The rate of photosynthesis as a function of leaf nitrogen content per unit leaf area for leaves that experience different irradiances within the vegetation. Open circles, points at which leaves in the 'optimal' stand operate. Filled circles, leaves of a stand in the evolutionarily stable situation at the end of the game, where SLA is such that PNUE at leaf level is maximized.

(Schieving, 1998). It is in line with conclusions by several authors (Fig. 1b). We stress that the absolute increases in LAI shown in Fig. 3a are higher than expected. Our model is a simple one and is not aimed at realistically mimicking the development over time of a vegetation, including the carbon economy of stems and roots and the water economy of the stand. Rather, it is a framework in which we focus on qualitative trends that follow from some simple relations. Given the results of this game theory model, we suggest in plants growing in dense stands a selective pressure towards a relatively high SLA.

Distribution of nitrogen and maximization of photosynthetic nitrogen-use efficiency

Another phenomenon related to the carbon gain of vegetation is the distribution of nitrogen. In plants growing in dense stands the distribution of nitrogen is more skewed (with higher content per unit leaf area at the top than at the bottom) than in plants in open

stands. Therefore, one would expect that in the end situation of the leaf area game, when LAI has increased compared with that of the 'optimal' stand, nitrogen distribution would be more skewed. However, this is not so. As in observations in analyses of stands growing naturally, nitrogen distribution is less skewed (Hirose & Werger, 1987; Pons *et al.*, 1989; cf. De Pury & Farquhar, 1997) (Fig. 1a). How is this possible?

For plants growing naturally, several hypotheses have been put forward to explain this phenomenon. It might be that higher concentrations of nitrogen in top layers are physiologically not achievable, as there must be an upper limit to the amount of chlorophyll and protein contained in 1 g of plant material (Pons *et al.*, 1989). It could also be that there is a limitation to decreasing nitrogen in bottom layers. Retranslocation might not be possible, or the costs for retranslocation might be high and not met by the relatively small benefits (Field, 1983). Sunfleck penetration into the vegetation could increase the return on investment lower in the canopy (Terashima & Hikosaka, 1995; but see De Pury & Farquhar, 1995). Moreover, with potential herbivory at the top of the canopy an extra layer of leaves lower down would be valuable (Gutschick & Wiegel, 1988), requiring for its function a certain amount of nitrogen; consequently, distribution cannot be skewed too much.

Another explanation follows from the model and requires some insight into the relationship between photosynthesis and leaf organic nitrogen. The ratio between the two is the photosynthetic nitrogen-use efficiency (PNUE) (Field & Mooney, 1986; Evans, 1989) and is for a leaf at depth *h* in the vegetation given by:

$$\Pi(i_a^h, n_a^h) = g(i_a^h, n_a^h)/n_a^h \qquad \text{Eqn 13}$$

Of interest in this respect is how a dilution of the available nitrogen over a range of leaf areas affects PNUE. Fig. 5a shows an example of gross photosynthesis as a function of leaf nitrogen content per unit leaf area, at an intermediate irradiance, combining Eqns 1–3. Mathematically, the leaf nitrogen content at which PNUE is maximal can be found by differentiating PNUE with respect to n_a:

$$\frac{\partial \Pi(n_a^h)}{\partial n_a^h} = \frac{1}{(n_a^h)^2}\left[n_a^h\frac{\partial g(i_a^h, n_a^h)}{\partial n_a^h} - g(i_a^h, n_a^h)\right]. \qquad \text{Eqn 14}$$

Eqn 14 becomes zero at the nitrogen content at which $n_a \partial g/\partial n_a = g$. From Fig. 5a it can be shown that this must be where the straight line (l) through the origin is tangent to the curve (c). This is the point at which PNUE is maximal, n_a^* indicating the leaf nitrogen content at which this occurs.

How do these equations relate to the leaf-area game described earlier? In the description of the model, it was stated that variation in SLA had no effect any more when $n_a \partial g/\partial n_a = g$ at every canopy

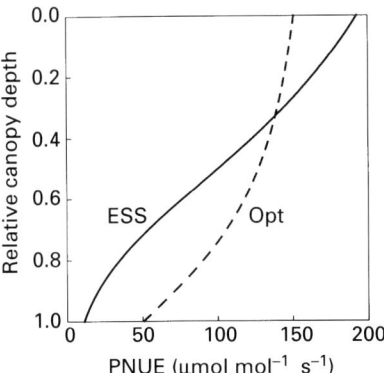

Fig. 6. *In situ* photosynthetic nitrogen-use efficiency (PNUE) as a function of the depth of the vegetation, for an optimal stand (Opt, broken line) and a stand in an evolutionarily stable situation at the end of the game (ESS, continuous line). For more detail see the Fig. 2 legend.

depth (Eqns 11, 12). This coincides exactly with the condition in which PNUE is at a maximum (Eqn 14). Therefore, within the model, a population that can resist invasion of leaf area players must have maximized PNUE for all its leaves.

A consequence of PNUE maximization is that nitrogen distribution of plants in these stands does not maximize the carbon gain of the canopy. A visual proof is given in Fig. 5b, where a family of curves, each representing a different irradiance, relates gross photosynthesis to nitrogen content per unit leaf area. Because leaves lower in the canopy experience low levels of light, photosynthesis curves saturate at low n_a levels, implying that the n_a^* for which PNUE reaches a maximum is lower for leaves at the bottom than at the top of the canopy (Fig. 5b, closed circles). However, it also implies that the slope $n_a \partial g / \partial n_a$ is different for the different layers. This is at variance with the condition formulated by Field (1983), that slopes should be equal at each depth in order to maximize canopy carbon gain. These operating points are indicated in Fig. 5b by the open circles. As shown by Hikosaka & Terashima (1995), differences between leaves that maximize PNUE and those that maximize carbon gain are very small at low irradiance but substantial at high irradiance. The consequence of the strategy of maximizing the PNUE of individual leaves is that the carbon gain of the stand is not maximal (Fig. 4b). This is the tragedy of the commons: the strategy that would maximize the carbon gain of all individual plants in the stand is not evolutionarily stable. Such a stand will be replaced by plants that play a different strategy, the end result being a stand in which total carbon gain is lower.

An implication of the lower carbon gain of the canopy at a given amount of leaf nitrogen per ground area is that the PNUE is lower in the evolutionarily stable situation at stand level. At first, this might be surprising, because we have concluded that the leaf area game is played up to the point at which PNUE

is maximized. However, it should be stressed that invaders maximize the PNUE of each leaf *for the light climate that they experience*. Stands in the end-game situation have higher LAI and consequently are characterized by lower light levels, especially low in the canopy (Fig. 2b). Consequently, at a given depth *h* the rate of photosynthesis of the lower leaves is less in the evolutionarily stable than in the optimal stand, as is PNUE (Fig. 6). This is not fully compensated for by higher PNUE in the upper part of the evolutionarily stable stand, thus PNUE of the whole stand decreases. The form of the photosynthesis–nitrogen relationship is crucial to this analysis. The family of curves shown in Fig. 5 has a threshold value for nitrogen content ($n_{a\,min}$), below which no photosynthesis would occur. If such a threshold did not exist, no SLA could be found for which PNUE is maximal and there would be a tendency to ever-increasing SLA. This is in line with simulations of Evans (1998), who showed that leaves with the lowest photosynthetic capacity per unit area and the highest SLA have the highest daily rate of photosynthesis per unit biomass.

Our model suggests a tendency for SLA and thereby PNUE of plants in dense vegetation to increase. Interestingly, in comparative experiments a strong correlation between SLA and PNUE is often found (Poorter & Evans, 1998); species with higher SLA also have higher PNUE. In plants grown in controlled conditions at intermediate light intensities the main reason is that nitrogen content per unit area in low-SLA species is high, resulting in high photosynthetic capacity. Because light is the limiting factor for photosynthesis in such an environment, part of the assimilatory machinery is not used, leading to low PNUE (Pons *et al.*, 1994; Poorter & Evans, 1998). In plants grown at high irradiance, additional factors play a role, but PNUE of low-SLA species would still increase if they were to spread their leaf biomass and nitrogen over a larger leaf area (Poorter & Evans, 1998).

Outlook

We stress that the model presented in this paper is a rather simplified representation of real vegetation. Although more complexity would not necessarily improve insight, we think that there are at least four points at which progress could be made. First, we assumed constant and uniform light conditions throughout the day. Changes in irradiance, the angle at which light penetrates the vegetation and the separation between direct and diffuse light can be of importance in determining the distribution of nitrogen within the canopy (Gutschick & Wiegel, 1988; Evans, 1998). Second, we did not consider differences in shoot height or leaf angle between the two species, factors known to have a strong effect on light competition (Barnes *et al.*, 1990; Hikosaka &

Hirose, 1997). Third, we assumed respiration to be directly dependent on nitrogen content, as found by Hirose *et al.* (1997), Anten *et al.* (1998) and Reich *et al.* (1998). However, at least part of night respiration is related to the amount of photosynthesis fixed during the preceding day (Ludwig *et al.*, 1975; Pons & Pearcy, 1994). Because leaves low in the vegetation operate close to the light-compensation point, the implication could be that their respiration rate is lower than would be expected on the basis of their nitrogen content. This in turn could imply that the net carbon gain of the canopy is underestimated at high LAIs (Ludwig *et al.*, 1965). A fourth point is that we focused on the 24-h carbon gain of a stand with a given amount of leaf nitrogen. Inclusion of the real growth process, including turnover of leaves, might yield further insights. However, the consequence of including these processes into the model would be that results could be obtained only by means of numerical simulations rather than by a relatively simple analytical solution.

CONCLUSIONS

The model formulated here shows the tragedy of the commons: plants that distribute their nitrogen at a given stand biomass in such a way that canopy photosynthesis is maximized are outcompeted by plants that follow a different strategy. Within the framework of the model, an evolutionarily stable strategy is achieved by increasing SLA to the point at which PNUE is maximized for each individual leaf. Consequences at stand level are that LAI is higher and nitrogen distribution less skewed than in a canopy in which carbon gain is maximized. Therefore, canopy carbon gain as well as carbon gain of the individual plants are lower.

ACKNOWLEDGEMENTS

We thank John Evans, Heinjo During, Lourens Poorter and Adrie van der Werf for increasing the quality of this paper by their critical but constructive comments, and the reviewers and John Farrar for their work on the manuscript.

REFERENCES

Aerts R, Chapin FS. 1999. The mineral nutrition of wild plants revisited: a re-evaluation of processes and patterns. *Advances in Ecological Research.* (In press).

Anten NPR, Schieving F, Medina E, Werger MJA, Schuffelen P. 1995a. Optimal leaf area indices in C_3 and C_4 mono- and dicotyledonous species at low and high nitrogen availability. *Physiologia Plantarum* **95**: 541–550.

Anten NPR, Schieving F, Werger MJA. 1995b. Patterns of light and nitrogen distribution in relation to whole canopy carbon gain in C_3 and C_4 mono- and dicotyledonous species. *Oecologia* **101**: 504–513.

Anten NPR, Werger MJA, Medina E. 1998. Nitrogen distribution and leaf area indices in relation to photosynthetic nitrogen use efficiency in savanna grasses. *Plant Ecology* **138**: 63–75.

Barnes PW, Beyschlag W, Ryel R, Caldwell MM, Flint SD. 1990. Plant competition for light analysed with a multispecies canopy model. III. Influence of canopy structure in mixtures and monocultures of wheat and wild oat. *Oecologia* **82**: 560–566.

Berendse F, Aerts R. 1987. Nitrogen-use efficiency: a biologically meaningful definition? *Functional Ecology* **1**: 293–296.

Corré WJ. 1983. Growth and morphogenesis of sun and shade plants. III. The combined effects of light intensity and nutrient supply. *Acta Botanica Neerlandica* **32**: 277–294.

De Pury DGG, Farquhar GD. 1997. Simple scaling of photosynthesis from leaves to canopies without the errors of big-leaf models. *Plant Cell and Environment* **20**: 537–557.

Evans JR. 1988. Acclimation by the thylakoid membranes to growth irradiance and the partitioning of nitrogen between soluble and thylakoid proteins. *Australian Journal of Plant Physiology* **15**: 93–106.

Evans JR. 1989. Photosynthesis and nitrogen relationships in leaves of C_3 plants. *Oecologia* **78**: 9–19.

Evans JR. 1993. Photosynthetic acclimation and nitrogen partitioning within a lucerne canopy. II. Stability through time and comparison with a theoretical optimum. *Australian Journal of Plant Physiology* **20**: 69–82.

Evans JR. 1998. Photosynthetic characteristics of fast- and slow-growing species. In: Lambers H, Poorter H, Van Vuuren MMI, eds. *Inherent variation in plant growth. Physiological mechanisms and ecological consequences.* Leiden, The Netherlands: Backhuys, 101–119.

Farquhar GD, Von Caemmerer S, Berry JA. 1980. A biochemical model of photosynthetic CO_2 assimilation in leaves of C_3 species. *Planta* **149**: 78–90.

Field CB. 1983. Allocating leaf nitrogen for maximization of carbon gain: leaf age as a control on the allocation program. *Oecologia* **56**: 348–355.

Field CB. 1988. On the role of photosynthetic responses in constraining the habitat distribution of rainforest plants. *Australian Journal of Plant Physiology* **15**: 343–358.

Field C, Mooney HA. 1986. The photosynthesis–nitrogen relationship in wild plants. In: Givnish TJ, ed. *On the economy of form and function.* Cambridge, UK: Cambridge University Press, 25–55.

Gutschick VP. 1988. Optimization of specific leaf mass, internal CO_2 concentration, and chlorophyll content in crop canopies. *Plant Physiology and Biochemistry* **26**: 525–537.

Gutschick VP, Wiegel FW. 1988. Optimizing the canopy photosynthetic rate by patterns of investment in specific leaf mass. *American Naturalist* **132**: 67–86.

Hikosaka K, Hirose T. 1997. Leaf angle as a strategy for light competition: optimal and evolutionary stable light-extinction coefficient within a leaf canopy. *Ecoscience* **4**: 501–507.

Hikosaka K, Terashima I. 1995. A model of the acclimation of photosynthesis in the leaves of C_3 plants to sun and shade with respect to nitrogen use. *Plant Cell and Environment* **18**: 605–616.

Hirose T, Ackerly DD, Traw MB, Ramseier D, Bazzaz FA. 1997. CO_2 elevation, canopy photosynthesis, and optimal leaf area index. *Ecology* **78**: 2339–2350.

Hirose T, Werger MJA. 1987. Maximizing daily canopy photosynthesis with respect to the leaf nitrogen allocation pattern. *Oecologia* **72**: 520–526.

Leuning R, Wang YP, Cromer RN. 1991. Model simulations of spatial distributions and daily totals of photosynthesis in *Eucalyptus grandis* canopies. *Oecologia* **88**: 494–503.

Ludwig LJ, Saeki T, Evans LT. 1965. Photosynthesis in artificial communities of cotton plants in relation to leaf area. I. Experiments with progressive defoliation of mature plants. *Australian Journal of Biological Sciences* **18**: 1103–1118.

Ludwig LJ, Charles-Edwards DA, Withers AC. 1975. Tomato leaf photosynthesis and respiration in various light and carbon dioxide environments. In: Marcelle R, ed. *Environmental and biological control of photosynthesis.* The Hague, The Netherlands: Junk, 29–36.

Maynard Smith J. 1982. *Evolution and the theory of games.* Cambridge, UK: Cambridge University Press.

Monk CD. 1966. An ecological significance of evergreenness. *Ecology* **47**: 504–505.

Monsi M, Saeki T. 1953. Über den Lichtfaktor in den Pflanzengesellschaften und seine bedeutingen für den Stoffproduktion. *Japanese Journal of Botany* **14**: 22–52.

Pons TL, Pearcy RW. 1994. Nitrogen reallocation and photosynthetic acclimation in response to partial shading in soybean plants. *Physiologia Plantarum* **92**: 636–644.

Pons TL, Schieving F, Hirose T, Werger MJA 1989. Optimization of leaf nitrogen allocation for canopy photosynthesis in *Lysimachia vulgaris*. In: Lambers H, Cambridge ML, Konings H, Pons SPB, eds. *Causes and consequences of variation in growth rate and productivity of higher plants*. The Hague, The Netherlands: Academic Publishing, 175–186.

Pons TL, Van der Werf A, Lambers H. 1994. Photosynthetic nitrogen use efficiency of inherently slow- and fast-growing species: possible explanations for observed differences. In: Roy J, Garnier SPB, eds. *A whole plant perspective on carbon–nitrogen interactions*. The Hague, The Netherlands: Academic Publishing, 71–77.

Pons TL, Van Rijnberk H, Van der Werf A. 1993. Importance of the gradient in photosynthetically active radiation in a vegetation stand for leaf nitrogen allocation in two monocotyledons. *Oecologia* **95**: 416–424.

Poorter H, De Jong R. 1999. Specific leaf area, chemical composition and leaf construction costs of plant species from productive and unproductive habitats. *New Phytologist* **143**: 000–000.

Poorter H, Evans JR 1998. Photosynthetic nitrogen-use efficiency of species that differ inherently in specific leaf area. *Oecologia* **116**: 26–37.

Poorter H, Garnier E. 1999. Ecological significance of inherent variation in relative growth rate. In: Pugnaire FI, Valladares F, eds. *Handbook of functional plant ecology*. New York, USA: Marcel Dekker 81–120.

Reich PB. 1998. Variation among plant species in leaf turnover rates and associated traits: implications for growth at all life stages. In: Lambers H, Poorter H, Van Vuuren MMI, eds. *Inherent variation in plant growth. Physiological mechanisms and ecological consequences*. Leiden, The Netherlands: Backhuys, 467–487.

Reich PB, Walters MB, Ellsworth DS, Vose JM, Volin JC, Gresham C, Bowman WD. 1998. Relationships of leaf dark respiration to leaf nitrogen, specific leaf area and life-span: a test across biomes and functional groups. *Oecologia* **114**: 471–482.

Schieving, F. 1998. *Plato's plant : on the mathematical structure of simple plants and canopies*. Leiden, The Netherlands: Backhuys.

Schieving F, Pons TL, Werger MJA, Hirose T. 1992a. The vertical distribution of nitrogen and photosynthetic activity at different plant densities in *Carex acutiformis*. *Plant and Soil* **142**: 9–17.

Schieving F, Werger MJA, Hirose T. 1992b. Canopy structure, nitrogen distribution and whole canopy photosynthetic carbon gain in growing and flowering stands of tall herbs. *Vegetatio* **102**: 173–181.

Spitters CJT, Aerts R. 1983. Simulation of competition for light and water in crop–weed associations. *Aspects of Applied Biology* **4**: 467–483.

Terashima I, Hikosaka K. 1995. Comparative ecophysiology of leaf and canopy photosynthesis. *Plant Cell and Environment* **18**: 1111–1128.

Werger MJA, Hirose T. 1991. Leaf nitrogen distribution and whole canopy photosynthetic carbon gain in herbaceous stands. *Vegetatio* **97**: 11–20.

Woodward FI. 1983. The significance of interspecific differences in specific leaf area to the growth of selected herbaceous species from different altitudes. *New Phytologist* **95**: 313–323.

APPENDIX I. A FORMAL DERIVATION OF THE PHOTOSYNTHETIC ADVANTAGE OF THE INVADING SPECIES I

The difference in net carbon gain between the plants of invading species I and the average plant in the stand is given by:

$$C_{TI}^{\delta} - C_{TA}^{\delta} = C_{TI}^{\delta} - \phi_R C_{TR}^{\delta} - \phi_I C_{TI}^{\delta}$$
$$= (1 - \phi_I) C_{TI}^{\delta} - \phi_R C_{TR}^{\delta}$$
$$= \phi_R (C_{TI}^{\delta} - C_{TR}^{\delta}) \qquad \text{Eqn 15}$$

Expanding the above expression gives:

$$C_{TI}^{\delta} - C_{TA}^{\delta} = \phi_R \int_0^{h_T} dh\, m^h \sigma^h \left[(1+\delta) g\left(i_a^h, \frac{n_a^h}{1+\delta} \right) \right.$$
$$\left. - g(i_a^h, n_a^h) \right] \qquad \text{Eqn 16}$$

Here σ^h and n_a^h are the SLA and nitrogen content per unit leaf area of species R plants. The nitrogen content of the I plants can be written as:

$$n_{aI}^h = n_a^h(\delta) = n_a^h/(1+\delta) \qquad \text{Eqn 17}$$

It should be noted that the respiration rates disappear from the expression $C_{TI} - C_{TA}$ and that the $(1+\delta)$ variation affects the total leaf area, the nitrogen content per unit leaf area and the irradiance at the leaf level.

Because δ is small, we can write:

$$C_{TI}^{\delta} - C_{TA}^{\delta} = \frac{d}{d\delta}(C_{TI}^{\delta} - C_{TA}^{\delta})|_{\delta=0} \cdot \delta \qquad \text{Eqn 18}$$

Interchanging differentiation and integration gives:

$$\frac{d}{d\delta}(C_{TI}^{\delta} - C_{TA}^{\delta})\bigg|_{\delta=0} = \phi_R \int_0^{h_T} dh\, m^h \sigma^h \times$$
$$\frac{d}{d\delta}\left[(1+\delta) g\left(i_a^h(\delta), \frac{n_a^h}{1+\delta} \right) - g(i_a^h(\delta), n_a^h) \right]\bigg|_{\delta=0}. \quad \text{Eqn 19}$$

Expanding the derivative $d/d\delta\,[..]$ in the integrand part of Eqn 19 gives:

$$\frac{d}{d\delta}[..] = g\left(i_a^h(\delta), \frac{n_a^h}{1+\delta} \right) + (1+\delta)\frac{d}{d\delta} g\left(i_a^h(\delta), \frac{n_a^h}{1+\delta} \right)$$
$$- \frac{d}{d\delta} g(i_a^h(\delta), n^h). \qquad \text{Eqn 20}$$

By the chain rule:

$$\frac{d}{d\delta} g\left(i_a^h(\delta), \frac{n_a^h}{1+\delta} \right) = \frac{\partial g^h}{\partial i_a^h}\frac{di_a^h}{dF^h}\frac{dF^h}{d\delta} - \frac{\partial g^h}{\partial n_a}\frac{n_a^h}{(1+\delta)^2},$$
$$\text{Eqn 21}$$

This gives for Eqn 20, evaluated at $\delta = 0$:

$$\frac{d}{d\delta}[..]|_{\delta=0} = g(i_a^h, n_a^h) + \frac{\partial g(i_a^h, n_a^h)}{\partial i_a^h}\frac{di_a^h}{dF^h}\frac{dF^h}{d\delta}\bigg|_{\delta=0}$$
$$- \frac{\partial g(i_a^h, n_a^h)}{\partial n_a} n_a^h - \frac{\partial g(i_a^h, n_a^h)}{\partial i_a^h}\frac{di_a^h}{dF^h}\frac{dF^h}{d\delta}\bigg|_{\delta=0}$$
$$= g(i_a^h, n_a^h) - \frac{\partial g(i_a^h, n_a^h)}{\partial n_a} n_a^h. \qquad \text{Eqn 22}$$

Substitution of the above expression into Eqn 19 gives:

$$\frac{d}{d\delta}(C_{TI}^{\delta} - C_{TA}^{\delta})|_{\delta=0} =$$
$$\phi_R \int_0^{h_T} dh\, m^h \sigma^h \left[g(i_a^h, n_a^h) - \frac{\partial g(i_a^h, n_a^h)}{\partial n_a} n_a^h \right], \quad \text{Eqn 23}$$

As pointed out in the main text, if this integral is positive, the invading plants will outcompete the resident plants.

New Phytol (1999), **143**, 213–219

Research review
The functional significance of leaf structure: a search for generalizations

MALCOLM C. PRESS

Department of Animal and Plant Sciences, University of Sheffield, Sheffield, S10 2TN, UK (tel +44 114 222 4111; fax +44 114 222 0002; e-mail m.c.press@sheffield.ac.uk)

Received 4 November 1998; accepted 8 April 1999

SUMMARY

The coupling between leaf structure and function is illustrated with reference to two examples, the C_4 photosynthetic pathway and leaf pubescence. A distinction is made between function and functional significance. The latter is defined as the role, significance or consequence of a structure, whereas the former is more simply the action that a structure is capable of performing. Using the two examples, four generalizations are made concerning the relationships between structure, function and functional significance: the functional significance of leaf structure is environment-dependent; the relationship between functional significance and structure is sometimes non-intuitive; functional equivalency means that there is often more than one 'solution' to the same 'constraint'; and the consequences of leaf structure can exert profound effects at levels of organization beyond those of the individual organism and may play a critical role in determining community structure and function, through interactions with other species and trophic levels. The importance of understanding the consequences in variation in leaf structure at the global scale is illustrated with reference to the issue of global climate change.

Key words: C_4 photosynthesis, function, functional significance, leaf pubescence, level of organization, plant–atmosphere relations, optimization.

INTRODUCTION

Leaves have evolved independently in at least six plant lineages, from liverworts to seed plant (Niklas, 1997). Almost all extant land plants possess leaves or have evolved photosynthetic stems or roots from leafy ancestors (Ingrouille, 1992). The primary role of leaves is to capture light and CO_2 and the basic molecular and biophysical processes are strongly conserved. Yet despite this, at higher levels of organization there is great diversity. Leaf arrangement and leaf morphology vary greatly between, and sometimes within, species, and there is substantial variation in structure at both microscopic and macroscopic scales. The current view is that this multiplicity of variation reflects trade-offs between the requirements of leaves to perform their primary functions and a suite of 'constraints', such as the need for mechanical support, prevention of water loss, regulation of energy exchange and defence (Parkhurst & Loucks, 1972; Gates, 1980; Givnish, 1986; Niklas, 1999). The Fourth *New Phytologist* Symposium on the causes and consequences of variation in leaf structure explored the function and functional significance of leaf structure at a number of different scales from photosynthetic performance

across mesophyll cell layers (Evans, 1999; Han *et al.*, 1999) to the extent to which specific leaf area affects global carbon cycling (D. J. Beerling & F. I. Woodward, unpublished).

In this review, a clear distinction is made between function and functional significance (Koehl, 1996; see also Calow, 1987). A function of a structure is 'simply' defined as a function that the structure is capable of doing. Functional significance, on the other hand, is interpreted as the role, significance or consequence of a structure, for a given species in a defined environment. In this paper, the coupling between structure and function is examined with respect to two important phenomena. The first is the C_4 photosynthetic pathway in leaves, where the biochemical operation of the pathway is highly dependent on anatomical and ultrastructural co-ordination. The second example pertains to the external structure of leaves and examines the effects of pubescence on energy budgets. The functional significances of the structure–function relations described are considered at higher levels of organization, specifically with regard to their physiological and ecological consequences.

An underlying theme of many of the papers in this special issue concerns the search for a series of generalizations or rules that describe the relationship between leaf structure, function and functional significance (Garnier *et al.*, 1999; Poorter & de Jong, 1999; Walters & Reich, 1999). The extent to which such generalizations can be made and their validity across species and environments is considered, with special reference to the two chosen examples.

C_4 PHOTOSYNTHESIS AND LEAVES

The structure–function relations

Photosynthetic pathways are excellent systems for the study of structure–function relations. The C_4 pathway has multiple evolutionary origins (Kellogg, 1999) and can be considered as a CO_2-concentrating mechanism that arose between 8–6 million yr ago, when atmospheric CO_2 concentrations were lower than those of today (Cerling *et al.*, 1997). The pathway is widespread in monocotyledonous plants but, in contrast, is both taxonomically uncommon and phylogenetically widely dispersed among dicotyledonous families (Ehleringer *et al.*, 1997).

Effective functioning of the C_4 photosynthetic pathway usually depends on the physical separation of the two carboxylation steps between the mesophyll and bundle sheath cells. Although many C_3 grasses have bundle sheath cells and many C_3 dicotyledonous plants possess paraveinal mesophyll cells, their arrangement and structure is distinct from the so-called Kranz anatomy of C_4 species. A second important structural feature of C_4 plants is the thickened wall of the bundle sheath cells, which prevents leakage of CO_2 back to the mesophyll cells (Hatch *et al.*, 1995). The distinctive combination of anatomy and biochemistry results in an estimated 3 to 20-fold increase in the concentration of CO_2 in the bundle sheath cells compared with the atmosphere.

Despite a strong degree of evolutionary convergence, there are distinct differences among the leaves of C_4 species, both at the biochemical level and with respect to cellular and subcellular organization (Leegood, 1997). Perhaps the most marked variation concerns the nature of the decarboxylating enzyme in the bundle sheath cells, with species employing either NADP-malic enzyme, NAD-malic enzyme or both NAD-malic enzyme and phospho*enol*pyruvate (PEP) carboxykinase (PCK-type species). Bundle sheath cells differ in leakiness to CO_2 between the subtypes because of differences in cell wall structure. Leakiness is greatest in the NAD-malic enzyme subtype and, in contrast to the other two subtypes, it also lacks suberized lamellae in the walls of the bundle sheath cells.

Although Kranz anatomy is widespread amongst C_4 plants, recent studies show that the structural arrangement already described is not always necessary for the operation of the pathway. Keeley (1998) reports ^{14}C and enzymological data that show the tight coupling of the C_3 and C_4 photosynthetic cycles in the juvenile leaves of *Orcuttia*, despite the absence of Kranz anatomy. There appears to be no cellular isolation of the phospho*enol*pyruvate carboxylase-(PEPC) and Rubisco-catalysed CO_2 assimilation pathways. *Orcuttia*, a member of the Orcuttieae tribe of the Poaceae, occurs in shallow rain-filled seasonal pools in the mediterranean climate region of California, USA. The genus forms juvenile foliage that remains submerged for periods of between 1 and 3 months before forming adult terrestrial foliage following the drying out of the pools. Keeley (1998) suggests that the very low diffusivity of CO_2 in water compared with air, together with the high velocity of the PEPC reaction, allow the C_4 photosynthetic pathway to operate in the absence of Kranz anatomy.

Thus, in aquatic environments, a different structural arrangement appears to allow the functioning of the CO_2-concentrating mechanism in angiosperms. Such modifications are also apparent in some species of algae, for example, in the marine macroscopic green alga *Udotea flabellum*, where C_4-like photosynthetic characteristics have been reported in the absence of the compartmentation typical of most angiosperms (Reiskind & Bowes, 1991). Other variations in structure and enzyme localization have been reported for both aquatic and terrestrial species (Ueno, 1996), confirming that Kranz anatomy is not a universal requirement for C_4 photosynthesis.

The functional significance of the C_4 pathway

The C_4 pathway has a number of physiological consequences that are well understood. First, the theoretical quantum yield (moles of CO_2 fixed per quantum of light energy absorbed) of C_4 photosynthesis is less than that of C_3 species under non-photorespiratory conditions, because extra ATP is required for the regeneration of PEP. However, the observed difference in quantum yield between C_4 and C_3 plants is temperature-dependent because of the variable costs of photorespiration in the latter. The high internal CO_2 concentration in the mesophyll cells of C_4 plants favours the carboxylase rather than the oxygenase reaction of Rubisco, thus virtually suppressing photorespiration. Second, the CO_2 concentrating mechanism allows C_4 plants to operate at lower stomatal conductances than their C_3 counterparts, thus improving water use efficiency. Third, the nitrogen (N) requirements of C_4 plants are also lower because the plants contain lower amounts of Rubisco as a consequence of the higher concentration of CO_2 at the site of fixation. This results in foliar N concentrations of up to 65% less than those in C_3 leaves and up to a doubling of the photosynthetic N-use efficiency (Long, 1999a).

Despite our understanding of the functional consequences of the C_4 pathway at the physiological level, most of the evidence that purports to demonstrate the functional significance of the pathway at the ecological level is crude. There are good correlations between the relative abundance of C_4 grasses in plant communities and climatic and environmental gradients. For example, the relative abundance of C_4 grasses is positively correlated with the minimum growing season temperature in Australia (Hattersley, 1983) and the Middle East (Vogel *et al.*, 1986), and negatively correlated with altitude in Kenya (Tieszen *et al.*, 1979). However, as Monson (1989) points out, it is not always clear whether the relative distributions of C_3 and C_4 grasses are more closely correlated with temperature gradients or the frequently accompanying gradients in water availability. There are also correlations between the relative abundance of C_4 subtype and environment, with the NADP-malic enzyme type increasing and the NAD-malic enzyme type decreasing with increasing annual precipitation (Henderson *et al.*, 1995), although the significance of these correlations is unclear. There are few ecological studies that explicitly explore which of the physiological attributes of the C_4 pathway confer competitive advantage over C_3 plants, and indeed, such advantages are not easy to demonstrate experimentally. For example, there are many studies that show that elevated concentrations of atmospheric CO_2 do not confer a competitive advantage on C_3 compared with C_4 species (Knapp & Medina, 1999), despite theoretical predictions (Cerling *et al.*, 1997).

Reproductive success must underlie the trends in abundance of C_3 and C_4 species and relationships between photosynthetic pathway, reproduction and environment are poorly explored, leading Henderson *et al.* (1995) to conclude that 'precisely which trait, or which combination of traits, is most relevant to the relative fitness of C_3 and C_4 species remains baffling'.

Although many C_4 species occur in warm, dry and high light environments, there are a small number of species with distributions in other habitats, such as in tropical rain forest understories and temperate coastal communities (Monson, 1989), in addition to the aquatic examples already described. This has caused some authors to question whether distribution is adaptive or whether it reflects phylogenetic inertia (Pearcy & Calkin, 1983; Pearcy *et al.*, 1987; Keeley, 1998). Indeed, one criticism of optimality theory is that organisms may not, at any given point in time, be optimally adapted to their environment (Gould & Lewontin, 1979). Long (1983) suggests that the positive correlation between C_4 species abundance and temperature may reflect the tropical and sub-tropical distribution of the progenitors of C_4 species, rather than their optimization to ambient environmental conditions. Keeley (1998), however, argues against phylogenetic inertia, citing molecular studies that show that relatively complex devel-

opmental and biochemical modifications in the C_4 pathway are controlled by a few regulatory genes, thus making evolutionary changes in pathway expression relatively easy (Ku *et al.*, 1996).

The lessons which I have attempted to illustrate here are twofold. First, although leaf anatomy and ultrastructure may be important for the operation of the C_4 photosynthetic pathway, function is not constrained by one structural solution. More than one solution has evolved, presumably because of the different relative constraints in different environments. Second, understanding the functional significance of leaf structure and function at the ecological level is complex and rarely straightforward or intuitive, a point which will be illustrated further in the following sections.

FUNCTIONAL SIGNIFICANCE OF LEAF PUBESCENCE

The study of coupling between leaves and the atmosphere is rich in examples that support the lessons already drawn. Multiple solutions to the same environmental constraints are illustrated by Gates in his classic text (Gates, 1980); for example, the different strategies that regulate energy exchange in the leaves of sympatric Australian trees. The interpretation of leaf pubescence serves to illustrate the problem of understanding the functional significance of structures and is considered in more detail (see also Gutschick, 1999).

A multitude of functions has been attributed to the presence of trichomes, hairs, thorns and bristles on

Table 1. *Some proposed functions and significances of leaf pubescence (from Gates, 1980)*

Function	Significance
Reflection of radiation, re-radiation and absorption of radiation (wavelength specific); shading of leaf surface	Protection from excess radiation at specific wavelengths (including UV-B radiation)
Reflection of radiation onto mesophyll cells	Increase light capture for photosynthesis
Increase gas diffusion pathway	Reduce water loss
Provision of moisture trap on leaf surface	Reduce water loss
Insulation	Protect against heat loss
Reduction of ion leaching	Nutrient retention
Movement of salts from within leaf to surface	Salinity tolerance
Defence	Protection from insect and vertebrate herbivores and pathogens
Stomatal protection	Reduces blockages by water and particulates and impedes pathogen entry
Increases boundary layer thickness	Reduces the impact of wind on energy budget

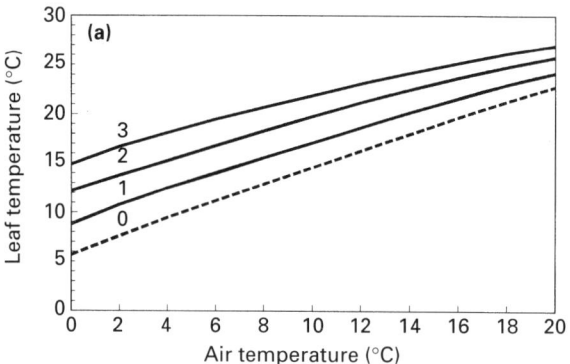

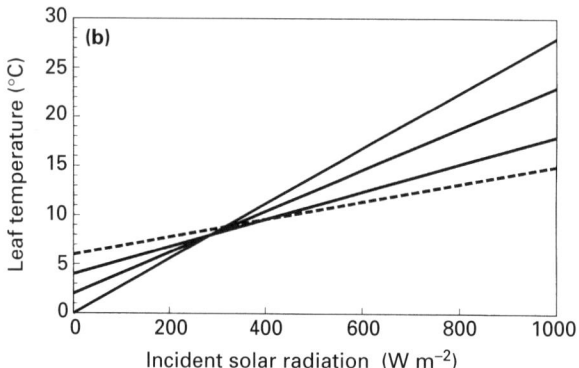

Fig. 1. (a) The calculated relationship between leaf and air temperature for the giant rosette plant *Espeletia timotensis* with 0, 1, 2 and 3 mm of pubescence under conditions of 800 W m^{-2} irradiance, 2 m s^{-1} wind speed, 150 s m^{-1} stomatal resistance and 5 g m^{-3} external humidity. (b) The calculated relationship between leaf temperature and incident solar radiation for leaves of *E. timotensis* with 0, 1, 2 and 3 mm of pubescence under conditions of 2 m s^{-1} wind speed, 8°C air temperature, 200 s m^{-1} stomatal resistance and 5 g m^{-3} external humidity. (Re-drawn, with permission, from Meinzer & Goldstein, 1985.)

the surfaces of leaves (Table 1) but their adaptive significance can depend on the environment in which they occur. It is commonly accepted that, in hot, arid environments, the primary role of leaf pubescence is to lower leaf temperature and transpirational water loss through lower absorption of solar radiation (Ehleringer & Mooney, 1978). However, at night, pubescence also serves to maintain tissue temperatures above those of the air. The cactus *Carnegiea gigantea* growing in northern Arizona has a mean apical pubescence 10 mm in thickness and spine coverage of 41% of plant surface area. Nobel (1988) calculated that these structures raise the minimum apical temperature by 6°C at night. Leaf pubescence also serves to increase tissue temperature for many arctic and alpine species and Körner & Larcher (1988) and Gauslaa (1984) show that the tissue temperatures of plants in these environments can exceed those of the air by >8°C. These generalizations demonstrate that the consequences of pubescence for tissue temperature are highly dependent on the environment.

Meinzer & Goldstein (1985) further illustrate this point in their studies of the caulescent giant rosette plant, *Espeletia timotensis*, which grows in the

northern Andes at altitudes of up to 4500 m. Leaf pubescence of up to 3 mm in thickness increases boundary layer thickness and resistance to convective and latent heat transfer, with purportedly only minor effects on solar radiation absorption. The extent to which leaf temperature is elevated is dependent on the air temperature and incident solar radiation (Fig. 1). For example, at an air temperature of 5°C, a wind speed of 2 m s^{-1} and an irradiance of 800 W m^{-2}, surface temperature with 2 mm pubescence was calculated to be 16°C, compared with 10°C for a non-pubescent leaf of the same size. At an air temperature of 20°C this difference was reduced from 6 to 3°C. Meinzer & Goldstein (1985) also demonstrate that the warming effect is greater at higher irradiances (Fig. 1), with pubescence resulting in leaf cooling below a critical irradiance, the value of which is largely a function of wind speed, air temperature and atmospheric humidity.

Attempting to interpret the functional significance of leaf structure in response to one environmental constraint may be unsafe. There is a large volume of literature that ascribes herbivore defence strategies to leaf hairs, spines, thorn and prickles (Grubb, 1992) and the evidence for their effectiveness in some species is certainly compelling, both for insect and vertebrate herbivores (Ezcurra *et al.*, 1987; Schoener, 1988). However, fewer studies have simultaneously addressed more than one of the putative significances of pubescence. One exception is that of Woodman & Fernandes (1991) on *Verbascum thapsus*, where it is demonstrated that pubescence serves at least a dual role, both providing some protection from leaf chewing insects and reducing transpiration loss from young leaves. However, the relative benefits were not estimated.

These examples attempt to illustrate that the functional significance of a structure cannot be interpreted in the absence of a detailed knowledge of the environment in which the organism is located and both organism-based and environmental complexity may preclude the quest for 'readily apparent cause–effect relationships' (Schuepp, 1993). It is perhaps not surprising, therefore, that such complexity has resulted in functional equivalency, as illustrated by discussion of the various mechanisms that could result in elevation of leaf temperature in tropical alpine giant rosette plants in the absence of leaf pubescence (Meinzer & Goldstein, 1985; Meinzer *et al.*, 1985), the conclusions drawn by Sandquist & Ehleringer (1997) on the alternative suites of characteristics employed by the desert plant *Encelia farinosa* that maximize carbon gain, and the brief review of the structure–function relations of C_4 grasses reported here.

FUNCTIONAL SIGNIFICANCE AT HIGHER LEVELS OF ORGANIZATION

So far, three generalizations have been drawn. First, the functional significance of leaf structure is environment-dependent; second, the relationship between functional significance and structure is sometimes non-intuitive; and third, functional equivalency means that there is often more than one solution to the same constraint. In these respects plants are no different from other taxa and the conclusions are consistent with those that arise from studies of the relationship between animal phenotypes and ecological success (Koehl, 1996).

Next, the consequences of leaf structure at higher levels of organization will be considered. In addition to questions pertaining to phylogenetic inertia, optimality theory has been further criticized for its tendency to consider individual characters in isolation at the expense of the larger scale picture, both at the organismal and community level (Parker & Maynard Smith, 1990; Pearcy & Valladares, 1999). Such considerations may often be necessary in order to gain a full appreciation of the functional significance of structure and, at this scale, complexity can be very great. Continuing with the theme of pubescence, the phenomenon may exert either positive or negative effects on neighbouring indi-

viduals, of either the same or different species, depending on the specific community. For example, in both hot desert and arctic ecosystems, pubescence of adults may assist in providing cooler and warmer environments, respectively, for seedling establishment. Indeed, facilitation in extreme environments is probably an under-recognized driving force in plant ecology (Callaway, 1995; Brooker & Callaghan, 1997).

The extent to which pubescence may defend individual species from herbivores may not only have profound effects on the plant itself but also on community composition, through shifts in competitive balance (Olff & Ritchie, 1998). Furthermore, an understanding of the significance of pubescence at the community level may require a knowledge of multispecies and multitrophic interactions, such as those involving plants, fungi (both mutualistic and pathogenic), herbivores, parasitoids and predators (Gange & Brown, 1997). For example, Marquis & Whelan (1996) argue that plant characters, such as pubescence, can influence (usually reduce) the accessibility of herbivores to predators and parasitoids. They suggest that selection might be acting on morphological traits that allow predators and parasitoids a greater probability of locating or gaining access to prey. Perhaps one of the better studied examples of a multispecies interaction is myrmecophy, where symbiotic ants live in modified plant organs, including leaves. The interactions are mostly considered to be mutualistic, with the plant trading food and nest sites for protection (Davidson & McKey, 1993). Thus, a fourth generalization can be added to the structure–function generalizations already expounded; that the consequences of leaf structure can exert profound effects at higher levels of organization and may play a critical role in determining both the structure and function of the community through interactions with other species and trophic levels.

FUTURE DIRECTIONS

Experimentation with transgenic and mutant plants provides an ideal way of elucidating the functional significance of changes in leaf structure, both internally and externally. Howell (1998) describes *Arabidopsis thaliana* genotypes that differ in a number of respects from their wildtype counterparts, as illustrated by the three examples that follow. First, over-expression of *knotted1*-like gene (*KNAT1*) leads to the production of transformed plants with lobed leaves, differing in shape and size of lobe as well as in the degree of serration (Chuck *et al.*, 1996). Second, a range of trichome mutants have been created by blocking trichome development at different stages, resulting in, for example, glabrous mutants or plants with non-uniform patterns of trichome distribution. Third, stomatal frequency, pattern and spacing can be altered (Yang & Sack, 1995). Such variations provide an unrivalled opportunity for understanding the functional significance of leaf structure, although if more than one gene has been altered, potential pleiotropic effects in the genome cannot be ignored.

A second approach that has gained appeal during the last decade is based on a recognition of the importance of phylogenetic relationships. The most widely used technique involves the construction of 'independent contrasts' from differences in trait values between species representative of independent evolutionary divergence events (Ackerly, 1999). By examining a trait across a wide enough range of species and genera, it is possible to separate adaptation from phylogeny, although the two may not necessarily be mutually exclusive. Thus, taxonomic information has the potential to play a greater role in understanding the functional significance of structural phenomena.

At the community level, assessing the relative importance of current constraints is difficult and only achievable by careful experimentation. Multispecies and multitrophic interactions are important in shaping ecosystem structure and function (Dobson & Crawley, 1994) and the recognition of these phenomena adds another dimension to the functional significance of leaf structure. The importance of soil processes is recognized (Cornelissen *et al.*, 1999), but less emphasis has been put on the emerging importance of leaf structure and composition on the links between soil biodiversity, element cycling and ecosystem function (Press, 1998; Bardgett *et al.*, 1998; Bardgett & Shine, 1999). Global climate change further adds to complexity at the community level and much attention is currently devoted to understanding how changes in leaf properties might affect interactions with other organisms under various global change situations (Bezemer & Jones, 1998; Peñuelas & Estiarte, 1998; Staddon & Fitter, 1998).

Long (1999b) provides an excellent example that illustrates that the feedback processes between leaf responses to climate change and climate are critical at the global scale. Elevated concentrations of atmospheric CO_2 are likely to result in mean reductions of stomatal conductance of approx. 20%, which in turn will lower transpirational cooling of leaves. Free air CO_2 exposure (FACE) studies suggest that lower transpirational cooling alone could result in a mean increase of 0.4°C of the earth's land surface temperature, with greater increases of up to 2–3°C during periods of maximum insolation (Sellers *et al.*, 1996), although it is important to recognize that scaling up from FACE studies may be problematic. Betts *et al.* (1997) provide further information on vegetation–climate feedback processes, suggesting that the importance of physiological effects can be partially offset by structural changes (increases in leaf area index and canopy height), but that their extent will be highly climate dependent. Further, Betts *et al.* (1997) point out that the timescales over which physiological and structural changes operate are likely to be different, further adding to complexity. These studies (Sellars *et al.*, 1996; Betts *et al.*, 1997) demonstrate the true global significance of variations in leaf structure and function.

ACKNOWLEDGEMENTS

I am grateful to Ian Woodward, Jenny Watling, Richard Leegood and David Lewis for their helpful comments on an earlier version of this article and to Eric Garnier and the co-organizers of the Fourth *New Phytologist* Symposium for inviting me to produce these 'after thoughts' and for giving me a free hand to rummage 'under the bonnet'.

REFERENCES

Ackerly DD. 1999. Comparative plant ecology and the role of phylogenetic information. In: Press MC, Scholes JD, Barker MG, eds. *Physiological plant ecology*. Oxford, UK: Blackwell Science. (In press.)

Bardgett RD, Shine A. 1999. Linkages between plant litter diversity, soil microbial biomass and ecosystem function in temperate grasslands. *Soil Biology and Biochemistry* **31**: 317–321.

Bardgett RD, Wardle DA, Yeates GW. 1998. Linking above ground and below ground interactions: how plant responses through foliar herbivory influence soil organisms. *Soil Biology and Biochemistry* **30**: 1867–1878.

Betts RA, Cox PM, Lee SE, Woodward FI. 1997. Contrasting physiological and structural vegetation feedbacks in climate change simulations. *Nature* **387**: 796–799.

Bezemer TM, Jones TH. 1998. Plant-insect herbivore interactions in elevated atmospheric CO_2: quantitative analyses and guild effects. *Oikos* **82**: 212–222.

Brooker RW, Callaghan TV. 1997. The balance between positive and negative plant interactions and its relationship to environmental gradients: a model. *Oikos* **81**: 196–206.

Callaway RM. 1995. Positive interactions among plants. *Botanical Review* **61**: 306–349.

Calow P. 1987. Towards a definition of functional ecology. *Functional Ecology* **1**: 57–61.

Cerling TE, Harris JM, MacFadden BJ, Leakey MG, Quade J, Eisenmann V, Ehleringer JR. 1997. Global vegetation change through the Miocene/Pliocene boundary. *Nature* **389**: 153–158.

Chuck G, Lincoln C, Hake S. 1996. *KNAT1* induces lobed leaves with ectopic meristems when overexpressed in *Arabidopsis*. *Plant Cell* **3**: 801–807.

Cornelissen H, Pérez-Harguindeguy N, Díaz S, Grime JP, Marzano B, Cabido M, Vendramini F, Cerabolini B. 1999. Leaf structure and defence control litter decomposition rate across species and life forms in regional floras on two continents. *New Phytologist* **143**: 191–200.

Davidson DW, McKey D. 1993. Ant–plant symbioses: stalking the Chuyachaqui. *Trends in Ecology and Evolution* **8**: 326–332.

Dobson A, Crawley MJ. 1994. Pathogens and the structure of plant communities. *Trends in Ecology and Evolution* **9**: 393–398.

Ehleringer JR, Cerling TE, Helliker BR. 1997. C_4 photosynthesis, atmospheric CO_2 and climate. *Oecologia* **112**: 285–299.

Ehleringer JR, Mooney HA. 1978. Leaf hairs: effects on physiological activity and adaptive value to a desert shrub. *Oecologia* **37**: 183–200.

Evans JR. 1999. Leaf anatomy enables more equal access to light and CO_2 between chloroplasts. *New Phytologist* **143**: 93–104.

Ezcurra E, Gómez JC, Becarra J. 1987. Diverging patterns of host use by phytophagous insects in relation to leaf pubescence in *Arbutus xalapensis* (Ericaceae). *Oecologia* **72**: 479–480.

Gange AC, Brown VK. 1997. *Multitrophic interactions in terrestrial systems.* Oxford, UK: Blackwell Science.

Garnier E, Salager J-L, Laurent G, Sonié L. 1999. Relationships between photosynthesis, nitrogen and leaf structure in 14 grass species and their dependence on the basis of expression. *New Phytologist* **143**: 119–129.

Gates DM. 1980. *Biophysical ecology.* New York, USA: Springer-Verlag.

Gauslaa Y. 1984. Heat resistance and energy budget in different Scandinavian plants. *Holarctic Ecology* **7**: 1–78.

Givnish TJ. 1986. *On the economy of plant form and function.* Cambridge, UK: Cambridge University Press.

Gould SJ, Lewontin RC. 1979. The spandrels of San Marco and the Panglossian paradigm - a critique of the adaptationist program. *Proceedings of the Royal Society of London, Series B* **205**: 581–598.

Grubb PJ. 1992. A positive distrust in simplicity - lessons from plant defences and from competition among plants and among animals. *Journal of Ecology* **80**: 585–610.

Gutschick VP. 1999. Biotic and abiotic consequences of differences in leaf structure. *New Phytologist* **143**: 3–18.

Han T, Vogelmann T, Nishio J. 1999. Profiles of photosynthetic oxygen evolution within leaves of *Spinacia oleracea*. *New Phytologist* **143**: 83–92.

Hatch MD, Agostino A, Jenkins CLD. 1995. Measurement of the leakage of CO_2 from bundle-sheath cells of leaves during C_4 photosynthesis. *Plant Physiology* **108**: 173–181.

Hattersley PW. 1983. The distribution of C_3 and C_4 grasses in Australia in relation to climate. *Oecologia* **57**: 113–128.

Henderson S, Hattersley P, von Caemmerer S, Osmond CB. 1995. Are C_4 pathway plants threatened by global climatic change? In: Schulze ED, Caldwell MM, eds. *Ecophysiology of photosynthesis.* Berlin, Germany: Springer-Verlag, 529–549.

Howell SH. 1998. *Molecular genetics of plant development.* Cambridge, UK: Cambridge University Press.

Ingrouille M. 1992. *Diversity and evolution of land plants.* London, UK: Chapman & Hall.

Keeley JE. 1998. C_4 photosynthetic modifications in the evolutionary transition from land to water in aquatic grasses. *Oecologia* **116**: 85–97.

Kellogg EA. 1999. Phylogenetic aspects of the evolution of C_4 photosynthesis. In: Sage RF, Monson RF, eds. *C_4 plant biology.* San Diego, CA, USA: Academic Press, 411–444.

Knapp AK, Medina E. 1999. Success of C_4 photosynthesis in the field: lessons from communities dominated by C_4 plants. In: Sage RF, Monson RF, eds. *C_4 plant biology.* San Diego, CA, USA: Academic Press, 251–283.

Koehl MAR. 1996. When does morphology matter? *Annual Review of Ecology and Systematics* **27**: 501–542.

Körner C, Larcher W. 1988. Plant life in cold climates. In: Long SP, Woodward FI, eds. *Plants and temperature.* Cambridge, UK: The Company of Biologists, 25–57.

Ku MSB, Kano-Murakami Y, Matsuoka M. 1996. Evolution and expression of C_4 photosynthesis genes. *Plant Physiology* **111**: 949–957.

Leegood RC. 1997. The regulation of C_4 photosynthesis. *Advances in Botanical Research* **26**: 251–316.

Long SP. 1983. C_4 photosynthesis at low temperatures. *Plant, Cell and Environment* **6**: 345–363.

Long SP. 1999a. Environmental responses. In: Sage RF, Monson RF, eds. *C_4 plant biology.* San Diego, CA, USA: Academic Press, 215–249.

Long SP. 1999b. Understanding the impacts of rising CO_2 - the contribution of environmental physiology. In: Press MC, Scholes JD, Barker MG, eds. *Physiological plant ecology.* Oxford, UK: Blackwell Science. (In press.)

Marquis RJ, Whelan C. 1996. Plant morphology and recruitment of the third trophic level: subtle and little recognised defences? *Oikos* **75**: 330–334.

Meinzer F, Goldstein G. 1985. Some consequences of leaf pubescence in the Andean giant rosette plant *Espeletia timotensis*. *Ecology* **66**: 512–520.

Meinzer FC, Goldstein GH, Rundel PW. 1985. Morphological changes along an altitude gradient and their consequences for an Andean giant rosette plant. *Oecologia* **65**: 278–283.

Monson RK. 1989. On the evolutionary pathways resulting in C_4 photosynthesis and Crassulacean acid metabolism (CAM). *Advances in Ecological Research* **19**: 57–110.

Niklas KJ. 1997. *The evolutionary biology of plants.* Chicago, IL, USA: University of Chicago Press.

Niklas KJ. 1999. A mechanical perspective on foliage leaf form and function. *New Phytologist* **143**: 19–31.

Nobel PS. 1988. Principles underlying the prediction of temperature in plants, with special reference to desert succulents. In: Long SP, Woodward FI, eds. *Plants and temperature.* Cambridge, UK: The Company of Biologists, 1–23.

Olff H, Ritchie ME. 1998. Effects of herbivores on grassland plant diversity. *Trends in Ecology and Evolution* **13**: 261–265.

Parker GA, Maynard Smith J. 1990. Optimality theory in evolutionary biology. *Nature* **348**: 27–33.

Parkhurst DF, Loucks OL. 1972. Optimal leaf size in relation to environment. *Journal of Ecology* **60**: 505–537.

Pearcy RW, Bjorkman O, Caldwell MM, Keeley JE, Monson RK, Strain BR. 1987. Carbon gain by plants in natural environments. *BioScience* **37**: 21–29.

Pearcy RW, Calkin HW. 1983. Carbon dioxide exchange of C_3 and C_4 tree species in the understory of a Hawaiian forest. *Oecologia* **58**: 26–32.

Pearcy RW, Valladares F. 1999. Resource acquisition by plants: the role of crown architecture. In: Press MC, Scholes JD, Barker MG, eds. *Physiological plant ecology.* Oxford, UK: Blackwell Science. (In press.)

Peñuelas J, Estiarte M. 1998. Can elevated CO_2 affect secondary metabolism and ecosystem function? *Trends in Ecology and Evolution* **13**: 20–24.

Poorter H, de Jong R. 1999. A comparison of specific leaf area, chemical composition and leaf construction costs of field plants from 15 habitats differing in productivity. *New Phytologist* **143**: 163–176.

Press MC. 1998. Dracula or Robin Hood? A functional role for root hemiparasites in nutrient poor ecosystems. *Oikos* **82**: 609–611.

Reiskind JB, Bowes G. 1991. The role of phosphoenolpyruvate carboxykinase in a marine macroalga with C_4-like photosynthetic characteristics. *Proceedings of the National Academy of Sciences, USA* **88**: 2883–2887.

Sandquist DR, Ehleringer JR. 1997. Intraspecific variation of leaf pubescence and drought response in *Encelia farinosa* associated with contrasting desert environments. *New Phytologist* **135**: 635–644.

Schoener TW. 1988. Leaf damage in island buttonwood, *Conocarpus erectus*: correlations with pubescence, island area, isolation and the distribution of major carnivores. *Oikos* **53**: 253–266.

Schuepp PH. 1993. Leaf boundary layers. *New Phytologist* **125**: 477–507.

Sellers PJ, Bounoua L, Collatz GJ, Randall DA, Dazlich DA, Los SO, Berry JA, Fung I, Tucker CJ, Field CB, Jensen TG. 1996. Comparison of radiative and physiological effects of doubled atmospheric CO_2 on climate. *Science* **271**: 1402–1406.

Staddon PL, Fitter AH. 1998. Does elevated atmospheric carbon dioxide affect arbuscular mycorrhizas? *Trends in Ecology and Evolution* **13**: 455–458.

Tieszen LL, Senyimba MM, Imbamba SK, Troughton JH. 1979. The distribution of C_3 and C_4 grasses and carbon isotope discrimination along an altitudinal and moisture gradient in Kenya. *Oecologia* **37**: 337–350.

Ueno O. 1996. Immunocytochemical localization of enzymes involved in the C_3 and C_4 pathways in the photosynthetic cells of an amphibious sedge, *Eleocharis vivipara*. *Oecologia* **199**: 394–403.

Vogel JC, Fuls A, Danin A. 1986. Geographical and environmental distribution of C_3 and C_4 grasses in Sinai, Negev and Judean Deserts. *Oecologia* **70**: 258–265.

Walters MB, Reich PB. 1999. Low-light carbon balance and shade tolerance in the seedlings of woody plants: do winter deciduous and broad-leaved evergreen species differ? *New Phytologist* **143**: 143–154.

Woodman RL, Fernandes GW. 1991. Differential mechanical defence: herbivory, evapotranspiration, and leaf hairs. *Oikos* **60**: 11–19.

Yang M, Sack FD. 1995. The *too many mouths* and *four lips* mutations affect stomatal production in *Arabidopsis*. *Plant Cell* **7**: 2227–2239.

Author Index

Root Dynamics and Global Change: An Ecosystem Perspective

The 5th *New Phytologist* Symposium and GCTE Focus 1 Workshop, 'Root Dynamics and Global Change: An Ecosystem Perspective' is being held in Townsend, TN, USA on October 19–22 1999.

Changes in the physiology and turnover of fine roots in forests and grasslands in response to rising atmospheric CO_2 concentrations, elevated temperatures, altered precipitation, or nitrogen deposition could be a key link between plant responses and longer-term changes in soil organic matter and ecosystem carbon balance. Observations of root dynamics and physiology to address ecosystem-scale questions are just starting to emerge in the global change research community. The Global Change and Terrestrial Ecosystems (GCTE) core project of the International Geosphere–Biosphere Programme identified this issue in its new implementation plan because of the opportunity to influence the direction of future research initiatives in this important component of global change research.

The meeting represents a significant achievement, as the fifth symposium since the *New Phytologist* Trust agreed to sponsor an annual meeting as part of its charitable aim in promoting plant science. The research area is one in which the journal has established the highest reputation, and the specific topic is particularly timely as global change research grows increasingly in importance. As the world's leading broad-spectrum plant journal, *New Phytologist* strives to publish the best papers, and thus we also hope that the location of the meeting – the first *New Phytologist* Symposium to be held in North America – will send a clear message about our international commitment.

Organization:

R. J. Norby	*Oak Ridge National Laboratory, TN, USA*
A. H. Fitter	*University of York, UK*
R. Jackson	*Duke University, NC, USA*

Invited participants include:

J. A. Arnone III	*Desert Research Institute, NV, USA*
O. Atkin	*University of York, UK*
H. BassiriRad	*University of Illinois – Chicago, IL, USA*
M. M. Caldwell	*Utah State University, UT, USA*
D. Eissenstat	*Pennsylvania State University, PA, USA*
J. F. Farrar	*University of Wales – Bangor, UK*
R. Hendrick	*University of Georgia, GA, USA*
J. D. Joslin	*University of Tennessee, TN, USA*
K. J. Nadelhoffer	*Marine Biology Laboratory, MA, USA*
E. G. O'Neill	*Oak Ridge National Laboratory, TN, USA*
K. Pregitzer	*Michigan Tech University, MI, USA*
H. Rogers	*National Soil Dynamics Laboratory, AL, USA*
P. Rygiewicz	*US Environmental Protection Agency, OR, USA*
D. Tingey	*US Environmental Protection Agency, OR, USA*
F. I. Woodward	*University of Sheffield, UK*
D. Zak	*University of Michigan, MI, USA*

Look out for details of the 6th New Phytologist *Symposium, 'Signalling in Plants', London, UK on September 18–22 2000 at www.journals.cup.org.*

Now available from *New Phytologist*

Fructose Polymers in Plants and Micro-organisms

Edited by J. F. Farrar, University of Wales Bangor, UK
and C. J. Pollock, Institute of Grassland and Environmental Research, Aberystwyth, UK

Polymers of fructose - fructans - are second only to starch as plant storage carbohydrates. In spite of their biological importance, many aspects of their metabolism and function are poorly understood. Current work promises great improvements in our knowledge as the tools of molecular biology are being used effectively for the first time. The increasing importance in the food industry is stimulating more work on the metabolism of fructans within both higher plants and micro-organisms, and their tissue-level localization is being determined. This volume presents a collection of edited papers from the Third International Symposium on Fructan. New insights into synthesis, the impact of molecular biology techniques and new physiological insights are presented in an accessible form.

£19.95   PB   0 521 62715 X   164 pp.   1997

Putting Plant Physiology on the Map

Genetical Analysis of Developmental and Adaptive Traits
Edited by H. Thomas,
Institute for Grassland and Environmental Research, Aberystwyth
and J. F. Farrar,
University of Wales, Bangor

This volume comprises a collection of case studies in which careful analysis of developmental and adaptive characters by plant physiologists and biochemists has been given a new and immensely fruitful dimension by assigning these traits to molecular maps. These edited papers were presented at the Second New Phytologist Symposium, held in 1997. The meeting captured the excitement of a time when disciplines are combining synergistically to revitalise whole areas of traditional plant science, and this volume is a timely record of the event.

£19.95   PB   0 521 64654 5   188 pp.   1998

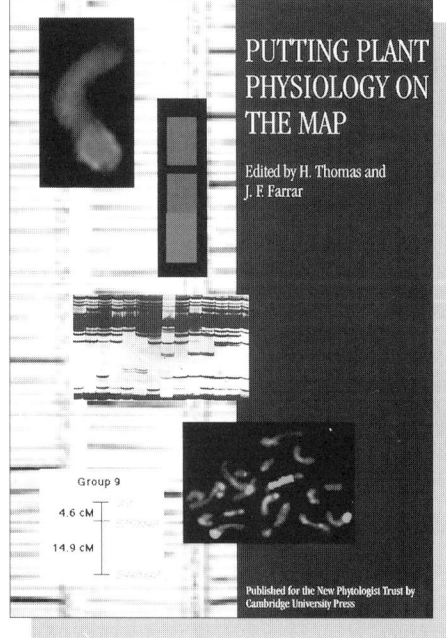

Cambridge books are available from good bookshops, alternatively phone UK +44 (0)1223 326070, fax UK +44 (0)1223 315052.
For further information, please e-mail Susan Chadwick on schadwick@cup.cam.ac.uk or browse our web server www.cup.cam.ac.uk

CAMBRIDGE UNIVERSITY PRESS   The Edinburgh Building
Cambridge, CB2 2RU, UK
www.cup.cam.ac.uk

The world's leading broad-spectrum plant journal

New Phytologist

International Journal of Plant Science

- **High impact** - the highest impact factor of all the world's broad-spectrum plant journals*
- **Wide ranging** - from intracellular processes to global environmental change
- **International** - Editors and Advisors from around the world
- **Independent** - the *New Phytologist* Trust is a charity dedicated to the promotion of plant science

Features include monthly Tansley Reviews – an unrivalled format marrying the personal views of experts with in-depth analysis – and Special Issues concentrating on current research themes.

With free online access to subscribers, no page charges, free colour and free offprints, why not take a closer look – also for free?

www.journals.cup.org

Recent Tansley Reviews

Calcium physiology and terrestrial ecosystem processes – *McLaughlin & Wimmer*

Biological clocks in *Arabidopsis* – *Millar*

Plant hybridization – *Rieseberg & Carney*

Recent Research articles

Parental environmental effects on life history traits in *Arabidopsis* – *Andalo* et al.

Emissions, biogenesis and metabolic utilization of chloromethane by potato tubers – *Harper* et al.

Colonization of wheat *para*-nodules by the N_2-fixing cyanobacterium *Nostoc* sp. strain 2S9B – *Gantar & Elhai*

Neither compatible nor self-incompatible pollinations of *Brassica napus* involve reorganization of the papillar cytoskeleton – *Dearnaley* et al.

Iron accumulation in root apoplasm of dicotyledonous and graminaceous species grown on calcareous soil – *Zhang* et al.

*Independent ISI data, 1997 impact factor 1.967, cited half-life 8.1 years

to contact the Journals Marketing Department –
North America: tel (914) 937 9600 x 154 *fax* (914) 937 4712 *email* journals_marketing@cup.org *web* www.cup.org
rest of the world: tel +44 (0)1223 325806 *fax* +44 (0)1223 315052 *email* journals_marketing@cup.cam.ac.uk *web* www.cup.cam.ac.uk

Take a closer look – FREE

Please send me a free sample copy of *New Phytologist*
Send this coupon to:
   Journals Marketing, Cambridge University Press,
   40 West 20th Street, New York, NY 10011-4211, USA
or The Edinburgh Building, Cambridge, CB2 2RU, UK
or send your request by fax or email using the
   information below

name

address

 **CAMBRIDGE** UNIVERSITY PRESS   The Edinburgh Building, Cambridge, CB2 2RU, UK
40 West 20th Street, New York, NY 10011–4211, USA

Now available from *New Phytologist*

Disturbance of the Nitrogen Cycle

Edited by T. A. Mansfield,
Lancaster University, UK,
K. W. T. Goulding, IACR-Rothamsted, UK
and L. J. Sheppard, Institute of Terrestrial
Ecology, Penicuik, UK

Some of the most formidable issues now
confronting environmental scientists are
the result of human intervention in the
global nitrogen cycle. Under natural
conditions the incorporation of
atmospheric nitrogen into compounds
that are usable in biological systems
relies on two main processes: fixation of
nitrogen by microorganisms, both
free-living and in symbiotic associations
(such as in legumes), and the
combination of nitrogen and oxygen in
the atmosphere during electrical storms.
Human activities now produce more 'fixed
nitrogen' than the global sum of these natural
processes, the main contributions coming from the
manufacture of nitrogenous fertilizers, and the
emissions of nitrogen oxides into the atmosphere
when fossil fuels are burnt. The resulting nitrogen
causes many different environmental problems -
acidification, climate change, eutrophication, ozone
destruction in the stratosphere and ozone formation
in the troposphere - which are common to most
countries. The transboundary nature of the problems
requires international dialogue and agreed action to
provide solutions.

The contributors to this book are scientists at the
leading edge of research on many important aspects of
the subject, and the articles represent their thoughts
on the problems. There is an emphasis on good
science and open debate about the complex issues
with which we are confronted.

The book is reprinted from *New Phytologist* 139
(1998), a Special Issue of the journal that followed the
3rd *New Phytologist* Symposium.

£19.95   PB   0521644267   248 pp.   1998

Cambridge books are available from good bookshops,
alternatively phone UK +44 (0)1223 326070, fax
UK +44 (0)1223 315052. For further information, please
e-mail Susan Chadwick on schadwick@cup.cam.ac.uk or
browse our web server www.cup.cam.ac.uk

The Tansley Review Collections 1

Leaf Development and Functions
Edited by A. M. Hetherington, Lancaster University, UK
The Tansley Reviews of *New Phytologist* are
commissioned in-depth reviews written from a
personal perspective by leading workers in plant
science. This selection of reviews includes discussions
of the use of *Arabidopsis* in the genetic and molecular
analysis of plant morphogenesis; molecular genetics of
cellular differentiation in leaves; chlorophyll as a
symptom and regulator of plastid development; auxin
action and auxin binding proteins; gene expression
during leaf senescence; control of organogenesis at the
shoot apex; diffusion of CO_2 and other gases inside
leaves and leaf boundary layers. They provide
stimulating and valuable contributions at the
forefront of contemporary plant science, for students
and researchers.

£9.95   PB   0 521 62716 8   216 pp.   1997

CAMBRIDGE
UNIVERSITY PRESS

The Edinburgh Building
Cambridge, CB2 2RU, UK
www.cup.cam.ac.uk